94.00
80€

INTERNATIONAL ASTRONOMICAL UNION
UNION ASTRONOMIQUE INTERNATIONALE

LUMINOUS STARS AND ASSOCIATIONS IN GALAXIES

PROCEEDINGS OF THE 116TH SYMPOSIUM OF THE
INTERNATIONAL ASTRONOMICAL UNION,
HELD AT PORTO HELI, GREECE, MAY 26–31, 1985

EDITED BY

C. W. H. DE LOORE

*Astrophysical Institute, Vrije Universiteit Brussel and
Rijksuniversitair Centrum Antwerpen, University of Antwerp, Belgium*

A. J. WILLIS

*Department of Physics and Astronomy,
University College London, U.K.*

and

P. LASKARIDES

University of Athens, Greece

D. REIDEL PUBLISHING COMPANY

A MEMBER OF THE KLUWER ACADEMIC PUBLISHERS GROUP

DORDRECHT / BOSTON / LANCASTER / TOKYO

Library of Congress Cataloging in Publication Data

International Astronomical Union. Symposium (116th : 1985 : Portokhèlion, Greece)
Luminous stars and associations in galaxies.

Includes index.
1. Stars—Magnitudes—Congresses.   2.   Galaxies—Congresses.   I. Loore, Camiel W. H. de.   II.   Willis, A. J. (Allan J.)   III.   Laskarides, Paul.
IV. Title.
QB799.I59   1985   523.8'22   86–10090
ISBN 90–277–2272–2
ISBN 90–277–2273–0 (pbk.)

*Published on behalf of
the International Astronomical Union
by
D. Reidel Publishing Company, P.O. Box 17, 3300 AA Dordrecht, Holland*

All Rights Reserved
© *1986 by the International Astronomical Union*

*Sold and distributed in the U.S.A. and Canada
by Kluwer Academic Publishers,
101 Philip Drive, Assinippi Park, Norwell, MA 02061, U.S.A.*

*In all other countries, sold and distributed
by Kluwer Academic Publishers Group,
P.O. Box 322, 3300 AH Dordrecht, Holland*

No part of the material protected by this copyright notice may be reproduced or utilized in any form or by any means, electronic or mechanical, including photocopying, recording or by any information storage and retrieval system, without written permission from the publisher

Printed in The Netherlands

TABLE OF CONTENTS

PREFACE   xiii

THE ORGANISING COMMITTEES   xv

LIST OF PARTICIPANTS   xvii

SESSION 1 - MAIN SEQUENCE AND SUPERGIANT STARS - I.

R. P. KUDRITZKI and D. G. HUMMER: Intrinsic Parameters of Hot Blue Stars   3

C. D. GARMANY: Luminous Blue Stars: Distribution and Numbers.   19

A. SANDAGE: Brightest Stars in Galaxies as Distance Indicators.   31

SESSION 2 - MAIN SEQUENCE AND SUPERGIANT STARS - II.

R. M. HUMPHREYS: The Cooler Supergiants (A to M): Crucial Signposts in the Lifecycles of Massive Stars.   45

W. L. FREEDMAN: The Distribution of Young Stars in Nearby Galaxies   61

POSTER PAPERS SESSIONS 1 and 2.

T. R. MANLEY: A Stellar Content Study of NGC 2403 with the automated Plate Scanner.   83

V. S. NIEMELA: Spectroscopic Binaries in the LMC.   85

E. ZSOLDOS: The Hypergiant Variable HR 8752=HD217476.   87

J. SMOLINSKI, J. L. CLIMENHAGA and J. M. FLETCHER: Activity in the Envelope of the G-type Hypergiant HD 217476.   89

D. J. STICKLAND: IRAS Observations of the Cool Galactic Hypergiants.   91

R. PRINJA and I. HOWARTH: O-Star Winds: Is Rotation Important?   93

D. H. MORGAN, K. NANDY and G. I. THOMPSON: The Absorption Feature at 1920 Å in the Spectra of Early-type Supergiants in the Small Magellanic Cloud. 95

L. MANTEGAZZA: The Light Variations of the Red Supergiant Mu Cephei. 97

G. R. GRIEVE and B. F. MADORE: Variability of Magellanic Cloud Supergiants. 99

M. KONTIZAS and E. KONTIZAS: Star Formation of Star Clusters in the SMC and their Adjoining Fields. 101

J. A. GRAHAM and R. M. HUMPHREYS: Red Supergiants in NGC 300. 103

L. GREGGIO: Simulations of the Brightest Stars in Galaxies as Distance Indicators. 105

C. DE JAGER, H. NIEUWENHUIJZEN and K. A. VAN DER HUCHT: The Dependence of the Stellar Rate of Mass Loss on Effective Temperature and Velocity 109

B. BOHANNAN, D. C. ABBOTT, S. A. VOELS and D. G. HUMMER: Steps to a New Calibration of the Spectral Type — Effective Temperature Relationship for Early-type Stars. 111

F. CASTELLI, C. MOROSSI and R. STALIO: Photospheric and "Chromospheric" CIV Lines in B-Main-Sequence Stars. 113

R. S. POLIDAN and R. STALIO: Voyager Observations of Hot Luminous Stars. 117

E. NASI: Synthetic HR Diagrams for Luminous Stars: A Test of Stellar Evolution Theory. 121

D. J. MacCONNELL, R. F. WING and E. COSTA: A Low-Dispersion, Near-Infrared Survey for Galactic M Supergiants. 123

DISCUSSION 125

SESSION 3 - LUMINOUS BLUE VARIABLES.

I. APPENZELLER: Instability in Massive Stars: An Overview. 139

B. WOLF: S Doradus Type; Hubble-Sandage Variables. 151

H. J. G. L. M. LAMERS: P Cygni Type Stars: Evolution and Physical Processes. 157

TABLE OF CONTENTS

SESSION 4 - OTHER LUMINOUS STARS WITH EMISSION LINES

N. R. WALBORN: Optical and Ultraviolet Spectral Morphology of Luminous OB Stars in the Galaxy and the Magellanic Clouds.   185

P. S. CONTI: Parameters of Wolf-Rayet Stars.   199

P. MASSEY: Wolf-Rayet Stars in Nearby Galaxies.   215

POSTER PAPERS - Sessions 3 and 4

P. HELLINGS: Bolometric Corrections and Magnitudes of WR+O Stars.   221

T. NUGIS: Two Possible Wind Models for Wolf-Rayet stars.   223

M. AZZOPARDI and N. MEYSSONNIER: A New Survey for $H\alpha$-Emission-Line Stars in the Small Magellanic Cloud.   225

B. BOHANNAN: A Large Sample of Emission-Line Stars in the Large Magellanic Cloud: Their Location in the Hertzsprung-Russell Diagram.   227

D. J. STICKLAND, C. LLOYD and A. J. WILLIS: Is AS 431 a Superluminous WR Star?   229

M. M. SHARA and A. F. J. MOFFAT: Wolf-Rayet Stars in M31.   231

A. F. J. MOFFAT, M. M. SHARA and W. SEGGEWISS: The Luminous Stellar Content of 30 Doradus and NGC 3603 - The Nearest Visible Giant HII Regions.   233

C. LEITHERER, I. APPENZELLER, G. KLARE, H. J. G. L. M. LAMERS, O. STAHL, L. B. F. M. WATERS and B. WOLF: The Massive Stellar Wind of the Hubble-Sandage Variable S Doradus.   235

D. VANBEVEREN: The WR/OB Number Ratio within 2.5 kpc from the Sun.   237

T. E. ARMANDROFF, P. MASSEY and P. S. CONTI: Wolf-Rayet Stars in M31's OB Associations.   239

K. A. VAN DER HUCHT, T. A. JURRIENS, F. M. OLNON, P. S. THE, P. R. WESSELIUS, P. M. WILLIAMS: IRAS Observations of Wolf-Rayet Stars.   243

O. STAHL, B. WOLF, M. DE GROOT and C. LEITHERER: High-Dispersion Spectroscopy of the Brightest Emission-Line Stars of the Magellanic Clouds   247

B. WOLF and O. STAHL: MWC 300: A Runaway Hypergiant. 248

R. VIOTTI, L. ROSSI, A. ALTAMORE, C. ROSSI and A. CASSATELLA:
   New Results on Eta Carinae. Evidence for an Asymmetric,
   Inhomogeneous Wind. 249

J. P. DE GREVE, P. HELLINGS and E. P. J. VAN DEN HEUVEL:
   On the Occurrence of WR+O Binaries. 253

C. DE JAGER and H. NIEUWENHUIJZEN: Stellar Atmospheric
   Instability in the Upper Part of the Hertzsprung-Russell
   Diagram. 255

P. J. McGREGOR, A. R. HYLAND and D. J. HILLIER: Infrared
   Spectroscopy of Southern P Cygni Stars. 257

A. J. WILLIS, P. S. CONTI, C. D. GARMANY and I. D. HOWARTH:
   Rapid Ultraviolet Spectral Variations in HD 50896
   (WN5+?). 259

D. J. HILLIER: The Formation of Nitrogen and Carbon Lines in
   HD 50896 (WN5). 261

F.-J. ZICKGRAF, B. WOLF, O. STAHL and C. LEITHERER:
   B(e)-Stars of the Magellanic Clouds. 265

A. J. WILLIS, I. D. HOWARTH, K. NANDY and D. H. MORGAN:
   The Mass Loss Rate of SK 80 (O7Iaf) in the Small
   Magellanic Cloud. 269

H. ZINNECKER: How to Form a 200 M_o Star. 271

G. MURATORIO, M. FRIEDJUNG and R. VIOTTI: FeII in the UV
   Spectrum of Luminous Emission Line Stars. 275

DISCUSSION 277

   SESSION 5 - MASSIVE STAR EVOLUTION.

A. MAEDER: Massive Star Evolution: Mass Loss and Mixing. 287

J. SILK: Physical Processes in Massive Star Formation. 301

C. CHIOSI: Effects of Convective Overshooting, Mass Loss
   (and Chemical Composition) across the HR Diagram. 317

J. P. DE GREVE and C. DE LOORE: Binary Evolution in the
   Upper HRD. 339

TABLE OF CONTENTS

SESSION 6 - LARGE STELLAR COMPLEXES IN GALAXIES

M. ROSA and S. D'ODORICO: The Exciting Stars of Giant Extragalactic HII Regions.   355

P. HODGE: Systems of Stellar Associations in Galaxies.   369

POSTER PAPERS - Sessions 5 and 6.

P. HELLINGS: The Average X-Ray Lifetime of Massive X-Ray Binaries.   383

S. R. SREENIVASAN and W. J. F. WILSON: Consequences of Rotational Mixing in late Type Massive Stars.   385

Y. N. EFREMOV, G. R. IVANOV and N. S. NIKOLOV: New Stellar Associations in M31.   389

P. Z. KUNCHEV and N. S. NIKOLOV: The Associations OB 110 and OB 112 in M33 Galaxy.   391

S. A. SILICH: The Dynamical Evolution of Expanding HI Shells and initial Mass Function of OB Associations.   393

R. L. PENNINGTON: Star Formation in NGC 5128.   395

G. BODIFEE: Oscillating Star Formation.   397

C. H. B. SYBESMA: The Evolution of Massive Close Binaries with Mass Loss and Overshooting. An Application to V729 Cyg (=BD+40 4220).   399

M. C. LORTET, M. HEYDARI-MALAYERI and G. TESTOR: CCD Observations of Young Stellar Associations and Multiple Systems in the Magellanic Clouds.   401

D. H. MORGAN and A. R. GOOD: Two Faint WC Stars near 30 Doradus.   403

E. KONTIZAS, A. DAPERGOLAS and M. KONTIZAS: Bright Stars in the SMC Clusters.   405

M. KONTIZAS, M. CHRYSOVERGIS, E. KONTIZAS and D. HADJIDIMITRIOU: Tidal Radii and Masses of the Clusters in the LMC.   407

E. KONTIZAS, E. THEODOSIOU and M. KONTIZAS: The Distribution of Bright Stars in the SMC Clusters   409

M. KAFATOS and R. McCRAY: Propagating Star Formation Induced by Superbubbles.   411

T. GEHREN, D. HUSFELD, R. P. KUDRITZKI, P. S. CONTI
    and D. G. HUMMER: Non-LTE Analysis of Massive
    Stars in the Magellanic Clouds.   413

P. DUBOIS: Metallicity Effect on Absolute Magnitude
    Determination.   415

E. KONTIZAS, E. XIRADAKI and M. KONTIZAS: Bright Stars
    in Five LMC Star Clusters.   417

G. R. IVANOV and P. Z. KUNCHEV: On the Arm Pattern of M33.   419

G. BERTELLI and A. BRESSAN: Convective Overshooting: The
    Upper Mass Limit for Stars undergoing Core He-Flash.   421

H. SCHILD: The Final Fate of Massive Stars.   423

DISCUSSION   425

SESSION 7 - LUMINOUS STELLAR CONTENT OF GALAXIES-INTEGRATED
            PROPERTIES - I.

J. G. HOESSEL: Luminosity Functions in Dwarf Irregulars.   439

J. M. SCALO: The Initial Mass Function of Massive Stars
    in Galaxies: Empirical Evidence.   451

J. C. WHEELER: Luminous Content of Galaxies: Inferences from
    Supernovae.   467

SESSION 8 - LUMINOUS STELLAR CONTENT OF GALAXIES:
            INTEGRATED PROPERTIES - II.

P. G. MEZGER: Luminous Stellar Content of the Galaxy:
    Inferences from Radio and Infrared Data.   479

POSTER PAPERS - Sessions 7 and 8.

A. W. CAMPBELL and L. J. SMITH: A Search for Wolf-Rayet
    Stars in Giant Extragalactic Bursts of Star Formation.   499

M. KAUFMAN, R. C. KENNICUTT and F. N. BASH: Giant HII
    Regions in M81.   503

J. MELNICK, R. TERLEVICH and M. MOLES: Warmers: Massive
    Stars in the Nuclei of Galaxies.   505

G. BERTELLI, A. G. BRESSAN, C. CHIOSI, E. NASI and
    L. PIGATTO: Convective Overshooting: New Integrated
    Colours vs. Age Relations for Star Clusters.   511

C. CHIOSI and L. PIGATTO: The Distance Modulus of LMC. 513

Y. H. CHU: NGC 2070 and NGC 3603. 515

DISCUSSION 517

    Summary of the Symposium.

J. A. GRAHAM and T. E. ARMANDROFF: What is happening to Massive Stars in Galaxies? 523

SUBJECT INDEX 529

PREFACE

The original ideas for this symposium were generated through discussions amongst several members of the Scientific Organising Committee attending IAU Symposium No. 99 (Wolf-Rayet Stars: Observations, Physics and Evolution) held in Mexico in 1981, and subsequently at IAU Symposium No. 105 (Observational tests of stellar evolutionary theory) at Geneva in 1983. It was recognised that with the advent of powerful new observing facilities, particularly with sensitive imaging and spectroscopic detectors on large telescopes, the study of luminous stars was being rapidly extended to objects beyond our own galaxy. Great advances were being made in such diverse, yet possibly related, phenomena as the exciting objects of giant HII regions (e.g. R136a), Eta Carina-like objects, Hubble-Sandage variables, as well as the most luminous 'normal' stars. Considerable progress was being made in advancing knowledge of the numbers and distributions of the most luminous objects in our own and neighbouring galaxies (spirals and irregulars) and exciting results emerging on star formation and the initial mass function of the upper end of the main-sequence in other galaxies. It was thus considered timely to arrange a symposium to bring together the 'stellar' observers and theoreticians to discuss recent progress and outstanding problems in the field. The overall aim of this symposium was to consider the properties and role of the luminous stars, stellar complexes and associations in the evolution of galaxies; to discuss objects found not only in the Galaxy but also in the Magellanic Clouds, the Local Group and beyond.

The organisation of the meeting relvolved around invited review talks, in eight sessions, reflecting the broad theme of the symposium and embracing a wide range of interrelated topics: main-sequence and supergiant stars; luminous blue variables and other stars with emission lines; massive stars evolution; large stellar complexes in galaxies; and the integrated luminous stellar content of galaxies. Advances in these fields on both observational and theoretical fronts were presented. In addition, many contributions were presented in the form of poster papers, highlighting exciting new results, and each day finished with a substantial and lively discussion session of the poster papers reflecting the theme of the day's review sessions.

An important aspect of all symposia are the discussions which follow the review papers and, of particular importance to the format of this symposium, of the poster discussion sessions. The editors have tried to faithfully record the content of these discussions in the proceedings. In the case of the reviews, participants were asked to record their questions and comments on paper, and the collecting and collating of these discussion sheets were capably handled at the meeting by several assistants to the Local Organising Committee, and their help is greatly appreciated. The poster discussions sessions were recorded on tape and these have been transcribed by the editors for inclusion in the proceedings. At the Vrije Universiteit Brussels the preparation and typing of the discussions was carried out by Rita Cardon whose hard work and assistance is greatly appreciated. The International Astronomical Union provided generous financial assistance to the Scientific Organising Committee, and these funds were used to enable several participants to attend the meeting. The Local Organising Committee received generous financial assistance from the Greek Ministry of Science and Culture which greatly facilitated the smooth running of the symposium and general local organisational matters. The Scientific Organising Committee is greatly indebted to Drs Paul Laskarides, Mary and Evangelos Kontizas and other members of the LOC whose untiring efforts both prior to and at the meeting resulted in a smooth and efficient local organisation. This was greatly appreciated by all participants and their guests. The location of the symposium at the Hotel 'Cosmos' (deemed an appropriate venue for an astronomical meeting!) in the small fishing village of Porto Heli in the southern Peleponese, proved an ideal setting for informal, yet highly constructive scientific interaction between the participants.

The Editors.

SCIENTIFIC ORGANISING COMMITTEE

P.S. Conti (Chairman), C. de Loore, S. d'Odorico, J. Graham, R. Humphreys, P. Laskarides, A. Maeder, S. van den Berg, A.J. Willis

LOCAL ORGANISING COMMITTEE

P. Laskarides (Chairman), E. Antonopoulou, E. Kontizas, M. Kontizas, P. Niarchos, E. Theodossiou, V. Tsikoudi

LIST OF PARTICIPANTS

H. ABLES, US Naval Observatory, Flagstaff, USA
M. ANDRACACOU, Dept. of Astronomy, University of Athens, Greece
E. ANTONOPOULOU, Dept. of Astronomy, University of Athens, Greece
I. APPENZELLER, Landessternwarte Königstuhl, Heidelberg, Germany
T. ARMANDROFF, Yale University Observatory, New Haven, CT, USA
M. AZZOPARDI, European Southern Observatory, München, Germany
G. BERTELLI, Istituto di Astronomia, Padova, Italy
G. BODIFEE, Astrofysisch Instituut, University of Brussels, Belgium.
B. BOHANNAN, University of Colorado, Boulder, USA
J. BREYSACHER, European Southern Observatory, München, Germany
A. CAMPBELL, Institute of Astronomy, Cambridge, England
C. CHIOSI, Istituto di Astronomia, Padova, Italy
M. CHRYSSOVERGIS, University of Athens, Greece
Y. CHU, Astronomy Dept., University of Illinois, Urbana, USA
P. CONTI, University of Colorado, Boulder, USA
A. DAPERGOLAS, National Observatory Athens, Greece
J.P. DE GREVE, Astrofysisch Instituut, University of Brussels, Belgium.
M. DE GROOT, Armagh Observatory, Northern Ireland
C. DE JAGER, Space Research Laboratory, Utrecht, the Netherlands
C. DE LOORE, Astrofysisch Instituut, University of Brussels, Belgium.
P. DUBOIS, Observatoire de Strasbourg, France
M. FEAST, South African Astronomical Observatory, South Africa
W. FREEDMAN, Mount Wilson and Los Campanos Observatory, Pasadena, California, USA
E. GAVRYUSEVA, Inst. for Nuclear Research, Acad. of Sciences, Moscow, U.S.S.R.
J. GRAHAM, Joint Institute for Laboratory Astrophysics University of Colorado and National Bureau of Standards, USA
L. GREGGIO, Istituto di Astronomia, Padova, Italy.
D. HATZIDIMITRIOU, National Observatory Athens, Greece
P. HELLINGS, Astrofysisch Instituut, University of Brussels, Belgium.
D. HILLIER, Mt. Stromlo Observatory, Woden P.O., Australia
P. HODGE, Astronomy Dept. University of Washington, Seattle USA
J. HOESSEL, Space Telescope Science Institute, Baltimore, USA
I. HOWARTH, Dept. of Physics and Astronomy, University College, London WCIE, England
D. HUMMER, Institut fur Astronomie und Astropysik, Universität München, Germany
R. HUMPHREYS, Dept. of Astronomy, Minneapolis, USA

LIST OF PARTICIPANTS

G. IVANOV, University of Sofia, Faculty of Physics, Department of Astronomy, Bulgaria
M. KAUFMAN, Ohio State University, Ohio, USA
M. KAFATOS, Department of Physics, G. Mason University, Fairfax, USA
E. KONTIZAS, National Observatory of Athens, Greece
M. KONTIZAS, University of Athens, Greece
P. KUDRITZKI, Institut fur Astronomie und Astrophysik, Universität München, Germany.
P. KUNCHEV, University of Sofia, Faculty of Physics, Department of Astronomy, Bulgaria
H. LAMERS, Space Research Labortory, Utrecht, the Netherlands
P. LASKARIDES, University of Athens, Athens, Greece
C. LEITHERER, Landessternwarte Königstuhl, Heidelberg, Germany
M.-C. LORTET, Observatoire de Meudon, Meudon, France
D. MacCONNELL, Michigan State University, Department of Physics and Astronomy, USA
B. MADORE, David Dunlap Observatory, University of Toronto, Ontario, Canada
A. MAEDER, Geneva Observatory, Sauverny, Switzerland
M.-L. MALAGNINI, Osservatorio Astronomico, Trieste, Italy
T. MANLEY, University of Minnesota, USA
L. MANTEGAZZA, Osservatory Astronomico di Brera, Milano-Merate, Italy
P. MASSEY, Kitt Peak National Observatory, Tucson, USA
P. McGREGOR, Mt. Stromlo and Siding Spring Observatory, Woden, Australia
J. MELNICK, European Southern Observatory, Chile
P. MEZGER, Max-Planck-Institut Fur Radio-Astronomie, Bonn, Germany
A. MOFFAT, Dept. of Physics, Universite de Montreal, Canada
D. MORGAN, Royal Observatory, Edinburgh, Scotland
X. MOUSSAS, Dept. of Astronomy, Sofia, Bulgaria
J. MURATORIO, Observatoire de Marseille, France
E. NASI-CHIOSI, Istituto di Astronomia, Padova, Italy
P. NIARCHOS, University of Athens, Athens, Greece
V. NIKOLOV, Astronomy Dept., University of Sofia, Bulgaria
V. NIEMELA, Instituto de Astronomia Y Fisica del Espacia, Buenos Aires, Argentina
T. NUGIS, W. Struve Tartu Astrophysical Observatory, Estonian, USSR.
R. PENNINGTON, University of Minnesota, USA.
L. PIGATTO, Osservatorio Astronomica, Padova, Italy
R. PRINJA, Dept. of Physics and Astronomy, University College London, England
A. RENZINI, Osservatorio Astronomica Universitario, Bologna, Italy
A. SANDAGE, Hale Observatories, Pasadena, California, USA
J. SCALO, Dept. of Astronomy, University of Texas, USA
M. SHARA, Space Telescope Science Institute, Baltimore, USA
S. SILICH, Main Astronomical Observatory, Academy of Sciences of the Ukraninian SSR, Kiev, USSR
J. SILK, University of California, Astronomy Department, Berkeley, California 94720, USA

LIST OF PARTICIPANTS

L.J. SMITH, Dept. of Physics and Astronomy, University College London, London, England
J. SMOLINSKI, Copernicus Astronomical Center, Torn, Poland
R. SREENIVASAN, University of Calgary, Dept. of Physics, Canada
O. STAHL, Landessternwarte Königstuhl, Heidelberg, Germany
R. STALIO, Osservatorio Astronomico, Trieste, Italy
D. STICKLAND, Rutherford Appleton Laboratories, England
C. SYBESMA, Astrofysisch Instituut, University of Brussels, Belgium.
E. THEODOSSIOU, University of Athens, Greece
V. TSIKOUDI, University of Ioannina, Dept. of Astronomy, Greece
B. TULLY, Institute for Astronomy, University of Texas, USA
D. VANBEVEREN, Physics Department, University of Brussels, Belgium
R. VIOTTI, Istituto Astrofisica Spaziale, Frascati, Italy
N. WALBORN, Space Telescope Science Institute, Baltimore, USA
C. WHEELER, University of Texas at Austin, Dept. of Astronomy USA
B. WOLF, Landessternwarte Königstuhl, Heidelberg, Germany
A.J. WILLIS, Department of Physics and Astronomy, University College London, England
E. XIRADAKI, Astronomy Institute, National Observatory of Athens, Greece
H. ZINNECKER, Royal Observatory, Blackford Hill, Edinburgh, England.
E. ZSOLDOS, Konkoly Observatory of the Hungarian Academy of Sciences Budapest, Hungary

SESSION 1.

MAIN SEQUENCE AND SUPERGIANT STARS - I.

Intrinsic Parameters of Hot Blue Stars

R.P. Kudritzki and D.G. Hummer[*]

Institut für Astronomie und Astrophysik der Universität
München

[*]Staff Member, Quantum Physics Division, National Bureau of
Standards; permanent address, Joint Institute for Laboratory
Astrophysics, University of Colorado and National Bureau of
Standards, Boulder, Colorado

I. Introduction

Advances in both theoretical understanding and observational capabilities in the past few years have made possible the determination of the effective temperature, surface gravity and chemical abundance of massive stars with unprecedented accuracy. These data are in turn important for the study of galaxies, as stars are important sources of information concerning the evolutionary state, past and present chemical composition, and distance of the parent galaxy. In addition to this diagnostic role, stars are crucial as sources of light, matter, and metals in the galaxy. Thus an improved understanding of massive stars makes possible a better determination of the physical conditions in a galaxy as well as a deeper understanding of how it functions.

The hottest stars in a galaxy fall into two distinct groups: 1) the young, massive, very luminous stars at the upper end of the main sequence with masses in the range 20-200 $M_\odot$; 2) the hot evolved stars with smaller masses and luminosities. Members of the first group, newly formed from the ISM, reflect the chemical composition of the galactic gas and provide information about the process of star formation. Their rapid evolution with strong mass loss, leading to Wolf-Rayet stars and Type II Supernovae, means that these stars in their various phases are responsible for the abundance of heavy elements in the galaxy. Because of their high luminosity, the stars can also be used as standard candles; a proposal for their exploitation in a new way appears at the end of this review.

The second group includes subdwarf O-stars, subdwarf B-stars and Extended Horizontal Branch objects, and central stars of planetary nebulae. These diverse objects have masses in the range 0.3 to 3 $M_\odot$ and are all in advanced stages of stellar evolution. As successors of the asymptotic giant branch, they exhibit abundance anomalies reflecting their evolutionary history rather than the composition of the ISM and so give less information on the current chemical content of the galaxy. On the other hand, as progenitors of White Dwarfs and probably of Type

I Supernovae, they play other roles in the development of galaxies. For
brevity we will discuss only the first group; for the second group we
refer to Kudritzki (1985).

II. The Photosphere of Hot Stars

The determination of stellar parameters from spectra requires a
detailed understanding of the stellar photospheres (and in some cases
of stellar winds), which is complicated by a severe failure of the
classical assumptions of Local Thermodynamic Equilibrium (LTE). This
has been suspected for many years and was clearly demonstrated by
Peterson and Scholz (1971) who showed that the LTE absorption profiles
of H I, He I and He II were much too weak. The reason is clear: because
of the intense radiation field the radiative transitions occur much
more rapidly than electron collisions. Consequently the rate equations
for the level populations and the state of ionization are dominated by
radiative processes, which are determined by solutions of the radiative
transfer equations. As these involve the excitation and ionization
state of the gas, the calculation of non-LTE model stellar atmospheres
awaited the development of powerful numerical algorithms, which were
first provided by Auer and Mihalas (1972, 1973). Detailed calculations
by these workers and by Kudritzki (1973, 1976, 1979) showed that non-
LTE absorption profiles are much stronger, leading to improved agreement with observation. The resulting relation between stellar parameters and the model profiles of the classification lines provided a new
calibration of spectral types as a function of effective temperatures
in the hands of Conti and co-workers (Conti and Alschuler, 1971; Conti,
1973; Conti and Frost, 1977).

However, for ζ Pup, the brightest O-star in the sky, attempts to
determine the effective temperature by different methods led to serious
discrepancies. The ratio of equivalent widths of He I and He II led to
a spectral type O4, corresponding to $T_{eff} \approx 50000$ K (Baschek and
Scholz, 1971; Conti, 1973; Conti and Frost, 1977). On the other hand,
$T_{eff} \approx 35000$ K was inferred from the angular diameter, given by
interferometry, together with the visual flux (Hanbury Brown, 1974;
Code et al., 1976; Davis et al., 1970), and also from fits of the
continuum energy distribution in the visual and UV to model atmosphere
fluxes (Holm and Cassinelli, 1977; Underhill et al., 1979; Remie and
Lamers, 1982). This outstanding descrepancy cast doubt on the adequacy
of the theory of model atmospheres, which was based on the so-called
"classical" assumptions: hydrostatic and radiative equilibrium, plane-
parallel geometry and no blanketing by metal lines. In order to resolve
this serious discrepancy a detailed analysis using NLTE model atmospheres and specially-obtained high-quality photographic spectra of
ζ Pup was carried out by Kudritzki et al. (1983), who compared calculations with observed profiles of Balmer and Pickering lines and equivalent width of He I λ 4471. The crucial point in their method is that
T_{eff}, log g and the helium abundance (y=He/(H+He) by numbers) are
determined simultaneously by constructing fit curves in the (log g,
log T_{eff})-plane, along which the calculated and observed equivalent

widths agree (see Fig. 1). In this way Kudritzki et al. (1983) determined for ζ Pup the parameters T_{eff} = 42000 K ± 2500 K, log g = 3.5 ± 0.15 and y = He/(H+He) = 0.14 ± 0.02. The calculated energy distribution for these parameters agrees satisfactorily with the UV measures of Brune et al. (1979), Code and Meade (1979), and Jamar et al. (1976), and with the visual data of Johnson and Mitchell (1975), although the differences among the UV observations and uncertainties in the (small) values of E_{B-V} precludes an exact comparison.

How can this wide range of temperatures be understood? In the first place, the usual calibration of spectral types vs. T_{eff} assumes that the ratio of equivalent width of HeI λ 4471 Å and HeII λ 4542 Å is primarily a function of temperature, whereas the models of Kudritzki et al. (1983) show that this ratio became very sensitive to gravity for smaller values of log g than were previously used in model calculations (see Fig. 2). Recent discussions by Tobin (1983) and by Abbott and Hummer (1985) show that methods using measured angular diameters and flux distributions for stars with T_{eff} > 40000 K are intrinsically subject to large errors; the latter authors and Kudritzki et al. (1983) stress also that the strong dependence of the flux on gravity leads to very large uncertainties in temperatures, since the gravity is undetermined. Methods relying on the energy

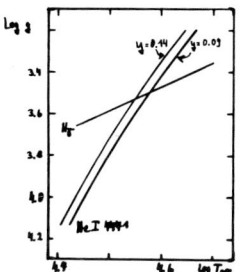

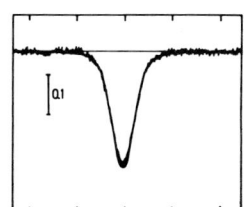

Fig. 1a. Fit diagram of Hγ and HeI λ 4471 for normal helium abundance y=0.09 and the enhanced value y= 0.14

Fig. 1b. Hγ profile compared with calculation at the intersection point (y=0.14) in Fig. 1a

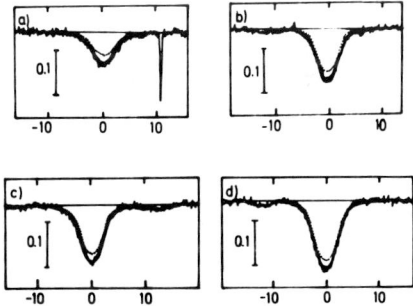

Fig. 1c. Profiles of He II lines compared with calculations at the intersections in Fig. 1a (dotted: y=0.09, full drawn: y=0.14)

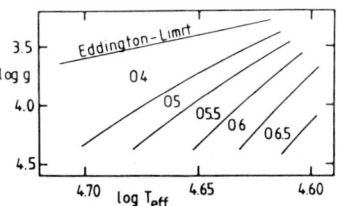

Fig. 2. The regions of the spectral types O4 to O6.5 in the log g-log T_{eff} diagram from NLTE models. The boundaries are given by W_λ(HeI4471)/W_λ(HeII4542) = 0.25, 0.35, 0.50, 0.63 and 0.79 from left to right.

distribution alone are even worse. In practice the determination of T_{eff} from the continuous energy distribution is further compromised by the uncertainties in de-reddening the spectrum.

A further check on the accuracy of the derived model comes from a comparison of the interferometrically measured angular diameter (allowing for the electron scattering envelope) of $(4.1 \pm 0.3) \times 10^{-4}$ arc sec with that obtained by comparing the measured and computed fluxes, $(3.8 \pm 0.3) \times 10^{-4}$ arc sec. Finally the He II lines in the UV agree well with Copernicus spectra except for He II λ 1640 Å which, along with the visual line He II λ 4686 Å, can be shown to be formed in the wind.

We believe that the critical case of ζ Pup shows clearly that the method of non-LTE analysis of the line spectrum gives reliable values for T_{eff}, log g and y. Additional physical processes must be taken into account to realize the full accuracy of the method. This is discussed in the following two sections.

III. Wind Blanketing

Photons scattered by a stellar wind back into the photosphere heat the surface layers and modify significantly the observed spectrum, as was first discussed by Hummer (1982), who pointed out that for this purpose the wind could be represented by a partial reflector with an albedo depending on frequency, and who estimated the strength of the effect based on analytical solutions of gray atmospheres. Observational evidence for wind blanketing was presented by Remie and Lamers (1982). Husfeld (1982) and Husfeld and Kudritzki (1983) used schematic albedos with full non-LTE atmosphere models. The calculations by Abbott and Lucy (1985) of a realistic albedo for ζ Pup by a Monte-Carlo solution of the transfer problem in the wind accounting for approximately 10^4 lines

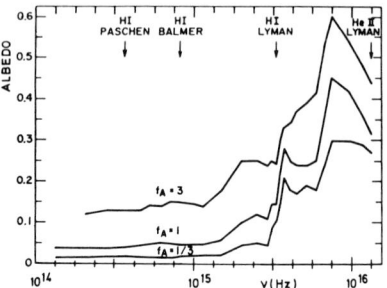

Fig. 3: Albedo vs. frequency for three values of mass loss rate. f_A is mass loss relative to that for ζ Pup ($\dot{M} = 5 \times 10^{-6} M_\odot$/yr)

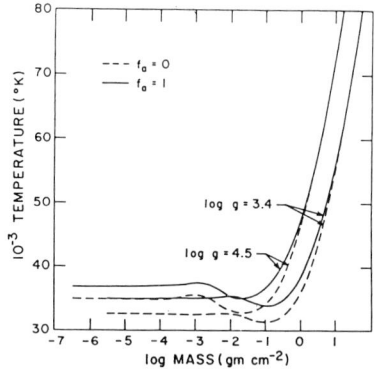

Fig. 4: Temperature vs. mass for T_{eff}=42000 K, log g=3.5. Curves are labeled with mass loss rate in units of that for ζ Pup

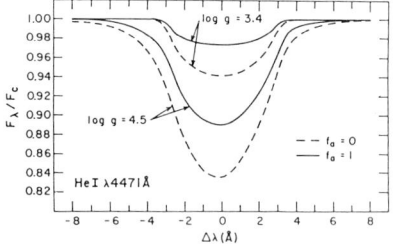

Fig. 5: Rotationally broadend line profile for He I 4471 Å, for models in Fig. 4

made possible the calculations by Abbot and Hummer (1985) of non-LTE models with realistic wind blanketing. In Figure 3 the albedo is given for three cases, labelled with $f_A = \dot{M}/\dot{M}(\zeta$ Pup). Figure 4 shows the run of temperature in models for which T_{eff} = 42000 K, log g = 3.5, y = 0.16, for the indicated mass loss rates; the surface temperature for f_A = 1 is about 2000 K above the unblanketed model and that for f_A = 3, about 6000 K higher.

The classification lines for early-type O-stars are also affected by wind blanketing, particularly He I λ 4471 Å, as can be seen from Fig. 5 for the models of Fig. 3. Thus the spectral type depends not only on T_{eff} and log g, but also on the mass loss rate $\dot{M}$! An analysis by Bohannan et al. (1985) of line profiles of ζ Pup, obtained with a CCD camera, in which the full profiles of a large number of H, He I and He II lines were fit by wind-blanketed models, yields T_{eff} = 41500 K ± 1500, log g = 3.5 ± 0.1 and y 0.17 ± 0.03. An analysis of the same data using unblanketed models gave a value of T_{eff} larger by some 4000 K! The agreement between the parameters obtained by Bohannan et al. from wind-blanketed models and those of Kudritzki et al. (1983) from unblanketed models is fortuitous and arises from a systematic difference between He I λ 4471 Å measures in the two sets of observations.

IV. Metal opacity and non-LTE line blanketing.

Nearly all of the non-LTE models at present in use contain only H and He; the atmospheric structure from these models is used to calculate the line profiles for metals in order to determine abundances, but the effects of the metals on the structure is usually neglected. One exception is the grid of models calculated by Mihalas (1972), which contain a "mean light ion" intended to simulate the continuum opacity of the ground states of C, N, O and their ions. Husfeld et al. (1984) included more detailed C, N and O opacities and concluded that these opacities were of minor importance to the atmospheric structure.

Another effect of great potential significance for the photospheres of hot stars is the blocking of outflowing radiation by a large number of lines, particularly in the UV. Although the calculation of line-blanketed models for cooler stars in which LTE is a valid assumption has been brought to a remarkable degree of elaboration, principally by Kurucz (1979), the correct treatment for hot stars requires a full non-LTE treatment of hundreds or even thousands of lines. Anderson (1985) has developed an ingenious procedure for carrying out these calculations efficiently. To date Anderson has been able to include all H and He lines and the most important lines and continua of C and its ions, as well as the bound-free transitons of N, O, Ne, Mg, Si and S, in a model with T_{eff} = 35000 K and log g = 4.0. The surface temperature is controlled by the resonance doublet of C λλ 1548 Å, which cools the gas from about 28500 K in an unblanketed H-He model to about 17300 K. However, from the region of the temperature minimum of the corresponding H-He model inwards the temperature is essentially unchanged, i.e.

no back warming is seen, contrary to what might have been expected. The combined effects of wind-blanketing, which heats the surface layers, and line-blanketing, which causes them to cool, are not yet known. As the heat capacity of the very low density gas at the top of a classical photosphere is so low, a small increase in the radiation field can cause a substantial change in temperature. Moreover, it seems plausible that the C IV resonance line in the wind may seriously reduce the ability of this transition to cool the photosphere.

The spectrum in the visual and UV regions including the lines of H I, He I and He II differs little from that of a standard non-LTE H-He model, but the flux at the confluence of the Lyman series, shortwards of approximately λ 921 Å, lies an order of magnitude below the standard model. Shortwards of the Lyman limit there are more significant discrepancies.

V. Radiation-driven Winds

There is now quite strong evidence that stellar winds in hot stars are driven by the force of radiation streaming from the photosphere, primarily through its interaction with the spectral lines, and that this phenomena is described, in its overall behaviour, by an improved version of the theory of Castor, Abbott and Klein (1975; =CAK), based on an earlier form of Lucy and Solomon (1970). Although controversy on this subject continues, very little in the way of quantitative production has been forthcoming from alternative theories.

A major step forward was taken by Abbott (1982), who calculated the radiative force using a list of approximately 10^4 lines. Although the physical representation of the line force was by far more realistic than with the original CAK line force, significant descrepancies still remained. In particular, $\dot{M}$ was systematically too large for OB-stars by a factor of 2-3 and too low for Wolf-Rayet stars by an order of magnitude, while the terminal velocity v_∞ was predicted to be roughly 1.3 v_{esc} instead of the observed value of 2-4.

A number of recent further improvements to the theory have led to significantly better agreement with observation. Pauldrach, Puls and Kudritzki (1986) have introduced into the CAK theory a factor accounting for the finite diameter of the photospheric disk, which results in much larger values of v_∞, in substantial agreement with observations. (This effect was also considered by Friend (1982)). These authors also solved the radiative transfer and gas dynamical equations using a sample of lines from Abbott's list, with strong, intermediate and weak lines weighted to reproduce Abbott's line force. This calculation is similiar to that of Weber (1981) but with technical and physical improvements. The agreement of v_∞ from the two methods, and with observation is remarkable, and in most cases the value of $\dot{M}$ also agrees with observations as can be seen in the following table.

Comparison of observed and theoretical wind properties

star	Sp.Type	10^{-4} T_{eff}	log g	$R/R_\odot$	$\dot{M}$ (obs)	$\dot{M}$ (MCAK)	$\dot{M}$ (CMF)	v_∞ (obs)	v_∞ (MCAK)	v_∞ (CMF)
					\multicolumn{3}{c}{$10^{-6} M_\odot/yr$}		km/sec	
P Cyg	B1Ia	1.8	2.0	68	20.	29.	23.	400	395	355
ε Ori	B0Ia	2.9	3.3	37	3.1	3.3	2.6	2010	1950	2055
ζ Ori	O9.5I	3.0	3.4	29	2.3	1.9	-	2290	2274	-
ρ Sgr	O4(f)V	5.0	4.1	12	4.0	4.0	3.8	3440	3480	3860
HD48099	O6.5V	3.9	4.0	11	0.63	0.64	-	3500	3540	-
HD42088	O6.5V	4.0	4.1	6.2	0.13	0.20	0.20	2600	2600	2720
λ Cep	O6ef	4.2	3.7	17	4.0	5.1	4.3	2500	2500	2590

MCAK = Modified CAK, CMF = Co-Moving Frame (solution of transport and dynamical equations)

This theory, used with correctly determined atmospheric parameters for the Wolf-Rayet star V-444 Cygni including an effective temperature of approximately 90000 K (Cherepashchuk, Eaton and Khaliullin, 1984), reproduces not only the observed mass loss and terminal velocity, but also the run of velocity and density in the W-R wind as inferred by Cherepashchuk et al. from multi-colour light curves (Pauldrach, Puls, Hummer and Kudritzki, 1985).

Another important result of this work is that v_∞/v_{esc} is not a constant but a function of g, T_{eff} and R. Consequently its mass loss rate will not be a unique function of the location in the H-R diagram, but will also depend on the evolutionary state of the star. This circumstance seems not to have been appreciated by critics of the radiation-driven wind theory. We believe that the time-averaged stationary wind properties of hot stars can be explained by this theory.

The direct numerical solution of the coupled radiation-gas dynamical equations allows the calculation of line force to be carried out in the subsonic region of the flow where the Sobolev approximation is unreliable. Thus the properties of the radiation-driven flow can be followed down into the photosphere. It turns out that the radiation pressure in the lines, which is usually neglected in calculating the photospheric radiation pressure, may be comparable to the usual continuum contribution for low gravity objects. Therefore the value of log g inferred by fitting to line profiles will be too small; for low gravity objects the effect will be on the order of 0.15 dex (Pauldrach, Puls and Kudritzki, 1986). Such calculations also provide the basis for a realistic calculation of extended photospheres, as appropriate to both high-luminosity O-B stars as well as to certain nuclei of planetary nebulae. Extended spherical model atmospheres with density distribution obtained in this way are now being developed in Munich.

VI. The H-R Diagram of the Most Massive Galactic O-Stars

The techniques described in section II for ζ Pup have been applied to the galactic O3-stars in the η Car region (Kudritzki, 1980; Simon et al., 1983). The results are given in the following table.

Parameters of the most luminous galactic O-stars

star	spectral type	T_{eff} 10^3 K	log g (cgs)	y	log L/L$_\odot$	M_V
HD 93250	O3V((f))	52	3.95	0.09	6.3±0.1	-6.4±0.25
93128	"	48	3.85	0.09	5.7±0.1	-5.2±0.25
303308	"	45	3.90	0.09	5.8±0.1	-5.5±0.25
93129A	O3If	45	3.60	0.09	6.2±0.1	-6.6±0.25
ζ Pup	O4f	42	3.50	0.14	6.0±0.4	-6.0±1.00

The accuracy for T_{eff}, log g and y is ±3000 K, ±0.15 and ±0.02, respectively. It is obvious from the table that most of the O3 stars are significantly cooler than the previously assumed value of 53000 K assigned to all O3 stars by Conti and Burnichon (1975). Moreover, they reveal a large spread in effective temperature, absolute magnitude and luminosity, which means that the detailed spectroscopic analysis technique as described above is needed to construct a reliable HR-diagram for these very luminous blue objects.

The loci of these five objects in the (log g, log T_{eff}) and (log L, log T_{eff}) forms of the H-R diagram are given in Figure 6, along with the newly calculated evolutionary tracks of Pylyser, Doom and de Loore (1985), which include the effects of mass loss, adjusted to yield the Humphreys-Davidson limit, and of overshooting according to the Roxburgh (1978) criterion. As the transformation between the two diagrams is trivial for the theoretical tracks, because the mass at each point is known, while that for the observed stars involves photometric and distance information, this way of comparing theory with observations gives more information than either form of the H-R diagram alone. In the first place, if the locus of a par-

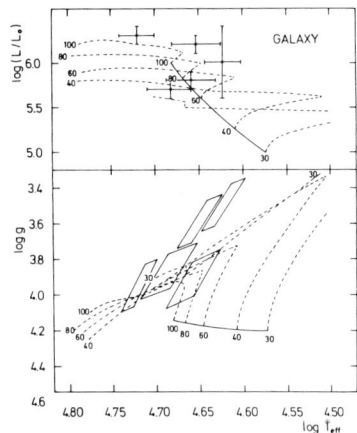

Fig. 6: (see text)

Fig. 7: (see text)

ticular star does not have a consistent relation to the track of a given mass, one is alerted to the certainty of error. Moreover the luminosity form gives better mass discrimination, while the log g form allows for a more direct test of evolution theory (but only if T_{eff} and log g are determined with sufficient precision).

In the present case we see that HD 93128 looks somewhat suspicious, although the extremes of the error bars include the same mass tracks. Also the two low-gravity stars HD 93129A and ζ Pup seem to lie outside the region of the tracks. It is therefore of interest to note the consequences of two effects discussed above (sections IV and V) which have not yet been included in the determination of these parameters: 1) wind blanketing can be expected to reduce T_{eff} by 4000-5000 K (except for ζ Pup) and 2) the effect of line radiation pressure in the photosphere can be expected to increase log g of the two low-gravity stars by approximately 0.15 dex. Thus the error boxes of these two stars will be expected to move down in the (log g, log T_{eff})-plane by 0.15 dex and the loci of all stars (except ζ Pup) will be shifted to the right by approximately .04 dex, as shown in Fig. 7. We stress that these changes are not based on detailed calculations for the individual objects. The agreement of the photometrically determined masses with the track masses seems to be considerably better in Fig. 7.

It is interesting that ζ Pup has both the largest luminosity and an enhanced helium content, indicating that it is substantially further evolved than the other four stars. Butler and Simon (1985) find in a preliminary analysis that its N abundance is enhanced by a factor between 6 and 10.

VII. Abundances in Massive O-stars

Non-LTE abundance determinations are based on statistical equilibrium calculations for the element in question using the appropriate model atmosphere to provide the run of temperatures, density and radiation field. Rotational broadening and collison broadening enter at this stage. The demands for atomic radiative and collisional data are enormous and frequently rather unreliable data must be used. Detailed abundance analyses have been made in this way by many workers, which cannot be reviewed here for lack of space. However, very recently Schönberner, Kudritzki and Simon (1984, 1985) have carried out an abundance study of He, C, N, O in 7 O-stars. Four of these objects are socalled "ON-stars", as their spectra show abnormally strong lines of nitrogen (Walborn, 1970, 1971, 1976). The other three objects are normal O-stars, which are used as standards for a differential study of the abundance of the strategic CNO-elements as an important prerequisite for a discussion of the evolutionary history of ON-stars.

The abundance analysis proceeds in two steps. First, the atmospheric parameters (T_{eff}, log g, y) are determined by the same techniques described in section II. This leads to a very interesting result which is displayed in the following table:

Atmospheric parameters

	normal stars				ON-stars			
name	$10^{-3} T_{eff}$	log g (cgs)	y	name	$10^{-3} T_{eff}$	log g (cgs)	y
τ Sco	33	4.15	0.1	HD 89137	30.0	3.25	0.26
10 Lac	38	4.25	0.09	θ Car	32.5	4.15	0.17
15 Mon	41	4.1	0.08	HD 14633	35.5	3.70	0.15
				HD 48279	37.5	4.00	0.15

Strikingly, all ON-stars are helium enriched, whereas the normal stars have normal helium abundance! The helium enrichment is strongest for HD 89137, which as a low gravity object has evolved away from the ZAMS.

An obvious question is whether the helium enrichment is due to the CNO-cycle. This is investigated in the second step where the C,N,O-lines (optical and UV high resolution spectra) are analysed on the basis of final non-LTE models and line-formation calculations (the latter carried out totally in a NLTE multi-level form for N III, but in LTE for the other ions). The results are given in the table:

CNO-Abundance of O and ON-stars

	normal stars						ON-stars				
object	C	N	O	N/C	N/O	object	C	N	O	N/C	N/O
τ Sco	-3.9*	-3.9	-2.9	1.0	0.2	HD 89137	-4.2	-2.3	-2.9	80	4
10 Lac	-3.2	-3.9	-2.7	0.2	0.06	θ Car	-4.7	-2.9	-3.1	65	1.5
15 Mon	-3.1	-4.0	-	0.1	-	HD 14633	-4.5	-2.7	-3.1	60	2.5
Sun	-3.36	-4.04	-3.12	0.2	0.1	HD 48279	-4.1	-2.6	-2.7	30	1.0

* columns C, N and O contain log ε, where ε is the number fraction: $\varepsilon = n_x / \Sigma n_x$

The abundances of the normal stars are essentially solar (except for carbon in τ Tau, which is well known already (Hardorp and Scholz, 1970)). However, for the ON-stars nitrogen is strongly enriched, carbon is strongly depleted and O remains about solar. This allows one to draw the firm conclusion that these stars are exposing nuclear processed matter at their surfaces. The elemental distribution corresponds to that of the incomplete CNO-cycle, where only the CN-equilibrium is well established (see Fig. 8). Fig. 9 shows the (log g, log T_{eff})-

diagram of the objects compared with evolutionary tracks by Maeder, 1982, which include convective mixing, overshooting and mass-loss. Qualitatively, the idea of photospheric enrichment by nuclear processed material as evolution proceeds is supported by Fig. 9. However, quantitativley the abundances are not in agreement. It is also not clear why the two ON-stars close to the ZAMS show CN-burned material. As they are binaries, the turbulent mixing due to shear instabilities is perhaps more efficient in carrying CN-burned matter to the surface.

VIII. Abundance Gradients in the Galactic Disk from Young Cluster B-stars

Observed abundance distributions in the disk set important constraints on theories of galactic evolution (Güsten and Mezger, 1983) and the abundance ratios (especially isotopic ratios) contain in principle information on the production processes (Talbot and Arnett, 1973). Analyses of emission lines of planetary nebulae and H II regions have yielded logarithmic abundance gradients of the order of -0.05 dex/kpc for N and O. Very recently this question has been investigated by Gehren et al. (1985), who determined the N and O abundances of a large sample of young B-stars in open clusters lying in the Galactic disk. As these stars are very young their abundances represent those of the gas from which they were formed. Moreover they are well described by stellar models, since convection is negligible and deviations from LTE are small. Using the Cassegrain echelle spectrograph (CASPEC)

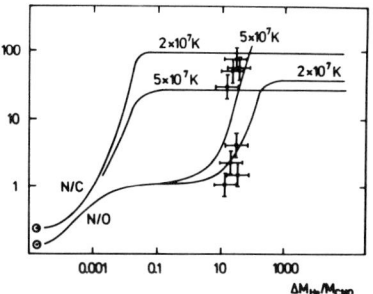

Fig. 8. Theoretical ratios of N/C and N/O as a function of newly created helium per total CNO-abundance (according to Caughlan, 1965). The observed ratios for the ON-stars are also plotted.

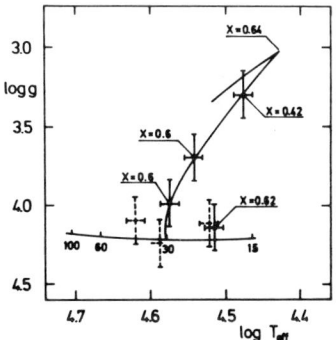

Fig. 9. The (log g, log T_{eff})-diagram of ON-stars (full drawn crosses) and normal O-star standard (dashed crosses). Evolutionary tracks by Maeder (1982) are also shown. For the ON-stars the hydrogen mass-fraction is indicated.

with a CCD detector on the ESO 3.6 m, high-resolution spectra have been obtained for remote open clusters, covering a range of galactocentric distances from 8.5 to 19.5 kpc.

A preliminary analysis of this material has been carried out using LTE model atmospheres, which should be adequate for a differential analysis in view of the low effective temperatures of these stars (T_{eff} < 30000 K). A grid of LTE models with solar He/H and metal/H abundances was calculated allowing for the Balmer and Lyman series as well as the

100 strongest metal lines. The resulting temperature stratifications were nearly identical with those obtained by Kurucz (1979). T_{eff} and log g were determined from fits to $H\beta$ and $H\gamma$ line profiles and from the relative strengths of Si II, Si III and Si IV lines. For each star 25 N II and 120 O II lines were used to determine the abundances. All abundances were determined relative to the standard star BS 2928, for which the abundances relative to the sun were found by Lambert (1978). This approach minimizes systematic errors arising from uncertain oscillator strengths and line broadening parameters, as well as small non-LTE effects (which may affect the sample as a whole). Figure 10 shows the abundance ratios N/H, O/H and N/O as functions of galactocentric distance. These results are consistent with a conclusion that no significant abundance gradient exists in the outer part of the Galaxy. Moreover, the scatter of roughly a factor of 2 in abundances for a given cluster is real.

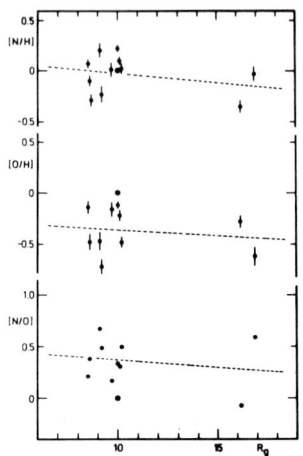

Fig. 10: Logarithmic N and O abundances and N/O ratios vs. galctocentric distance R_g. 2σ error bars refer to rms. error of single stars and include systematic errors. Dashed lines are least square fits.

IX. Future Developments

In addition to the recent developments in the theory of stellar atmospheres and winds described above, a very important contribution is now being made by the London-Belfast-Boulder opacity project, which is producing photoionization cross sections, oscillator strengths, damping coefficients and electron scattering cross sections of unprecedented accuracy and completeness for all atoms and ions of astrophysical interest.

These developments are matched by dramatic improvements in observational techniques. Capabilities for UV and IR spectroscopy continue to increase by leaps and bounds. Modern spectrographs and solid state detectors of high quantum efficiency and (hopefully) linear response make possible the spectroscopy of a wide range of stars, with characteristics summarized in the following Table:

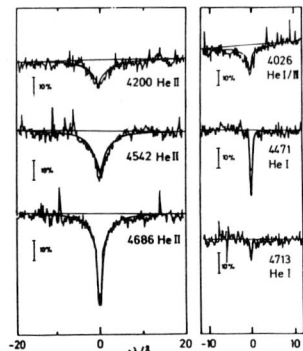

Fig. 11: He I and He II lines of ROB 162, and profiles calculated from final models.

m_V	S/N	$\Delta\lambda/\lambda$	t_{exp}
$\leq 7^m$	100-1000	10^4-10^5	< 60 min
7^m-14^m	30-100	2×10^9	< 90 min
14^m-20^m	10-50	2×10^3	120 min

Consequently accurate quantitative spectroscopy on very distant as well as nearby stars is now feasible. Thus we can investigate in detail the evolution of stars in all stages, which in turn allows us to study the structure and chemical evolution of galaxies.

An example of recent work illustrates the current situation. Figure 11 shows the He lines of ROB162, the only hot sdO star (m_V = 13.3) in the metal-deficient globular cluster NGC 6397. These spectra, with S/N≈ 30 and $\lambda/\Delta\lambda \approx 2\times10^4$, were obtained with 60 minute exposures on the ESO 3.6 m using CASPEC with a CCD detector. Heber and Kudritzki (1984) have determined the parameters T_{eff} = 51000 K ± 2000 K, log g = 4.5 ± 0.2 and n_{He}/n_H = 0.1 ± 0.02; calculated profiles for models with these parameters are given in the Figure.

We and collaborators in Munich and Boulder have underway a long-term project studying the evolution of massive stars in the Galaxy and the Magellanic Clouds, using spectra obtained with CASPEC on the ESO 3.6 m. We plan to obtain UV spectra with the High Resolution Spectrograph when Space Telescope becomes available. By means of wind-blanketed model atmospheres we can analyze these spectra, as outlined above, to determine T_{eff}, log g, abundances and M for a large number of stars. Thus we will obtain reliable H-R diagrams as well as information on chemical abundances and their spatial distributions for three galaxies of substantially different metallicities. These data provide meaningful constraints on many aspects of stellar evolution theory: main sequence location and width, the evolution of massive stars, the effects of mass loss and turbulent mixing, CNO-production and the evolution to WR stars and Type I Supernovae, all as functions of galactic metallicity. Moreover various aspects of chemical evolution, such as abundance gradients and IMF, can be studied as functions of Z. We are also interested in the role of hot stars as sources of ionization of H II regions: are the properties and abundances of ionized regions consistent with those of the ionizing stars? A first step in this program is reported by Gehren et al. (1986).

A further, and possibly more far reaching, application of the developments can now be considered - the use of massive stars as distance indicators. The idea is simple: by determining T_{eff} and log g from spectral lines by means of non-LTE model analysis and observing the terminal velocity v_∞, one has three relations among the three fundamental quantities mass, luminosity and radius. Specifically, from the relation between v_∞ and v_{esc} given above, and knowing v_∞, T_{eff} and g, one finds R and thus L or M_V. From a preliminary analysis, it appears that one can determine g to sufficient accuracy from observations of the Balmer lines with a FWHM resolution of < 3Å, or perhaps even 5Å if good S/N can be obtained. High accuracy in T_{eff} is not needed. For faint objects, Space Telescope will make possible the observations of UV lines to find v_∞.

We can see how well this idea works by using it on four of the five stars discussed in Section VI for which v_∞ is known from IUE observations. These data are summarized in the following table:

Star	v_∞	$r/R_\odot$	M_V (derived)	M_V (literature)
HD 93250	3300 km/s	19	-6.5	-6.4
93129	3950	27	-7.1	-6.6
303308	3200	14	-5.7	-5.5
ζ Pup	2400	19.5	-6.3	-6.0

These results encourage us to believe that we can provide new information on distances of galactic OB stars and central stars of planetary nebulae. Can this purely spectroscopic method be used for extragalactic objects? As an example, consider Space Telescope with the Faint Object Camera or Faint Object Spectrograph, using the long slit. With $\lambda/\Delta\lambda \approx$ 1000-2000, the limiting magnitude is $m_V \approx 23$ for both blue and V spectral regions. Thus with $M_V \approx -6$, we have $m-M_V \approx 29^m$.

D.G. Hummer acknowledges a "Senior U.S. Scientist Award" from the Alexander von Humboldt Stiftung.

References

Abbott, D.C., 1982, Ap.J. 281, 774
Abbott, D.C. and Hummer, D.G., 1985, Ap.J., 294, 286
Abbott, D.C. and Lucy, L.B., 1985, Ap.J., 288, 679
Anderson, L.S., 1985, Ap.J., 15 Nov.
Auer, L.H and Mihalas, D., 1972, Ap.J.Suppl., 24, 153
Auer, L.H and Mihalas, D., 1973, Ap.J.Suppl., 25, 433
Baschek, B. and Scholz, M., 1971, Astron. Astrophys., 15, 285
Bohannan, B., Abbott, D.C, Voels, S.A. and Hummer, D.G., 1985, these Proceedings
Brune, W.H., Mount,G.H. and Feldman, P.D., 1979, Ap.J. 227, 884
Butler, K. and Simon, K.P., 1985, Proc. ESP Workshop on CNO Abundances, ed. I.J. Danziger
Castor, J.I., Abbott, D.C. and Klein, R.I., 1985, Ap.J. 195, 157
Cherepashchuk, A.M., Eaton, J.A. and Khaliullin, K.F., 1984, Ap.J. 281, 744
Code, A.D., Davis, J., Bless, R.C, and Hanbury Brown, R., 1970, MNRAS 189, 601
Code, A.D and Meade, M.R., 1979, Ap.J.Suppl. 39, 195
Conti, P.S., 1973, Ap.J. 179, 181
Conti, P.S. and Alschuler, W.R., 1971, Ap.J. 170, 325
Conti, P.S. and Frost, S.A., 1977, Ap.J. 212, 728

Davis, J., Morton, D.C., Allen, L.R., and Hanbury Brown, R., 1970, MNRAS **150**, 45
Friend, D.B., 1972, Ph.D.Thesis, Univ. of Colorado
Gehren, T., Husfeld, D., Kudritzki, R.P., Conti, P.S. and Hummer, D.G., 1986, these Proceedings
Gehren, T., Nissen, P.E., Kudritzki, R.P. and Butler, K., 1985, Proc. ESO Workshop on CNO Abundances, ed. J.J. Danziger
Güsten, R. and Mezger, P.G., 1983, Vistas in Astronomy **26**, 159
Hanbury Brown, R., Davis, J. and Allen, L.R., 1974, MNRAS **167**, 121
Hardorp, J. and Scholz, M., 1970, Ap.J.Suppl **19**, 193
Heber, U. and Kudritzki, R.P., 1984, Mitt. Astron. Ges. **62**, 249
Holm, A.V. and Cassinelli, J.P., 1977, Ap.J. **211**, 432
Hummer, D.G., 1982, Ap.J. **257**, 724
Husfeld, D., 1982, Diplomarbeit, Univ. Kiel
Husfeld, D. and Kudritzki, R.P., 1983, Mitt. Astron. Ges. **60**, 306
Husfeld, D., Kudritzki, R.P., Simon, K.P. and Clegg, R.E.S., 1984, Astron. Astrophys. **134**, 139
Jamar, C., Macau-Hercot, D., Monfils, A., Thompson, G.I., Hoziaux, L., and Wilson, R., 1976, Ultraviolet Bright-star Spectrophotometric Catalogue (ESA SR-27)
Johnson, H.L. and Mitchell, R.I., 1975, Rev. Mex. Astrf. Af. **1**, Nr. 3
Kudritzki, R.P., 1973, Astron. Astrophys. **28**, 103
Kudritzki, R.P., 1976, Astron. Astrophys. **52**, 11
Kudritzki, R.P., 1979, Proc. of 22nd Liege conf., 295
Kudritzki, R.P., 1980, Astron. Astrophys. **85**, 174
Kudritzki, R.P., 1985, Proc. ESO Workshop on CNO Abundances, ed. I.J. Danziger
Kudritzki, R.P., Simon, K.P. and Hamann, W.R., 1983, Astron. Astrophys. **118**, 245
Kurucz, R., 1979, Ap.J.Suppl. **40**, 1
Lambert, D.L., 1978, MNRAS **182**, 249
Lucy, L.B. and Solomon, P., 1970, Ap.J. **159**, 879
Maeder, A., 1982, Astron. Astrophys. **105**, 149
Mihalas, D., 1972, Non-LTE Model Atmospheres for B and O Stars, NCAR-TN/STR-76, National Center for Atmospheric Research, Boulder
Pauldrach, A., Puls, J., Hummer, D.G. and Kudrizki, R.P., 1985, Astron. Astrophys. **148**, L1
Pauldrach, A., Puls, J. and Kudritzki, R.P., 1986, Astron. Astrophys., to be submitted
Peterson, D.M. and Scholz, M., 1971, Ap.J. **163**, 51
Pylyser,E., Doom, C. and de Loore, C., 1985, Astron. Astrophys., in press
Remie, H. and Lamers, H.J.G.L.M., 1982, Astron. Astrophys. **105**, 85
Roxburgh, I.W., 1979, Astron. Astrophys. **65**, 281
Schönberner, D., Kudritzki, R.P. and Simon, K.P., 1984, Proc. 4th Europ. IUE Conf., ESA SP-118, p. 267
Schönberner, D., Kudritzki, R.P. and Simon, K.P., 1985, Astron. Astrophys., to appear
Simon, K.P., Jonas, G., Kudritzki, R.P. and Rahe, J., 1983, Astron. Astropys. **125**, 34
Talbot, R.J. and Arnett, W.D., 1973, Ap.J. **186**, 51

Tobin, W., 1983, Astron. Astrophys. 125, 168
Underhill, A.B., Divan, L., Prevot-Burnichon, M.L., Doazan, V., 1979, MNRAS 189, 601
Walborn, N.R., 1970, Ap.J. 161, L149
Walborn, N.R., 1971, Ap.J. 164, L67
Walborn, N.R., 1976, Ap.J. 205, 419
Weber, S.V., 1981, Ap.J. 243, 954

Discussion : KUDRITZKI & HUMMER

SREENIVASAN :

I am happy to learn of the improvements you have made to stellar wind-theory. Have you also included a realistic energy conservation statement in the theory? CAK assumed radiative equilibrium and used a temperature profile in accordance with that assumption.

KUDRITZKI :

We have performed test calculations, which show that the temperature stratification in the wind is not crucial for the dynamics of the wind. We have used constant temperatures of different values as well as radiative equilibrium stratifications. That had nearly no effect on the mass loss rate as well as on V_{inf}.

LAMERS :

The observations show that the ratio V_{inf}/V_{esc} decreases with temperature from 3.5 at $T_{eff} \geq 30.000K$ to 1 at T_{eff} 10.000 K. This relation is important in your proposed methods for distance calibration. Do your calculations agree with this observed relation?

KUDRITZKI :

Yes, at least down to T_{eff} 20.000 K (P Cygni). We did not yet calculate winds for cooler stars like, for instance, alpha Cygni (T_{eff} 9000 K), where it is not clear whether our theory still works. The advantage of these cooler objects is that their absolute V-magnitude is brighter. However V_{inf} is propably too small to be measured with sufficient accuracy.

LUMINOUS BLUE STARS: DISTRIBUTION AND NUMBERS

Catharine D. Garmany
Joint Institute for Laboratory Astrophysics
University of Colorado and National Bureau of Standards
Boulder, Colorado 80309

I. INTRODUCTION

Because the term "luminous blue stars" is relative, let me begin by delineating what region of the H-R diagram I am considering. For the purpose of this talk, I will discuss primarily O-type stars, with $T_{eff} > 30,000$ K and $M_{bol} > -7$, but I will also mention their evolved descendants the B supergiants with $30,000$ K $> T_{eff} > 10,000$ K and $M_{bol} > -8$, and their even later evolutionary form, the Wolf-Rayet (W-R) stars with $T_{eff} > 30,000$ K and uncertain M_{bol}. Although these stars are not among the visually brightest stars, they are the bolometrically most luminous as well as the hottest stars. They are also the most massive. They are an important channel in the metal enrichment in a galaxy through the action of mass loss via stellar winds and their ultimate disruption as supernovae (Maeder 1981). They also contribute in a major way to the energy balance of the interstellar medium (Abbott 1982).

If we are to understand galactic evolution in detail, it is vital that we understand stellar evolution and star formation rates. Two fundamental questions, posed by Scalo (1984) and Freedman (1985) among others, are: how does the star formation rate (SFR) vary among galaxies, and can all star forming clouds produce the initial mass function (IMF) each time? It is natural to start seeking answers to these questions at the massive end of the spectrum and to expect that variations in the SFR or the IMF will affect the entire top of the H-R diagram.

W-R stars are currently observable in many Local Group galaxies, and their distribution provides tantalizing clues (Massey, this volume). However, understanding these clues requires the total evolutionary picture for massive stars. At present, such data are only attainable for our Galaxy and the Magellanic Clouds. This talk will deal with massive stars within 3 kpc of the Sun and in the LMC and SMC. Section II will cover what observations are required to resolve these stars and the calibrations used to construct an H-R diagram. Section III deals with our current knowledge of massive stars in the solar vicinity and Sec. IV with massive stars in the Magellanic Clouds.

II. PARAMETERS FOR MASSIVE STARS

To answer questions about massive star evolution, SFR and the IMF, we must start with a data set. Perhaps the most universally useful is a "theoretical H-R diagram," a term used by Humphreys (1983), and consisting of M_{bol} vs. T_{eff} for a volume limited sample of stars. The sample could consist in principle of a single cluster at a known distance, but because the mass function tapers off so rapidly for massive stars, a cluster invariably suffers from small number statistics. It also ignores the issue of the field O stars. Hence a number of recent studies have used a large volume of space and attempted to count all the stars within that volume and assign an M_{bol} and T_{eff} to each star.

Let us first consider problems connected with determining M_{bol} and T_{eff}. The observations generally include apparent magnitude and colors, usually V, (B-V) and (U-B), a spectral type which varies in quality from a low dispersion objective prism type to a high quality MK type, and information on cluster membership. Distances for cluster members usually rely on cluster fitting for less massive but more numerous B stars. These cluster stars have then been used to define the absolute magnitude calibration as a function of spectral type for the O and B supergiants. This calibration has been thoroughly discussed by Blaauw (1963); it has been reexamined over the years by Walborn (1972), Conti et al. (1983) and Humphreys and McElroy (1984), as more cluster stars have become available. However, it is well to remember that these basic calibrations still make use of the old distance modulus for the Hyades! The difference between the old and new Hyades distance modulus is 0.2 in M_V, but before we rush to increase the M_V calibration for O stars, it should be noted that Blaauw (1963) also discussed an independent determination of the M_V calibration from the Scorpio-Centaurus association and confirmed the results based on the (old) Hyades calibration. Until a careful reexamination of cluster fitting from the Hyades to h and χ Persei is made, we should keep in mind that the systematic uncertainty in the M_V calibration possibly exceeds 0.3 mag. The current state of M_V vs. spectral type is shown in Fig. 1.

The radical difference between an H-R diagram based on M_V rather than on M_{bol} is illustrated by Massey (1985, Fig. 3). The bolometric correction is usually computed as a function of T_{eff}. As temperature determinations are discussed by Kudritzki (this volume), I will say nothing further. Figure 2 shows bolometric correction vs. T_{eff} as recently computed by different authors.

For stars that are apparently not cluster members, distance must be computed from the relation:

$$V - M_V = 5 \log D - 5 + 3.1 \, [(B-V)-(B-V)_0)]$$

In addition to the uncertainty in the M_V calibration, this equation contains assumptions about the ratio of total to selective absorption and about the intrinsic color of the star. How well are the intrinsic colors known? As there are no nearby unreddened O stars, these numbers are based on a large degree of extrapolation (FitzGerald 1970).

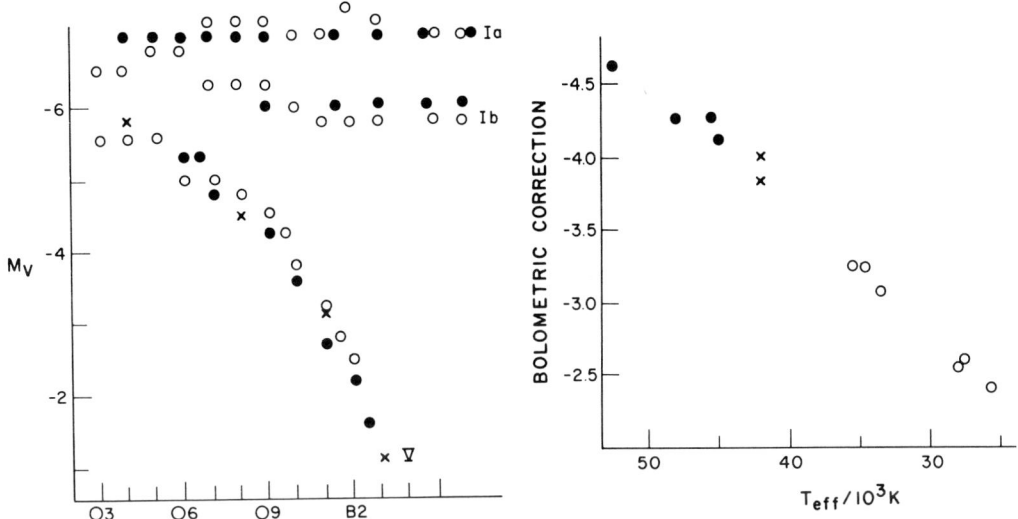

Fig. 1. The M_V-spectral type calibration by different authors. Walborn 1972 (●), Garrison 1978 (×), Humphreys and McElroy 1984 (∇).

Fig. 2. Bolometric correction vs. temperature. Kudritzki et al. (1983) (●), Remie and Lamers 1982 (o), Abbott and Hummer 1985 (×).

Furthermore, as Massey (1985) has illustrated, the intrinsic UBV colors for O stars are degenerate in the sense that there is little difference between the $(U-B)_0$ or $(B-V)_0$ for an early or a late O star. Recently, Massa and Savage (1985) have found from a study of Galactic stars that none of the O stars have UV colors much bluer than normal BO V stars. We will discuss further evidence for intrinsically redder colors for O stars in Sec. IV.

III. GALACTIC STARS

How complete are existing catalogs of massive stars? In the past 10 years there have been repeated updates of such lists: Cruz-Gonzales et al. (1974) (600 O stars), Humphreys (1978, Galactic associations), Garmany, Conti, and Chiosi (1982, 780 O-stars), Humphreys and McElroy (1984, O and B stars). Garmany et al. felt the O-star were essentially complete to 2.5 kpc from the Sun; Humphreys and McElroy felt the O and B stars were complete to 3 kpc for stars brighter than $M_{bol} = -8$. However, a complete list of stars is an elusive goal, as we have found in updating our list of O stars (Garmany 1984). We currently have a computer list of 1088 O stars, although data on some of these stars are not complete. This is a 40% increase over our list in 1982, and some of the new additions lie within 2.5 kpc of the Sun.

The arguments for completeness within a given volume of space rely on the time-honored technique of star counts, which in the Galaxy should increase with the square of the distance if the stars are confined to the plane. That they are so situated is shown in Fig. 3, which gives the distribution in Galactic Z of the O stars in our Catalog and the luminous B stars from Humphreys and McElroy. These stars all lie in the plane of the Galaxy, but Fig. 3 also shows that the plane is warped. The midplane is negative in the third and fourth quadrant and positive in the first and second.

Now, if we examine a plot of the log (number of O and B stars) vs. distance we find that the slope of the line is just about 2 out to 3 kpc, and then becomes flat. To zero order, this suggests that our most current list of O and B stars is basically complete to 3 kpc. But should these stars follow a uniform star count relation? We are surveying a region containing three Galactic spiral arms, and the total number of O and B stars within the solar circle is significantly greater than the number outside the solar circle. Indeed, a plot of log (number stars) vs. distance for the region within the solar circle has a slope greater than 2, reflecting the interarm gap between the Sun and the Carina-Sagittarius arm. The same plot for the region outside the solar circle has a slope less than 2, which probably reflects the gap in the arms between the local arm (or spur) and the Perseus arm. See Fig. 4.

The nonuniform distribution of O and B stars is even more striking if we consider only stars that have initial masses greater than 40 $M_\odot$ (Fig. 5) according to evolutionary models. The major

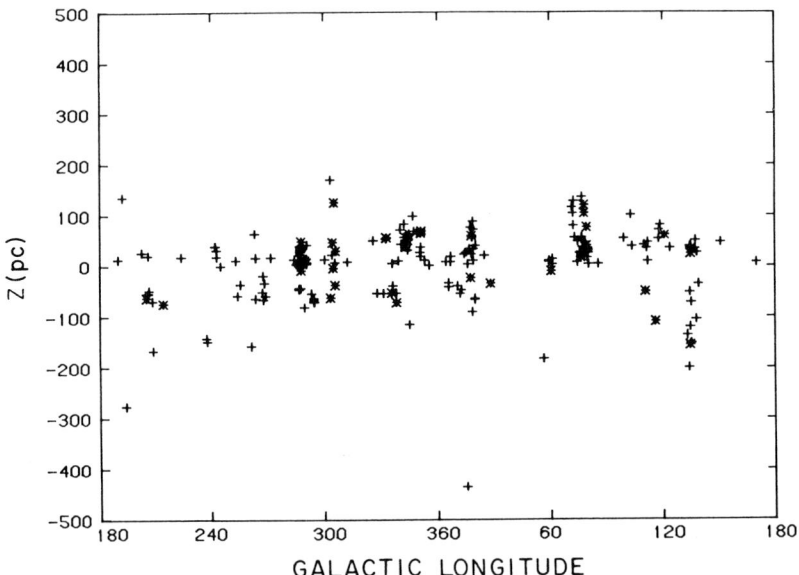

Fig. 3. The O and luminous B stars in the plane of the Galaxy.

LUMINOUS BLUE STARS: DISTRIBUTION AND NUMBERS

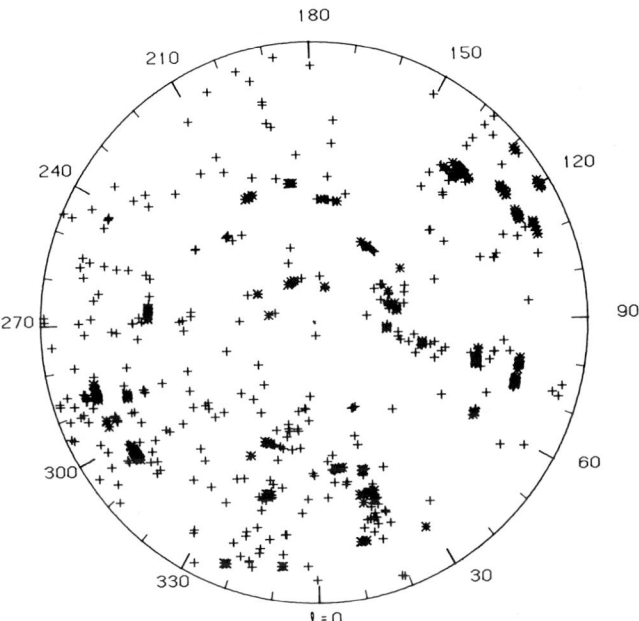

Fig. 4. The O and luminous B stars projected onto the Galactic plane within 3 kpc of the Sun.

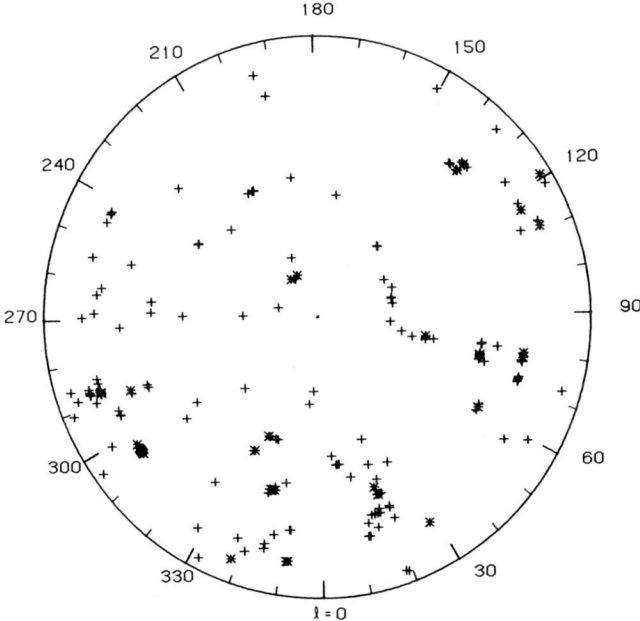

Fig. 5. Stars with initial mass greater than 40 $M_\odot$ projected onto the Galactic plane within 3 kpc of the Sun.

spiral arms are defined, but what is more interesting, there is a very strong gradient in the space density of stars. As pointed out by Conti et al. (1983) the same gradient is observed in the distribution of W-R stars: O and B stars of 40 $M_\odot$ and greater, as well as W-R stars, are preferentially found in the Carina-Sagittarius arm and the Cygnus arm, but not in the Perseus arm or Orion spur. This offers further indirect evidence that W-R stars are the descendants of very massive O stars, and suggests that probing the W-R population in other galaxies will tell us about the uppermost end of the mass function. (See Conti, this volume, Fig. 6 for the distribution of W-R stars in the plane.)

The greater space density of massive stars and W-R stars within the solar circle is incontrovertible; the cause of this is subject to debate. Garmany et al. (1982) argued that the data on O stars indicated a real difference in the upper end of the initial mass function (IMF) between the inner and outer region of the Galaxy. Humphreys and McElroy (1984), using data on the O and B stars, have concluded that only the star formation rate (SFR) is different, and that differences in the slope of the IMF can be explained by missing late-type O stars in the inner region of the Galaxy. (It should be emphasized here that the region of the Galaxy under consideration extends only 3 kpc from the Sun. By comparison, surveys of giant molecular clouds (Cohen et al. 1985; Solomon et al. 1985) extend 10-15 kpc from the Sun.)

I have reexamined this issue, using our current O-star list which contains 777 stars within 3 kpc and also the Humphreys and McElroy (1984) list of B stars brighter than $M_{bol} = -7$ within the same distance, for a total of 1,041 stars. Table 1 shows the distribution by mass, using evolutionary tracks by Bressan et al. (1981), for these stars divided into the region interior to and exterior to the solar circle. Indeed, the ratio of stars interior to the Sun to those exterior to the Sun is 1.8 (671/370) but the ratio of stars more massive than 40 $M_\odot$ inward vs. outward is 3.0 (128/42). If the difference is entirely caused by missing stars in the mass range 20-40 $M_\odot$, then we have to conclude that all these stars have been detected outside the solar circle, but only 55% of them have been detected inside the solar

Table 1. Number of stars within 3 kpc interior and exterior to the solar circle.

Mass interval	Interior	Exterior
>100	2	0
80 - 100	4	1
60 - 80	27	5
40 - 60	95	36
30 - 40	174	80
20 - 30	369	248

LUMINOUS BLUE STARS: DISTRIBUTION AND NUMBERS

circle. Why should there be such a disparity? One suggestion is that the missing stars are hidden in molecular clouds. However, opinions differ on whether giant molecular clouds might be preferentially hiding early O-type stars or late O stars.

IV. MASSIVE STARS IN THE MAGELLANIC CLOUDS

The Magellanic Clouds provide an ideal laboratory for comparison of massive stars. They are close enough to observe normal main sequence O stars, their reddening is very low, and they are far enough out of the Galactic plane that foreground contamination of blue stars is not an issue. In addition, there are a number of fascinating contrasts between the LMC and the SMC. The LMC is more massive than the SMC by a factor of 4 to 5 (Fujimoto 1979). The difference in metallicity (based on H II regions) has been studied by Dufour (1984) who finds that compared to the Galaxy, the LMC is deficient by 2-4 and the SMC by 5-20 for CNO.

The indications of massive star formation all point toward differences between the LMC and SMC. Davies et al. (1976) found that H II regions are richer and larger in the LMC than the SMC. Dark nebulae in the SMC are about half the size of those in the LMC (Hodge 1974). Stellar associations of blue stars are much richer in the LMC than in the SMC (Lucke 1974; Hodge 1985). Although surveys of CO are incomplete, it appears that CO clouds are much more widespread in the LMC than the SMC (Rubio, Cohen and Montani 1984). Humphreys (1983) found that luminous blue stars are deficient in the SMC compared to the LMC and the Galaxy. The W-R population is dramatically different: there are 105 known W-R stars in the LMC (Breysacher 1981; Hutchings et al. 1984; Conti and Garmany 1983) but only 8 in the SMC (Azzopardi and Breysacher 1979).

There have been a number of studies of the luminosity function (LF) of the luminous stars in the Clouds, which are reviewed by Freedman (1984). As the IMF can, in principle, be derived from the luminosity function, these studies should reflect the current state of the massive star population. In general, it has been found that the slope of the LF, or the IMF, is similar in the Clouds, and similar to the Galaxy (Lequeux et al. 1980, Humphreys and McElroy 1984, Freedman 1984). What can we conclude from these studies about the O-star population in the Clouds? The answer seems to be not very much, in large part because much of the available stellar data include only photometry (UBV) and objective prism spectroscopy. As pointed out in Sec. II, photometric determinations of spectral type carry a large uncertainty. Slit spectra are required to resolve early from late O, or early B stars.

Conti, Massey and I began a program a few years ago to obtain slit spectra at classification dispersion for candidate O stars in the Clouds. So far we have taken 175 spectra in the LMC and 134 in the SMC with the CTIO 4-meter, image tube and IIIa-J plates at a dispersion of 47 Å/mm. Our observing lists have been extracted from catalogs of early-type stars in the Clouds. In the LMC this includes

Rousseau et al.'s (1978) catalog of 1822 stars, from which we chose candidates having objective prism classifications, many by Sanduleak (1969). In the SMC we have used the catalog by Azzopardi and Vigneau (1982) of 524 stars. Of course, some stars in these catalogs have already been classified from slit spectra by Walborn (1977), Crampton (1979), Crampton and Greasley (1982), and Humphreys (1983); our observing list does not include these stars except for comparison.

Combining our new data with published data that include both spectral types and photometry confirms our feeling that photometry is useful in choosing candidates, but cannot identify O stars unambiguously. Figure 6 is a color-color plot for LMC stars classified as O6.5 or earlier, either from published work, on our new classification. The line defining unreddened class V stars (FitzGerald 1970) is also indicated, and an arrow shows the magnitude and slope of the average foreground reddening (McNamara and Feltz 1980). Examination of this figure suggests either a much steeper reddening curve applies in the LMC or the unreddened colors are too blue.

From the UV, there is evidence that the reddening law is steeper and the extinction is much higher than in the Galaxy (Nandy et al. 1981, Fitzpatrick and Savage 1984). Although the reddening law might be somewhat steeper in the optical region, it could not explain the position of the stars in Fig. 6. Figure 7 is a similar color-color plot, but contains stars classified O6.5 through O9. There is little difference in the position of these stars and the early O stars, suggesting that the intrinsic colors of O stars may be redder than generally assumed. As mentioned earlier, Massa and Savage (1985) have found from the UV that the colors of O stars are not any bluer than a normal B0 V star. (A detailed study of this matter is in preparation.)

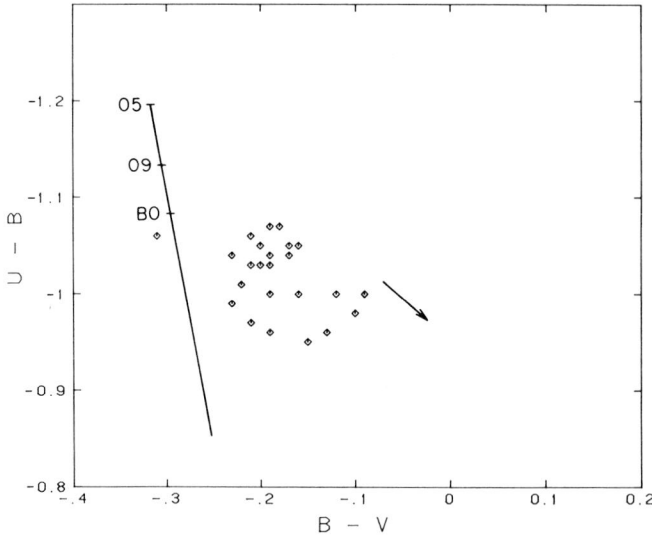

Fig. 6. Color-color plot for stars in the LMC earlier than O6.5.

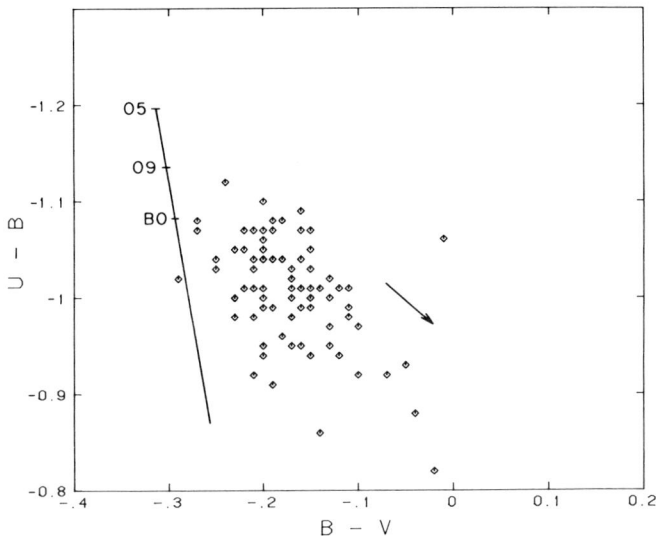

Fig. 7. Color-color plot for stars in the LMC classified as type O6.5 through O9.

How much progress has been made in identifying a major portion of the O stars in the LMC? Although we have observed most of the candidates, there are a great number of stars not catalogued by Rousseau et al. (1978). These include the stars in the associations studied photographically by Hodge and Lucke (1970). We have obtained spectra for 37 stars in four of these associations and discovered 29 new O stars. Eastwood and Massey have begun a program to obtain CCD frames of the Lucke-Hodge associations; this will tell us a great deal more about massive stars in the LMC. An H-R diagram of the LMC based on all available data represents mainly field O and B stars, so it is premature to compare it with the Galaxy.

Our search for massive stars in the SMC seems closer to completion than that in the LMC: not only are we dealing with a smaller galaxy, but the number of O stars is very small. Azzopardi and Vigneau (1982) estimate that at least 80% of the SMC members brighter than B = 14.3, and outside the central part of clusters or nebulae, are included in their catalog. Based on colors and objective prism classification, spectral types have been published for 75% of the candidate O stars, and unlike the LMC there do not seem to be scores of unidentified O stars in associations.

Hodge (1985) has completed a study of the associations in the SMC and finds that, compared to the LMC, bright members are scarce. Not counting the two richest clusters, NGC 346 and NGC 330, there are an average of five stars brighter than 14.2 in each association. Many of these stars already have been classified, especially the brighter ones. Thus it is now appropriate to examine the H-R diagram for massive

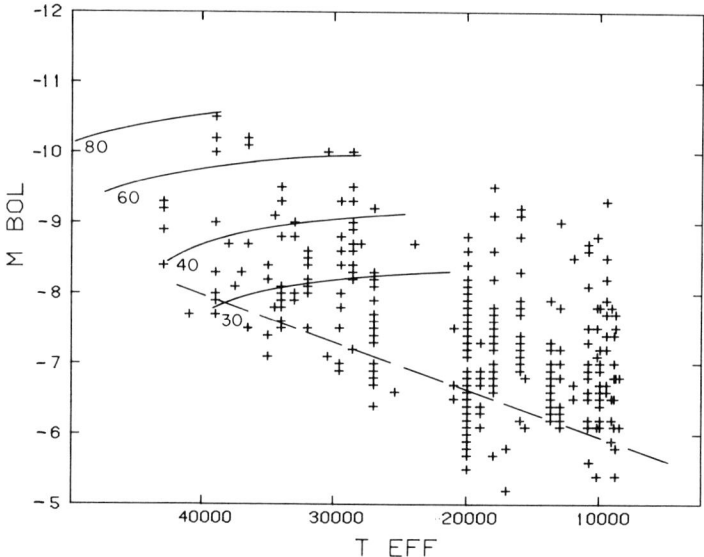

Fig. 8. The H-R diagram for stars in the SMC classified from slit spectra. The data are incomplete below the dashed line.

stars in the SMC, and expect it to be complete enough to be signif-iciant.

Figure 8 contains all stars with spectra in the SMC. Evolutionary tracks by Bressan et al. (1981) are shown, and the region where the data are incomplete is cross hatched. One thing is clear in Fig. 8: there are very few stars above 40 $M_\odot$. Recall that there are only eight stars classified as W-R in the SMC, and two of these look more like Of stars on spectra taken by P. Massey and myself. The situation in the SMC is reminiscent of the outer region of our Galaxy, and if we compute an IMF for stars above 30 $M_\odot$ in the SMC, the points lie within root N error bars of the points outside the solar circle.

At present, it appears that the SMC has produced remarkably few massive O stars in recent stellar generations, and consequently has very few W-R stars. On the other hand, the LMC is rich in W-R stars and it appears to be equally rich in massive O stars.

This work has been done in collaboration with Drs. P. Conti and P. Massey, and I gratefully acknowledge their help in preparing this paper. I especially thank Dr. Conti for presenting this paper on very short notice when illness prevented me from attending IAU Symposium #116. This research has been supported by NSF Grant AST83-12964.

REFERENCES

Abbott, D. C. 1982, Astrophys. J. **263**, 723.
Abbott, D. D. and Hummer, D. G. 1985 (preprint).
Azzopardi, M. and Breysacher, J. 1979, Astron. Astrophys. **75**, 120.
Azzopardi, M. and Vigneau, J. 1982, Astron. Astrophys. Suppl. **50**, 291.
Blaauw, A. 1963, Basic Astronomical Data, ed. K. Aa. Strand, Stars and Stellar Systems, Vol. III, University of Chicago, p. 383.
Bressan, A. G., Bertelli, G. and Chiosi, C. 1981, Astron. Astrophys. **102**, 25.
Breysacher, J. 1981, Astron. Astrophys. Suppl. **43**, 203.
Cohen, R. S., Grabelsky, D. A., May, J., Bronfman, L., Alvarez, H., Thaddeus, P. 1985, Astrophys. J. (Letters) **290**, L15.
Conti, P. S. and Garmany, C. D. 1983, Pub. A. S. P. **95**, 411.
Conti, P. S., Garmany, C. D., de Loore, C., and Vanbeveren, D. 1983, Astrophys. J. **274**, 302.
Crampton, D. 1979, Astrophys. J. **230**, 717.
Crampton, D. and Greasley, J. 1982, Pub. A. S. P. **94**, 31.
Cruz-Gonzales, C., Recillas-Cruz, E., Costero, R. Peibert, M., Torres-Peimbert, S. 1974, Rev. Mex. Astron. Astrof. **1**, 211.
Davies, R. D., Elliott, K. H. and Meaburn, J. 1976, Mem. Roy. Astron. Soc. **81**, 89.
Dufour, R. 1984, Structure and Evolution of the Magellanic Clouds, eds. S. van den Bergh and K. S. de Boer, Reidel, p. 353.
FitzGerald, M. P. 1970, Astron. Astrophys. **4**, 234.
Fitzpatrick, E. L. and Savage, B. 1984, Astrophys. J. **279**, 578.
Freedman, W. L. 1984, Ph.D Thesis, University of Toronto.
Freedman, W. L. 1985 (preprint).
Fujimoto, M. 1979, in The Large-Scale Characteristics of the Galaxy, ed. W. B. Burton, Reidel, p. 557.
Garmany, C. D. 1984, Pub. A. S. P. **96**, 779.
Garmany, C. D., Conti, P. S. and Chiosi, C. 1982, Astrophys. J. **263**, 777.
Garrison, R. F. 1978, IAU Symp. 80 The HR Diagram, eds. by A. J. Philip and D. S. Hayes, Reidel, p. 147.
Hodge, P. W. 1974, Pub. A. S. P. **86**, 236.
Hodge, P. W. 1985 (preprint).
Hodge, P. W. and Lucke, P. B. 1970, Astrophys. J. **75**, 933.
Humphreys, R. M. 1978, Astrophys. J. Suppl. **38**, 309.
Humphreys, R. M. 1983, Astrophys. J. **265**, 176.
Humphreys, R. M. and McElroy, D. 1984, Astrophys. J. **284**, 565.
Hutchings, J. B., Crampton, D., Cowley, A. P. and Thompson, I. B. 1984, Pub. A. S. P. **96**, 811.
Kudritzki, R. P., Simon, K. P. and Hamann, W.-R. 1983, Astron. Astrophys. **118**, 245.
Lequeux, J., Martin, N., Prevot, L., Prevot-Burnichon, M. L., Rebeirot, E. and Rousseau, J. 1980, Astron. Astrophys. **85**, 305.

Lucke, P. B. 1974, Astrophys. J. Suppl. **28**, 73.
Maeder, A. 1981, Astron. Astrophys. **101**, 385.
Massa, D. and Savage, B. D. 1985, Astrophys. J. (in press).
Massey, P. 1985, Pub. A. S. P. **97**, 5.
McNamara, D. H. and Feltz, K. A. 1980, Pub. A. S. P. **92**, 587.
Nandy, K., Morgan, D. H., Willis, A. J., Wilson, R. and Gondhalekar, P. M. 1981, Mon. Not. R. Astr. Soc. **196**, 955.
Remie, H. and Lamers, H. J. G. L. M. 1982, Astron. Astrophys. **105**, 85.
Rousseau, J., Martin, N., Prevot, L., Rebeirot, E., Robin, A. 1978, Astron. Astrophys. Suppl. **31**, 243.
Rubio, M., Cohen, R. and Montani, J. 1984, Structure and Evolution of the Magellanic Clouds, eds. S. van den Bergh and K. S. de Boer, Reidel, p. 399.
Sanduleak, N. 1969, Cerro Tololo Inter-American Obs. Contrib. No. 89.
Scalo, J. 1984, Fundamentals of Cosmic Physics (in press).
Solomon, P. M., Sanders, D. B., Rivolo, A. R. 1985, Astrophys. J. (Letters) **292**, L19.
Walborn, N. R. 1972, Astron. J. **77**, 312.
Walborn, N. R. 1977, Astrophys. J. **215**, 53.

Discussion : GARMANY.

HUMPHREYS :

I have 2 comments.
1. It is well known that spiral structure diagrams for individual stars show a lot of scatter due to the uncertainties in the luminosities of individual stars. Use of associations and stars clusters cleans up the diagram considerably.
2. I agree that the number of massive stars is greater inside the solar circle but I do not think there is a significant difference in slope because of incompleteness in the counts of massive stars in our galaxy which is probably more of a problem for the inner region because it is affected more by observations from the southern hemisphere and higher extinction toward the galactic center.

BRIGHTEST STARS IN GALAXIES AS DISTANCE INDICATORS

Allan Sandage
Carnegie Institution of Washington, Department of Astronomy
813 Santa Barbara St., Pasadena, CA 91101-1292

ABSTRACT

The current status of the absolute magnitude calibration of the brightest blue and red supergiants in galaxies of different absolute magnitudes shows trends of M(star) with M(parent galaxy). Red supergiants show a more shallow correlation than the blue stars for galaxies brighter than $M_B = -14$. For fainter galaxies, the red supergiant method appears to become totally degenerate.

Four areas of application of the brightest star data are discussed as (1) determining M_B of the Eddington limit to be ~ -10 for blue supergiants, (2) calibration of $M_B(max) = -20.0\pm0.4$ for type I supernovae, leading to a Hubble constant of $H_o = 43\pm10$ km s^{-1} Mpc^{-1}, (3) detection of the deceleration of the cosmological expansion by the Local Group leading to a Local Group mass of 4×10^{11} M$_\odot$ and a mass-to-blue light ratio of 3, and (4) use of the brightest stars to map the Virgo cluster velocity perturbation of the Hubble flow.

1. INTRODUCTION

Persistently, one of the most difficult problems in astronomy has been the determination of accurate distances to external galaxies. The only two fundamental methods we know depend either on (1) properties of the stellar content such as luminosities of variables, brightest stars, normal novae, supernovae, linear sizes of clusters, associations, H II regions, etc., or (2) properties of the dynamics via Öpik's (1922) method as applied in its restricted form by Tully and Fisher.

For distances beyond the Local Group and outward into the general field where the cosmological redshift dominates over local streaming motions, Hubble depended only on brightest "stars." His first calibration (Hubble 1926) of M_B(stars) was essentially his last (Hubble 1936) at M_B(brightest) = -6.1 based on the brightest resolved stars in only a few of the nearest galaxies (M31, M33, LMC, SMC, and NGC 6822--and strangely M101 for incorrect reasons) whose distances he already had determined from Cepheids.

In 1950, as Baade was revising Hubble's M31 distance outward by a factor of ∼2, Hubble remarked how ironic it had turned out that from 1924 to the late-1940's his distance scale was thought to be much too large by stellar spectroscopists everywhere; no one had believed that they had ever seen stars brighter than Mpg ≃ -3 spectroscopically. The calibration of distances in our own galaxy has also undergone a drastic revision over the past 30 years, just as has the extragalactic distance scale. The brightest Galactic stars are now indeed put near $M_B \simeq -10$, which is the current value from the external galaxies as well.

Hubble's work that ended in 1936 gave an expansion constant of $H_o = 530$ km s^{-1} Mpc^{-1} which is 10 times (5 magnitudes) larger than the current value. His problems were:

a) use of $M_B = -6$ rather than -10 for the brightest blue stars,
b) incorrect magnitude scales in the Selected areas such that stars listed as $m_{pg} \simeq 18-20$ are actually at B ≃ 20-22 (Stebbins, Whitford, and Johnson 1950; Sandage 1983a, Fig. 5, 1983b, Fig. 6),
c) real stars in his galaxies start near B ≃ 22 (in Virgo) rather than his 19.5, due partly to item (b) and partly because the objects he identified as brightest stars in remote galaxies are not stars but are rather compact H II regions or small associations.

2. CALIBRATION OF THE BRIGHTEST STARS

Beginning in 1950 when the Hale 5m reflector was put into routine use, a long-range program of faint photometry was begun in nearby galaxies as the next step. The aim has been to establish an accurate distance scale to calibrate the cosmological expansion rate. The work has proceeded along two separate routes. (1) Relative distances of clusters of galaxies over a redshift range of $0.003 < z \lesssim 0.4$ were established by measuring total magnitudes of the brightest E galaxies in them.
(2) Measurement of absolute distances to highly resolved nearby galaxies was attempted via Cepheids and brightest stars so as to walk along the way to the Virgo cluster.

The Cepheid P-L relation was recalibrated using Cepheids in galactic clusters (Sandage and Tammann 1968) and then applied to NGC 2403 directly (Tammann and Sandage 1968) and M81 and M101 indirectly. In the summary of the brightest star calibration as it existed from this early work (Sandage and Tammann 1974), the brightest red supergiants were isolated as a new distance indicator. From the data available at that time we concluded that $M_V(1) = -7.9\pm0.1$ mag for these red stars, independent of the parent galaxy. We had only 8 calibrators (the Galaxy, LMC, SMC, NGC 6822, IC 1613, NGC 2403, IC 2574, and Ho II) of which only 5 were based on Cepheids.

In studying the blue supergiants we also produced a calibration based on 12 galaxies in which $<M_B(1)>$ varies from -8 to -9.5 as the parent galaxy changed from Mpg = -14.4 to -19, an effect first discussed extensively by Holmberg (1950).

The work since 1974 has been aimed at improving these values by increasing the sample with Cepheids, and also by strengthening the internal photometry. Cepheids have been found and worked up in the faint Im dwarfs of Sextans A and Sextans B, and WLM (Sandage and Carlson 1985a,b). Holmberg IX, the companion to M81 at a common distance of m-M = 28.8, has a color-magnitude diagram (Sandage 1984). Cepheids have been found in the Pegasus dwarf, in Leo A, NGC 3109, and IC 5152 and are currently under study in the latter two. Cepheids in NGC 300 have been studied by Graham (1984), and photometry of the brightest red and blue stars is nearly completed (see also Graham and Humphreys, this volume for their independent value of $<M_V(4)> = -7.9$ for four red supergiants).

TABLE 1
Summary of the Absolute Magnitudes of the
Brightest Red and Blue Stars

Name	M_B(parent)	B(3)	$(m-M)_{AB}$	$M_B(3)$	V(3)	$(m-M)_{AV}$	$M_V(3)$
(1)	(2)	(3)	(4)	(5)	(6)	(7)	(8)
M81	-22.8	19.1	28.8	-9.7	19.7	28.7	-9.0
M101	-21.3	19.3	29.2	-9.9	20.5	29.2	-8.7
NGC 2403	-19.5	18.1	27.9	-9.8	27.8	19.4	-8.4
M33	-19.0	15.4	25.3	-9.9	16.6	25.0	-8.4
Solar Neigh	-18.5	....	....	-8.8	....	....	-8.0
LMC	-18.4	9.5	18.9	-9.4	11.1	18.8	-7.7
SMC	-16.8	10.6	19.3	-8.7	11.8	19.3	-7.5
NGC 6822	-15.8	16.9	25.0	-8.1	16.9	24.8	-7.9
IC 1613	-14.7	16.7	24.6	-7.9	17.0	24.5	-7.5
Ho IX	-14.5	20.1	28.8	-8.7	20.6	28.7	-8.1
Sextans A	-14.3	18.7	26.2	-7.5	18.3	26.2	-7.9
Sextans B	-14.3	19.4	26.2	-6.8	18.9	26.2	-7.3
WLM	-13.5	17.7	24.9	-7.2	17.9	24.9	-7.0
GR8	-11.3	(18.4)	25.2	(-6.8)	(21.0)	25.2	(-4.2)

The calibration of the brightest red and blue stars now available is set out in Table 1 and plotted in Figures 1 and 2, taken from the summary plots given elsewhere from the last paper of the series (Sandage and Carlson 1985b). The principle features of the diagrams are (1) the luminosity of the blue stars progressively decreases with M(parent) when the parent galaxies are fainter than $M_B \simeq -19$. (2) In the brightest four galaxies in the sample (M81, M101, NGC 2403, M33) the brightest stars

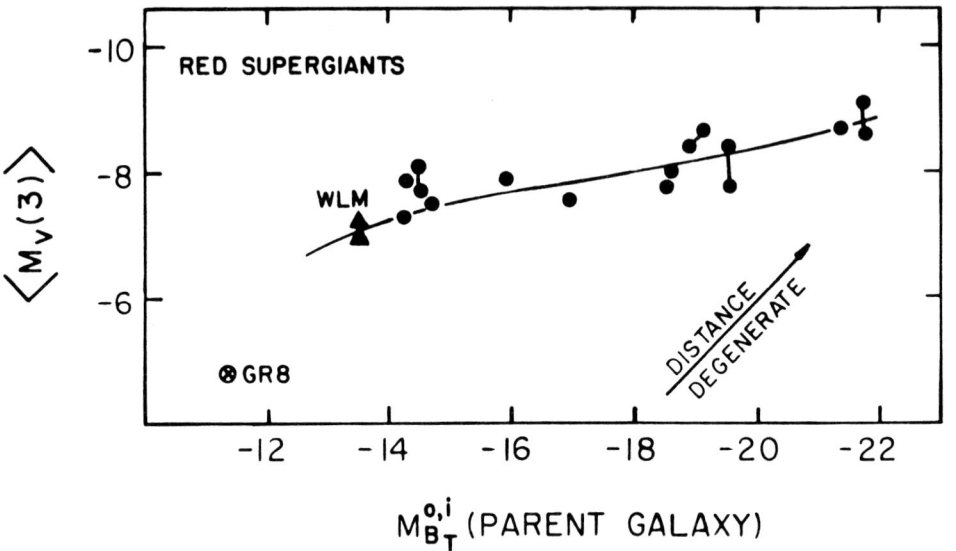

Figure 1. Calibration of the mean absolute visual magnitude of the three brightest red supergiants in the calibrating galaxies in Table 1.

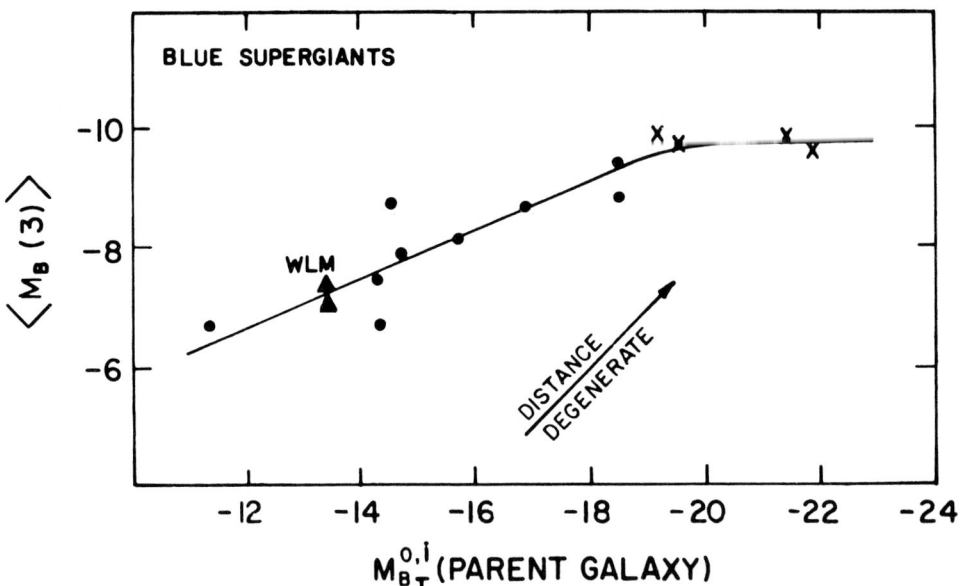

Figure 2. Same as Figure 1 for the B absolute magnitude of the three brightest blue supergiants.

are irregular blue variables that have characteristic amplitudes of $\Delta B \simeq$ 1 mag and characteristic time scales of variability of $\sim$10 to 30 years (Hubble and Sandage 1953). (3) The rate of change of M_B(star) with M_B(galaxy) is less steep than the distance-degenerate line along which there would be no distance information. It is the intersection of the line of slope 1 with the intrinsic calibration line from which the distance to an unknown galaxy can be determined. (4) However, the scatter of the points for the 13 calibrating galaxies about the mean line is large. Such a scatter is expected for the fainter galaxies due to stochastic sampling problems of the luminosity function (Greggio this volume and 1986), amounting to at least ±0.7 mag. This stochastic problem makes use of the brightest blue stars as distance indicators less attractive than the red supergiants shown in Figure 2. (5) The red supergiant calibration shows less dependence of $<M_V(3)>$ on M_B(parent) than for the blue stars. Nevertheless, there is a dependence with a mean slope of dM_V(star)$/dM_B$(parent) $\simeq 0.2$. (6) If the data for GR8 are correct, the red star relation becomes entirely degenerate for M_B(parent) fainter than $\simeq -14$ because the intrinsic line that connects WLM with GR8--not shown in Figure 2 has the degenerate slope of 1. This is the problem we faced in Leo A and the Pegasus dwarf (Sandage 1986), showing that for dwarf galaxies fainter than $M_B \simeq -14$ the brightest red supergiants may not be distance indicators at all.

3. APPLICATION OF THE BRIGHTEST STAR DATA

Four areas of application of the data on the brightest stars and the distances to their parent galaxies can be mentioned.

3a. Determining M_B for the Eddington Limit

If the data for M81, M101, NGC 2403, and M33 are representative, the brightest blue stars may define the upper limit of stability for stars as they have just moved off the main sequence. The two suggestive reasons are: (1) the stars are variable as mentioned before, and (2) there is no correlation of M_B(star) with M_B(parent galaxy) for them as there is for fainter galaxies (Figure 2). The explanation of the slope in Figure 2 for M_B(parent galaxy) fainter than -19 is undoubtedly the purely statistical effect of sampling the stellar luminosity function at fainter magnitudes with the condition that $\phi(M) = 1$, as the normalization constant for $\phi(M)$ becomes smaller for fainter parent galaxies (Holmberg 1950; Sandage and Tammann 1974). The leveling off of the M_B(star) = $f[M$(galaxy)$]$ correlation brighter than M_B(galaxy) = -19 then requires $\phi(M)_{star}$ to have a vertical cut off at $M_B \simeq -10$. The most reasonable explanation is an instability, which is the Eddington limit where the radiation pressure equals the gas pressure. As these variables are of spectral type F with $T \simeq 10,000°K$, the bolometric correction is small, suggesting $M_{bol} \simeq -10.5$ as this limit.

3b. Calibration of $M_B(max)$ for SNe I as a Distance Indicator

Evidence is growing that most supernovae of type I are standard explosions with a tightly fixed absolute luminosity at maximum. The main points (Cadonau, Sandage, and Tammann 1985) are (1) the Hubble diagram is well defined with a scatter in magnitude about the linear velocity-distance relation that can be due entirely to observational error (Kowal 1968; Sandage and Tammann 1981, Fig. 20), (2) near identity of the light curve shapes at any given time after maximum (Tammann 1978; Elias et al. 1981), and (3) near identity of the spectra at any given phase (cf. Wheeler 1985). Although there may be a small subset ($\sim 10\%$) of anomalous type I SNe (Panagia 1985), these can be identified and removed from the sample.

To determine the Hubble constant from the observed redshift-apparent magnitude relation requires calibration of $M_B(SN)$ at maximum. The purely astronomical method is to determine distances to galaxies which have had well observed SNe I. This has been done (Sandage and Tammann 1982) for two galaxies (NGC 4214 and IC 4182) using the brightest resolved stars and the calibrations of Figures 1 and 2 here, and revised for internal absorption in the two galaxies (Cadonau et al. 1985) to give $M_B(max) = -20.1 \pm 0.4$. This is in good agreement with the value of -19.8 ± 0.7 from the expansion parallaxes (Arnett 1982; Branch 1982) and with theoretical calculations from an increasingly realistic model of the explosion (Sutherland and Wheeler 1984), giving $M_B = -20.0 \pm 0.2$. If $M_B(max) = -20.0 \pm 0.4$, the magnitude-redshift relation requires the Hubble constant to be $H_o = 43 \pm 10$ km s^{-1} Mpc^{-1}.

3c. Deceleration of the Cosmological Expansion by the Local Group

The mass of the Local Group must retard the cosmological expansion, causing the effective Hubble constant to increase outward over distances where the effect is non-negligible. Calculations (Sandage 1986) show that the deceleration caused by the Local Group should be measurable within a distance of ~ 5 Mpc provided that the distances to the test galaxies can be determined with high accuracy. Using the galaxies that have gone into the calibration of the brightest stars in Table 1 gives a suggestive result that the deceleration has been found. The best-fit mass of the Local Group obtained by comparing the family of decelerating velocity-distance relations with the observations is 4×10^{11} M⊙, with a firm upper limit of less than 3×10^{12} M⊙. This gives a best-fit mass-to-light ratio of 3 (in solar blue units) and an upper limit of M/L < 20.

3d. Use of Brightest Stars to Map the Virgo Perturbation

Precise distances to galaxies strategically placed in angle and distance from Virgo are needed to map the perturbation of the cosmological expansion field caused by the overdensity of the Virgo cluster complex. Brightest stars are expected to be an important distance indicator for galaxies beyond the reach of the Cepheids. Lists and three atlases of galaxies closer than ∼50 Mpc have been prepared for use in planning observing programs with Space Telescope (Sandage 1985a,b,c).

To map the velocity perturbation it is required to determine the distances to many galaxies by a fundamental (nonredshift) method, and then to compare the observed redshifts with those predicted from a nonperturbed cosmological velocity field. Particular directions and distances from Virgo are advantageous to maximize the perturbation signal. Figure 3 illustrates this problem using a version of the Tonry-Davis (1981) diagram calculated for an "infall" velocity toward Virgo of 220 km s^{-1}. Dotted lines are the unperturbed Hubble flow velocities; solid lines are the contours of the perturbed velocity due to Virgo. Shown as crosses are galaxies from the first atlas where resolution into stars is expected to be easy, even from the ground in some cases.

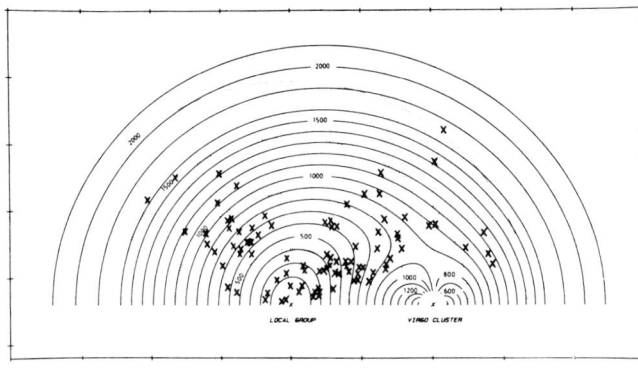

Figure 3. A Tonry-Davis diagram for a Virgo "infall" velocity of 220 km s^{-1}. Crosses are galaxies from the first atlas of candidates for resolution into brightest stars.

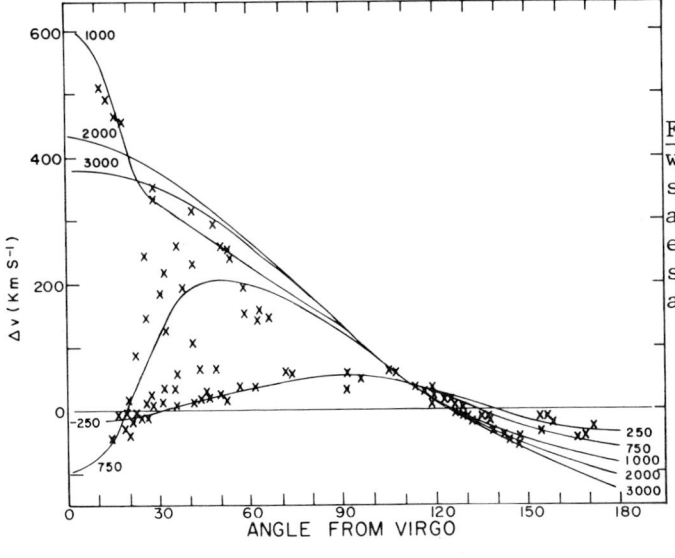

Figure 4. A Kraan-Korteweg diagram showing the same perturbation effect as Figure 3 in a different representation. Crosses are the same galaxies as in Figure 3.

Figure 4 illustrates the same effect in a different representation due to Kraan-Korteweg (1986). Shown as ordinate is the velocity difference between the ideal Hubble flow and the perturbed field as a function of angle from Virgo and the observed velocity. The same galaxies as in Figure 3 are plotted.

More distant galaxies which are more difficult to resolve but whose brightest stars can be studied with Space Telescope are shown in Figure 5.

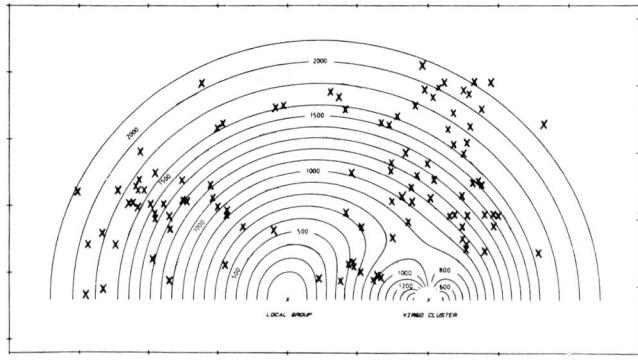

Figure 5. Same as Figure 3 but showing galaxies from the second atlas of candidates. These galaxies are more distant those in Figure 3.

Finally, Figure 6 is a cautionary note showing the very small signal which must be detected in order to see the perturbation field. What is measured by any method of distance determination is the percentage signal, $\Delta r/r = \Delta v/v$. This, calculated from Figure 4, is plotted in Figure 6 as a function of angle from Virgo and of the measured velocity.

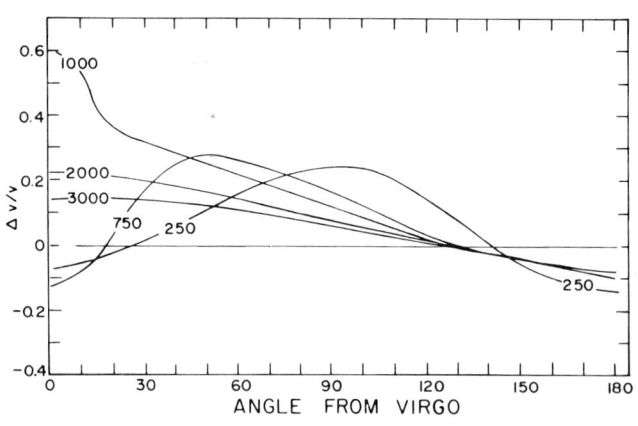

Figure 6. The percentage signal $\Delta v/v$ that must be detected to measure the perturbation, calculated from Figure 4. The photometric difference to be detected is Δmag = 2.17 $\Delta v/v$.

This signal $\Delta v/v$ translates into a photometric difference of $\Delta\text{mag} = 2.17 \Delta v/v$, showing that we must detect differences of at most only ~ 0.5 mag in order to map the perturbation using photometric methods. The scatter in Figures 1 and 2 shows that the job using brightest stars will be difficult.

REFERENCES

Arnett, W.D.: 1982, in Supernovae: A Survey of Current Research, eds. M. J. Rees and R.J. Stoneham (Doredrecht: Reidel), p. 221.
Branch, D.: 1982, in Supernovae: A Survey of Current Research, eds. M.J. Rees and R.J. Stoneham (Dordrecht: Reidel), p. 267.
Cadonau, R., Sandage, A., and Tammann, G.A.: 1985, in Lecture Notes in Physics, No. 224, ed. N. Bartel (Berlin: Springer), p. 151.
Elias, J.H., Frogel, J.A., Hackwell, J.A., and Persson, S.E.: 1981, Ap. J. Letters, 251, L13.
Graham, J.A.: 1984, A.J., 89, 1332.
Greggio, L.: 1986, Astron. Astrophys., in press.
Holmberg, E.: 1950, Medd. Lunds Obs. Ser. 2, No. 128.
Hubble, E.: 1926, Ap. J., 64, 321.
Hubble, E.: 1936, Ap. J., 84, 270.
Hubble, E., and Sandage, A.: 1953, Ap. J., 118, 353.
Kowal, C.T.: 1968, A.J., 73, 1021.
Kraan-Korteweg, R.: 1986, Astron. Astrophys., in press.
Öpik, E.: 1922, Ap. J., 55, 406.
Panagia, N.: 1985, in Lecture Notes in Physics No. 224, ed. N. Bartel (Berlin: Springer), p. 14.
Sandage, A.: 1983a, in IAU Coll. No. 100, Kinematics, Dynamics, and Structure of the Milky Way, ed. W.L.H. Schuter (Dordrecht: Reidel), p. 315.
Sandage, A.: 1983b, A.J., 88, 1108.
Sandage, A.: 1984, A.J., 89, 621.
Sandage, A.: 1985a,b,c, A.J., 90, in press (October issue).
Sandage, A.: 1986, Ap. J., in press.
Sandage, A. and Carlson, G.: 1985a, A.J., 90, 1019.
Sandage, A. and Carlson, G.: 1985b, A.J., 90, 1464.
Sandage, A. and Tammann, G.A.: 1968, Ap. J., 151, 531.
Sandage, A. and Tammann, G.A.: 1974, Ap. J., 191, 603.
Sandage, A. and Tammann, G.A.: 1981, in Astrophysical Cosmology, Proc. of the Study Week on Cosmology and Fundamental Physics, ed. H.A. Brück, G.V. Coyne, and M.S. Longair (Rome: Specola Vaticana), p. 23.
Sandage, A. and Tammann, G.A.: 1982, Ap. J., 256, 339.
Stebbins, J., Whitford, A.E., and Johnson, H.L.: 1950, Ap. J., 112, 469.
Sutherland, P.G. and Wheeler, J.C.: 1984, Ap. J., 280, 282.
Tammann, G.A.: 1978, Mem. Soc. Astron. Italiana, 49, 315.
Tammann, G.A. and Sandage, A.: 1968, Ap. J., 151, 825.
Tonry, J.L. and Davis, M.: 1981, Ap. J., 246, 680.
Wheeler, J.C.: 1985, in Lecture Notes in Physics, No. 224, ed. N. Bartel (Berlin: Springer), p. 34.

Discussion : SANDAGE.

HUMPHREYS :

It is important to understand where the difference in the luminosity calibration of the M supergiants comes from. We use different methods to derive the luminosities. Sandage uses the apparent modulus with the apparent magnitude of the brightest red supergiants to derive their M_V's. I use the true distance modulus corrected for reddening with the visual magnitudes of the stars corrected for the individual reddening affecting each star. If the reddening were uniform then one might expect the two methods to give very similar results; however in M33 the reddening is both significant and variable.

Comparison of the M_V's with the M_K and M_{Bol} measured from infrared photometry shows that when a reddening correction is <u>not</u> applied the resulting $(V-K)_o$ and BC_V (the bolometric correction at $\overline{V}$) are too blue and too small respectively. $(V-K)_o$ and BC_V derived from Sandage's method for M_V are not physically meaningful for M supergiants.

SANDAGE :

There is much less difference than it seems at first between Humphrey's reply here and the calibration of $\langle M_V(3)\rangle$ vs M (parent galaxy) for the red supergiants given in the main report. My main point is that $\langle M_V(3)\rangle$ is <u>not</u> constant at $+7.9 \pm 0.1$ as we originally thought and as Humphreys has now adopted. Rather, $\langle M_V(3)\rangle$ for the red supergiants becomes fainter as M (parent galaxy) becomes fainter over the entire range of M_B (parent) between -22 and -13.5. Humphreys obtains the same relation even with her individual adopted correction for A_V. If she were to plot her M101 point in the $\langle M_V(3)\rangle$ vs M_B (parent) correlation at her value of $\langle M_V(3)\rangle = -9.0$ and if she were to decide to use the M81 apparent modulus of $(m-M)_{AB} = 28.8$, giving $\langle M_V(3)\rangle = -8.8$, our respective diagrams would be nearly the same. There is, then, no argument except over the value of m-M = 28.8 for M81.

There is no question that $(m-M)_{AB} = 27.55$ used by Humphreys is too small by approximately 1 mag. The Cepheids and novae show that the apparent modulus of M81 is considerably fainter than that for NGC 2403. One will of course obtain an incorrect value of $\langle M_B(3)\rangle$ for the blue stars, and $\langle M_V(3)\rangle$ for the red stars, by using the same modulus for both M81 and NGC2403.

This then brings us to the question of the absorption. If the absorption is the same in the mean between the Cepheids (from which the apparent modulus is derived) and the red supergiants, the correct M_V(RSG) is always obtained by combining the <u>apparent</u> modulus $(m-M)_{AV}$ with the apparent magnitude of the RSG <u>no matter how high Av might be</u>. An error is introduced only if $\langle A_V(\text{Cepheids})\rangle$ is different from $\langle A_V(\text{RSG})\rangle$, i.e. in the presence of <u>differential</u> absorption. In M33 there is no evidence for large differential absorption between Hubble's

Cepheids near the center and our new Cepheids in Field 25. Hence the question of absorption (the main point of disagreement between Humphreys and myself) is not the issue.

Fainter galaxies, such as IC 1403, Sextans A, Sextans B, WLM, (etc) where the Av values themselves are expected to be very small, clearly show $\langle M_V(3) \rangle$ for the RSG to be fainter than -7.9. In WLM at M_B(parent) $= -13.5$, $\langle M_V(3) \rangle = -6.7$. Hence, there is a faintward trend of the RSG in fainter galaxies.

In summary, absorption is not the cause of the apparent difference between Humphreys and Sandage. Rather it is caused by (1) Humphreys' use of the small modulus of 27.55 for M81 for which there is no observational justification, (2) her neglect of the M101 point at $\langle M_V(3) \rangle = -9.0$, M_B(parent) $= -22$, and (3) her neglect of the data for WLM, Sextans A and Sextans B which give $\langle M(3) \rangle$ near -7.0 rather than -7.9.

Finally one should note that if M81 were put at $(m-M)_{AB} = -27.5$ then the 3 brightest blue irregular variables in M81 would each be approximately 1.3 magnitude fainter than -9.9 and they would not fit the blue supergiant calibration where now in M101, M81, M33 and NGC2403 $\langle M_B(3) \rangle = -9.9 \pm 0.1$ using the distances listed in the contribution. One would then have to say that the brightest blue irregular variables in any given galaxy are not at, or very near the de Jager-Humphreys-Davidson upper limiting line in the HRD. This, of course, would be unfortunate. But that conclusion is not necessary if $(m-M)_{AB} = 28.8$ for M81 as the Cepheid and novae data require.

SESSION 2.

MAIN SEQUENCE AND SUPERGIANT STARS - II.

Chairman : C. de Loore.

1. R. HUMPHREYS: The Cooler Supergiants (A to M): Crucial Signposts in the Lifecycles of Massive Stars.

2. W. FREEDMAN: The Distribution of Young Stars in Nearby Galaxies.

THE COOLER SUPERGIANTS (A to M): CRUCIAL SIGNPOSTS IN THE LIFECYCLES OF MASSIVE STARS

Roberta M. Humphreys
University of Minnesota

The intermediate and late-type supergiants are the visually brightest stars. They are among the first stellar objects observed in other galaxies and provide our first clues to the conditions of massive star evolution in galaxies of different types. They are not as massive as the hottest and most luminous stars in the upper left of the HR diagram. Nevertheless, these somewhat lower mass stars ($\approx$20-50 $M_\odot$) with relatively cool temperatures play a major role in our efforts to understand massive star evolution. These supergiants are usually considered to be post hydrogen burning stars, and their relative numbers in the HR diagram provide essential comparisons with models for the later stages of massive star evolution. Most importantly, the most luminous cooler supergiants define the stability limit for massive stars in the HR diagram.

1. Fundamental Properties

For the purpose of this review paper, the intermediate and late-type supergiants will be the stars with spectral types A through M. For the essential comparisons with stellar structure models and evolutionary tracks we must know the effective temperatures and total luminosities, which are determined from the total energy distribution of the stars. The effective temperatures for these supergiants range from about 10000°K to 3000°K. With the increased use of infrared observations, the measurement of these fundamental quantities has been greatly improved for the cooler supergiants (Lee 1970; Ridgway et al. 1980; Elias, Frogel and Humphreys 1985). In this paper I am using the summary by Flower (1977) with the bolometric corrections for M supergiants by Elias et al. (1985).

We are of course very interested in the luminosities of these stars both for stellar evolution studies and to evaluate their potential as distance indicators. Nearly all of these stars are too distant for direct measurement of their distances and thus their luminosities. In our galaxy, their visual and bolometric luminosities are derived from membership in clusters and associations with known distances (e.g.,

Humphreys 1978). A fundamental reference for this procedure is Blaauw (1965). Recent luminosity calibrations by Walborn (1972, 1973), Schmidt-Kaler (1983) and Humphreys and McElroy (1984) show small differences for some of the groups of stars, but overall there is little change even with the much more extensive data available now. These luminosity calibrations typically have a standard deviation of 0.5 mag for an individual luminosity and standard errors of the mean of 0.1 to 0.2 mag. For a star in a stellar aggregate, the luminosities are better determined with typical errors of ± 0.25 mag due to uncertainties in the distance moduli.

In any study of the stellar populations and stellar evolution in other galaxies we must know the distances. For the most part, we rely on published distances from the Cepheid period-luminosity relation, and the uncertainties for Local Group galaxies are ± 0.2 to ± 0.5 mag.

Another fundamental parameter for massive stars is of course mass loss, which has figured significantly in recent model calculations for massive star evolution. As a result of observations in both the ultraviolet and infrared, we now realize that mass loss is very likely occurring to some degree in all stars in the upper part of the HR diagram. Reviews by Hutchings (1978), Barlow (1978, 1981), and Zuckerman (1980) summarize the situation for the hot and cool supergiants and recent papers by Hagen, Humphreys and Stencil (1981), Lambert, Hinkle and Hall (1981), and Kunasz and Morrison (1982, 1983) discuss the mass loss rates for the intermediate type supergiants.

In Table 1 I have summarized the range of physical parameters for the luminous stars of different spectral types or temperatures based on observations in our galaxy and the Magellanic Clouds.

Table 1

	Hot	Intermediate	Cool
Spectral Types	O,B	A,F,G	K,M
Effective Temperature(°K)	50000-12000	11000-4000	<4000
Luminosity Range($L/L_\odot$)	10^4-5×10^6?	10^4-8×10^5	10^4-5×10^5
Mass Range($M/M_\odot$)	15-200 or 300	15-60	15-50
Size Range($R/R_\odot$)	10-200	30-1000	300-2000
Mass Loss($\dot{M}$)	10^{-7}-10^{-5}	10^{-7}-10^{-5}(10^{-4}:)	10^{-7}-10^{-4}

Although the cooler, more evolved supergiants are not the most massive or most luminous stars, they are without doubt the largest!

2. Observations of the Cooler Supergiants in Our Galaxy and Others

For our studies of massive star evolution in different galaxies we want to know how their basic properties, luminosity and mass, may depend on their environment and whether they vary from galaxy to galaxy. Comparisons of stellar models with observations hinge on the HR diagram, for which we need spectra and accurate photometry for the individual stars. However, there are numerous uncertainties in the resulting HR diagrams. How universal are the colors, the effective temperatures and bolometric corrections and the interstellar extinction law? The best evidence to date suggests that the differences, if any, are probably small, but we must know the true distances to galaxies if we want to compare the star formation rates, the luminosity and initial mass functions and study the effects of morphological type on stellar evolution. Many of the differences we are looking for may be rather small effects, and errors in the distance can lead to erroneous conclusions about the factors influencing stellar evolution.

Our galaxy and the Magellanic Clouds provide an excellent comparison of massive star evolution in different environments and with different metallicities. The HR diagrams are an efficient way to compare the properties of the luminous stars in these three galaxies. The M_{Bol} vs. log T_{eff} diagrams are shown in Figures 1, 3 and 4 in Humphreys and McElroy (1984). The luminosities for the nearly 2300 galactic supergiants are from their membership in 91 associations and clusters. The basic data for the Magellanic Cloud stars come from many sources in the literature cited in Humphreys and McElroy (1984). Their luminosities, corrected for extinction, are derived from the adopted true distance moduli of 18.6 mag and 19.0 mag for the LMC and SMC, respectively, based on Cepheids and RR Lyrae stars.

Comparison of the galactic and LMC HR diagrams reveal similar populations of luminous stars in both galaxies. The LMC and our solar region have essentially the same upper envelopes to their stellar luminosities. The HR diagram for the SMC shows some differences. The hottest, most luminous stars are fewer in number and are noticeably less luminous than stars of comparable temperature in the solar region and LMC, but the large scale features of the three HR diagrams are similar. The upper luminosity boundary for the late supergiants is the same in all three galaxies, although there are no known high luminosity yellow supergiants (FGK) in the SMC.

Figure 1 shows the composite HR diagram (M_{Bol} vs. log T_{eff}) for the stars with $M_{Bol} \leq -8.5$ mag in the Galaxy and Magellanic Clouds. The evolutionary tracks from models with mass loss by Maeder (1981b, 1983) are also shown for comparison. The effects of mass loss have been discussed by numerous authors (de Loore et al. 1978; Chiosi et al. 1978; Stothers and Chin 1979; Maeder 1980, 1981a,b, 1983; de Loore and de Greve 1981; Chiosi 1981; Falk and Mitalas 1981, 1983; Sreenivasan and Wilson 1982; Brunish and Truran 1982a,b). For this paper we are especially interested in the post main-sequence evolution for comparison with the observations of intermediate and late-type supergiants.

In 'conservative' evolution, with no mass loss, the stars leave the main sequence and end their lives as M supergiants (Stothers and Chin 1976; Lamb 1976). In these models the 15 to 50 $M_\odot$ stars begin helium burning as B supergiants and thus have very short lifetimes as red supergiants which is in sharp disagreement with the observed numbers of M supergiants in our galaxy and the Clouds.

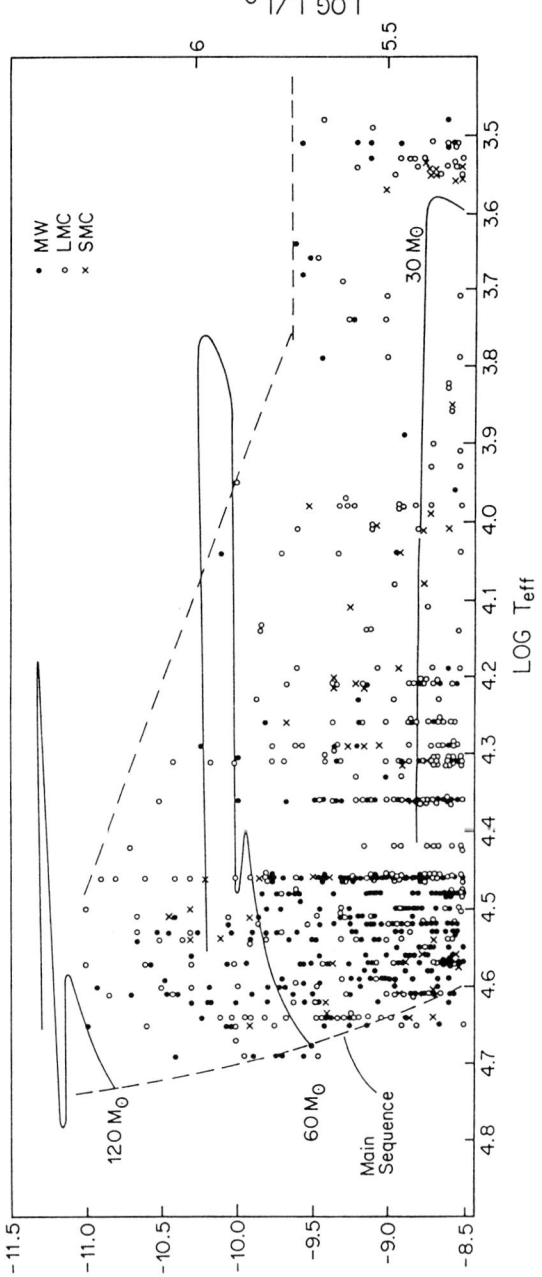

Figure 1 — The M_{Bol} vs. log T_{eff} diagram for luminous stars ($M_{Bol} \leq -8.5$ mag) in the Galaxy, LMC and SMC. The upper luminosity boundary (dashed line) is based on the distribution of the most luminous normal stars in these three galaxies.

Mass loss is observed in all of the luminous supergiants and their O star progenitors. Therefore, inclusion of mass loss in the models is not only physically necessary but also produces better agreement with the observations. The length of time in the red supergiant stage is increased by factors of 4 to 5 for 25 to 50 $M_\odot$ stars. High mass stars >50 to 60 $M_\odot$ never reach the red supergiant stage with mass loss at the observed rates. This result also agrees with our observations (see Figure 1) for the Galaxy and the Clouds, and explains the observed upper luminosity boundary for the intermediate and late-type supergiants.

A few evolutionary tracks for massive stars have been computed with the chemical abundances of the LMC (Maeder 1980; Brunish and Truran 1982b) and SMC (Hellings and Vanbeveren 1981; Brunish and Truran 1982b). The principal difference is that the models are bluer and slightly more luminous at comparable evolutionary stages.

Blue to red supergiant ratios are often used as indicators of the relative lifetimes of the massive stars in different stages for comparison with the models of stellar evolution. Counts of blue and red stars are also considered possible indicators of metallicity variations. A much more complete data set is now available for the massive stars in our galaxy and the Clouds (see Table 9 in Humphreys and McElroy 1984). The B/R ratios in these three galaxies show two phenomena: 1) little or no variation with luminosity when the data is corrected for incompleteness, and 2) a gradient with location in our galaxy and a difference between our galaxy and the Clouds, which are attributed to metallicity variations. The models with moderate mass loss give much better agreement between the expected and observed ratios than do the older, non-mass loss calculations.

The later evolution of the cooler supergiants has been discussed by Chiosi et al. (1978) and Maeder (1981a,b). They suggest that the high mass loss during the red supergiant stage favors the formation of Wolf-Rayet stars. To determine the range of initial masses of stars that evolve to WR stars, Schild and Maeder (1984) and Humphreys, Nichols and Massey (1985) have studied the WR stars in young clusters and associations and find that most WR stars had initial masses greater than 40 to 50 $M_\odot$ which is larger than the initial masses of most M supergiants. Thus only the most luminous M supergiants are potential progenitors of WR stars. A small group of very luminous late-type supergiants with extensive circumstellar dust shells, known as supergiant OH/IR sources, are potential candidates for subsequent evolution to WR stars. Alternatively, they may indeed be the final evolutionary stages for 30 to 50 $M_\odot$ stars. These objects include stars like VY CMa (M4-M5Ia), VX Sgr (M4-M8I) and IRC+10420 (F8Ia-0). They have luminosities right at the upper luminosity boundary and mass loss rates $\approx 10^{-4}$ $M_\odot$/yr. Infrared observations have revealed M supergiants of this type in other galaxies. Terry Jones and I are investigating the properties of these stars for eventual evolution to the WR stage.

We know very little about the stellar content of M31, the most massive, most luminous galaxy in our Local Group. Because of its large size, high tilt angle, and tightly wound arms there are no extensive surveys for the blue and red stars. Our information has essentially been limited to observations of stars in Baade's Field IV, the only region for which a color-magnitude diagram exists.

For this reason Elly Berkhuijsen (MPFR, Bonn), Michael Newberry (U. Michigan) and I initiated a stellar content survey of M31 using the Automated Plate Scanner at the University of Minnesota. This survey has been used to generate a preliminary list of the brightest blue and red stars in a region centered on Baade's Field III and NGC 206. Spectra for classification have been obtained for 11 candidate blue supergiants in this field plus 9 additional bright stars in prominent stellar associations elsewhere in M31. Only two of these stars were confirmed to be supergiants. Nine suspected red supergiants were observed in the near infrared (6800-9000Å). The near-infrared CaII triplet is a very strong luminosity indicator, and dwarfs and supergiants can be readily separated even on low resolution spectra (10-15Å). The paper in this volume on the M supergiants in NGC 300 by Graham and Humphreys illustrates the value of the near-infrared spectroscopy. Three of the red stars in M31 are confirmed M supergiants.

The visual luminosities of these confirmed supergiants, determined from the apparent distance modulus of 24.5 mag with no correction for reddening, are significantly less luminous than the brightest stars of similar spectral type in our galaxy, M33 and the Clouds. This is in contrast to what we might initially expect in a more massive, more luminous galaxy. Although all of M31 has not been surveyed yet, the area studied is comparable in surface area to M33. These results are very preliminary and further spectroscopy is planned. If these results are confirmed in other spiral arm regions in M31, they will have very important implications for the factors influencing massive star evolution, such as morphological type.

Very little work has been done on the intermediate and late-type supergiants in galaxies beyond our Local Group. Even though these supergiants are the visually brightest stars, they are quite faint in these distant galaxies and very little spectroscopic or photometric work has been done on them (Humphreys 1980b). We (Humphreys, Aaronson, Lebofsky, McAlary, Strom and Capps 1985) have just recently finished a program of near-infrared spectroscopy and JHK photometry of candidate red supergiants in M101 (Humphreys and Strom 1983; Sandage 1983b), NGC 2403 (Sandage 1984b) and M81 (Sandage 1984a) using the KPNO 4-meter, NASA's IRTF on Mauna Kea and the MMT in Arizona. Marc Aaronson and I have also observed spectra for classification of the brightest early-type stars in these same galaxies with the reticon scanner on the MMT. The blue spectra cover 1000Å at a resolution of 1Å and are excellent for classification.

The initial results for NGC 2403 are much as we might expect for a galaxy very similar to M33 and NGC 300. The brightest M supergiants have M_V near -8 mag while the visually brightest A-type stars are $\simeq -9.5$ mag as in M33. The results for M81 are surprising in the same sense as for M31. Adopting the distance to the M81 group from NGC 2403 the brightest red stars are near $M_V \simeq -8$, as in the other spirals and Magellanic type irregulars, but the brightest early-type supergiants identified so far are less luminous than similar stars in other spirals. These preliminary results for M31 and M81 may be our first indication of a dependence of the massive star population and the upper end of the IMF on morphological type. The luminosities of the HII regions in each of these two galaxies also suggest that there may be fewer of the most massive stars in M31 and M81 (Kennicutt 1984, 1985). The well known relation between the brightest blue supergiant (A-type) and the luminosity of the galaxy has been determined for irregulars and Sc-type spirals, but our first results for the brightest A-type supergiants in the Sb-type spirals M31 and M81 suggest they are less luminous than would be expected for the luminosities of their galaxies.

In contrast, the blue and red supergiants in M101 are over-luminous. Using a distance modulus of 29.2 mag from Sandage and Tammann (1974b) the brightest stars are a magnitude or more brighter both visually and bolometrically than the brightest known stars of the same spectral types in other galaxies (see Fig. 2). The very high luminosities near $M_{Bol} \simeq -11$ mag correspond to initial masses of ≥ 100 $M_\odot$. The uncertainties in the apparent bolometric magnitudes are very small. The bolometric magnitudes for the M supergiants determined from the infrared photometry are known to ± 0.1 mag and are independent of metallicity effects and uncertainties in the extinction and intrinsic colors. These luminosities lead to serious inconsistencies with our present understanding of the physics of massive star evolution and the effects of mass loss.

Modern models for massive star evolution including the effects of mass loss (see for example models by Chiosi, Nasi, Sreenivasan 1978; Maeder 1981a,b; Stothers and Chin 1983) show very clearly that stars with initial masses $>50-60$ $M_\odot$ do not evolve to the red supergiant part of the HR diagram. But the adopted distance modulus of M101 (29.2 mag from Sandage and Tammann 1974b) leads to luminosities that imply much greater masses. One can always argue that the evolutionary models and the effects of mass loss are grossly in error, and M supergiants with initial masses of ≈ 100 $M_\odot$ exist in M101, but there will also be serious problems with the stability of evolved stars of such high luminosity and high initial mass. Or, the proposed distance is too large; <u>a closer distance by $\approx .8$ mag would be required to produce agreement with the physics</u>.

3. The Stability Limit for the Photospheres of the Most Massive Stars

Humphreys and Davidson (1979) first drew attention to the lack of evolved very massive stars in the upper HR diagram. We proposed an empirical upper luminosity boundary based on the observed distribution

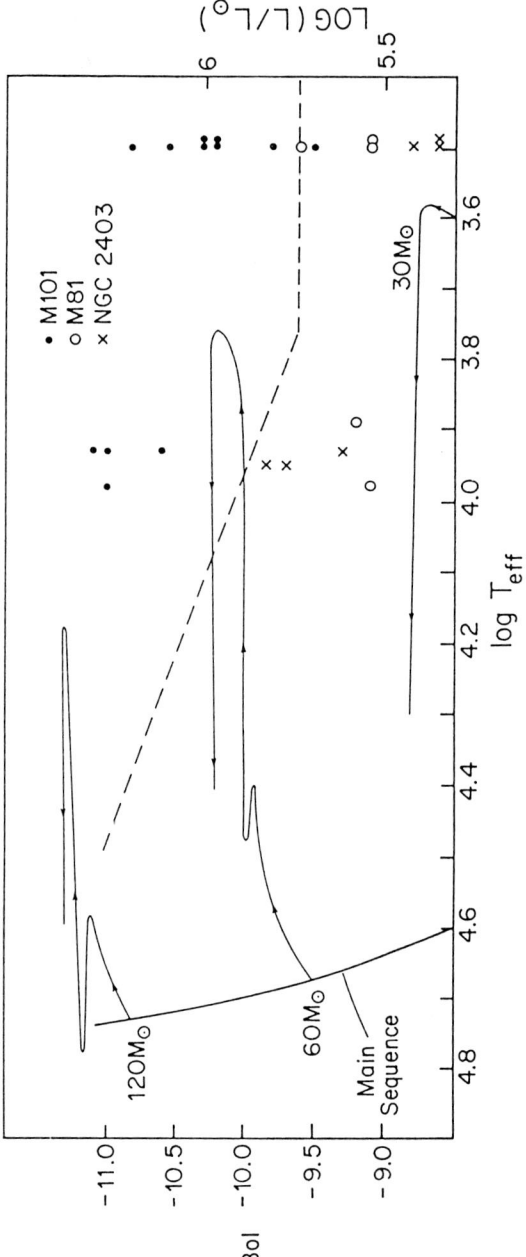

Figure 2 – The M_{Bol} vs. log T_{eff} diagram showing the positions of the confirmed supergiants in NGC 2403, M81 and M101. The dashed line is the upper luminosity boundary from Figure 1.

of the most massive 'normal' stars in our galaxy and the LMC. But what prevents a very massive star from evolving into a highly luminous cooler supergiant? We suggested an intuitive answer involving episodes of enhanced mass loss. We were motivated by the observations of unstable stars like η Car, P Cyg and other luminous blue variables, many of which lie near the critical boundary. Many of these stars are known to suffer spectacular episodes of mass ejection. We suggested that as the very massive stars evolve to cooler temperatures, they encounter a stability limit and suffer high mass loss, which prevents further evolution to cooler temperatures.

What is the cause of the instability? Several alternatives have been suggested, (see the review by Stothers and Chin 1983) including an internal vibrational instability, surface radiation pressure or a pressure gradient due to turbulence. In his comprehensive book, The Brightest Stars, (1979) de Jager proposed that a turbulent pressure gradient develops in the atmospheres of the cooler supergiants due to the dissipation of mechanical energy. The stability limit or de Jager limit is reached when the turbulent pressure gradient equals the acceleration due to gravity. Maeder (1983) showed that the de Jager limit for the most massive stars halts further evolution to the right and is accompanied by enhanced mass loss. After the mass ejection the stars reverse their evolution back to higher temperatures at essentially constant luminosity, in agreement with observations of the luminous blue variables between maximum and minimum (Appenzeller and Wolf 1981, Humphreys et al. 1984).

The onset of the stability limit for massive stars very likely corresponds to the observed upper luminosity boundary for the cooler supergiants. In a recent paper de Jager (1984) has shown that there is a stability limit to the atmospheres of hypergiants (spectroscopically class Ia+, Ia-0, and 0). The turbulent motions from the dissipation of the mechanical flux tend to destabilize the atmosphere. For more luminous stars the dissipation occurs deeper in the photosphere and leads to increasing mass loss. At the observed luminosity boundary ($M_{Bol} \approx$ -9.5 mag) for the cooler supergiants the mass loss rates are already near 10^{-4} $M_\odot$/yr. and de Jager has shown that these stars are nearly unstable. A star a magnitude brighter would be losing mass at nearly a factor of 10 times higher rate. This greatly alters the evolutionary tracks for more massive stars. As they evolve to cooler temperatures they approach the stability limit for their photospheres with increasing mass loss. According to Maeder (1983), the increasingly high mass loss rates increase the mass fraction of the core and the star's evolution reverses to warmer temperatures. Thus stars greater than 50-60 $M_\odot$ do not evolve across the HR diagram to cooler temperatures.

The recognition of a stability limit in the upper HR diagram has been one of the most important recent developments in studies of massive stars. This stability limit is very likely defined by the most luminous cooler supergiants which de Jager has shown are near the limits for the stability of their photospheres. The higher resolution observations required for further study of the atmospheres of the stars all along the upper luminosity boundary will soon be possible with the larger ground-based telescopes and Space Telescope.

References

Appenzeller, I. and Wolf, B.: 1981, in "The Most Massive Stars," ed. D'Odorico, Baade, and Kjar (Garching: ESO), p. 131.
Barlow, M.J.: 1978, in "Mass Loss and Evolution of O Stars," IAU Symposium #83 (Dordrecht, Reidel), p. 119. ed.P.Conti,C.deLoore
Barlow, M.J.: 1981, in "Wolf Rayet Stars: Observations, Physics, Evolution," IAU Symposium #99 (Dordrecht, Reidel), p. 149.
Blaauw, A.: 1965, in "Stars and Stellar Systems, Vol. 3, Basic Astronomical Data," ed. K. Aa. Strand (Chicago: Univ. of Chicago Press), p. 383.
Brunish, W.M. and Truran, J.W.: 1980a, Astrophys. J. 256, p. 247.
Brunish, W.M. and Truran, J.W.: 1980b, Astrophys. J. Suppl. 49, p. 447.
Chiosi, C.: 1981, in "The Most Massive Stars," ed. D'Odorico, Baade, and Kjar (Garching, ESO) p. 27.
Chiosi, C., Nasi, E., and Sreenivasan, S.R.: 1978, Astron. Astrophys. 63, p. 103.
de Jager, C.: 1980, "The Brightest Stars," (Reidel, Dordrecht), pp. 11-14.
de Jager, C.: 1984, Astron. Astrophys. 138, p. 246.
de Loore, C., de Greve, J.P., Vanbeveren, D.: 1978, Astron. Astrophys. 67, p. 373.
de Loore, C. and de Greve, J.P.: 1981, in "The Most Massive Stars," ed. D'Odorico, S., Baade, and Kjar (Garching, ESO), p. 85.
Elias, J.H. and Frogel, J.A.: 1985, Astrophys. J. 289, p. 141.
Elias, J.H., Frogel, J.A., and Humphreys, R.M.: 1985, Astrophys. J. Suppl. 57, p. 91.
Elias, J.H., Frogel, J.A., Humphreys, R.M., and Persson, S E.: 1981, Astrophys. J. (Letters) 249, p. 55.
Falk, H.J. and Mitalas, R.: 1981, Mon. Not. Roy. Astron. Soc. 196, p. 225.
Falk, H.J. and Mitalas, R.: 1983, Mon. Not. Roy. Astron. Soc. 202, p. 19.
Flower, P.J.: 1977, Astron. Astrophys. 54, p. 31.
Freedman, W.: 1984, Ph.D. Dissertation, University of Toronto.
Graham, J.A. and Humphreys, R.M.: 1985, in preparation.
Hagen, W., Humphreys, R.M., and Stencel, R.E.: 1981, Pub.A.S.P. 93, p. 567.
Hellings, P. and Vanbeveren, D.: 1981, Astron. Astrophys. 95, p. 14.
Humphreys, R.M.: 1978, Astrophys. J. Suppl. 38, p. 389.
Humphreys, R.M.: 1979, Astrophys. J. 231, p. 384.
Humphreys, R.M.: 1980a, Astrophys. J. 241, p. 587.
Humphreys, R.M.: 1980b, Astrophys. J. 241, p. 598.
Humphreys, R.M.: 1983, Astrophys. J. 269, p. 335.
Humphreys, R.M., Blaha, C., D'Odorico, S., Gull, T.R., and Benvenuti, P.: 1984, Astrophys. J. 278, p. 124.
Humphreys, R.M. and Davidson, K.: 1979, Astrophys. J. 232, p. 409.

Humphreys, R.M., Jones, T.J., and Sitko, M.L.: 1984, Astron. J. 89, p. 1155.
Humphreys, R.M. and McElroy, D.B.: 1984, Astrophys. J. 284, p. 565.
Humphreys, R.M., Nichols, M., and Massey, P.: 1985, Astron. J. 90, p. 101.
Humphreys, R.M. and Strom, S.E.: 1983, Astrophys. J. 264, p. 458.
Hutchings, J.B.: 1978, in "Mass Loss and Evolution of O Stars," IAU Symposium #83 (Dordrecht, Reidel), p. 3.
Kennicutt, R.C.: 1984, Astrophys. J. 287, p. 116.
Kennicutt, R.C.: 1985, private communication.
Kunasz, P.B. and Morrison, N.D.: 1982, Astrophys. J. 263, p. 226.
Kunasz, P.B., Morrison, N.D., and Spressant, B.: 1983, Astrophys. J. 266, p. 739.
Lamb, S.A., Iben, I., and Howard, W.M.: 1976, Astrophys. J. 207, p. 209.
Lambert, D.L., Hinkle, K.H., and Hall, D.N.B.: 1981, Astrophys. J. 248, p. 638.
Lee, T.A.: 1970, Astrophys. J. 162, p. 217.
Madore, B.F., McAlary, C.W., McLaren, R.A., Welch, D.L., Neugebauer, G., and Matthews, K.: 1985, Astrophys. J., in press.
Maeder, A.: 1980, Astron. Astrophys. 92, p. 101.
Maeder, A.: 1981a, Astron. Astrophys. 99, p. 97.
Maeder, A.: 1981b, Astron. Astrophys. 102, p. 401.
Maeder, A.: 1983, Astron. Astrophys. 120, p. 113.
McGregor, P.J.: 1981, Ph.D. Dissertation, Australian National University, Canberra.
McGregor, P.J. and Hyland, A.R.: 1981, Astrophys. J. 250, p. 116.
McGregor, P.J. and Hyland, A.R.: 1984, Astrophys. J. 277, p. 149.
Ridgway, S.T., Joyce, R.R., White, N.M., and Wing, R.F.: 1980, Astrophys. J. 235, p. 126.
Sandage, A.: 1983a, Astron. J. 88, p. 1108.
Sandage, A.: 1983b, Astron. J. 88, p. 1569.
Sandage, A.: 1984a, Astron. J. 89, p. 621.
Sandage, A.: 1984b, Astron. J. 89, p. 630.
Sandage, A. and Tammann, G.A.: 1974a, Astrophys. J. 191, p. 603.
Sandage, A. and Tammann, G.A.: 1974b, Astrophys. J. 194, p. 215.
Schild, H. and Maeder, A.: 1984, Astron. Astrophys. 136, p. 237.
Schmidt-Kaler, Th.: 1983, in "Landolt-Bornstein, New Series," Group 6, Vol. 2, Part B.
Sreenivasan, S.R. and Wilson, W.J.F.: 1982, Astrophys. J. 254, p. 287.
Stothers, R. and Chin, C.W.: 1976, Astrophys. J. 204, p. 472.
Stothers, R. and Chin, C.W.: 1979, Astrophys. J. 233, p. 267.
Stothers, R. and Chin, C.W.: 1983, Astrophys. J. 264, p. 583.
van den Bergh, S.: 1976, in "Redshifts and the Expansion of the Universe, IAU Symposium #37 (Paris: CNRS), p. 13.
Walborn, N.R.: 1972, Astron. J. 72, p. 312.
Walborn, N.R.: 1973, Astron. J. 78, p. 1067.
Zuckerman, B.: 1980, in "Ann. Rev. Astron. Astrophys." 18.

Discussion : HUMPHREYS.

LAMERS :

There seems to be some confusion about the nomenclature of the luminosity upper limit, which is sometimes called Humphreys-limit; de Jager-limit, Humphreys-Davidson limit or de Jager-Humphreys limit.
I propose that we give proper credit to those who discovered or predicted the different effects. I propose that we call the Observed Luminosity Upper Limit the "Humphreys-Davidson limit" (Humphreys and Davidson, 1979, Astrophys. J. 232, 409) and The Predicted Turbulent Instability Limit the "De Jager limit" (de Jager, 1980, The Brightest Stars: Reidel, Dordrecht; de Jager, 1984, Astron. Astrophys, 138, 246). The interesting astrophysical question is : is the De Jager-limit equal to the Humphreys-Davidson limit?

(This proposal was accepted by the participants).

LORTET :

Did you take into account the possibility that many bright stars may be member of a (undiscovered) tight cluster?

HUMPHREYS :

Yes. The spectra allow me to distinguish when the stars are composite. For example one of the supergiants in M31 had a composite spectrum, one of the blue stars in M101 is very likely more than one star and several of the blue star candidates in M101, M81 and NGC 2403 were found to be HII regions.

SREENIVASAN :

You suggested that morphological features might be a factor determining stellar evolution. Could you indicate how the physics would be different if it were the case?

HUMPHREYS :

Our first look at the luminous supergiants in M31 and M81 suggests that the most massive star progenitors are fewer numerous in the Sb type spirals. Presumably the the massive star formation rate is lower in the Sb spirals than in Sc spirals, like M33 and Magellanic Irregulars (i.e. LMC). This suggests that the morphological type may influence star formation. Perhaps the shock associated with the density wave or whatever mechanism initiates star formation is not so strong and does not produce as many massive stars in Sb type spirals.

RENZINI :

How were these M-type supergiants picked up? I mean, in which bands were the surveys done? Surveys limited to BV bands are indeed likely to miss bolometrically bright stars of late M spectral type.

HUMPHREYS :

You are correct. Humphreys and Strom used BVRI plates to select the candidate M supergiants in M101. The candidate red supergiants from Sandages work are found from blinking B and V plates.

THE DISTRIBUTION OF YOUNG STARS IN NEARBY GALAXIES

W. L. Freedman
Mount Wilson and Las Campanas Observatories
813 Santa Barbara St.
Pasadena, California 91101.

1. INTRODUCTION

Although luminous stars are relatively rare, they can potentially be studied out to large distances. In our own Milky Way, this advantage is offset by obscuration due to dust in the plane of the Galaxy. In addition, distances to these individual stars are extremely difficult to determine. The study of external galaxies allows a panoramic view of the system and its individually brightest stars which are all at a common distance. The spatial distribution of star forming regions is immediately apparent, and the effects of obscuration are minimized. Nearby resolved galaxies therefore provide a rich resource for examining the properties of the intrinsically brightest stars and their relation to other components of the galaxy.

Spectroscopy of stars in external galaxies will be covered in other reviews in this volume. Therefore, this review will concentrate on studies of photometry in galaxies near enough that stars can be resolved in them. Studies of luminosity functions, spatial distributions of young stars, and comparison with the distribution of gas, and radial distributions of stars in external galaxies will be discussed.

2. STELLAR PHOTOMETRY OF BRIGHT STARS IN NEARBY GALAXIES

Much of the work on stellar photometry in external galaxies has recently been reviewed by Scalo (1985a) and Freedman (1984), and will not be repeated here. Instead, below are tabulated studies in which photometry of bright stars in external galaxies have been obtained. Included also for reference, is the type of study (photographic (pg), photoelectric (pe), or CCD).

Table 1

Nearby Galaxies with Stellar Photometry

Galaxy	Type	References	Study
M31	Sb	Reddish (1962)	pg
		Baade and Swope (1963)	pg
		Freedman (1986)	CCD
		Humphreys (1986)	pg
M81	Sb	Sandage (1984a)	pg
		Freedman (1984, 1986)	pg, CCD
M33	Sc	de Vaucouleurs (1961)	pg
		Madore (1970, 1978)	pg
		Reddish (1978)	pg
		Humphreys and Sandage (1980)	pg
		Freedman (1984)	pg, CCD
		Freedman et al. (1986)	pg, CCD
NGC 2403	Sc	Tammann and Sandage (1968)	pg
		Sandage (1984b)	pg
		Freedman (1984, 1986)	pg, CCD
M101	Sc	Sandage (1983)	pg
		Humphreys and Strom (1983)	pe
IC 1613	Irr	Sandage and Katem (1976)	pg
		Hodge (1978, 1980)	pg
		Freedman (1986)	CCD
NGC 6822	Irr	Kayser (1967)	pg
		Hodge (1980)	pg
		Freedman (1986)	CCD
Sextans A	Irr	Reddish (1978)	pg
		Hoessel, Schommer and Danielson (1983)	CCD
		Freedman (1984, 1986)	CCD
Ho I, II	Irr	Hoessel and Danielson (1985)	CCD
Ho IX	Irr	Sandage (1984)	pg
		Freedman (1984, 1986)	CCD
LMC	Irr	Shapley (1931)	pg
		de Vaucouleurs (1955, 1956)	pg
		Westerlund (1961)	pg
		Hodge (1961)	pg

			Lucke (1972)	pg
			Butcher (1977)	pg
			Rousseau et al. (1978)	pe
			Hardy (1978)	pg
			Stryker (1981)	pg
			Stryker and Butcher (1982)	pg
			Hardy et al. (1984)	pg
SMC	Irr		Ardeberg and Maurice (1977)	pe
			Hardy and Durand (1984)	pg

3. LUMINOSITY FUNCTIONS IN NEARBY GALAXIES

Until recently, the data on luminosity functions in nearby galaxies was inhomogeneous, and, in many cases, photometry was available for only small samples of stars, often obtained for purposes other than the construction of luminosity functions. This problem is beginning to be solved with the advent of fast plate-measuring machines, and software analysis programs for obtaining photometry in crowded fields.

It must be stressed however, that the study of luminosity functions in even the nearest galaxies is limited to the very brightest stars alone; therefore no direct information is acquired about the low-mass end of the luminosity function outside of our own Galaxy. Furthermore, no information is obtained on possible variations of the luminosity or mass function with time. However, such studies do offer the advantage of allowing a comparison of the bright end of the luminosity function in a wide range of environments, having differing metallicities, kinematics, amounts of present star formation, total mass of system, etc.

3.1 Comparison of Luminosity Functions

There are conflicting interpretations of the data in the literature concerning whether or not real variations exist in the slope of the luminosity function comparing one galaxy to another. The first systematic comparison of published luminosity functions was recently undertaken by Scalo (1985a). Scalo finds that there is a remarkable agreement between the luminosity functions of the stars in M33, M31, NGC 6822, IC 1613, the LMC and SMC. Many cases where differences have been previously claimed are shown to be marginal, especially in view of the magnitude of the statistical uncertainties and the variety of methods used to obtain and reduce the data in the various studies. He concludes that the available data for $M_v < -3$ mag is consistent with a galaxy-wide universal luminosity function.

Freedman (1984) obtained UBV prime focus plates at the Canada-France-Hawaii telescope (CFHT) for M33, NGC 2403, and M81, and at Cerro Tololo for NGC 300. Positions, magnitudes, and colors were measured for several thousand images over the face of each of these galaxies using the Automatic Plate Measuring (APM) machine (see Kibblewhite et al. 1984). In addition, BVRI CCD data were obtained at the Kitt Peak 4m and the CFHT 3.6m for several fields in each of M33, NGC 2403, Ho IX, Sextans A, and Leo A.

The upper end of the main-sequence luminosity function is populated by hot, luminous, blue stars. Thus, apparent luminosity functions based on stars of all colors may be relatively insensitive to differences in the upper end of the mass function. Therefore, luminosity functions were constructed with a sample of only the bluest stars in each galaxy, as determined from their U-V and B-V colors. An additional advantage to restricting the sample to the bluest stars is that foreground contamination by stars in our own Galaxy is virtually eliminated. V, B and U (when available) luminosity functions were obtained.

Figure 1 presents V luminosity functions for the blue stars in 10 nearby galaxies. The data for M33, NGC 2403, M81, NGC 300, Holmberg IX, Sextans A, and Leo A are from Freedman (1984). Data for the LMC, SMC and NGC 6822 are from Rousseau et al. (1978), Ardeberg and Maurice (1977), and Kayser (1967), respectively. The data are displayed on an arbitrary number scale, as a function of absolute visual magnitude, M_v.

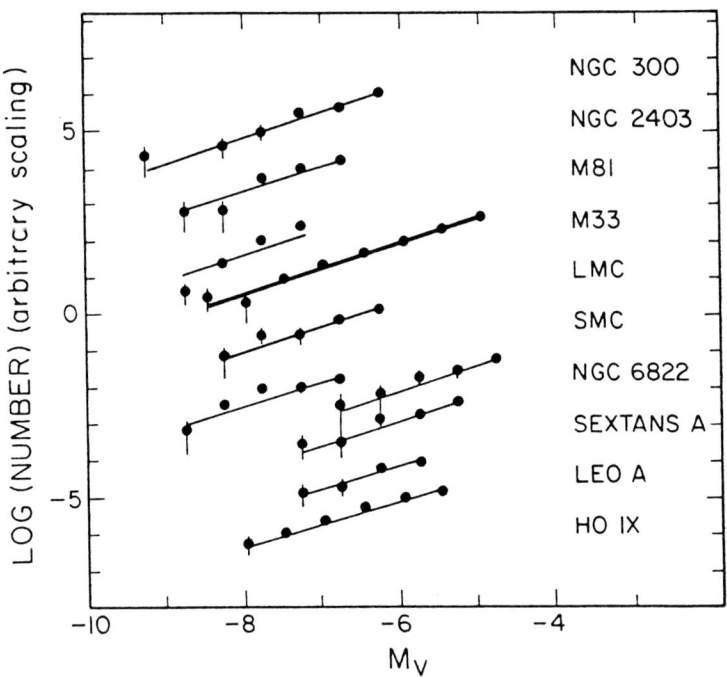

Figure 1 - The slope of the upper end of the luminosity function for a sample of ten nearby galaxies.

From this plot, the similarity of the slopes of the luminosity functions for this sample of galaxies is immediately evident over the range of absolute magnitude studied ($-9 < M_v < -5$ mag). The slope of the upper end of the luminosity function shows little variation from one galaxy to another. The largest deviations from universality occur at the brightest end where the numbers of stars are low. In addition, caution should be exercised in interpreting any of the apparent upper magnitude cut-offs because of uncertainties in the adopted distances to these galaxies. Due to the small numbers however, random errors are expected to be much larger at this end, and the data remain consistent with a universal slope for the upper end of the luminosity function.

It is also found that the slopes of the B and U luminosity functions also show little change from galaxy to galaxy. Although in general, it might be expected that the B and U luminosity functions might more accurately reflect differences in the mass function of the hottest stars, it should be recalled that all of the V luminosity functions shown here are based on a sample of the bluest stars in each galaxy. Thus these V luminosity functions are not for the visually most luminous stars, but for the bluest luminous stars.

3.2 The Slope of the Luminosity Function with Radius in M33

It is of interest to compare the slope of the luminosity function as a function of radius, particularly in galaxies with measured abundance gradients. Terlevich and Melnick (1983) have presented evidence based on a study of giant extragalactic HII regions that the slope of the mass function is a function of metallicity.

Conflicting evidence has been presented for the case of the change in the slope of the initial mass function as a function of galactocentric radius in our Galaxy. Garmany, Conti, and Chiosi (1982) have claimed that the slope of the galactic initial mass function is steeper in the region at distances beyond the distance of the sun. Based on a study of young clusters, Burki (1977) concludes also that the initial mass function in our Galaxy varies as a function of radius; however he finds the gradient in the opposite sense to that claimed by Garmany, Conti and Chiosi (1982). Scalo (1985a) has reanalyzed the data from the Garmany, Conti and Chiosi O-star catalog and concludes that the catalog is incomplete at the faint end. However, correcting for incompleteness in the Galaxy remains a difficult problem. Humphreys and McElroy (1984) have recently made an attempt to correct the counts, and find that there is no evidence for a change of slope with radius in the Galaxy. Further, they conclude that the slopes of the stellar mass functions are similar for the Galaxy, the LMC, and the SMC.

Berkhuijsen (1983) determined the slope of the luminosity functions for a number of Humphreys and Sandage (1980) associations in M33. She found factors of two variation in the slope, suggesting that the luminosity function within M33 is varying as a function of radius, and therefore, by implication, as a function of metallicity.

3.4 Problems Remaining and Future Work

The sample of galaxies now needs to be extended to include earlier Hubble types. The numbers of stars measured in many of the galaxies also needs to be increased to improve the statistics and further investigate the limits to the universality of the slope of the observed luminosity function. A larger, statistically reliable, sample of stars needs to be obtained for other galaxies with observed gradients in addition to M33, (such as M101) in order to investigate whether or not such galaxies might exhibit gradients in the slopes of their luminosity functions. A CCD study of this kind is now underway. Further, statistically reliable samples of stars within individual associations in the nearest galaxies should be obtained to conclusively determine whether or not variations exist on relatively small spatial scales within galaxies.

Within the uncertainties, the slope of the upper end of the luminosity function shows little variation. However, the uncertainties are such that 15 percent variations in the slope from galaxy to galaxy may have gone undetected. It will therefore be of considerable interest to obtain data on more galaxies, and to obtain larger samples of stars for many of the galaxies which have been discussed here in order to improve the statistics and decrease the uncertainties. In particular, the reliability of the M81 calibration must be checked, since no faint magnitude sequence was available for this galaxy. Such programs are now being undertaken.

At present, the result remains that within the uncertainties, there is no evidence for significant departures from universality for the slope of the upper end of the luminosity function.

4. RADIAL DISTRIBUTIONS IN EXTERNAL GALAXIES

It has been known for some time that the total disk light of spiral galaxies can be well represented by an exponential distribution (e.g., de Vaucouleurs 1959, Freeman 1970). More recently, the CO distributions in late-type spirals have been found to follow the exponential profiles of the integrated blue light (Young and Scoville 1982a,b), and the Hα and integrated light distributions also exhibit a close correspondence (DeGioia-Eastwood et al. 1981, Hodge and Kennicutt 1983).

Late-type Scd galaxies, which have been mapped in CO, show a close correspondence in their radial distributions with the optical blue surface-brightness distributions. The HI distribution displays a distinctly flatter radial distribution for these same systems (e.g., M101, M51, NGC 6946 and IC 342). For the Galaxy and M31, a ring of CO is observed, as well as a hole in the HI distribution. In NGC 2841 and M81, central HI holes are also present. To date, the galaxies displaying central holes tend to be early-type systems which have larger bulges than later-type galaxies, indicating that a causal connection may be present, perhaps through the resulting differences in their rotation curves and/or plane thicknesses induced by the bulge.

Freedman (1984) compared the radial distributions of blue light, HII regions, OB stars, Wolf-Rayet stars, supernova remnants, and neutral hydrogen in M33. Wevers (1984) obtained integrated light and neutral hydrogen distributions for a sample of 16, mainly late-type galaxies.

The results of all of the above studies illustrate that the neutral hydrogen distributions are significantly flatter than the distributions of recent star formation constituents (e.g., OB stars, HII regions and Wolf-Rayet stars), the molecular gas distributions, as well as the integrated light distributions.

The integrated blue light of a galaxy contains a contribution both from the young stellar population and an old disk giant population with an age of 1-3 billion years. The correspondence between the OB star, HII regions, Wolf-Rayet stars, supernova remnants and the integrated disk light suggests that the star formation rate at present does not differ from the past rate, and specifically, that star formation has not proceeded at a rate proportional to the mean gas density with some power greater than unity. If star formation had proceeded at such a rate, then one would expect that the total gas content at the center of the galaxy, (where the mean density of gas is greatest), would be depleted due to star formation, at a faster rate than in the outer regions of the galaxy where the mean gas density is lower. In such a case, the young stellar population would not be reflecting the distribution of the old disk light, but would exhibit a flatter radial profile. In fact, the HI displays a distribution that one might expect to see if the star formation at the center were proceeding at a higher rate than in the outer regions.

The apparent discrepancy between the implications of a relatively flat neutral hydrogen gas component and the steeper radial distributions of recent star formation constituents and of molecular gas late-type spirals is a challenge to present theories of galaxy evolution. Young and Scoville (1982a,b) conclude, on the basis of the similarity of the CO and blue light distributions, that the star formation rate per nucleon is a constant. However, this hypothesis cannot simply account for the strong galactic abundance gradients observed in galaxies. A closed model where the rate of star formation proceeds linearly with the gas density predicts a linear increase of the abundance with time, and no change with radius. That abundance gradients are observed implies that star formation cannot have proceeded linearly with gas density as a function of time, over the extent of the disk, for a simple closed model.

Four possible alternatives which might explain this discrepancy are listed below.

1. The yield is not constant as a function of radius. Gusten and Mezger

(1983) consider a bimodal star formation model which could produce a variable yield and thus explain the observed abundance gradients for the case where star formation proceeds linearly with gas density. Larson (1985) has also recently discussed a bimodal star formation model, in order to account for the unseen mass in the solar neighborhood.

2. Radial gas flow or infall. For a recent review, see Lacey and Fall (1985). Lacey and Fall find that for our Galaxy, they can fit the observed age-metallicity relation, the radial abundance gradient, star formation rate, and total surface density of gas (HI plus H_2) and stars, when radial flows at velocities of less than about 1 km/sec are assumed.

3. The CO/H_2 ratio may vary as a function of radius in galaxies. Bhat et al. (1985) and Harding and Stecker (1985) have recently presented evidence based on studies of the gamma-ray distribution in the Galaxy which suggests that the CO/H_2 ratio increases toward the galactic centre, and that the mass in molecular gas has been significantly overestimated. However, even if, in the extreme, the HI and H_2 distributions turn out not to be too significantly different, the flatter distribution of gas with respect to recent tracers of star formation (HII regions, OB stars, Wolf-Rayet stars) still remains to be explained.

4. The conversion of HI into H_2 proceeds faster at the center of the galaxy.

Two apparently contradictory alternatives seem to be present, both of which leave unexplained observations. 1) The total gas content has a distribution similar to the recent star formation, and the star formation rate proceeds linearly with gas density. This explains the observed radial distributions of all constituents except the HI, and also fails to account for abundance gradients. 2) The efficiency of conversion of gas into stars is higher in the center relative to the outer regions. This naturally explains the abundance gradients, but first begs the question as to why the efficiency is higher. Second, if the total gas content is being depleted at a faster rate in the center, then why don't the newly formed stars follow the flatter, depleted, total gas distribution, but instead exhibit a steeper gradient, similar to the integrated light.

In summary, recent observations of the radial distributions of gas, young objects and the integrated light from an older population in late-type galaxies reveal that these distributions are all very similar. The implication is that the star formation histories in these galaxies have been the same for the last few billion years. Spiral galaxies exhibit abundance gradients, indicating that more processing per unit mass has occurred toward the center. The causal relationship between the neutral hydrogen distribution and other young Population I tracers is still far from clear. These issues remain at present, as

known) does not solve the problem since the resolution of the data is not improved compared to the HI surveys.

ACKNOWLEDGEMENTS. WLF would like to thank the American Astronomical Society for an International Travel Grant which provided partial support to attend this meeting.

REFERENCES

Ardeberg, A., and Maurice, E. 1977, Astr. and Ap., 30, 261.

Baade, W., and Swope, H. 1963, A. J., 68, 435.

Berkhuijsen, E. M. 1983, Astr. and Ap., 127, 395.

Bhat, C.L., Issa, M. R., Houston, B. P., Mayer, C. J., and Wolfendale, A. W. 1985, Nature, 314, 511.

Blair, W. P., and Kirshner, R. P. 1985, Ap. J., 289, 582.

Burki, G. 1977, Astr. and Ap., 57, 135.

Butcher, H. 1977, Ap. J., 216, 372.

DeGioia-Eastwood, K., Grasdalen, G. L., Strom, S. E. and Strom, K. M. 1984, Ap. J., 278, 564.

de Vaucouleurs, G. 1955, A. J., 60, 126.

de Vaucouleurs, G. 1956, Irish A. J., 4, 13.

de Vaucouleurs, G. 1959, Handbuch der Physik, 53, 275.

de Vaucouleurs, G. 1961, Year Book Amer. Phil. Soc., p. 268.

Diaz, A. I. and Tosi, M. 1984, M. N. R. A. S., 208, 365.

Einasto, J. 1972, Ap. Letters, 11, 195.

Emerson, D. J. 1974, M. N. R. A. S., 169, 607.

Freeman, K. C. 1970, Ap. J, 160, 811.

Freedman, W. L. 1984, PhD thesis, University of Toronto.

Freedman, W. L. 1985, *Ap. J.*, Nov. 1, in press.

Freedman, W. L. 1986, in preparation.

Freedman, W. L., Kibblewhite, E., Irwin, M., Bunclark, P., Bridgeland, M., and Madore, B. 1986, in preparation, Paper I.

Garmany, C. D., Conti, P. S., and Chiosi, C. 1982, *Ap. J.*, 263, 777.

Guibert, J., Lequeux, J. and Viallefond, F. 1978, *Astr. Ap.*, 68, 1.

Gusten, R., and Mezger, P. 1982, *Vistas in Astron.*, 26, 159.

Hamajima, K. and Tosa, M. 1975, *Publ. Astr. Soc. Japan*, 27, 561.

Harding, A. K., and Stecker, F. W. 1985, *Ap. J*, 291, 471.

Hardy, E. 1978, *Ap. J.*, 223, 98.

Hardy, E., and Durand, D. 1984, *Ap. J.*, 279, 567.

Hardy, E., Buonanno, R., Corsi, C.E., Janes, K.A. and Schommer, R.A. 1984, *Ap. J.*, 278, 592.

Hartwick, F. D. A. 1971, *Ap. J.*, 163, 431.

Hodge, P. W. 1961, *Ap. J. Suppl.*, 6, 235.

Hodge, P. W. 1978, *Ap. J. Suppl.*, 37, 145.

Hodge, P. W. 1980, *Ap. J.*, 241, 125.

Hodge, P. W., and Kennicutt, R. C., 1983, *A. J.*, 88, 296.

Hoessel. J. G., and Danielson, G. E. 1984, *Ap. J.*, 286, 159.

Hoessel, J. G., Schommer, R. A., and Danielson, G. E. 1983, *Ap. J.*, 274, 577.

Humphreys, R. M. 1986, in preparation.

Humphreys, R. M., and McElroy, D. B. 1984, *Ap. J.*, 284, 565.

KAUFMAN :

In M33 the radial distribution of the number of HII regions of all luminosities agrees with the blue light distribution; however the radial distribution of giant radio HII regions is much more sharply peaked than the overall distribution of blue light.

FREEDMAN :

Yes, that is correct. But if one compares either the number distribution or the flux from all HII regions, the radial distributions as compared to the stars are very similar. I am not quite sure how to interpret the steeper distribution of bright HII regions. It is difficult to rule out a selection effect of finding faint HII regions toward the center. There will also be a crowding problem due to the higher surface density of HII regions toward the center. Some bright HII regions may actually be aggregates of several smaller HII regions.

MOFFAT :

You stated that the ILF slopes at the bright end are constant for many different galaxies. Can you give us a more quantitative estimate of the slope and the slope of the more fundamental IMF (and their variations)?

FREEDMAN :

The slope of the upper end of the luminosity function for blue stars is about 0.7 ± 0.1. Scalo (see review of this volume) has tabulated the slope of the IMF for the present and several other recent determinations.

Discussion : FREEDMAN.

ZINNECKER :

You showed that the radial distribution of the integrated blue light in M33 is very similar to the radial distribution of the young stars, and from this you concluded that the star formation rate hasn't changed since about 1 Gyr ago. I think your conclusion depends somewhat on what fraction the young stars alone contribute to the total blue light. Can you estimate their percentage?

FREEDMAN :

What you say is very true, the conclusion depends strongly on the shape of the IMF adopted, upper and lower mass cutoffs, and the star formation role as a function of time. The point that I meant to illustrate however, was that contrary to some previous studies which assumed that <u>all</u> of the blue light is due to young stars, for a "standard" (Salpeter) IMF with a slowly decreasing star formation rate, about half of the blue light comes from stars up to a few times 10^9 years.

MASSEY :

What magnitude range does your luminosity function cover (answer : 4). Over that sort of range, all that the constancy of the luminosity function in V means is that the exponent of the IMF does not vary by 3 or so (see my PASP 97,5 figure). Differences such as that suggested by Garmany et al. would go completly undetected in a luminosity function, even in V.

FREEDMAN :

The uncertainty in the sensitivity of the slope of the mass function derived from luminosity functions depends on the accuracy of the measured luminosity function. Thus, in order to detect changes in the slope of the mass function, at the level suggested by Garmany et al., the original sample of stars must be complete, otherwise, as you point out, small differences or errors in the luminosity function will be magnified in the mass function determination. As discussed by Scalo, and also by Humphreys and McElroy (1984, ApJ <u>284</u>, 565), the Galactic catalogues appear to be incomplete. Their analysis suggests that, within the uncertainties, no trend of the slope with radius is evident in the Galaxy.

In other words, as you suggest, very small changes (or errors) in the luminosity function will give rise to much larger changes (or errors) in the mass function.

Sandage, A. R., and Katem, B. ,1976, A. J., 81, 743.

Sanduleak, N. 1969, A. J., 74, 47.

Scalo, J. 1985a, Fundamentals of Cosmic Physics, in press.

Scalo, J. 1985b, A. J., submitted.

Schmidt, M. 1959, Ap. J., 129, 248.

Schmidt, M. 1962, in IAU Symposium No. 15, Problems of Extragalactic Research, ed. G. C. McVittie, (New York: MacMillan), p 170.

Schmidt, M. 1963, Ap. J., 137, 758.

Shapley, H. 1931, Harvard Bull., No. 881.

Stothers, R., and Chin, C. 1980, Ap. J., 240, 885.

Stryker, L. C. 1981, PhD thesis, Yale University

Stryker, L. C., and Butcher, H. R. 1982, in IAU Colloquium No. 61, Astrophysical Parameters for Globular Clusters, ed. A. G. D. Philips and D. S. Hayes, (Schenectady: Davis Press), p. 255.

Talbot, R. J., Jr. 1971, Ap. Lett.s, 8, 111.

Talbot, R. J., Jr. 1980, Ap. J., 235, 821.

Tammann, G. A., and Sandage, A. R. 1968, Ap. J., 151, 825.

Terlevich, R., and Melnick, J. 1983, ESO Preprint No. 264.

Tosa, M. and Hamajima, K. 1975, Publ. Astr. Soc. Japan, 27, 501.

Westerlund, B., 1961, Uppsala. Astron. Obs. Ann., 5, 1.

Wevers, B. M. H. R., PhD thesis, University of Groningen.

Young, J., and Scoville, N. 1982a, Ap. J., 258, 467.

Young, J., and Scoville, N. 1982b, Ap. J. (Letters), 260, L11.

Humphreys, R. M., and Sandage, A. R. 1980, **Ap. J. Suppl.**, **44**, 319.

Humphreys, R. M., and Strom, S. E. 1983, **Ap. J.**, **264**, 458.

Kayser, S. 1967, **A. J.**, **72**, 134.

Kibblewhite, E. J., Bridgeland, M. T., Bunclark, P., and Irwin, M. 1984, in **Proceedings of the Astronomical Microdensitometry Conference**, NASA Conf. Publ. 2317, p.277.

Lacey, C. G., and Fall, S. M. 1985, **Ap. J**, **290**, 154.

Larson, R. B. 1977, in The Evolution of Galaxies and Stellar Populations, ed. B. M. Tinsley, and R. B. Larson, (New Haven: Yale University Observatory), p. 97.

Larson, R. B. 1985, preprint.

Lucke, P. B. 1972, PhD thesis, University of Washington.

Madore, B. F. 1970, Master's thesis, University of Toronto.

Madore, B. F. 1977, **M. N. R. A. S.**, **178**, 1.

Madore, B. F. 1978, **Observatory**, **98**, 169.

Mathis, J. S. 1959, **Ap. J.**, **129**, 259.

Newton, K. 1980, **M. N. R. A. S.**, **190**, 689.

Reddish, V. C. 1962, **Zeit. Ap.**, **56**, 194.

Reddish, V. C. 1978, **Stellar Formation**, (New York: Pergamon), p. 87.

Rousseau, J., Martin, N., Prevot, L., Rebeirot, E., Robin, A., and Brunet, J. P. 1978, **Astr. and Ap. Suppl.**, **31**, 243.

Salpeter, E. E. 1955, **Ap. J.**, **121**, 161.

Sandage, A. R. 1983, **A. J.**, **88**, 1569.

Sandage, A. R. 1984a, **A. J.**, **89**, 621.

Sandage, A. R. 1984b, **A. J.**, **89**, 630.

POSTER PAPERS - SESSIONS 1 and 2.

Chairman : A.J. Willis.

1. T.R.MANLEY: A Stellar Content Study of NGC 2403 with the Automated Plate Scanner.

2. V.NIEMELA: Spectroscopic Binaries in the LMC.

3. E. ZSOLDOS: The Hypergiant Variable HR 8752=HD217476.

4. J.SMOLINSKI, J.L.CLIMENHAGA and J.M.FLETCHER: Activity in the Envelope of the G-type Hypergiant HD 217476.

5. D.J.STICKLAND: IRAS Observations of the Cool Galactic Hypergiants.

6. R.K.PRINJA and I. HOWARTH: Is Rotation Important?

7. D.H.MORGAN, K.NANDY and G.I.THOMPSON: The Absorption Feature at 1920 A in the Spectra of Early-type Supergiants in the SMC.

8. L.MANTEGAZZA: The Light Variations of the Red Supergiant Mu Cephei.

9. G.R.GRIEVE and B.F.MADORE: Variability of Magellanic Cloud Supergiants.

10. M.KONTIZAS and E.KONTIZAS: Star Formation Rate of Star Clusters in the SMC and their Adjoining Fields.

11. J.A.GRAHAM and R.M.HUMPHREYS:
    Red Supergiants in NGC 300.

12. L.GREGGIO: Simulations of the Brightest Stars in Galaxies as Distance Indicators.

13. C. de JAGER, H.NIEUWENHUIJZEN and K.A. VAN DER HUCHT:
    The Dependence of the Stellar Rate of Mass Loss on Effective Temperature and Luminosity.

14. B.BOHANNAN, D.C.ABBOTT, S.A. VOELS and D.G.HUMMER:
    Steps to a New Calibration of the Spectral Type-Effective Temperature Relationship for Early-type Stars.

15. E.CASTELLI, C.MOROSSI and R.STALIO:
    Photospheric and "Chromospheric" CIV lines in B-Main-Sequence Stars.

16. R.POLIDAN and R.STALIO:
    Voyager Observations of Hot Luminous Stars.

17. E.NASI: Synthetic HR Diagrams for Luminous Stars. A Test of Stellar Evolution Theory.

18. D.J.McCONNELL, R.F. WING and F. COSTA:
    A Low Dispersion Near Infrared Survey for Galactic M-Supergiants.

A STELLAR CONTENT STUDY OF NGC 2403 WITH THE AUTOMATED PLATE SCANNER

Thomas R. Manley
University of Minnesota

In this poster we illustrate how the Minnesota Automated Plate Scanner (APS) is being used to study the luminous stellar content of NGC 2403. Presented are a brief description of the APS, examples of photometric calibration, separation of stellar and non-stellar images, and a preliminary color-magnitude diagram. The eventual goal is to study the evolution of massive stars via color-magnitude diagrams, luminosity functions and star formation rates.

The APS is a very high speed measuring machine. Two plates are mounted on a moving table with accurately calibrated lead screws controlled by an M6809 microprocessor. A laser beam is passed through a rotating octagonal prism, is then split and focussed to a 10 micron spot on each of the two plates as well as on a reticle used to determine position. The light passing through the plates is detected by semiconductor photodetectors. A PDP 11/60 computer records the positions of the plates where the density crosses either of two (soon to be four) threshold levels. The data are then sorted by the PDP 11/60 into images and parameterized with ellipses. The ellipse parameters are the X and Y positions of the center, the diameter, ellipticity and orientation of the ellipse, and FUZ, which is a measure of the goodness of fit of the ellipse to the transit endpoints. A catalog of the ellipse parameters is made for all of the images on each plage. Plate pairs can be matched to produce a catalog of images that are on both plates. All programming is done in FORTH, and results are displayed on a Grinnell image processor.

The scanning is done in strips 12mm wide at rates of 1.5, 3 or 6 mm/sec. Due to its very high speed, the APS is best at problems where a large number of images need to be processed.

The scans of the plates are calibrated photometrically with photoelectric and CCD observations of nearby field stars in the standard U, B, V and R magnitudes. To get more accurate photometric calibrations, color equations were determined to convert the CCD photometric system to the plate/filter system. This was done by taking a series of spectra of main sequence stars from B0 to M5, multiplying by the published spectral response curves for the plate/filter or CCD/filter combination, and integrating over wavelengths to get a relative response in a passband. For example, the color term for V is the slope of a plot of $V_{pg}-V_{CCD}$ as function of $(B-V)_{CCD}$. Once this is done the image diameter-magnitude calibrated can be well determined for each plate/filter combination.

Two of the ellipse parameters are used to separate stellar from non-stellar images, ellipticity and FUZ. Plots of ellipticity vs. diameter and FUZ vs. diameter can be used to separate stellar and non-stellar images.

Preliminary plots of V vs. V-R for two areas of approximately the same size, one inside NGC 2403 and one of a nearby star field are discussed. Only images classified as stellar from FUZ and ellipticity criteria are considered. The number of stars is about the same in both plots, but there are many more faint stars on the galaxy plot, which reflects the dominant contribution from stars in the galaxy. Adopting an apparent distance modulus of $(m-M)_V = 27.8$, we are seeing stars in the galaxy down to an absolute magnitude of about -7 on the V vs. V-R diagram.

SPECTROSCOPIC BINARIES IN THE LMC

Virpi S. Niemela [1,2]
Instituto de Astronomia y Fisica del Espacio
Buenos Aires, Argentina

Preliminary results are presented of an observing programme aimed to obtain estimates of the stellar masses from studies of spectroscopic binary systems in the Large Magellanic Cloud. These are the first steps with the final purpose to determine an empirical mass-luminosity relation in a galaxy other than our own.

The observations have been carried out at CTIO, Chile, with the image tube spectrograph attached to the 1 m reflector. The spectrograms have a reciprocal dispersion of 45 A/mm, and were obtained on III aJ emulsion, baked in forming gas. Three double lined systems have been observed, namely HV2543, HV2241 and NS105-67[. The preliminary orbital elements of these systems are listed in Table 1, and their radial velocity variations are shown in Figure 1.

Table 1. Orbital parameters for 3 double-lined binaries.

Parameter	HV2543	HV2241	NS105-67[
Period(days)	4.829	4.343	3.301
V_o(km/s)	+ 290	+ 285	+ 270
K_1(km/s)	160	157	224
K_2(km/s)	273	261	355
$M_1 \sin^3 i (M_o)$	26	21	41
$M_2 \sin^3 i (M_o)$	15	12	26

Notes to Table 1: the orbits were assumed to be circular. Data for NS105-67[are from Niemela and Morrell (1986, in preparation).
The approximate spectral types of the components are O9I:+O8V; O7V:+O8V and O4f+O6V, for HV2543, HV2241 and NS105-67, respectively.

---
1. Visiting astronomer, CTIO, NOAO.
2. Member of Carrera del Investigador, CIC, Prov.BsAs, Argentina

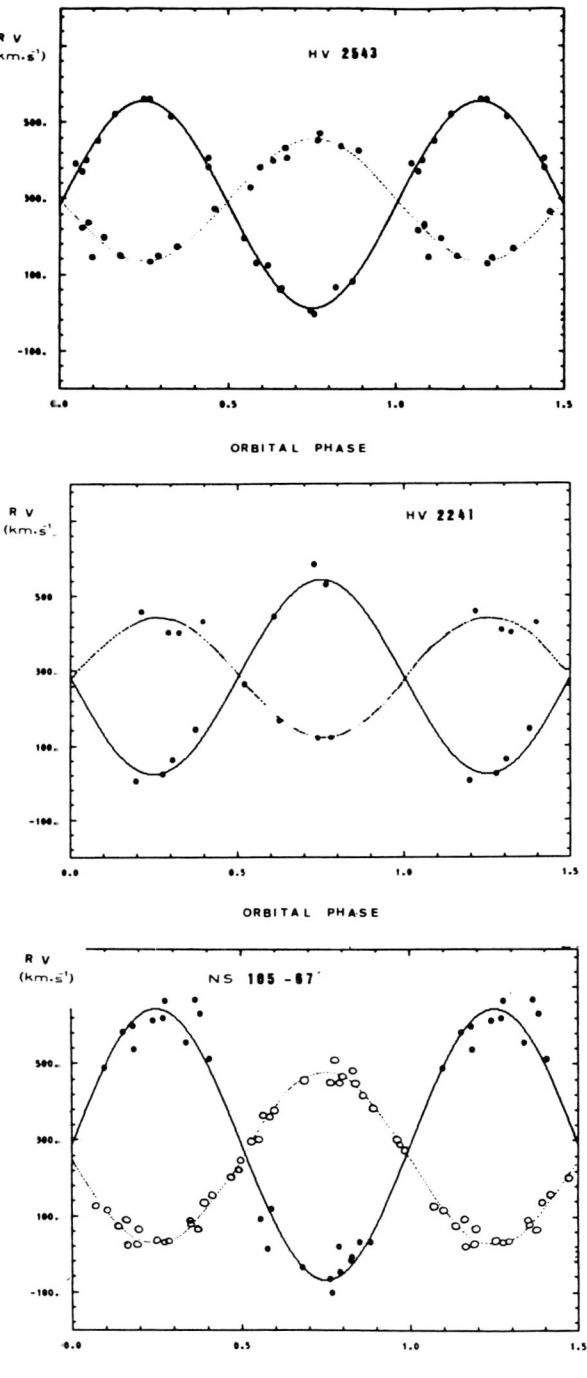

Fig. 1. Radial-velocity curves derived from the computed orbits of the three binaries, with the observed velocities plotted.

THE HYPERGIANT VARIABLE HR 8752 = HD 217476

E. Zsoldos
Konkoly Observatory
Budapest XII, P. O. Box 67
H-1525, Hungary

HR 8752 is one of the brightest stars of the Galaxy. Humphreys (1978) gives its absolute visual and bolometric magnitudes between -9.1 and -9.5, and between -9.2 and -9.5, respectively. The star received its variable star name, V509 Cas, in 1972 (Kukarkin et al., 1972). Radial velocity variations were first noticed by Harper (1923) and by Campbell and Moore (1928). In Harvard Ann. Vol. 44 (Pickering, 1899) there are indications for light variation at the end of the 19th century.

The light variation of HR 8752 is quasi-periodic (Zsoldos and Olah, 1985). A semi-period of 400 days can be assigned to the star. The theoretical models of Lovy et al. (1984) make it possible to derive the physical parameters of HR 8752 (assuming that it pulsates in the radial fundamental mode). Adopting a value of -9.3 for the absolute bolometric magnitude, the effective temperature can be obtained from the P-L-C relation for cool stars of Lovy et al. (1984)

$$\log P_o = -0.275\ M_{bol} - 3.918 \log T_{eff} + 14.543 \ . \qquad (1)$$

The pulsation constant can be derived from the following equation (Lovy et al., 1984)

$$\log Q_o = -0.054\ M_{bol} - 0.864 \log T_{eff} + 1.635 \ . \qquad (2)$$

The radius is easily obtainable from the $Q=P\sqrt{\rho}$ relation.

The observed and derived parameters of HR 8752 are as follows:

Semi-period	400 days
$\log (L/L_\odot)$	5.62
$\log T_{eff}$	3.70
Mass	30 $M/M_\odot$
$\log Q_\odot$	-1.060

$\log (R/R_o)$ 2.93

These quantities are based upon the assumption of radial pulsation. The values are in good agreement with those of Arellano Ferro (1985).

I wish to thank Drs. K. Olah and B. Szeidl for their invaluable help.

REFERENCES

Arellano Ferro, A. 1985, preprint
Campbell, W. W. and Moore, J. H. 1928, Publ. Lick Obs. Vol. 16
Harper, W. E. 1923, Publ. Dominion Astrophys. Obs. $\underline{2}$,189
Humphreys, R. M. 1978, Astrophys. J. Suppl. $\underline{38}$,309
Kukarkin, B. V.; Kholopov, P. N.; Kukarkina, N. P. and Perova, N. B. 1972, Inf. Bull. Var. Stars No. 717
Lovy, D.; Maeder, A.; Noels, A. and Gabriel, M. 1984, Astron. Astrophys. $\underline{133}$,307
Pickering, E. C. 1899, Harvard Ann. $\underline{44}$,1
Zsoldos, E. and Olah, K. 1985, Inf. Bull. Var. Stars No. 2715

ACTIVITY IN THE ENVELOPE OF THE G-TYPE HYPERGIANT HD 217476

J.Smoliński[1], J.L.Climenhaga[2], and J.M.Fletcher[3]
[1]Copernicus Astronomical Center, Toruń, Poland
[2]Department of Physics, University of Victoria, Victoria, Canada
[3]Dominion Astrophysical Observatory, Victoria, Canada

Yellow hypergiant, HD 217476, (HR 8752) is known to have extensive circumstellar envelope. Its unusual spectrum makes it enigmatic object. On the H-R diagram this hypergiant lies among the most luminous stars known in six Local Group galaxies (Humphreys et al.,1984). We have observed it with the same equipment during last fifteen years at the Dominion Astrophysical Observatory. Our present results are based upon several hundreds of high dispersion spectra (6 and 10 Å/mm). They indicate the following model:

1) Hypergiant HD 217476 is the binary system with about 620-days period as derived from the radial velocity curve. The secondary is the B1 V-type star according to the ultraviolet observations by Stickland and Harmer (1978). This binary system is contained in the common expanding envelope and surrounded by the H II region observed at the radio wavelengths (Smoliński, Feldman, and Higgs, 1977) and in the [N II]lines (Sargent 1965).

2) The mentioned expanding envelope is very active, what is seen in the line splitting and in the changes of the intensities of the components in H_α and in the neutral metallic lines, especially in the Fe I lines. Basing on the three blue-shifted components of neutral metallic lines (with the radial velocities reduced in respect to the binary system, i.e. about 35, 54 and 84 km/s) one can consider three regions or shells in this envelope. We have found that these components are visible as absorption or emission or even disappear depending on the time. However, their radial velocities are rather stationary. As an example, these kinds of activity of the expanding envelope are shown for Fe I λ 6430.8A line in Fig.1 and Fig.2.

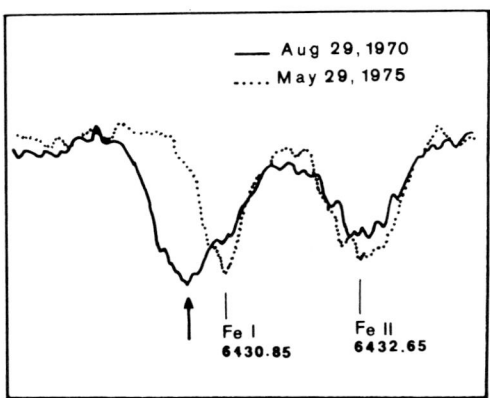

Fig.1 The characteristic changes of the envelope component (from the absorption to disappearance) as shown by arrow.

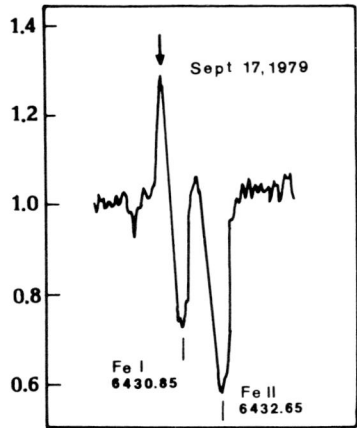

Fig.2 The same as Fig.1 but the change into emission.

It is worth to notice that the proposed model for the hypergiant HD 217476 is supported by the majority of the observational facts obtained from the ultraviolet up to the radio wavelengths.

REFERENCES

Humphreys, R.M., Blaha, C., D´Odorico, S., Gull, T.R., Benvenuti, P., 1984, Astrophys. J., 278, pp. 124 - 136
Sargent, W.L.W., 1965, Observatory, 85, pp. 33 - 35
Smoliński, J., Feldman, P.A., Higgs, L.A., 1977, Astron. Astrophys.,60, pp. 277 - 280
Stickland, D.J., Harmer, D.L., 1978, Astron. Astrophys., 70, pp. L53 - L56

IRAS OBSERVATIONS OF THE COOL GALACTIC HYPERGIANTS

D. J. Stickland
Space and Astrophysics Division, Rutherford Appleton Laboratory, Chilton, Didcot, Oxfordshire, U.K.

It is still a matter for some debate as to how far the most massive stars ever get towards the right-hand side of the HR diagram during the course of their evolution. The apparent absence of very luminous red stars suggests either that stars never reach this region at all, or that if they do, their sojourn there is very brief indeed. If the latter is the case, one might expect to see an increase in mass-loss activity with luminosity among the brightest stars that we actually do see — the cool galactic hypergiants. An attempt has been made to search for one of the signatures of mass-loss — thermal reemission from ejected dust — using recently released data from the InfraRed Astronomical Satellite (IRAS).

In the first phase of this study, a list of stars with $M_{BOL} < -8.0$ mag. and with spectral type F or later has been compiled, largely from the catalogue of Humphreys (1978). Optical observations have been combined with ground-based infrared measurements from the Catalog of Infrared Observations (Gezari et al. 1984) and colour-corrected results from the IRAS Point Source Catalogue. The fluxes were all dereddened according to their colour excesses; the reddening law of Savage and Mathis (1979) was employed (but with A_λ / E_{B-V} taken as 2.15 mag at 0.7μ).

The dereddened fluxes were then compared with black-body energy curves for temperatures appropriate to the spectral types, with emphasis on good fitting in the optical and near infrared. Any residual flux in the infrared was extracted and characterised by a broad distribution taken to be due to isothermal dust, and a spike at 10μ. For the dust cloud, both the temperature and cross-section on the sky compared with the disc of the parent star were determined.

The main conclusion is that the present sample must be extended to cover a wider range of luminosities. Among the present limited sample (of 10 F,G,K type stars and 11 M type stars) it is apparent that infrared excesses only exist for stars cooler than mid-late G type, and for these, the diameters of the dust clouds seem frequently to be ~ 10^4 $R_\odot$. The size of the 10μ feature is probably linked to the size of the general dust emission, at least for the M type stars, but may also be a function of dust temperature.

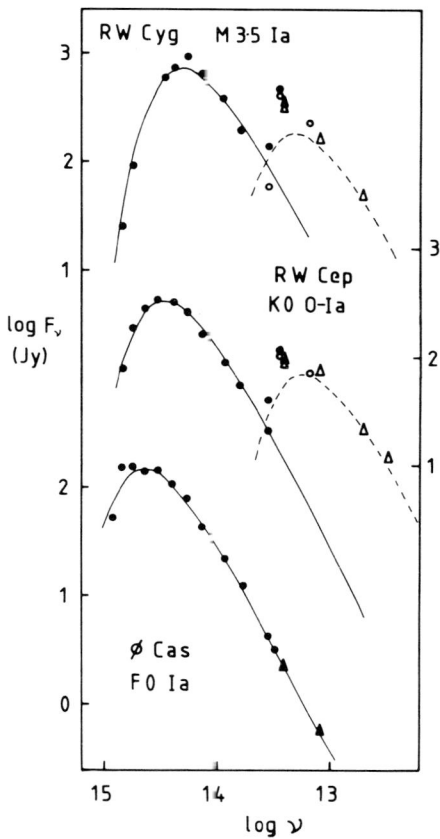

Figure 1. Some representative energy curves of cool hypergiants. Circles represent ground-based observations and triangles IRAS data after correction for reddening. Filled symbols depict the observed points, open symbols the IR excess after subtraction of a black-body curve (continuous line). The dashed curve indicates the dust cloud energy distribution.

REFERENCES

Gezari,D.Y., Schmitz,M., Mead,J.M.:1984, NASA Ref.Publ.1118
Humphreys,R.M.:1978, Astrophys.J.Suppl.,38,309
Savage,B.D., Mathis,J.S.:1979, Ann.Rev.Astron.Astrophys.,17,73

O-Star Winds: Is Rotation Important?

Raman Prinja and Ian Howarth,
Dept. of Physics & Astronomy,
University College London,
Gower St., London WC1E 6BT.

The inclusion of rotation as an ingredient in radiation pressure driven stellar wind models is a nontrivial undertaking. Those bold enough to attempt an investigation of its likely importance include Castor (1979), Abbott (1980), and Marlborough & Zamir (1984), whose work shows that the critical point is expected to move away from the stellar surface with increasing rotation. Observationally, the surface mass flux is predicted to be insensitive to rotation, but the wind acceleration is expected to be less than in a nonrotating star and the terminal velocity smaller.

We are working on an ultraviolet study of O star winds, and have preliminary mass-loss rates and terminal velocities for about 150 stars; here we examine the dependence of these data on rotation rate. As an empirical handle on the last quantity we adopt Conti & Ebbets' (1977) measures of $v_e \sin i$, which we assume is related in a simple way to the projected equatorial rotation velocity.

Figure 1 shows the residuals from a linear fit of $\log_{10}(\dot{M})$ vs. M_{BOL}, plotted against $v_e \sin i$. Rotation appears not to be a significant additional parameter, as expected. Figure 2 shows the column measured for the N^{4+} narrow components — discrete features in the absorption troughs of P Cygni profiles — as a function of $v_e \sin i$. A lower boundary is sketched in to emphasise a possible trend, *i.e.* rapid rotators may always have strong narrow components (although strong components are not exclusively associated with high values of $v_e \sin i$), but this is obviously not a well established result.

Figures 3 and 4 show the terminal velocities against $v_e \sin i$. If rotation does have a detectable rôle in radiation pressure driven winds we might hope to see its effects here, since they are predicted to be quite large (*e.g.* figure 2 of Castor 1979) and v_∞ can normally be measured directly and unambiguously from the high resolution spectra we have examined. Unfortunately, as can be seen from the figures, no significant trends are evident in our data. This result, while perhaps unexpected, may be due to the lack of stars in our sample rotating very near to break-up velocity; certainly we do not regard it as casting significant doubt on radiation pressure as an important mechanism in O star winds, especially given the state of development of the models.

References:

Abbott, D.C., 1980. *Astrophys. J.*, **242**, 1183.
Castor, J.I., 1979. *IAU Symp. 83: Mass Loss and Evolution of O-type Stars*, ed. P. Conti and C.W.H. de Loore (Dordrecht: Reidel), p. 175.
Conti, P., and Ebbets, D., 1977. *Astrophys. J.*, **213**, 438.
Marlborough, J.M., and Zamir, M., 1984. *Astrophys. J.*, **276**, 706.

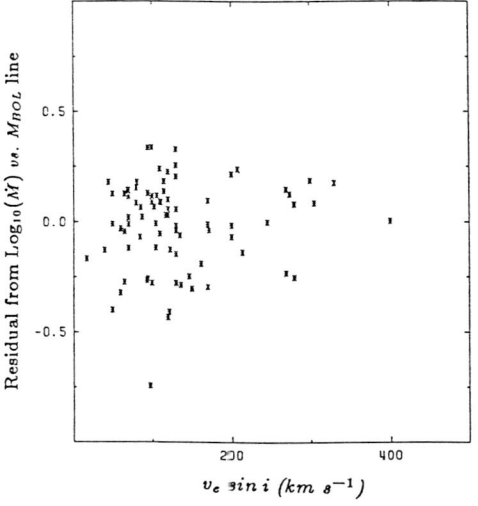

Fig. 1 — Residuals from $\log_{10}(\dot{M})$ vs. M_{BOL} line versus $v_e \sin i$.

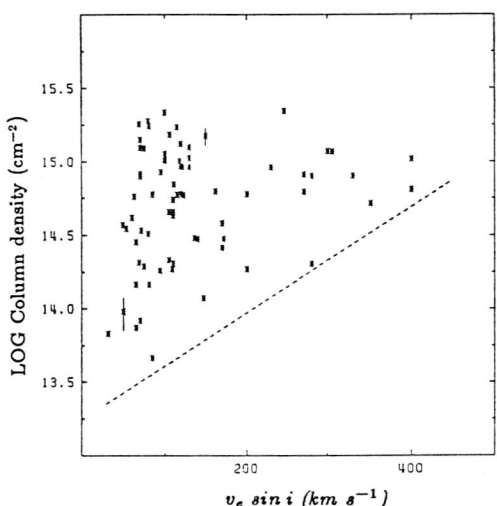

Fig. 2 — Log of the narrow component column density versus $v_e \sin i$.

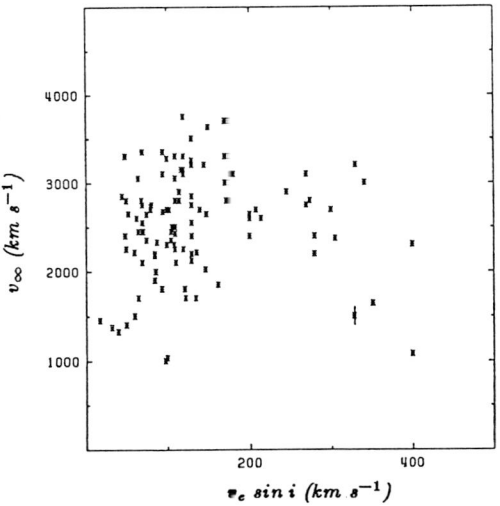

Fig. 3 — Terminal velocity versus $v_e \sin i$.

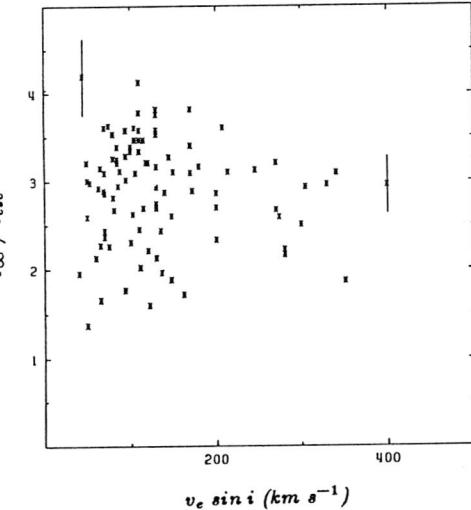

Fig. 4 — The ratio of terminal velocity to escape velocity verus $v_e \sin i$.

The absorption feature at 1920Å in the spectra of early-type
supergiants in the Small Magellanic Cloud.

D.H. Morgan, K. Nandy and G.I. Thompson
Royal Observatory, Edinburgh.

The 1600-2500Å spectra of many early-type supergiants are dominated by
a broad depression of the continuum near 1920Å. This depression,
often called the 1920Å feature, was detected in S2/68 (TD-1) spectra
of early B stars (Thompson et al. 1974) and was shown to be strongly
luminosity dependent. It was also shown that its profile could be
matched by the cumulative effects of FeIII absorption lines.

The 1920Å feature is also prominent in LMC early-type supergiants. In
their studies of interstellar extinction in the LMC, Nandy et al.
(1981), by separating the 1920Å feature from the 2200Å interstellar
extinction maximum, found the profile of the 1920Å feature to be the
same for LMC stars as for galactic stars and found its strength to be
dependent on both spectral type and luminosity, being strongest for
stars of spectral type B0-B3 with $-8.5 < M_V < -6.5$.

Early-type supergiants in the SMC have metal absorption lines
generally weaker in strength than those of their LMC counterparts
(Crampton and Greasley 1982). Is the 1920Å feature, attributed to
FeIII lines, also weak in the SMC stars?

The observations used to answer this question are some 24 IUE low
resolution spectra. The strength of the 1920Å feature is defined as
$S(1920) = -2.5 \log [2F^o_{1920}/(F^o_{1700}+F^o_{2200})]$ where F^o_λ is the
dereddened stellar flux in a band 30Å wide centred at wavelength λ.
This is the same as the definition of S(1920) used by Nandy et al.
(1981) for the LMC stars. The fluxes were dereddened using the
appropriate extinction laws and colour excesses (Prévot et al. 1984).
The reddening of many of the stars in question is small.

The results are shown in Figure 1 where S(1920) is plotted against
absolute magnitude M_V. M_V was calculated using (B-V) colours and
spectral types quoted by Azzopardi and Vigneau (1982) and Prévot et
al. (1984) and an assumed distance modulus of 19.1 (Crampton and
Greasley 1982). The chief source of error lies in the spectral
classification: an error of ±1 subclass is assumed here. The results

for the LMC stars are those presented by Nandy et al. (1981).

It is clear that S(1920) is on average weaker in the SMC B0-B3 stars than in comparable LMC stars and as with the LMC data, has a considerable spread. The SMC O stars, like those in the LMC, do not show significant 1920Å features. Similarly, SMC stars of type B5-B6 (not shown in Figure 1) also show the 1920Å feature but at a lower strength than the B0-B3 stars.

The profile of the 1920Å feature was measured from the five SMC stars with the greatest S(1920) and good signal-to-noise ratios. The profile of the 1920Å feature was then obtained by fitting an appropriate continuum and subtracting a suitably scaled Lorentz profile (see Howarth 1983) representing the weak 2200Å feature. The average 1920Å profile normalized to 0.1 mag at 1920Å is shown in Figure 2 along with the profiles for the LMC and galactic supergiants taken from Nandy et al. (1981). It is clear that the profiles of the features are the same in all three galaxies.

The 1920Å feature, having a large equivalent width, can in principle be used as a distance indicator. However its difference in strength in the two Magellanic Clouds coupled with its strong spectral class dependence prevents its use as a reliable luminosity class indicator.

References
Azzopardi M. and Vigneau J., 1982. Astron. Astrophys. Suppl.,50, 291.
Crampton D. and Greasley J., 1982. Publs. astron. Soc. Pacif.,94, 31.
Howarth I.D., 1983. Mon. Not. R. astr. Soc., 203, 301.
Nandy K. et al., 1981. Mon. Not. R. astr. Soc., 196, 955.
Prévot M.L. et al. 1984. Astron. Astrophys., 132, 389.
Thompson G.I. et al. 1974. Astrophys. J., 187, L81.

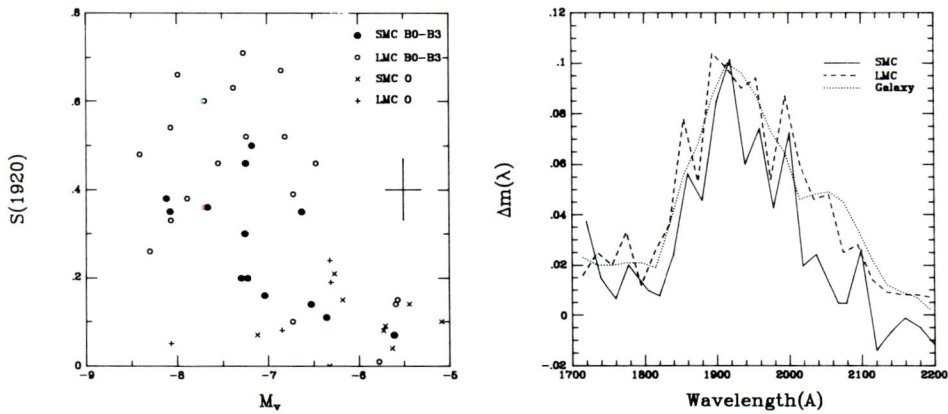

Fig.1 S(1920) vs absolute magnitude   Fig.2 profile of 1920Å feature

THE LIGHT VARIATIONS OF THE RED SUPERGIANT MU CEPHEI

Luciano Mantegazza
Osservatorio Astronomico di Brera
Via E. Bianchi 46
22055 Merate
Italy

The causes of light variations in red supergiants are still discussed. In particular it is not clear if these variations result from deterministic or narrow-band stochastic (pseudo-periodic) processes or from a mixture of both types.

Mu Cephei is among these objects that with the greatest deal of observations as it has been under surveillance since 1848. Its visual observations from 1881 to 1935 has been discussed in several papers. These data are quite reliable because they are due to the largest part to only one observer. The most recent analysis of them (Mantegazza,1982; hereafter cited as paper I) showed that they can be interpreted as the result of the superimposition of two waves with periods of 4700 and 873 days and of their non-linear couplings. However this fact, even if it tends to suggest the presence of deterministic processes (e.g. multimode pulsations), is not sufficient to rule out the possibility of pseudo-periodic ones. In fact the observations cover only a few cycles of the longer wave and besides the visual data cannot be very accurate. These random processes could be connected for example to the star's convective motions or to the stellar rotation.

In order to verify the previous results and to improve the knowledge of the phenomena regarding mu Cephei I decided to collect all the observations of this star following 1938. Not all of these data are reliable, therefore after an accurate selection I adopted the photoelectric measurements published by: Larsson-Leander(1962), Coyne and Kruszewski (1968), Polyakova (1975) and references therein, Polyakova (1978), and Abramyam (1982). These data span over 7800 days, thus they permit a resolution in frequency that is about 3 times inferior to that of the visual data analysed in paper I. Because of the limited resolution it was not possible to fit on these data the model of paper I (in fact the least-squares solution is unstable). On the other hand the gap of 17 years between the two data sets prevented from extrapolating that model. Therefore it was necessary to analyse the photoelectric data separately. Fig. 1 shows the autocorrelation function estimated from these data. It clearly shows two peaks centred at about 900 and 4700 days, values which correspond to the periods derived in paper I. Successively the data were analysed with the least-squares power spectrum tecnique adopted in paper

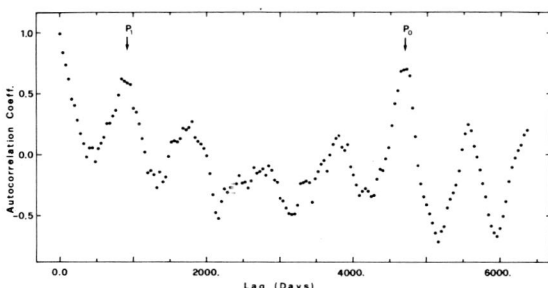

Fig. 1 - Autocorrelation function

I. Due to the limited resolution it was not possible to identify all the terms found in paper I. However by calculating successive spectra in which the previously identified terms were introduced as known-constituents (see paper I) it was possible to single out 7 terms: $f_0 (=1/P_0)$, $2f_0$, $3f_0$, $4f_0$, $f_1 (=1/P_1)$, $2f_1$, f_0+f_1. Finally the values of f_0 and f_1 were iteratively refined in such a way that the 7 term model as a whole gave the best least-squares fit. The values found were 4650 and 920 days for P_0 and P_1 respectively. These values are coincident within their uncertainties with those found in paper I. It is possible to see that the terms of the model of paper I that are not present in this solution are too close to the terms above listed in order to be resolved. The synthesized light curve computed with this model is shown in fig.2. In this figure the dots represent forty days averages of the photoelectric data.

In conclusion the analysis of the photoelectric data furnish an independent confirmation that in the light curve of mu Cephei are present two cyclic phenomena and their non-linear couplings, and that there is a substantial agreement with the model derived from the analysis of the visual data. However a much longer set of observations is needed in order to verfy if these terms are due to purely deterministic processes and in order to identify the physical processes that generate them.

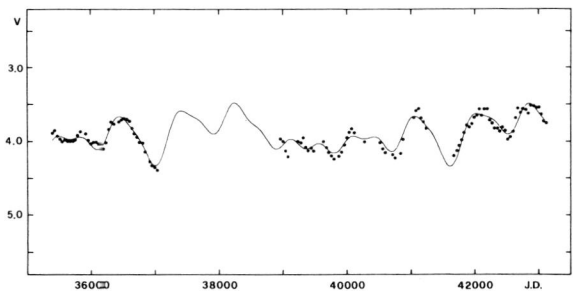

Fig. 2 - Dots: 40-day means of the photoelectric data
solid line: synthesized light curve

REFERENCES

Abramyam,G.V.: 1932, Bull. BYURAKAN Obs. 53, 3;
Coyne,G.V., Kruszewsky,A.: 1968, Astron.J. 73, 20;
Larsson-Leander,G.: 1962, Arkiv Astron. 3, n.21, 285;
Mantegazza,L.: 1932, Astron.Astrophys. 111, 295;
Polyakova,T.A.:1975, Variable Stars 20, 75;
Polyakova,T.A.:1978, Variable Stars 20, 529.

VARIABILITY OF MAGELLANIC CLOUD SUPERGIANTS

G. R. Grieve
Department of Geophysics and Astronomy
University of British Columbia
Vancouver, British Columbia
Canada

and

Barry F. Madore
David Dunlap Observatory
Department of Astronomy
University of Toronto
Toronto, Ontario
Canada   M5S 1A1

ABSTRACT. Statistics on the variability of 127 high-luminosity, intermediate spectral-type supergiants in the Magellanic Clouds are presented.

From a new photoelectric survey of intermediate-type supergiants in the Magellanic Clouds (Grieve and Madore 1986) thirty-eight stars in the LMC and twenty-three stars in the SMC have been judged to be variable by more than 0.05 mag. This sample represents about fifty percent of the stars monitored for variability.

For most stars, the magnitude variation detected fell in the range from 0.05 to about 0.20 mag. There are a few examples of larger variations, but these tend to be special cases (e.g., K supergiants). None of the stars studied show extreme brightness variations (of several magnitudes) as seen in the *Hubble-Sandage* variables.

The distribution of variables (filled circles) and the constant stars (open circles) are compared in Figures 1 and 2 for the LMC and SMC, respectively. Also shown are the positions of the brightest long-period Cepheids (crosses) bounded by the approximate edges of the cepheid instability strip. In general the more luminous supergiants are more likely to be variable than the fainter stars. This statement is quantitfied in Table 1 and gives support to the findings of Maeder (1980), who concludes that for galactic supergiants the amplitude of variation is an increasing function of luminosity.

References:

Grieve, G. R., and Madore, B. F. 1986, *Ap. J. Suppl.*, (in preparation).
Madore, B. F. 1985, in *I.A.U. Colloquium No. 82*,
 Cepheids: Theory and Observations, ed. B. F. Madore,
 (Cambridge: Cambridge University Press), p. 166.
Maeder, A. 1980, *Astron. Astrophys.*, 90, 311.

STAR FORMATION OF STAR CLUSTERS IN THE SMC AND THEIR ADJOINING FIELDS.

M. Kontizas* and E. Kontizas**
* University of Athens, ** National Observatory.

Photometric and recent spectroscopic studies of the SMC have shown that the differences observed in the SMC clusters and those of our Galaxy could be attibuted to differences in metallicity, star formation rate and/or the Initial Mass Function (IMF) (Humphries, 1983). The studied clusters NGC152 and KRON3 are located at the west side of the bar of the SMC and their adjoining fields represent the halo population of this galaxy.

From photographic plates taken with the 3.8 m AAT telescopes (Stewart, 1980) the LFs were found (Kontizas et al; 1984) and compared to theoretical models (Stryker, 1984). The bright part of the LF and spectral classification of the bright stars (Kontizas et al, 1985) in these clusters has provided the observed ratio of C/M stars, that was compared to the theoretical ratios derived by Miller and Scalo (1982).

From the above investigation it was found that :
1) The faint part of the observed LF of both fields (Fig 1a) seems to fit the theoretical LF Stryker (1984) implying constant star formation, low metallicity stars, masses of $0.5-3M_\odot$ and ages younger than 5×10^9 with a strong component of intermediate age stars.

2) The cluster's LFs (Fig. 1b) fit the models by Stryker (1984) implying intermediate age stars of low metallicity and constant birth-rate.

The comparison of the observed C/M ratios with the theoretical models (Miller and Scalo, 1982) implyies that either the metallicity is very low or their masses are high. Considering that the LF supports the low metallicity assumption it can be concluded that the field and clusters are rather intermediate age stellar population of low metallicity, constant birth rate and continuous star formation for about 5×10^9 years.

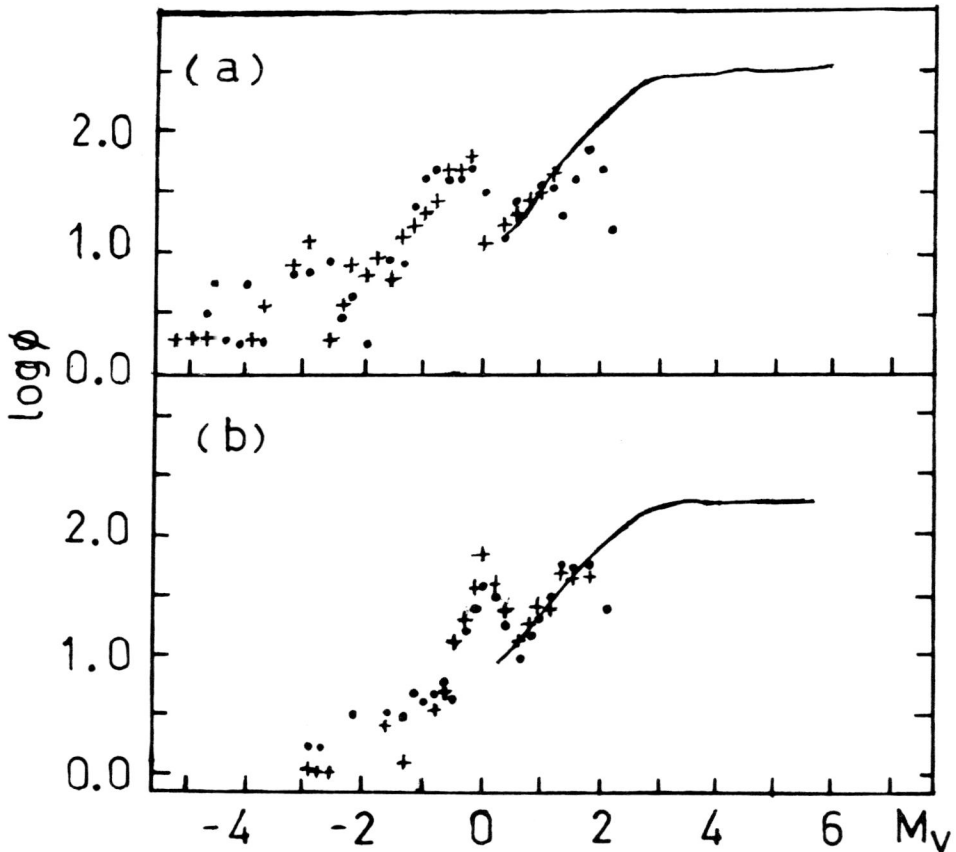

Figure 1. LFs of (a) the clusters NGC152 (crosses), KRON3 (dots) and (b) their adjoining fields. The solid line represents the theoretical model (Stryker, 1984).

REFERENCES

Humphries, R. M. : 1983, Ap. J., 265, 176.
Kontizas, M., Kontizas, E. and Stewart, N. : 1984, IAU, Symp No 105, p. 121.
Kontizas, E., Dapergolas, A. andKontizas, M. : 1985, submitted.
Miller, G.E., and Scalo J. M : 1982, Ap. J. 263, 259.
Stryker, L. L. : 1984, IAU Symp. No 108, p. 79.
Stewart, N. : '980, Ph. D. Thesis Edinburgh University.

RED SUPERGIANTS IN NGC 300

J.A. Graham
Cerro Tololo Inter-American Observatory
National Optical Astronomy Observatories
Casilla 603, La Serena, Chile.

Roberta M. Humphreys
Astronomy Department
University of Minnesota
Minneapolis, MN 55455

Red supergiant stars that are members of the nearby Sc galaxy NGC 300 have been isolated by spectrophotometry and infrared photometry from a blink comparator survey of red stars in the general field of this highly resolved stellar system. Our spectroscopic criterion is the strength of the CaII triplet near $\lambda 8500$ which is more than twice as strong in red supergiants as in red dwarfs. Red supergiants and dwarf stars also occupy distinct regions in a two-color (J-H) vs (H-K) diagram. Both methods clearly segregate the supergiants which, with visual magnitudes close to 19, must be members of NGC 300. Of the 18 red stars we examined, only 6 are probable members of NGC 300. The other 12 are likely foreground stars belonging to the Galaxy. A principal finding is that, unless special care is taken, contamination by foreground stars can be an important factor when the red stars of an external galaxy are selected only on the basis of their colors in the visual and near infrared spectral region.

Photometry of 4 M supergiants indicates absorption corrections in V averaging $0^{m}\!.7$. Adopting a true distance modulus of $26^{m}\!.1$ for NGC 300 (Graham, A. J. **89**, 1332), a mean absolute visual magnitude of $-7^{m}\!.9$ is found for these stars. The IR photometry leads to a mean absolute magnitude at K of $-11^{m}\!.1$ and a mean bolometric absolute magnitude of $-8^{m}\!.5$.

Detailed results will be published in a paper now in preparation.

SIMULATIONS OF THE BRIGHTEST STARS IN GALAXIES AS DISTANCE INDICATORS

Laura Greggio
Dipartimento di Astronomia
CP 596, I-40100 Bologna, Italy

The brightest stars in galaxies have been often considered useful as distance indicators. However, one may expect that the absolute magnitude of the brightest stars in galaxies should depend on the IMF and on the present SFR, and that statistical effects may be important, although different in different photometric bands, due to the different numbers of objects effectively sampled.

To investigate on this problem I have performed a number of numerical simulations of the absolute magnitude of the brightest objects in a population of stars more massive than 15 $M_\odot$, in four different bands: two UV bands, centered at 1400 Å and 1750 Å, the B and the V bands. Each synthetic population is constructed by performing a number N of random extractions of pairs (mass,age), according to a given IMF and a constant SFR. The location in the theoretical HR Diagram for each pair is found by linear interpolation on the evolutionary tracks with moderate mass loss computed by Maeder (1981a, 1981b, 1983), while the magnitudes are determined by using model atmospheres avaliable in the literature. Different values of N are explored, namely N= 50,100,200,500,1000,2000,4000,8000, while the corresponding number of simulations are 1200,800,600,500,400,300,250 and 100. In each run, only the three brightest stars were retained. Two different slopes of the IMF, namely 2.5 and 3, and two different values of the upper mass cut off (150 and 200 $M_\odot$) were tested.

The main results are the following:
i) the UV bands always sample the MS stars and are therefore most effective for selecting the brightest stars in galaxies with ongoing star formation. The statistical effects are found to be very important for the less luminous galaxies, as the theoretical distributions can be as wide as 2 mag. As the total number of objects sampled increases, the probability of sampling the upper end of the mass distribution increases and, correspondingly, the magnitude of the brightest stars decreases and

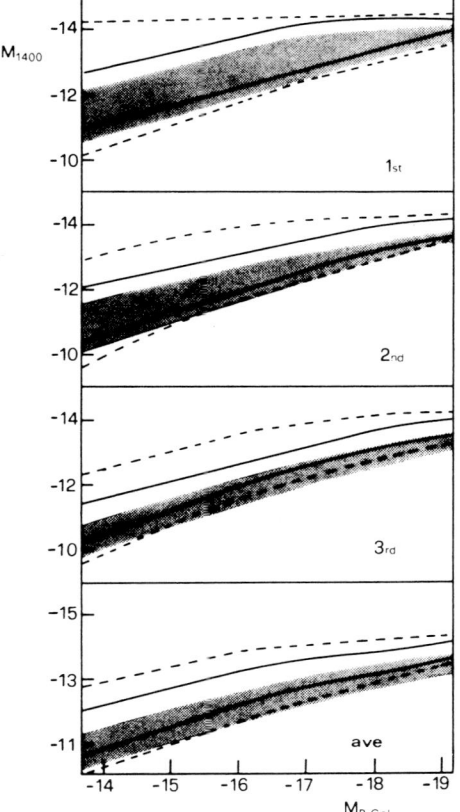

Fig. 1 : HMFW of the distributions (solid line) and 99% probability level for the brightest UV stars as function of the B magnitude of the parent galaxy. Shaded area: HMFW for s=3.

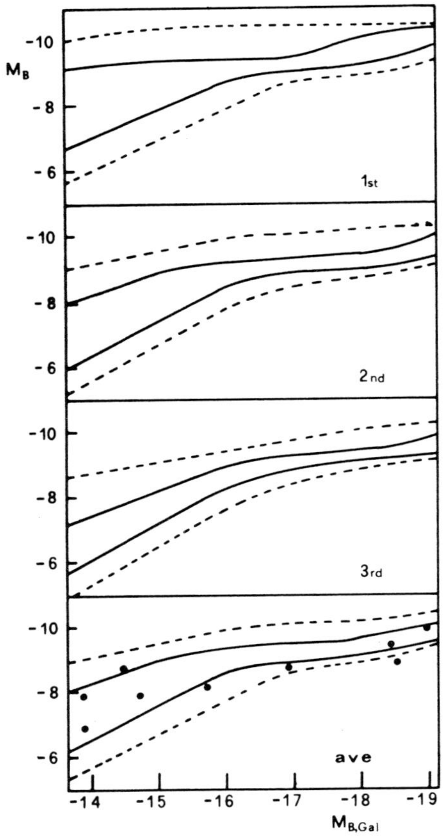

Fig. 2 : The same as Fig. 1 for the B magnitude. The dots are the observations reported by Sandage (1984).

the distributions get narrower (see Fig. 1). Therefore, for the most luminous galaxies, the brightest UV stars can be used to derive distances, although their magnitudes are dependent on the IMF. Infact, it is found that when the IMF slope s varies from 2.5 to 3, the magnitudes of the brightest stars get get larger by ∼0.3−0.5 mag, while when the upper mass cut off varies from 200 to 150 $M_\odot$, they vary by 0.5−0.7 mag, for the most luminous galaxies.

ii) Contrary to the UV bands, the optical bands are more likely to sample stars in advanced evolutionary stages. However, for the less luminous galaxies, the brightest stars turned out to be MS objects in ∼45% of the cases, due to the fact that the evolutionary lifetime of the core H-burning phase is much longer than that of the core He-burning stage, and consequently, for low N's the number of evolved stars is very small. Since the B magnitude of MS stars is appreciably lower than that of the A supergiants, the distributions turn out to be very wide. The same result was found by Schild and Maeder (1983). As the total number of objects increases, the fraction of A supergiants among the brightest stars increases rapidly: the low luminosity tail of the distribution disappears and the HMFW gets as narrow as ~ 0.4 mag. For the brightest galaxies, however, the upper end of the mass distribution becomes well populated and the fraction of MS stars among the brightest objects increases. Correspondingly, the distributions tend to get wider again (see Fig. 2).

In all the explored bands, the distributions for the magnitude of the 3rd brightest stars, or for the average magnitude of the three brightest stars are the narrowest and then, they can be used with more confidence for the estimate of the distance modulus of the parent galaxy. As a general conclusion, one may suggest the use of observations in different photometric bands of the brightest stars in large samples of galaxies, all at the same redshift. This should minimize the statistical effects and allow a good estimate of the most probable value of the absolute magnitude of the brightest stars.

REFERENCES

Maeder, A. 1981a, Astron. Astrophys. 99, 97
Maeder, A. 1981b, Astron. Astrophys. 102, 401
Maeder, A. 1983, Astron. Astrophys. 120, 113
Sandage, A. 1984, Astron. J. 89, 630
Schild, H., Maeder, A. 1983, Astron. Astrophys. 127, 238

The dependence of the stellar rate of mass loss on effective temperature and velocity

C. de Jager, H. Nieuwenhuijzen and K.A. van der Hucht
Astronomical Observatory and Laboratory for Space Research, Utrecht, The Netherlands

From the existing literature data have been collected on the stellar rate of mass loss for 189 stars with known or derived values of the effective temperature T_{eff} and luminosity L. It appears that $\dot{M}$ depends only on T_{eff} and L for the O- through M-type stars brighter that about 3×10^3 $L_\odot$. This is shown in Figure 1, where we have plotted for each star the value of $-\log(-\dot{M})$ with $\dot{M}$ expressed in solar masses per year. An interpolation formula has been derived for the $\dot{M}(T_{eff}, L)$ dependence. If we define $X = \log T_{eff}-4$, and $Y = \log(L/L_\odot)-5$, then

$-\log(-\dot{M}) = a_1 + a_2X + a_3X^2 + a_4X^3 + a_5Y + a_6Y^2 + a_7Y^3 + a_8XY + a_9X^2Y + a_{10}XY^2$ (1).

Values for a_1 through a_{10} are given in De Jager et al. (1985). A histogram of the deviations of $-\log(-\dot{M})_{obs}$ with respect to the values calculated with Equation (1) is shown in the insert to Figure 1. It yields a sigma value of 0.53, part of which is probably of cosmic origin. It appears that the Of stars are not deviating from Equation (1). Some groups of stars do not fit to our interpolation formula. These are the Wolf-Rayet and the C-stars. Figure 2 shows the rates of mass loss for WR stars, the C and S star, the Be stars and the nuclei of planetary nebulae. The histograms of deviations of $-\log(-\dot{M})$ with respect to Equation (1) are given in the insert to Figure 2 for the WR and the C-stars. For these we found:

WR-stars: $-\log(-\dot{M})_{observed} + \log(-\dot{M})_{calc} = 1.12 \pm 0.03$
C-stars : $= 1.24 \pm 0.03$

Their rate of mass loss is, hence, a factor of about 15 times larger than that for 'normal' stars with the same luminosity and T_{eff}. A comparison is not possible for Be-stars and nuclei of planetary nebulae because we have no reference data.

Reference:
De Jager, C., Nieuwenhuijzen, H., and van der Hucht, K.A., 1985, Astron. and Astrophys., submitted

to as wind-blanketing (see Abbott and Hummer 1985 for details). The effect of wind-blanketing on a stellar photosphere appropriate to zeta Pup is to warm the layers where the helium lines arise so that lines of equivalent strength are formed at an effective temperature some 4,000K° cooler than in a wind-free atmosphere.

For zeta Puppis, the relation between spectral type and effective temperature is changed from earlier spectroscopic analysis roughly 20% by accounting for gravity, 10% by using higher precision observations, and 10% from wind-blanketing (Tables 1 and 2).

TABLE 2: STELLAR PARAMETERS FOR ZETA PUPPIS

	$T_{effective}$	log g	[Y]
Kudritzki et al. (1983)	42,000±2,500	3.5±0.15	0.14±0.03
unblanketed (this analysis)	46,500	3.6	0.16
blanketed (this analysis)	42,000±1,500	3.5±0.1	0.17±0.03

We determine the same temperature with our wind-blanketed analysis as did Kudritzki et al. (1983) from their unblanketed study because of a systematic difference in measured equivalent widths (this analysis − Kudritzki et al. = 0.070±0.093Å). Kudritzki et al. co-added line profiles from spectra obtained with Kodak type IIIa-J photographic emulsion for an estimated signal-to-noise ratio of 100:1. They would have found a higher temperature similar to our unblanketed result with line profiles equivalent to ours.

The emergent flux of luminous stars is very sensitive to gravity. For zeta Puppis, an uncertainty of 0.2% in the wing of H gamma represents a change of 0.1 in log g. The gravity must be measured with this precision to define with realistic accuracy bolometric corrections and the flux of ionizing photons. The RCA CCD camera we used for these observations is limited in its photometric precision by calibration of fringing effects. For this analysis we have minimized residual fringing effects by adding together spectra taken at slightly different grating settings and thus different fringe patterns.

References:

Abbott, D. C. and Hummer, D. G. 1985, Astrophys. J. 179, 286.
Conti, P. S. 1973, Astrophys. J. 179, 161.
Kudritzki, R. P., Simon, K. P., and Hamann, W. R. 1983, Astron. Astrophys. 118, 245.
Mihalas, D. 1972, Non-LTE Model Atmospheres for B and O Stars (NCAR-TN/STR-76; Boulder, National Center for Atmospheric Research).
Panagia, N. 1973, Astrophys. J. 178, 929.

PHOTOSPHERIC AND "CHROMOSPHERIC" CIV LINES IN B-MAIN-SEQUENCE STARS

Fiorella Castelli[*], Carlo Morossi[*] and Roberto Stalio[**]
(*)Osservatorio Astronomico, Trieste
(**)Dipartimento di Astronomia, University of Trieste

The presence in the far-UV spectra of early-type stars of spectral lines of superionized atoms is argument of controversial debate among astronomers. Presently there is agreement on the non-radiative origin of these ions but not on the proposed mechanisms for their production nor on the proposed locations in the stellar atmosphere where they are abundant. Cassinelli et al. (1978) suggest that the Auger mechanism is operative in a cool wind blowing above a narrow corona to produce these ions; Lucy and White (1980) introduce radiative instabilities growing into hot blobs distributed across the stellar wind; Doazan and Thomas (1982) make these ions to be formed in both pre- and post-coronal, high temperature regions at low and high velocity respectively.

The definition of superionization and the identification of the superionized species are also object of debate. If we define as superionized those ions which cannot be formed in a stellar photosphere because of the too large radiative energy necessary, then OVI is always superionized and NV is "almost" always superionized, being possibly present in the photospheres of the hottest stars. On the contrary the definition of the superionization character of the CIV resonance lines requires some further analysis also because CIV has been widely observed on a very large number of early-type stars and its presence or absence may give some clue for the understanding of the atmospheric structure of these stars.

Previous literature reports on the presence of CIV lines in main sequence B stars (these are the best to study because there is little effect on the CIV profiles from thick stellar winds) at approximately photospheric velocity are: 1) by Underhill (1982) who reports that CIV lines are observed to be moderately strong in B1 main sequence stars and display blue extended wings; they are instead weak and sharp at B3 and B5; 2) by Doazan et al. (1984, 1985) who show the presence of variable CIV at spectral type B7 in an ex-Be star, ϑ CrB; 3) by Molaro et al. (1982) who observed CIV absorption lines in HD119921, A0 V.

No photospheric model, constructed under standard equations, can predict the existence of CIV ions over the temperature range indicated by the observations. Thus the question is: at what spectral type in a main sequence star photospheric CIV ceases to contribute to the observed profile? In order to answer this question we compared IUE spectra of a set of B main sequence stars with synthetic photospheric spectra (Castelli 1985) obtained from the solar abundance model atmospheres reported in Table 1. The spectral range investigated was from 1541 to 1553 A. The results are summarized in Figures 1, 2 and 3.

Figure 1 compares the observed spectrum of γ Peg (B2 IV, V sin i $\sim$ 10 km/s) with the synthetic spectrum obtained from model A and of ι Her (B3 IV, V sin i = 11 km/s) with the spectrum from model B. Both synthetic and observed spectra show sharp and deep lines of CIV at stellar velocity, indicating its photospheric origin. Note that the 1550 line is blended with a strong Fe III line (at 1550.2). Figure 2 compares two observed spectra of ϑ CrB, an ex-Be star (B6 V, V sin i = 393 km/s) taken at different epochs with model C. The strongest contributors to the absorption lines indicated by the dots correspond to lines of Fe II-1547.8, Fe II-1548.7, Fe III-1550.2, Fe II-1550.9, Fe II-1551.2. Thus there is no photospheric CIV at the effective temperatures of a B6 star. We note the variability of the observed profiles (this point is more extensively discussed in Doazan et al., 1985) which suggest a change from a one component, low velocity absorption profile to a two components at low and high velocity profile. Figure 3 compares the observed spectrum of HD119921 (A0 V, V sin i > 400 km/s; 220 km/s according to Molaro et al.)) with model D. The same arguments as for ϑ CrB can be raised to infer the non-photospheric character of the observed CIV.

We conclude that the boundary at which photospheric CIV disappears is between spectral types B3 and B6 in main sequence stars. After B6 it is fully superionized. The observations (we refer in particular to ϑ CrB which we are monitoring since several years) seem to support Doazan and Thomas (1982) model of chromospheric CIV. In ϑ CrB, and possibly in HD119921, it is formed both in a pre-coronal, solar-like chromosphere, at low velocity, and in a post- coronal "chromosphere" at high velocity. Variable mass loss makes the profile components to vary in strength and eventually in velocity.

REFERENCES

Cassinelli, J.P., Olson, G., Stalio, R.: 1978, Astrophys. J., 220, 85.
Castelli, F.: 1985, Astronet Documentation Facility, no. 6.
Doazan, V., Thomas, R.N.: 1982, in "B-Stars with and without Emission Lines", ed. A.B. Underhill and V. Doazan, NASA SP-456
Doazan, V., Morossi, C., Stalio, R., Thomas, R.N., Willis, A.J.: 1984, Astron. Astrophys., 131, 210.
Doazan, V., Morossi, C., Sedmak, G., Stalio, R., Thomas, R.N.: 1985, Astron. Astrophys., in press.
Lucy, L.B., White, R.L.: 1980, Astrophys. J., 241, 300.
Molaro, P., Morossi, C., Ramella, M., Franco, M.L.: 1983, Astron. Astrophys., 127, L3.
Underhill, A.B.: 1982, in "B-Stars with and without Emission Lines", ed. A.B. Underhill and V. Doazan, NASA SP-456.

Table 1: The set of models

Model	Teff	log g	V sin i	Model	Teff	log g	V sin i
A	21780	4	0	C	13500	4	220
B	17460	4	10	D	10000	4	250

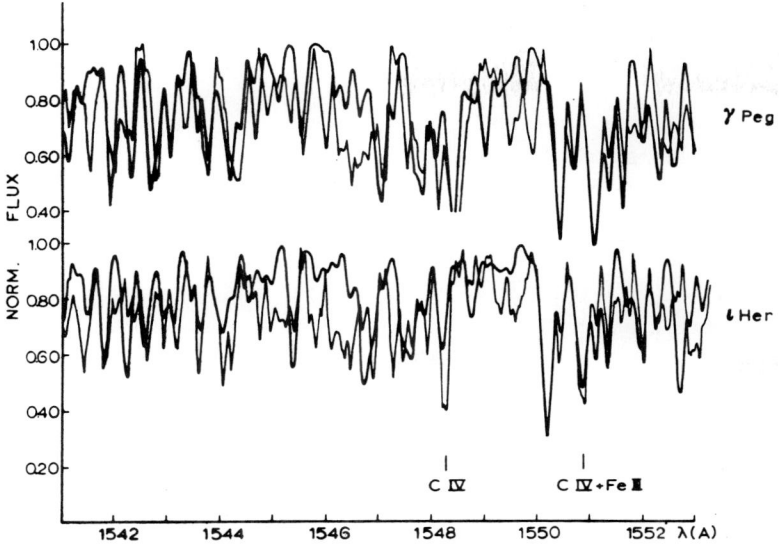

Figure 1: Observed (thin line) and synthetic (thick line) spectra of γ Peg and ι Her.

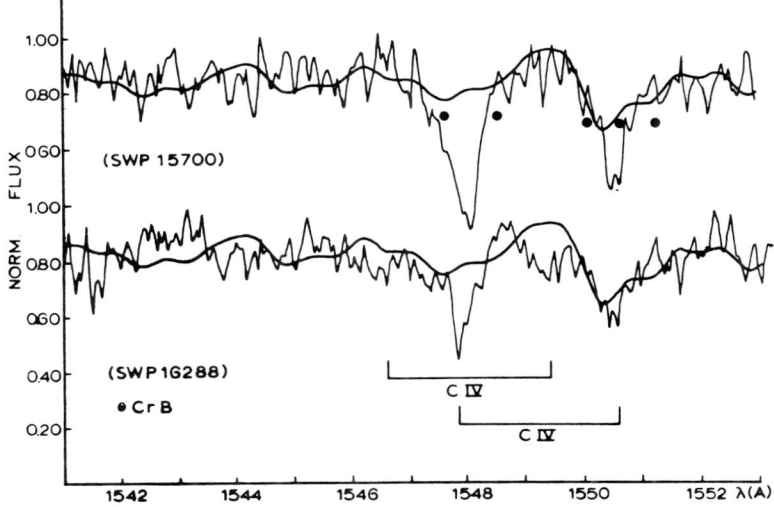

Figure 2: Same symbols as in Fig. 1; the IUE spectra of ϑ CrB were taken on Dec. 10, 1981 and Feb. 8, 1982.

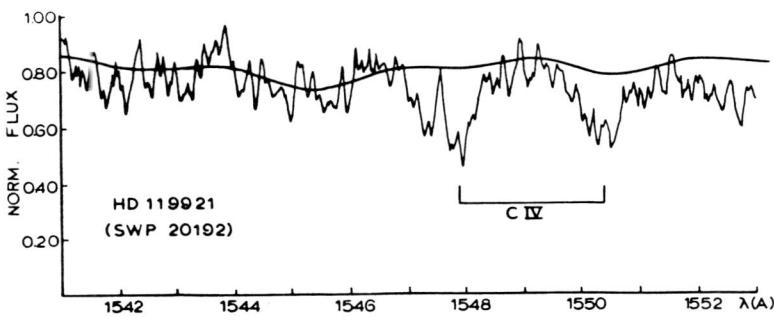

Figure 3: Same symbols as in Fig. 1.

VOYAGER OBSERVATIONS OF HOT LUMINOUS STARS

Ronald S. Polidan and Roberto Stalio
Lunar and Planetary Laboratory, University of Arizona, Tucson
Dipartimento di Astronomia. University of Trieste, Italy

This is a progress report on a study of the far-UV spectra of hot stars observed with the Voyager ultraviolet spectrometers (UVS); we discuss UV flux distribution and variability of a set of early type high luminosity stars.

The UVS's are objective grating spectrometers sensitive in the 500-1700 A region; the reciprocal dispersion is 9.26 A per detector channel yielding an effective resolution of about 20 A (approximately 2 channels). More information on the UV spectrometers is given in Broadfoot et al. (1977). The adavantages of their use for stellar work are discussed in Polidan, Stalio and Peters (1985, Paper I); the same authors give an evaluation of the mean error in flux per channel under typical observational circumstances: shortward of 1200 A the error is approximately 3%, longward of 1200 A it becomes of the order of 7%. Systematic errors due to uncertainties in the star position on the slit and to the effects of time variable instrument sensitivity are found to be negligible. The Voyager UVS calibration is discussed in Holberg et al. (1982).

In Paper I we have studied the problems of flux distributions, gravity darkening and flux variability in a set of main sequence B and Be stars. In particular we have shown that the far-UV flux of main sequence B stars is highly sensitive to stellar effective temperature. In this section we discuss the far-UV flux distribution of high luminosity stars of spectral type ranging from O4 to B8. Table 1 gives the basic information about the stars in study and Figure 1 shows the far-UV spectra which have been dereddened (see Paper I) and normalized at 1400 A. From these relative flux distributions one sees that the far-UV gives good definition of the spectral type of B-supergiants. Viceversa the relative fluxes of two supergiants (ζ Pup and ζ Ori), which lie at the lower and upper extremes of the O spectral class, are less separated in their relative far-UV fluxes. The likely reason for this smaller temperature dependence is that at the high effective temperatures of the O-stars one sees in the far-UV the rising (towards the extreme UV) part of the spectrum, not the region where the maximum flux occurs as is observed in the B stars. In addition, when comparing the relative fluxes of main sequence and supergiants B-stars of the same spectral type (see also Figure 1 in Paper I) one notices that the supergiants appear cooler due to either real differences in effective temperature and line blocking from both photosphere and stellar wind.

Five of the seven program stars were investigated for flux variability in the far-UV by intercomparing Voyager data when multiple observations were available. We have (1) measured integrated 950-1150 A fluxes, and (2) compared the absorption features at 980 and 1030 A which are mostly due to resonance lines of CIII and OVI ions respectively. Table 2 shows that no significant changes in the 950–1150 A continuum have been detected in most of the stars. There are probable variations in γ^2 Vel which could be ascribed to the binary character of the star. In the spectra of γ^2 Vel we also noticed a change of spectral shape shortward of 1100 A; with spectral shape changes there may be some line change as well. Figure 2 reports variations in the CIII and OVI profiles observed in ζ Ori. The line flux changes are of the order 7 sigmas and probably reflect a change in the rate of mass loss from the star.

We have presented the first results from a program to investigate the far-UV spectra of hot, luminous stars with the Voyager UV spectrometers. Two results have been obtained :
1) the far-UV range is a better discriminator for spectral types in the B-supergiant class than it is among the O-supergiants;
2) a remarkable "episode" of line flux variability has been detected in the spectrum of ζ Ori which we tentatively ascribe to variable mass loss.

REFERENCES

Broadfoot, A.L., et al.: 1977, Space Sci. Rev., 21, 183.
Holberg, J.B., Forrester, W.T., Shemansky, D.E., Barry, D.C.:1982, Astrophys. J., 257, 656.
Polidan, R.S., Stalio, R., Peters, G.J.: 1985, Astrophys. J., in press.

Table 1: The stars in study

Star	Spectral Type	V	E(B-V)	Observation Date(1)
γ^2 Vel	WC8+O7.5e	1.78		See Table 2
ζ Pup	O4 If	2.22	0.05	38, 1980
ζ Ori	O9.7 Ia	2.05	0.09	289, 1981
ϵ Ori	B0 Ia	1.70	0.06	202, 1981
ϵ CMa	B2 II	1.50	0.00	5, 1982
η CMa	B5 Ia	2.45	0.00	328, 1981
β Ori	B8 Ia	0.12	0.00	53, 1978

(1) The observation date reported refers to the spectra of Figure 1; for the other observation dates see Table 2.

Table 2: 950–1150 A flux ratios

γ^2 Vel	80/030:81/282:82/005=1.18:1.00:1.18
ζ Pup	80/038:81/328:82/005=1.08:1.00:0.96
ζ Ori	81/202:81/282:82/005=0.97:0.93:1.00
ϵ CMa	81/282:82/005=1.00:0.93
η CMa	81/328:82/005=1.00:1.00

VOYAGER OBSERVATIONS OF HOT LUMINOUS STARS

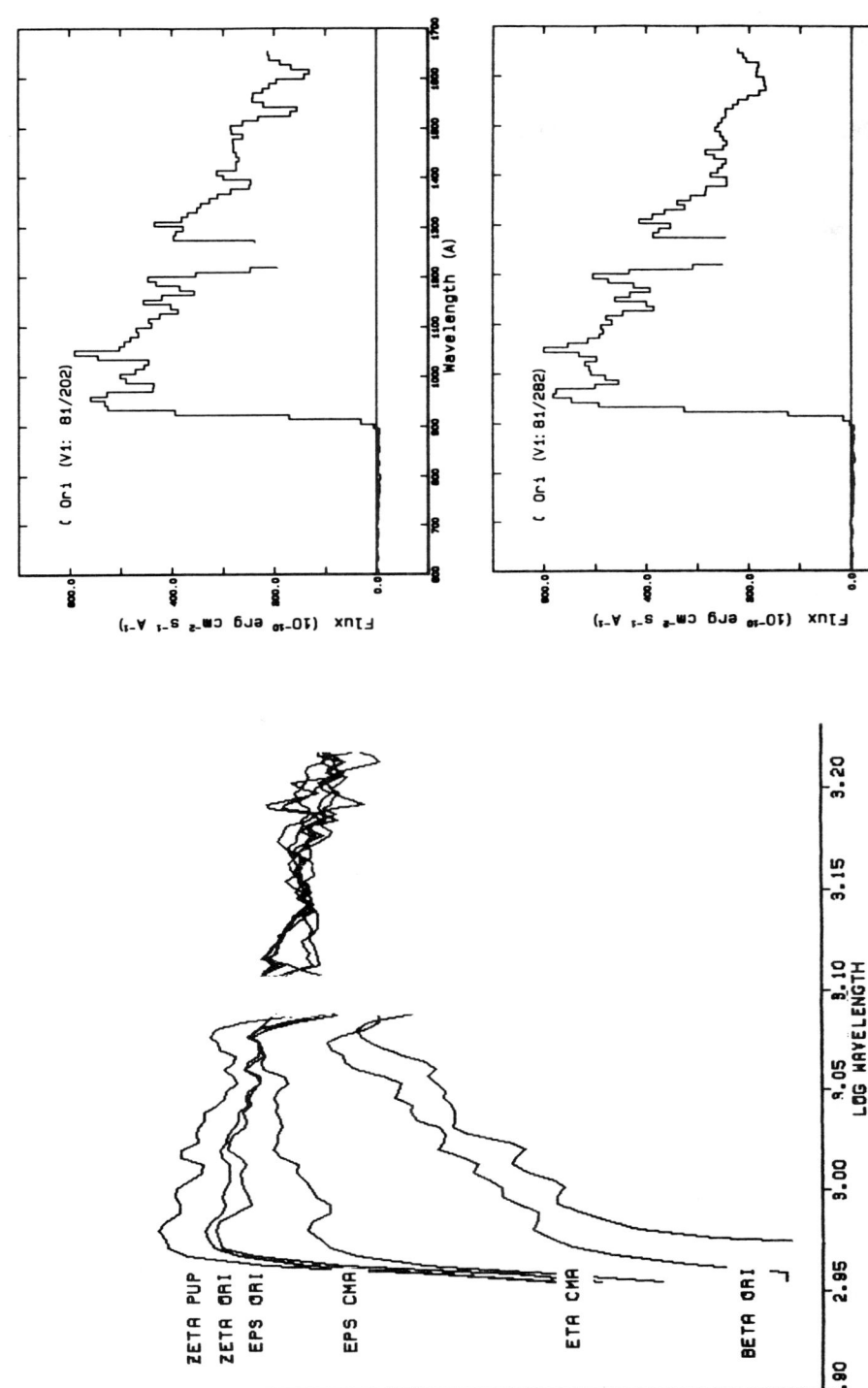

Figure 1: Voyager spectra of the hot, luminous stars identified in Table 1 (except ζ Vel). All spectra have been dereddened and normalized at 1400 Å.

Figure 2: Variable OVI and CIII profiles in ζ Ori.

SYNTHETIC HR DIAGRAMS FOR LUMINOUS STARS : A TEST OF STELLAR EVOLUTION THEORY

EMMA NASI
Osservatorio Astronomico
Padova   Italy

The composite HR diagram for luminous stars in our galaxy by Humphreys and McElroy(1984) is the only source which provides enough data for the comparison with the results of massive stars theoretical models .
The luminosity limit for completeness of Humphreys' and McElroy's catalog(1984) is about $M_{Bol}=-8$ mag, when restricted to 3 Kpc from the Sun . Inside these limits there are 378 luminous stars in galactic associations and clusters, whose distances are known with sufficient reliability .
The correspondent observational HR diagram is reproduced in Fig. 1 .
Synthetic HR diagrams for massive stars were constructed with the aim of reproducing the observed features for galactic luminous stars .
By means of a Montecarlo technique, stars were randomly distributed in the M_{Bol}-Log T_{eff} plane, weighted on the Salpeter's initial mass function and with a constant star formation rate . The different evolutionary scenarios considered for the comparison with observations are :
a) Evolution with mass loss by stellar wind all across the HR diagram as for Case C by Maeder (1981a,b) shown in Fig 2 ;
b) Evolution with mass loss by stellar wind and convective overshooting from the central cores as for $\lambda=1$ by Bressan et al.(1981) in Fig 3 ;
c) Evolution with mass loss, convective overshooting and modified CNO opacity as in Bertelli et al. (1984) in Fig 4 .
The number of stars in Figures 2 , 3 and 4 is the same as in Figure 1 .
The comparison of the relative number of stars in different regions of the observed HR diagram with the distribution expected from models with mass loss (Fig 2) and with mass loss and overshooting (Fig 3) evidences an excess of A type stars and a main sequence theoretical width not sufficient to account for the observed distribution . As discussed in Bertelli et al.(1984), if on the observational ground the catalog incompleteness is the more serious problem, theoretical models give results in good agreement with the observed distribution only with a suitable combination of mass loss ,overshooting and a moderate increase of standard radiative opacities in the region of the CNO ionization . The synthetic HR diagram in Fig 4 shows the best agreement with the observed one(Fig 1) ,thus supporting the idea that standard theory must be modified .
A modification of the opacities is actually one way to improve the agreement between theory and observations .

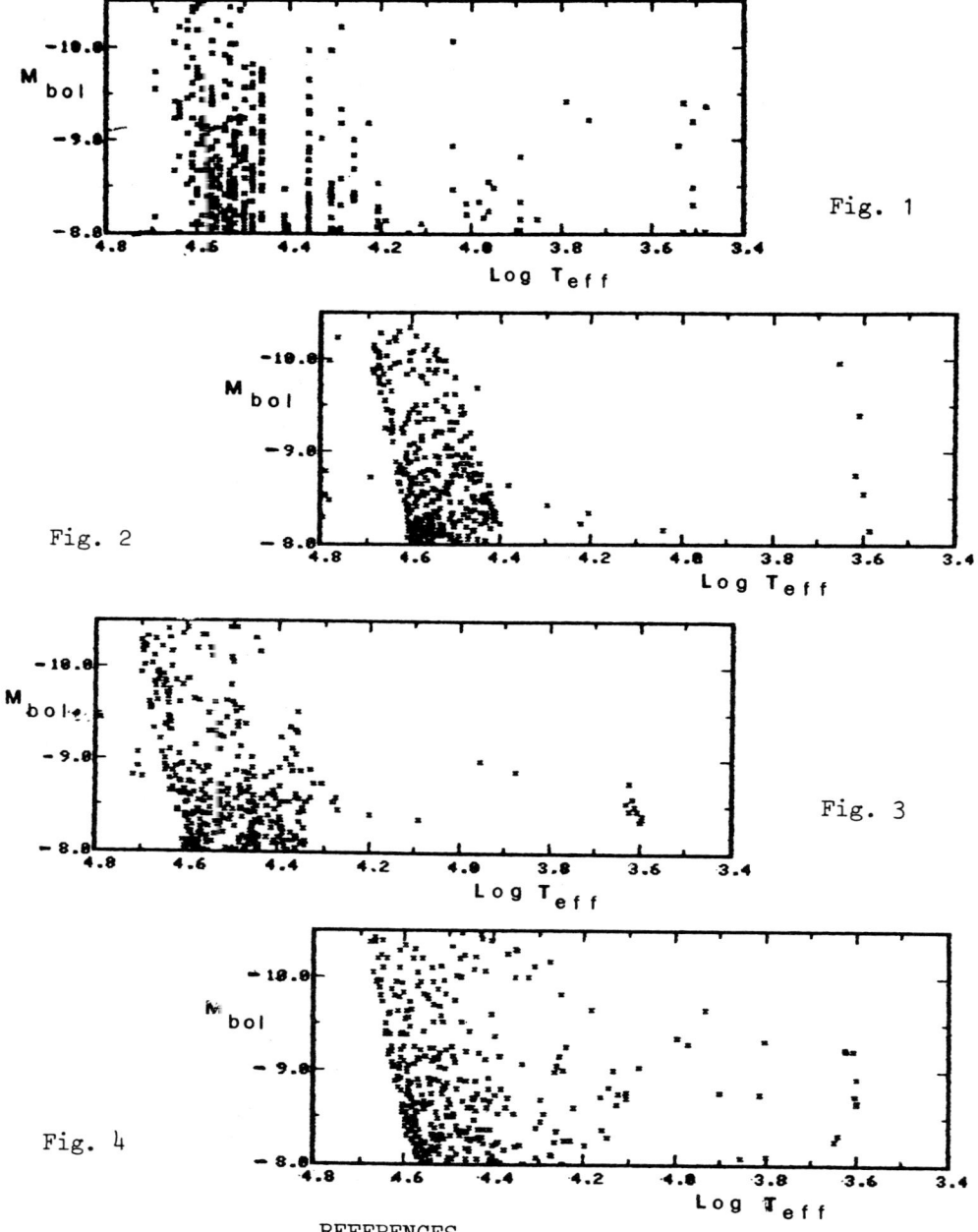

Fig. 1

Fig. 2

Fig. 3

Fig. 4

REFERENCES

Bertelli,G.,Bressan,A.G.and Chiosi,C. :1984,Astron.Astrophys. 130 ,279
Bressan,A.G.,Bertelli,G.and Chiosi,C. :1981,Astron.Astrophys. 102 , 25
Humphreys,R.M.and McElroy,D.B. :1984,Astrophys.J. 284 ,565
Maeder,A. :1981a, Astron. Astrophys. 99 , 97
Maeder,A. :1981b, Astron. Astrophys.102 ,401

A LOW-DISPERSION, NEAR-INFRARED SURVEY FOR GALACTIC M SUPERGIANTS

D. J. MacConnell
Michigan State University, East Lansing, MI USA
R. F. Wing
Ohio State University, Columbus, OH USA
E. Costa
Universidad de Chile, Santiago, Chile

I. Introduction

While there have been many surveys for luminous, blue galactic stars and their numbers can be considered somewhat complete, such is not the case for red supergiants (see e.g. Humphreys and McElroy 1984). One result of this incompleteness is that the ratios B/R and WR/R, often used as diagnostics for evolutionary models of massive stars and the variation of the ratios with galactocentric distance, are not well known for the Galaxy. In an attempt to improve the statistics, the first author began an objective-prism survey within 6 deg of the southern galactic plane using I-N plates. The dispersion is 3400 A/mm at the A-band, and the spectra cover the range 6800-8800 A; the deepest plates reach ir mag ~13. The detection of possible M supergiants on such plates was first discussed by Nassau, et al. (1954) and depends on the presence of TiO at 7054 A and a spectrum sharply tapered to the blue. For supergiants, this shape results from integration of interstellar dust over a long path-length, but any sample of red stars with tapered spectra contains M giants in heavily-obscured regions and S stars; thus follow-up observations of the candidate stars are necessary.

II. Status of survey

To date, all plates on 33 centers (596 sq deg) have been searched and several hundred candidate stars have been marked, however only those on 74% of this area have had coordinates determined and thus are available for follow-up observations. The plates searched are in two regions of nearly equal area: A: $208 \leq$ long ≤ 248 (224 sq deg) and B: $288 \leq$ long ≤ 304 (219 sq deg); the number of known M supergiants in these regions are 3 and 31, respectively (Egret 1980).

III. Follow-up observations

A. Eight-color photometry. In March 1984 we began a program of 8-color measures in Wing's system (Wing 1971) which White and Wing (1978) have shown to be effective for the classification of red supergiants. The observations, made at Cerro Tololo, are continuing, and over 100 program stars have been measured; however, to date, only measures for 39 candidates in region B have been reduced. The photometry indicates that nearly 50% of the best candidates are new K/M supergiants.

B. CCD spectroscopy. In March 1985 we obtained CCD spectra of 57 program stars in both regions and of 31 revised MK standards of later type. The spectra cover the range 6362-8842 A at a resolution of 8 A. Several approaches have been employed to classify the program stars using the PANDCRA/ROO package of programs (developed at Mt. Stromlo and set up by S. Simkin). Spectral types were determined by quasi-photometry of the digital spectra using the band-passes of Wing's filters 1 and 2. Similar photometric and equivalent-width measures of features known to be luminosity-sensitive in late-type stars were made: the Ca II triplet (Jones, et al. 1984), Fe I 8688.6 A (Humphreys 1971), and the (0,2) CN-band system in the region 7870-7970 A (Sharpless 1956). None of these indicators is ideal; there are some inconsistencies among them, and the assignment of divisions within class I is more difficult than in the blue region as Humphreys (1971) stated. We have also experimented with pixel-by-pixel division of the spectra of program stars by those of the standard stars and with a correlation method, but the spectra are too noisy for effective use of these techniques.

IV. Summary

We have classified 27 new, red supergiants in an area of 443 sq deg along the southern galactic plane where 34 M supergiants were previously known. Most of the new stars are in the range 11 < V < 15, and the faintest suffer 5 or 6 mag of visual absorption. We are continuing the survey over the remainder of the plates available (1700 sq deg) and will pursue follow-up observations. We expect to about double the number of late supergiants.

Acknowledgements. We wish to thank the staff of the Cerro Tololo Interamerican Observatory for support of the observing program, and DJM thanks Susan Simkin for help with the digital spectra and gratefully acknowledges the American Astronomical Society for the award of a Small Research Grant.

REFERENCES
Egret, D. 1980, Catalogue of Supergiant Stars, C.D.S. 5017.
Humphreys, R. M. 1971, in Proc. Conf. Late-Type Stars, ed. by G. W. Lockwood and H. M. Dyck (KPNO Contrib. No. 554), p. 41.
Humphreys, R. M. and McElroy, D. B. 1984, Ap. J. 284, 565.
Jones, J. E., Alloin, D. M., and Jones, B.J.T. 1984, Ap. J. 283, 457.
Nassau, J. J., Blanco, V. M., and Morgan, W. W. 1954, Ap. J. 120, 478.
Sharpless, S. 1956, Ap. J. 124, 342.
White, N. M. and Wing, R. F. 1978, Ap. J. 222, 209.
Wing, R. F. 1971, in Proc. Conf. Late-Type Stars, ed. by G. W. Lockwood and H. M. Dyck (KPNO Contrib. No. 554), p. 145.

POSTER PAPERS - SESSIONS 1 and 2.

Chairman : A. WILLIS.

Willis:

I thought I would give a brief overview of a selection of the poster papers and hopefully inspire some informal discussion. We have seen today many HR diagrams and evolutionary tracks and it seems to me that one must ask which tracks to use and what do we believe regarding the physics that goes into generating these models. The poster paper by Nasi on synthetic HR diagrams for luminous stars uses various model evolutionary schemes involving mass loss and other effects and tries to reproduce the HR diagrams that are observed now. They conclude that one can only reconcile the observed HR diagrams with models incorporating overshooting and a CNO opacity bump. I am sure the stellar evolution pundits here will comment on this.

Maeder:

There has been a debate about the CNO opacity bump which started with the Carson opacity model. Recently there has been a joint paper from the Los Alamos and Carson groups concluding that there is no opacity bump. The origin of this bump is in some trick in the Carson model, so that the current status from atomic physics is that there is no bump, and that its assertion is hypothetical. There is another thing that is not hypothetical and that is the change in the C12 alpha-gamma O16 rate. Usually in stellar evolution when one is changing the energy production rate there is no great influence on the lifetime, but in this case during the He burning phase, traditionally one was converting He to C and a bit of O was formed. Now with this new rate you convert He to C and then most of C is converted to O. Moreover as the rate is larger, the core is larger so that the He burning lifetime is greater and this can explain the larger number of stars outside the MS. So I would say in summary that the present status is that the opacity bump does not exist, but there is some effect in the cross sections that considerably increases the number of stars outside the MS.

Chiosi:

The synthetic HR diagrams produced by Nasi are meant to test how closely current models for massive stars may reproduce the observational HR diagrams. Usually the comparison is made counting stars in different spectral groups and comparing those stellar counts with theoretical lifetimes. Broadly speaking, the groupings of spectral type correspond to MS, post MS and WR stages and/or stars. This may give only rough indications, while the synthetic HR diagrams are a more sophisticated tool. There are of course problems with completeness, Sp:BC:Teff scales, non hydrostatic effects on stellar radii etc.; nevertheless it seems to me that the simulations in question strongly indicate that models evolved at constant mass as well as in presence of mass loss hardly can match the observational data. The scheme with mass loss and overshooting is much better and indeed the closest to reality. Now comes the problem of the revised reaction rate. It is true that the novel rate increases the lifetime of the core helium burning phase as compared to the core hydrogen burning phase, but it is also true that adopting the mass loss rates that everybody believes in we are producing models which preferentially locate stars at the left part of the HR diagram, which have probably to be identified with WR objects. So there is a group of stars (A to M) which is not easily accounted for by any type of models. In this context comes the suggestion that CNO opacity is still underestimated. The recent paper by the Los Alamos group is not conclusive in this respect. In fact, suppose that the Los Alamos and Carson groups are reconciled as far as the past divergence in their CNO calculation is concerned, this does not imply that real CNO opacities may not be slightly higher than usually estimated. This is the spirit of our suggestion.

Willis:

Roberta mentioned the new techniques of IR surveys of late supergiant stars, with recent results for the Carina region almost doubling their numbers. This is clearly a powerful new technique and the results have serious implication for say the number ratio of WR/RSG.

Humphreys:

I think it is very impressive that Jack MacConnel has been able to get very good IR spectra demonstrating the very powerful use of the IR CaII triplet with CCD spectra on such faint objects.

Willis:

The paper by Graham and Humphreys using the IR CaII triplet to disentangle foreground stars and real members in NGC 300 and thereby get correct statistics gives a (J-H) vs (H-K) diagram in which is stated evidence for a composition gradient, and I was not sure how this was arrived at.

Humphreys:

The evidence for the composition gradient comes from the HII regions which have to be taken into account. You can see in the 2-colour diagram that some of the stars do have compositions more like LMC values than galactic values.

DISCUSSION

McGregor:

I would like to ask if you have classified the spectra in the red to determine their types, whether they are K or M?

Humphreys:

Yes the types are given in the paper.

Zinnecker:

Was there any attempt to measure CO in these objects?

Humphreys:

No.

Willis:

Roberta's review included some mention of instabilities and mechanical energy deposition in supergiant stars and two posters show observational data on HR 8752 and HD 217476 which may provide evidence for such effects. David Stickland pointed out to me that these papers are on the same star. The first paper by Zsoldos suggests variability with a 400 day period caused by pulsations, whilst the second by Smolinski et al. suggests a binary system with a period of 620 days. There appears to a be common envelope with at least three shells with considerable line activity. Obviously we need to consider the cause of such variability, and this star (and others) is clearly going to be providing the kind of observational material that de Jager would like for his models of pulsation induced by mechanical energy deposition. I know Stickland has observed this star and may wish to comment.

Stickland:

The only question I have at the moment concerns the periodicities we find in this star. Dave Lambert has been studying this star and finds a period of about one year in the lines. We have periods of 400 and 600 days, and I wonder if we are dealing with a semi-irregular pulsation or whether it is really a binary.

Smolinski:

Well, our observations cover at least 15 years at high dispersion and there is no doubt that the velocities of the lines are changing. In addition there is a common envelope which sometimes has the same velocity as the star, sometimes is not seen and at other times is split into several components. If the star and the envelope velocities are the same at some times they will be superimposed, so we really only see one line. The mistake of others is not to use high dispersion and be unable to distinguish the star and envelope components. For this reason it is very difficult to obtain the RV. With our data of over 200 high dispersion spectra we are able to distinguish the stars component lines and according to our present results it seems clear that it is a binary. There may be additional evidence for pulsations. However, I think perhaps that the optical light variations may result from increased line blanketing when multiple components are seen, which may explain some of the photometric variations being interpreted as due to pulsations.

de Jager:

I would like to know how sure we are about the binary hypothesis - did you try to calculate the elements of the system?

Smolinski:

At the moment we have some elements but these are only preliminary.

Viotti:

There is also radio emission from HR 8752, is that not so?

Smolinski:

We are not sure of the origin of the radio emission but I think it comes from the HII region in which this star is just inside, which is excited by a hot B star nearby.

Willis:

We heard in Rolf's talk this morning about mass loss in OB and WR stars with radiation pressure being highlighted as the mechanism for initiating and driving the mass loss in hot stars - there is a poster by de Jager, van der Hucht & Nieuwenhuijzen which studies all the rates available in the literature and on looking at correlations with stellar parameters concludes that the mass loss rates are a sole function of L and T(eff) for all types except WR and C stars where the observed rates appear too high by a factor of ten or so. I think we know the reason why the rates seem too high for the WR stars, since the radiative

DISCUSSION

luminosities for these stars are taken using older values of T(eff) of say 30000 K rather than the more recently deduced higher values of say 100000 K. The higher values based on 1985 results for temperatures will give higher L and thus higher predicted mass loss rates, which are likely to come into agreement with observations. How the C stars can be explained I do not know, and we have to ask what is the cause of the discrepancy in this case.

Lamers:

I am not sure that their analysis shows that mass loss depends on temperature and L alone, the paper assumes such a dependence and then it fits. Whether this is true remains to be seen. If you take the example of the Be stars, which are known to be rapid rotators, and which have the same L and T(eff) as normal B stars with considerably lower mass loss rates, there we know that there is another effect. In this case either the mass loss is due to something else or it is enhanced by some other mechanism than radiation pressure.

de Jager:

I agree of course. Our intention was to test possible correlations of mass loss rates and stellar parameters in order to see if a fairly simple parametrisation could be found which could be used in stellar evolution models. This appears to be the case for all normal types of stars (O to M).

Chiosi:

Did you compare the observed mass loss rates with the predictions of the Reimers formula in the domain of intermediate mass loss? I remember the rates predicted by Reimers are somewhat lower than observed in the range of the bright supergiants.

de Jager:

We made some rough comparisons but not systematically. There are in the literature about a dozen formulas which describe the mass loss in dependence on the radius, mass and L, but these refer only to parts of the HR diagram. Reimers formula relates to the cool stars and not the hot stars. Other formulae apply only to hot stars. But I did not really make a solid comparison. This is something we should do.

US Voice:

I believe there was a paper by Wayne Waldron in ApJ letters last year which found a similar relationship. Also it is not clear to me that the relationship demonstrates that radiation pressure is in fact the significant mechanism in late type stars.

de Jager:

We did not discuss the mechanism of mass loss but with regard to the late type stars I am sure that mechanical turbulence plays the significant role. Radiation pressure is unimportant for stars later than B-type.

Moffat:

I do not think it is quite fair to say that these are "1985" WR temperatures of 100000K. These values have been around for a while, it's just that some people have not accepted them. The Russians for instance have published these values in 1975.

Willis:

That is certainly a fair comment. In talking about mass loss in cool stars, and in addressing activity on the supergiant region of the HR diagram - if there is sudden and extensive mass loss in these cool stars we might expect to see dust forming in the shed material. Dave Stickland has a poster on IRAS observations of cool hypergiants more luminous than M(Bol) = -8. Apparently there is direct evidence for dust emission around stars later than G-type, but not around the earlier F hypergiants.

Humphreys:

The circumstellar dust feature around F, G, K and M supergiants has been known since the early 1970's. One of the problems we are all concerned with is what direction those stars are going in the HR diagram - are they approaching the supergiant region or on their way out of it. Do you have any intuition on that subject?

Stickland:

No, I have no ideas about that at all. I think that the fact that the brighest F stars do not show dust emission maybe means that they are moving from left to right, and one will see the dust produced later on when they go through some period of mass loss. On the other hand, maybe they have dispersed it and they are going backwards. I just don't know. The important thing is to extend the survey because there are rather a limited range of absolute mags here, and in act, if you

DISCUSSION

look a little bit later to 89 Her you do see the dust, but it is hot dust and also it is a peculiar star. Its IR excess is rather similar to Eps Aur, with a temperature of about 700 K. What I was really trying to do (apart from get a ticket to this conference!) was to see if the amount of dust emission is greater in the more luminous stars, and the present sample does not show that. The fact that the sample is limited is historical: the IRAS Point Source catalogue came up at RAL a few months ago, and there was not another in the UK. I thought 'what should I do with this', so I looked at my favourite star, HR 8752, but it did not seem to be very exciting in the IR. Then I looked at the other "official" hypergiants and they showed a rather strange collection of results, so I extended it using your paper of 1978. Finally I remembered the horrible BC's for M stars, and so rapidly threw in some of those too. I ended up totally confused, and maybe you are confused as well.

Lamers:

I want to add a word of warning for those who use the IRAS Point Source Catalogue, particularly in the two long wavelength bands of 60 and 100 microns. A large number of the fluxes quoted in the catalogue are not very reliable. If you look at the real tracings of the data you can see that many of the Point Source data are in fact due to some background. So one has to be very wary of the 60 and 100 microns data.

Stickland:

I quite agree and reiterate that warning. In fact most of the stars in my paper have only 12 and 25 micron data, a few have longer wavelength observations.

Willis:

I omitted reference to another paper dealing with variability in these cool supergiants, which refers to the star mu Cep and its light variations which seems to come up with two periods of 4700 days and 873 days and the author believes these are due to multi-mode pulsations. Here again we may be dealing with observational evidence for mechanical energy deposition.

de Jager:

I do not think that there may be any difference between a star like this and the Sun where we have now discovered many hundreds of periods by refined observation. I guess that this kind of seismology applied to these supergiants may really give us profound information on the internal constitution of these stars.

Willis:

I would draw attention to the important paper by Hummer, Bohannan and Abbott on the recalibration of T(eff) for hot stars taking into account the recognition that wind blanketing is so important for properly determining the photospheric parameters. Their analysis of Zeta Pup provides the solution to the long standing temperature question, and most excitingly I think, the He and N abundance enhancements. My question to the authors is that when you say that the derived abundances are consistent with CNO burning products, have you looked into the question of getting that stuff out into the atmosphere from the nuclear burning regions?

Hummer:

Of course not!

Willis:

Thank you David. The point I am getting at is that if we believe the WR stars are chemically evolved in which we have stripped off all the outer atmosphere during the H-burning phase, then one can automatically expose the interior burning products. Here clearly one still has a very large H atmosphere content, and not therefore necessarily peeled down to interior regions, so there may be a problem.

Bohannan:

I think the next step is to look at a variety of Of stars. When you look at Zeta Pup you realise it is slighly more evolved than other O stars.

Kudritzki:

I want to make a general comment. Evolutionary tracks taking into account mass loss and turbulent mixing in general produce this type of chemical enrichment in the photosphere. Whether the agreement is really quantitative in this place in the HR diagram, depends on the mass loss rates before. We cannot be sure of a steady mass loss rate for example. Qualitatively there does appear to be good agreement.

DISCUSSION

Maeder:

A general comment. For a long time in stellar evolution only the HR diagram was used as a constraint on models. Now it is clear that the C/N and O/N ratios are also very strong constraints on models and must be taken into account in order to give a good description of stellar evolution.

Bohannan:

One of the things that stellar astrophysicists get critisised about is that they just hack away trying to get better and better stellar parameters. For instance is changing the temperature of Zeta Pup 4000K significant? The answer is that it is. That change makes a big difference when you consider the amount of ionising radiation and M(Bol), so one would be quite wrong to say that a change of only ten percent in T(eff) is insignificant. It does have profound effects in the most luminous and most massive stars in the Galaxy.

Willis:

We also have discussion opportunities for todays review talks.

Conti:

I would like Alan Sandage to respond to Roberta's 5-minute comments after his talk - what is your reaction to that attack.

Sandage:

I dit not see it as an attack. The difference between Roberta Humphreys and me about the calibration of the red supergiants is not as great as it might seem. Because it all depends on the absorption question and I am not concerned about three tenths of a magnitude in M33. What I am concerned about is whether the absolute magnitudes of the RSG's is a function at all of the absolute magnitude of the parent galaxy. I think from what Roberta showed this afternoon, we are in essential agreement because she puts for M101 a value of -9 and if that was put on her diagram it would be the highest point and essentially the brightest galaxy and the other points slope down through -8 to something like -6.7 for WLM. I think we both agree that the absorption problem in galaxies like the SMC or even fainter like Sextans A, Sextans B, IC 1613 and especially WLM which is -13.5 is not present. So the question of the internal absorption is not present fainter than -16 to any great extent. I would like to stress that if you use the apparent distance modulus and the apparent magnitude, if there is a veil of absorption no matter how strong, if it is constant you get the right absolute magnitude. So the

Wendy Freedman and I have got spectra of one of the blue stars there and John Hutchings and I have got IUE UV spectra of it. It is a pretty normal late O supergiant. We also have frames looking for WR stars and it seems to be full of them. Even though it's only one small region, if it is representative of all of the area I do not believe the answer for the lack of very luminous supergiants to be the lack of suitable progenitors - they do seem to be there, at least in NGC 206.

SESSION 3.

LUMINOUS BLUE VARIABLES.

Chairman : R. Sreenivasan.

1. I.APPENZELLER: Instability in Massive Stars: An Overview.

2. B.WOLF: S Doradus Type ; Hubble-Sandage Variables.

3. H.LAMERS: P Cygni Type Stars: Evolution and Physical Properties.

INSTABILITY IN MASSIVE STARS: AN OVERVIEW

I. Appenzeller
Landessternwarte
D-6900 Heidelberg
Federal Republic of Germany

ABSTRACT: Dynamical, vibrational, and thermal instabilities of massive blue stars are discussed as possible mechanisms for the observed brightness variations of such objects. Relaxation oscillations (on local thermal time scales) due to dynamical instabilities of the stellar wind flows appear to be the most likely mechanism, at least for the S Dor variables. Very massive main-sequence stars with $M > 10^3\ M_\odot$ should be violently vibrationally unstable and therefore should differ significantly from stable main-sequence stars of lower mass.

1. INTRODUCTION

Except for main-sequence O stars all very bright blue stars are variable (cf. e. g. Maeder 1980). The amplitudes of the visual brightness fluctuations vary between a few hundredths of a magnitude in the less luminous stars and several magnitudes in some of the "Hubble-Sandage" or "S Doradus" variables (cf. B. Wolf's contribution to this volume). So far there exists no consistent and established theory of these light variations. But it is usually assumed that the variations are caused by some type of stellar instability (see e. g. Stothers and Chin, 1983). Therefore, in the following the known types of instabilities which may play a role in massive blue stars are reviewed and discussed.

In the astrophysical literature (as in daily life) the term "instability" is used with different meanings. In order to avoid misunderstandings I would like to emphasize that throughout this paper "instability" will be used in the conventional physical sense to describe equilibrium configuration only. An equilibrium state will be called "stable" if an (infitesimally) small perturbation does not result in a significant deviation from the unperturbed state. It will be called "unstable" if a small perturbation can lead to a substantial deviation (from the initial state) which is growing with time. Such an instability may lead to a disasterous runaway (as in the case of the dynamical instability which causes the supernova events at the end of a massive star's evolution), or the effects of an instability may be limited by non-linear effects (as in the case of the vibrational instability of the δ Cep stars). In the latter case the instability may cause no more than a minor modification of the equilibrium configuration. The above definition of "instability" does not cover non-equilibrium states. Hence, theories, which for massive stars (or their outer layers) assume the permanent absence of an equilibrium will not be discussed here.

2. HYDROSTATIC AND DYNAMIC EQUILIBRIUM

In the textbooks of stellar structure stars are defined as gaseous spheres in hydrostatic equilibrium. Hence, in order to study stellar stability we have to analyze the behaviour of the stellar hydrostatic equilibrium in the presence of various types of perturbations. Mathematically the stellar hydrostatic equilibrium is expressed by the hydrostatic equation

$$\frac{1}{\rho} \frac{dP}{dr} = - \frac{G M_r}{r^2} \qquad (1)$$

where P is the total pressure, M_r the mass inside a sphere of radius r around the star's center, G the gravitational constant, and ρ the gas density. P may contain several different significant contributions, e. g.

$$P = P_G + P_R + P_T + P_M \qquad (2)$$

where P_G is the gas pressure, P_R the radiation pressure, P_T the turbulent pressure, and P_M the magnetic pressure. Although magnetic fields may play an important role in the outer layers of luminous blue stars (cf. e. g. Friend and Mac Gregor 1984), too little is known about the strength and geometry of such fields to allow a meaningful discussion. Therefore, in the following the last term in Equ. (2) will always be neglected. P_T is negligible in the deep interior of all stars (cf. e. g. Cox and Giuli 1968). However, P_T may become significant in the outermost layers of stars with outer convection zones. P_R contributes significantly to the total pressure in all layers of massive stars. For $M \gtrsim 130 \, M_\odot$ main-sequence stars P_R usually is the dominating term in Equ. (2). In the outermost layers of massive stars P_R may be more important than P_G even at lower mass values.

For the atmospheric layers (which form the outer boundary of a star) Equ. (1) is often rewritten into the form

$$\frac{1}{\rho}\frac{dP_G}{dr} = -g_{eff} = -(g-g_R-g_T-\ldots) \qquad (3)$$

where

$$g_R = -\frac{1}{\rho}\frac{dP_R}{dr} = \frac{4\pi}{c}\int \kappa_\nu \pi F_\nu d\nu \qquad (4)$$

and

$$g_T = -\frac{1}{\rho}\frac{dP_T}{dr} = -\frac{\alpha}{\rho}\frac{d}{dr}(\overline{\rho v^2_T}) \qquad (5)$$

and where the various symbols have the following meaning: $g = GM/r^2$ = gravitational acceleration, c = velocity of light, κ_ν = monochromatic mass absorption coefficient, F_ν = monotonic flux of frequency ν, α = numerical factor, and v_T = velocity of turbulence elements.

For all realistic stellar models we have $dP_G/dr < 0$. Hence from Equ. (3) follows for fully hydrostatic atmospheres $g_{eff} > 0$. However, as pointed out first by Parker (1958), instead of Equ. (3) stars may also have non-static boundary conditions, which in the strictly stationary case can be written

$$v\frac{dv}{dr} + \frac{1}{\rho}\frac{dP_G}{dr} = -g_{eff} = -(g-g_R-g_T-\ldots) \qquad (6)$$

where $v = dr/dt$. (Equ. (6) can be extended to the more general non-stationary case by adding on the lefthand side the expression dv/dt). Physically Equ. (6) means that stationary equilibrium solutions with stellar winds, mass loss, or mass accretion are possible. At least the great majority of all stars show winds and some stars are known to show mass accretion. Hence Equ. (6) (and not Equ. (3)) should be considered the normal boundary condition of stars. In contrast to Equ. (3) Equ. (6) allows solutions with $g_{eff} < 0$, provided $dv/dr > 0$. Hence, contrary to claims in some textbooks, stationary stars can exist even for $g_R > g$, although for $g_R \gg g$ the mass loss $\dot{M}$ tends to become excessive.

3. RELATIVISTIC DYNAMICAL INSTABILITY

As noted above, with increasing mass P_R becomes the dominant pressure term in main-sequence stars. If $P = P_R$ the righthand and lefthand sides of Equ. (1) show exactly the same relative variation for adiabatic radius changes. Hence, we have <u>indifferent</u> hydrostatic equilibrium, independent of the radius, and a small velocity perturbation will result in a displacement growing linearly with time.

Stothers and Chin (1983), who found that under certain conditions in massive post-main-sequence stars unstable hydrogen-shell burning or core hydrogen flashes may occur. However, the computations of Stothers and Chin also showed that if these types of instabilities are present in massive stars, they are not expected to be observable, since they would cause only very small light variations (<0.05 mag) on very long time scales ($\gtrsim 10^3$ years). Therefore, according to our present knowledge thermal instabilities of the nuclear sources also can be excluded as the cause of the observed variability of massive luminous stars.

6. MASS LOSS INSTABILITIES

The wind flows from luminous blue stars are known to be unstable to Rayleigh-Taylor and related instabilities (see e.g. Nelson and Hearn 1978, Kahn 1981). These instabilities may be very important for heating the wind flows. But because of their small-scale nature they probably contribute little to the visual brightness variations.

Radiatively driven winds can also become dynamically unstable if a random increase of the wind results in an increase of the driving term g_R, which enters into Equ. (6). Any increase of g_R and the corresponding decrease of g_{eff} results in an expansion of the massive star's outflowing envelope and an increased optical depth of the flow. Hence the increase of the mass loss rate will be accompanied by a decrease of the effective temperature. For a constant luminosity $L = 4\pi\sigma R^2 T_{eff}^4$ the effective radius and the gravitational acceleration at a given optical depth are expected to vary approximately according to $g \sim R^{-2} \sim T_{eff}^4$. As the effective acceleration can be written as $g_{eff} = g - g_R - ... = g(1 - g_R/g - ...)$, the stability of the wind against $\dot{M}$ or T_{eff} fluctuations will depend on whether g_R varies (compared to g) more rapidly or less rapidly with T_{eff}. If g_R also decreases $\sim T_{eff}^4$ (or more rapidly), the wind is expected to be stable. If g_R decreases less rapidly, a temperature (or $\dot{M}$) fluctuation may lead to a slightly decreased T_{eff}, followed by an increased g_R/g, a consequently decreased g_{eff}, an increased $\dot{M}$ and consequently a rapidly growing further decrease of T_{eff}. Hence, for g_R decreasing less rapidly than $\sim T_{eff}^4$ radiatively driven winds are expected to be unstable.

An accurate qualitative study of the functions $g_R(T_{eff})$ require a grid of detailed non-static atmospheric models of massive luminous stars. So far only very fragmentary data of this type exist. However, qualitatively some information on the behaviour of g_R as a function of T_{eff} can perhaps be obtained from the static (LTE) computations of Kurucz and Schild (1976) reproduced in Figure 1. (While LTE

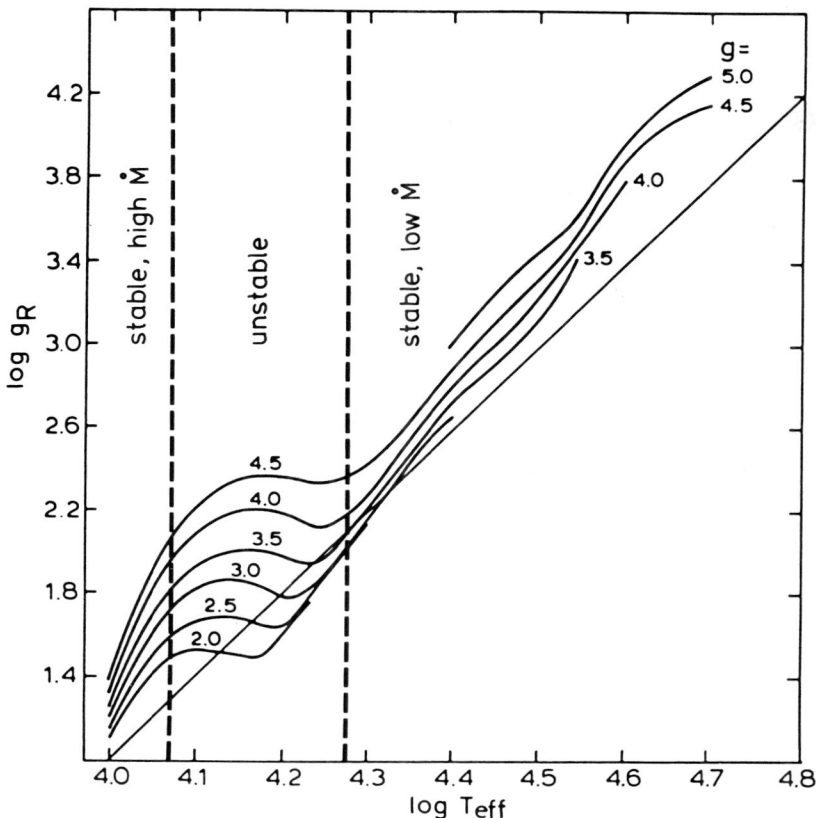

Figure 1. The radiative acceleration g_R (at a Rosseland optical depth of 10^{-4}) as a function of the effective temperature T_{eff} for different values of the gravitational acceleration g (Kurucz and Schild 1976). The solid straight line indicates the relation $g_R \sim T_{eff}^4$.

is probably a poor approximation, the static approximation may be justified, as in the temperature range discussed below merging UV lines form a quasi-continuum.) According to Figure 1 for most values of T_{eff} g_R is expected to decrease (with decreasing T_{eff}) more rapidly than T_{eff}^4. However, for $10^4 \lesssim T_{eff} \lesssim 2\ 10^4$ g_R decreases less rapidly and for some values of T_{eff} we have even $dg_R/dT_{eff} < 0$. This "bump" in the g_R-T_{eff} relations is caused by an increased κ due to the recombination of higher ions of the iron-group elements. In this range of effective temperatures the mass loss from sufficiently luminous stars (where g_R is important in Equ. (6)) should be unstable

The instability outlined above can possibly explain the observed variability of the S Dor variables. As described in detail in Dr. B. Wolf's contribution to this volume, S Dor variables are massive and luminous post-main-sequence stars. On their path in the HR diagram towards lower effective temperatures these objects are expected to encounter the upper T_{eff} limit of the "instability strip" of Figure 1. If the instability outlined above exists, at this effective temperature the mass loss rate will rapidly increase (on a dynamical time scale of the order of a few months), the effective radius will increase as well, and the effective temperature will decrease, until a new stable equilibrium is reached at the "red" boundary of the instability strip. However, this (now again dynamically stable) "maximum state" (characterized by high mass loss, low effective temperature, and - because of the lower bolometric correction - high visual brightness) very likely is not in thermal equilibrium: Strong blanketing by the now optically thick expanding envelope will result in a slow heating of the outermost stellar layers until (after a time interval determined by the <u>thermal</u> time scale of the stellar surface layers, which may be of the order of years or decades) g_R/g becomes too small to support the expanded envelope. The envelope will then collapse and after some oscillations settle back to the high temperature, low mass-loss equilibrium state. But, now the blanketing has disappeared. Hence, the photospheric and subphotospheric layers will start to cool again and a new cycle of the g_R-energized "relaxation oscillations" can begin.

In the "stable" domain the g_R-T_{eff} relation seems to be relatively close to indifferent equilibrium. Therefore, if the $g_R(T_{eff})$ relation is not a smooth curve the above mechanism may possibly also cause smaller-amplitude variations as observed in the "normal" very luminous blue stars not belonging to the S Dor class. On the other hand, it must also be kept in mind that Figure 1 is based on static LTE computations which give only poor approximations for the conditions in the envelopes of luminous blue stars. Fully reliable computations are hardly possible at present. Hence, so far we cannot realy prove that the necessary conditions for the mechanism outlined above do actually exist. On the other hand, the dense forest of absorption lines observed in IUE spectra of S Dor variables during maximum state makes the existence of the wind instability very likely.

A related type of relaxation oscillation scenario for the variability of the S Dor stars was suggested by Maeder (1983), who pointed out that the turbulent contribution g_T to the effective acceleration may cause a recurring instability. Maeder's mechanism is based on de Jager's (1980, 1984) suggestion that the observed absence of very luminous cool stars (Humphreys and Davidson 1979) is due to turbulence-pressure induced mass loss, which increases with decreasing effective temperature. In fact, as shown by Maeder (1980), the extent of the outer convection zone of supergiant stars increases rapidly with decreasing effective temperature. If g_T is dynamically important it is expected to increase with decreasing T_{eff}, which obviously could lead to an instability. Unfortunately convection in stars and, in particular, in their outermost layers is a poorly understood topic and very little is known on convection in non-static atmospheres. (Practically all stellar evolution and interior structure computations are based on hydrostatic models.) Moreover, at present it is not clear, how a g_T-instability of the wind in massive stars would be limited. Nevertheless, g_T-driven relaxation oscillations also seem to be a viable possibility for explaining the variations of massive luminous stars.

7. CONCLUSIONS

Of the various instabilities discussed in the literature the conventional dynamical, thermal, and vibrational instabilities all seem to be inadequate to drive the observed light variations of luminous blue stars. On the other hand, mass loss instabilities may possibly explain the observations. In the case of the S Dor variables the observed amplitudes and time scales of the observations would be consistent with the existence of g_R-or g_T-relaxation oscillations of the mass loss rates. These oscillations would be caused by dynamical instabilities, but governed by thermal time scales. If main-sequence stars of $M > 10^3 M_O$ exist, such objects are expected to be violently vibrationally unstable. They are expected to show surface properties which are quite different from the equilibrium main-sequence stars of lower mass.

REFERENCES

Appenzeller, I.: 1970a, Astron. Astrophys. 5, 355
Appenzeller, I.: 1970b, Astron. Astrophys. 9, 216
Appenzeller, I., Fricke, K.: 1971, Astron. Astrophys. 12, 488
Appenzeller, I., Tscharnuter, W.: 1973, Astron. Astrophys. 25, 125
Appenzeller, I., Tscharnuter, W.: 1974, Astron. Astrophys. 30, 423
Cox, J. P., Giuli, R. T.: 1968, Principles of Stellar Structure, Gordon and Breach, New York 1968, Vol. 1, Chapter 14
de Jager, C.: 1980, The Brightest Stars. D. Reidel Publ. Company, Dordrecht 1980
de Jager, C.: 1984, Astron. Astrophys. 138, 246
Fricke, K.: 1973, Astrophys. J. 183, 941
Friend, D. B., Mac Gregor, K. B.: 1984, Astrophys. J. 282, 591
Humphreys, R. M., Davidson, K.: 1979, Astrophys. J. 232, 409
Kahn, F. D.: 1981, Monthl. Not. R. Astron. Soc. 196, 641
Kurucz, R. L., Schild, R. E.: 1976, IAU Symp. 70, 377
Ledoux, P.: 1941, Astrophys. J. 94, 537
Maeder, A.: 1980, Astron. Astrophys. 90, 311
Maeder, A.: 1983, Astron. Astrophys. 120, 113
Nelson, G. D., Hearn, A. G.: 1978, Astron. Astrophys. 65, 223
Papaloizou, J. C. B.: 1973, Monthl. Not. R. Astron. Soc. 162, 169
Parker, E. N.: 1958, Astrophys. J. 128, 664
Schwarzschild, M., Härm, R.: 1959, Astrophys. J. 129, 637
Stothers, R.: 1976, Astrophys. J. 204, 853
Stothers, R., Chin, C.: 1983, Astrophys. J. 264, 583
Stothers, R., Simon, N.: 1970, Astrophys. J. 160, 1019
Talbot, R.: 1971, Astrophys. J. 165, 121
Weigelt, G., Baier, G.: 1985, Astron. Astrophys. (in press)
Ziebarth, K.: 1970, Astrophys. J. 162, 947

Discussion : APPENZELLER.

LAMERS :

I agree with the suggestion that radiation pressure due to FeII etc. might play a role in the instabilities of B and A hypergiants. In my paper I show that the acceleration of the wind of P Cygni is probably due to singly ionized metal lines.

STALIO :

The atmospheric structure of HS variables resulting from your presentation can be schematically seen as consisting of a star, a stellar wind and an expanding shell around. Do you have any estimate of the mass of the shell? What happens to the shell when an outburst occurs? Is it disrupted or does it act as a barrier for the newly ejected material?

KUDRITZKI :

1) Would it not be better to use Abbott's force multiplier for the calculation of g_{rad}, since it takes into account the velocity field?
2) What amplitudes and periods do you expect for the vibrational instability of O3-stars at the ZAMS?

APPENZELLER :

1) Abbott's results were obtained assuming relative high expansion velocities. For the low expansion velocity in SDor type stars the static computations of Kurucz and Schild may still be a better approximation.
2) For the most massive O-stars we would expect pulsation amplitudes (limited by nonlinear effects) of the order 0.1 mag. on timescales of hours.

DE JAGER :

My limit assumes that stellar photospheres become unstable and produce a large rate of mass loss because $g_{eff} < 0$. Actually, g_{eff} consists of several terms: $g_{eff} = g_{grav} + g_{rad} + g_{turb} + g_{rot}$. In hot stars g_{rad} is important (the classical Eddington case) but in cooler stars dissipation of mechanical energy (most probably the energy of convective motions and non-radial pulsatation) may cause g_{turb} to become important. There is some recent evidence supporting this latter assumption. Nieuwenhuyzen and I (proceedings this symposium) found that, if other things remain the same, the rate of mass loss is positively correlated with the photospheric value of v_{turb}.

S DORADUS TYPE; HUBBLE-SANDAGE VARIABLES

B. Wolf
Landessternwarte
Königstuhl
6900 Heidelberg
West Germany

ABSTRACT. The luminous S Dor variables or Hubble-Sandage variables have gained considerable interest during the past few years for their exceptional (if compared to normal luminous hot stars) stellar wind properties. They seem to represent a short intermezzo in the evolution of the very massive stars prior to becoming WR stars. More recent results on S Dor variables (mainly of the LMC) are briefly summarized. Since most of the data have been or are going to be published elsewhere only an outline will be presented here.

1. INTRODUCTION

S Dor variables or Hubble-Sandage variables belong to the most luminous stars ($M_{pg} \approx$ -7 to -11.5) in the universe. They are characterized by irregular photometric variations of more than one magnitude on timescales of years to decades. They show a strong UV-excess ($(U-B) \leq$ -0.8) and are redder at visually bright phases. First low dispersion spectrograms of five objects of this class in M31 and M33 were discussed by Hubble and Sandage (1953). More detailed spectroscopic studies of the LMC-S Dor variables (cf. e. g. Thackeray (1974) and literature quoted therein) have shown that their highly variable spectra are characterized by strong Balmer emission lines and pronounced forbidden lines ([Fe II], [N II] etc.) which dominate the maximum and minimum spectra, respectively.

The last review on S Dor variables or Hubble-Sandage variables has been represented during the IAU Symposium No. 67 by Sharov (1975). Since that time considerable technical advances have been made allowing to obtain spectrograms of high resolution of the brightest S Dor variables in the optical range (e. g. with CASPEC attached to the 3.6 m ESO telescope) and in the satellite UV with IUE. In addition infrared data in the JHKL(M)(N) passbands have been obtained for a number of these variables. Some of the more recent results on LMC-S Dor variables are summarized in the following.

2. S DOR VARIABLES OF THE LMC

Three S Dor variables are known to be members of the LMC: the prototype S Dor, R71 (Thackeray, 1974) and R127 (Stahl et al. 1983). Since these LMC objects are only slightly reddened and comparatively close, they can be studied in more detail than S Dor variables in other galaxies. In a series of papers (Wolf et al., 1980, Wolf et al. 1981, Stahl and Wolf, 1982, Appenzeller and Wolf, 1982, Stahl et al., 1983, Stahl et al., 1985) our group has discussed detailed photometric (UBVRIJHKL) and spectroscopic (ground-based and IUE) observations of these three LMC-S Dor variables. The most important results of these investigations can be summarized in the following way: S Dor variables are hot OB supergiants which during maximum are surrounded by cool ($T_e \approx$ 8000 - 10 000 K), dense ($N_e \approx 10^{11}$ cm^{-3}), slowly expanding ($v_{exp} \approx$ 100 - 200 km s^{-1}) envelopes. The mass-loss rate during maximum is of the order of 10^{-4} $M_\odot yr^{-1}$; it is by about a factor 100 lower during minimum phases. On the other hand the bolometric luminosity of the S Dor variables seems to remain almost constant. The observed brightness variations in the visual and photographic range are regarded as a consequence of the variable mass-loss rate and a correspondingly variable redistribution of the spectral flux.

A particularly detailed discussion of the envelope characteristics of the prototype S Dor has been carried out on the basis of extensive coordinated observations by our group quite recently (cf. Leitherer et al.; this volume and Leitherer et al. 1985).

From the data a model was derived according to which a hot stellar core (which can only be observed in the minimum phase) is surrounded by a cool envelope of highly variable extent and with physical conditions similar to the atmospheres of late B or early A supergiants. This "pseudo-photosphere" appears essentially static although it shows radial velocity variations of small amplitude ($\approx$10 km s^{-1}) indicative of large scale pulsation-like motions. From the pseudo-photosphere the strong wind ($\dot{M} \approx 5 - 10 \cdot 10^{-5}$ $M_\odot yr^{-1}$) starts at low velocity ($v_0 \approx v_{sound}$) and gets slowly accelerated approaching its terminal velocity ($v \approx$ 130 km s^{-1}) only at large distances ($r > 10$ R_*). The flat velocity law is similar to that of P Cygni (Waters, 1984). Possible mechanisms which may drive the unusual wind (compared to the wind of normal hot stars) of S Dor variables are discussed elsewhere in this volume (Appenzeller, 1986).

3. EVOLUTIONARY STATUS

During the past few years the S Dor variables have gained considerable interest in connection with current theories of the evolution of very massive ($M \gtrsim 50\ M_\odot$) stars with mass loss. It has been suggested (Sterken and Wolf 1978, Humphreys and Davidson 1979, Wolf et al. 1980) that the S Dor variables which are located around the upper envelope of known stellar luminosities (Fig. 1) represent a short-lived phase as immediate progenitors of the massive WR stars. Presumably during this S Dor phase of enhanced mass loss matter is rapidly removed from the surface of massive stars until the processed matter characterizing the spectra of late WN stars becomes visible.

This earlier suggestion got further support from Maeder's (1983) computation of evolutionary tracks of very massive stars including the S Dor phase. The best witness for this scenario, however, appears to be the recently discovered S Dor variable R127 of the LMC. This star had earlier been classified by Walborn (1982) as a transition Of/WN9-10 star before an S Dor-type outburst was observed by Stahl in 1982 (cf. Wolf and Stahl, 1983). For a detailed description of this key object for the understanding of the evolutionary status of S Dor variables see Stahl et al. (1983) and Stahl and Wolf (1985).

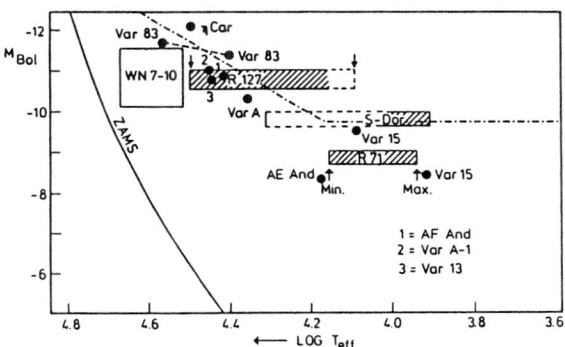

Fig. 1. Location of the LMC-S Dor variables R71, S Dor, and R127 in the HRD. Also included in the figure are the S Dor variables or Hubble-Sandage variables of M31 and M33 (Humphreys et al., 1984) and the upper envelope (dashed-dotted line) of known stellar luminosities (Humphreys and Davidson, 1979). The approximate position of the late WN-type stars is also given.

REFERENCES

Appenzeller, I., Wolf, B.: 1982, in ESO Workshop "The Most Massive Stars", p. 131, eds. S. D'Odorico, D. Baade, K. Kjär
Appenzeller, I.: 1986 (this volume)
Hubble, E., Sandage, A.: 1953, Astrophys. J. 118, 353
Humphreys, R. M., Davidson, K.: 1979, Astrophys. J. 232, 409
Humphreys, R. M., Blaha, C., D'Odorico, S., Gull, T. R., Benvenuti, P.: 1984, Astrophys. J. 278, 124
Leitherer, C. et al.: 1986 (this volume)
Leitherer, C., Appenzeller, I., Klare, G., Lamers,H.J.G.L.M. Stahl, O., Waters, L.B.F.M., Wolf, B.: 1985, Astron. Astrophys. (in preparation)
Maeder, A.: 1983, Astron. Astrophys. 120, 113
Sharov, A. S.: 1975, IAU Symposium 67, Variable Stars and Stellar Evolution, eds. V. E. Sherwood and L. Plant. D. Reidel, Dordrecht, p. 275
Stahl, O., Wolf, B.: 1982, Astron. Astrophys. 110, 272
Stahl, O., Wolf, B., Klare, G., Cassatella, A., Krautter, J., Persi, P., Ferrari-Toniolo, M.: 1983, Astron. Astrophys. 127, 49
Stahl, O., Wolf, B., de Groot, M., Leitherer, C.: 1985, Astron. Astrophys. Suppl. (in press)
Stahl, O., Wolf, B.: 1985, Astron. Astrophys. (in preparation)
Sterken, C., Wolf, B.: 1978, Astron. Astrophys. 70, 641
Thackeray, A. D.: 1974, Mon. Not. R. Astron. Soc. 168, 221
Walborn, N. R.: 1982, Astrophys. J. 256, 452
Waters, L. B. F. M.: 1984, Astron. Astrophys. (in press)
Wolf, B., Appenzeller, I., Cassatella, A.: 1980, Astron. Astrophys. 88, 15
Wolf, B., Stahl, O.: 1983, The Messenger No. 33, pg. 11
Wolf, B., Appenzeller, I., Stahl, O.: 1981, Astron. Astrophys. 103, 94

Discussion : WOLF.

MASSEY :

If Jay Gallagher were here, he'd point out that the distribution of H-S variables in M33 doesn't match that of WR stars, i.e., they are not found where they should be if they are very massive stars. Couldn't these objects be lower mass (30 M_o) binaries? What is their distribution in the LMC?

WOLF :

There is little evidence that the S Dor phenomenon is due to binarity. Due to the position of the S Dor variables in the HRD close to the Humphreys-de Jager limit we regard the evolutionary status outlined above as more plausible. The prototype S Dor is situated in the HII region N119. R127 is close (but clearly outside) to the giant HII region 30 Dor. R71 is a problem; it is not a member of any association.

WALBORN :

The distribution of typical Hubble-Sandage variables indicates that they are somewhat older than the youngest HII complexes. On the other hand, Eta Carinae may be considered an extreme Hubble-Sandage variable, and it is located in the Carina Nebula in association with O3 and WN-A stars. It has a mass probably between 100 and 200 Mo.

KUDRITZKI :

I think, the question where the HS-variables come from and what they are at the moment can be settled by looking at the progenitors. I promise to pay a high prize in the forms of beer or champagne to anybody who can give me calibrated spectra of R127 in the pre-HS stage, when it was still an Of-star.

WOLF :

We do not have calibrated spectra of R127 at quiescence. But there are several related stars, classified by Walborn (1977), with spectra very similar to R127 at quiescence. From one of these stars (Sk-67°266) we have CASPEC spectra of high quality. However it seems to me that the spectra of these stars are too peculiar for fine analytic abundance determinations. In this respect the more moderate S Dor variable R71 seems more promising. This object was also observed by us with CASPEC during a quiet phase and shows a rather normal early B-supergiant spectrum.

GRAHAM :

1) About <u>how</u> long is the period of accelerated mass loss activity in S Dor variables likely to last?
2) Is it possible that such stars are progenitors of massive WR stars which have ring nebulae around them?

WOLF :

1) The period of enhanced mass loss characterizing the S Dor-phase is supposed to be of the order of 10^3 to 10^4 years.
2) In fact it is suggested by us that the S Dor-variables are progenitors of massive WR stars. We found narrow (NII) 6548, 6583 lines (total width of zero intensity about 120 kms^{-1}) in the three LMC-S Dor variables indicative of ring nebula as observed around the galactic S Dor variable AG Car. In the case of R127 Stahl from a deep CCD image found clear evidence for extended structures presumably due to a nebulosity.

VIOTTI :

The observations seem to suggest at least two mechanisms which could be responsible for the large variability in these very luminous objects :

(1) changes in the structure of the atmosphere, like in the MC stars mentioned by B. Wolf and in AG Carinae (this galactic object is presently varying from 6 to 8^m at constant M_{Bol}), and
(2) dust formation, around stars, as in the case of Eta Car. This latter one could well be responsible for the large variability of the Hubble-Sandage variables in other galaxies, but one needs extreme IR monitoring of these faint stars to know more about these problems.

It is anyhow clear that in the theoretical HR diagram, most of these stars (maybe all) are moving right and left at constant bolometric luminosity. Since the temperature variations are short time changes, related to changes in the stellar structure, not to evolutionary effects, one should not use this parameter to describe the evolutionary stage of the Hubble-Sandage variables.

BOHANNAN :

In looking at the bright emission-line stars in the LMC, I find 9 stars that are spectroscopically similar to S Dor, 10 if you count R127. They are found in emission regions or associations at about the same frequency as other early-type supergiants. One can get a handle on the lifetime of the S Dor phase by allowing 2 to 3 times as many Wolf-Rayet stars. The S Dor-like stars I've identified in the LMC span nearly 5 magnitudes in luminosity.

P CYGNI TYPE STARS: EVOLUTION AND PHYSICAL PROCESSES

Henny J. G. L. M. Lamers[*]
Center for Astrophysics and Space Astronomy
University of Colorado
Boulder, Colorado 80309 USA

ABSTRACT. The class of P Cygni Type (PCT) stars is defined and the brightest members in our galaxy, LMC, M31 and M33 are identified. They are located near the upper luminosity limit in the HRD in the range of $8500 \lesssim T_{eff} \lesssim 27,000$ K and $-11.0 \lesssim M_{bol} \lesssim -7.8$. We suggest that all PCT stars are S Dor variables, and that the reverse may also be true. The basic parameters of the PCT stars are derived and compared with those of normal supergiants: the effective gravity is a factor 3 to 10 lower, the mass loss rate is a factor 3 to 10 higher, the terminal velocity is about a factor 10 lower. This results in a wind density which is a factor 30 higher and thus produces the P Cygni lines in the vi*ual spectrum. The history and the physical processes in the star P Cygni are discussed. The photometric variability and the shell ejections on a timescale of about a month are probably due to non-radial pulsations. The acceleration of the wind is due to radiation pressure by numerous ($\sim 10^3$) metal lines in the Balmer continuum. This can also explain the S Dor type of variability. The proposed mechanism will automatically lead to a P Cygni phase in the evolution of the most massive stars and to the existence of an upper limit in the HR diagram. The expected lifetime of the PCT phase is about 10^4 to 10^5 years. Some unresolved problems and recommendations for future research are discussed.

1. THE DEFINITION OF P CYGNI TYPE STARS

P Cygni type stars are stars which, in their visual spectrum, have lines with so-called P Cygni profiles, i.e. an emission component

---

[*]On leave from Space Research Laboratory and Sonnenborgh Observatory, Utrecht, The Netherlands.

which is either undisplaced or slightly red-shifted and a blue-shifted absorption component. They are called after their prototype: P Cygni (HD 193237). This star appeared in the sky in August 1600 as a nova and after a century of irregular brightness variations it has now settled down as a B1 Ia$^+$ hypergiant of V ≃ 4.9 magnitude.

Although most astronomers seem to agree which stars should be called "P Cygni-stars," there is no clear definition. In his classical study, Beals (1950) called them "stars with P Cygni profiles in the visual spectrum." This definition, however, would also include some Of stars, Be-stars during occasional phases and some B-supergiants which one would normally not call P Cygni stars. Other definitions or descriptions can be found, e.g. in de Jager (1980) and Underhill (1982). A clear but restricted definition is given by Underhill (1960): "Stars with a spectrum similar to P Cygni." This would limit the class to stars with spectral types early-B, whereas there might also be P Cygni type stars of types late-B or A.

The problem of the definition of the P Cygni type stars arises because the P Cygni <u>spectral characteristic</u>, namely the presence of P Cygni profiles in the visual spectrum, is due to a certain <u>astrophysical phenomenon</u> (i.e. a very large mass loss) which may not be restricted to a certain <u>class of objects</u>. For instance, if a Be-star would temporarily develop P Cygni profiles in its visual spectrum, would we call it a P Cygni star? Obviously not. On the other hand, the P Cygni characteristic is usually seen in the spectra of the most luminous B- and A-type supergiants, which do indeed form a certain class of objects. Therefore, I propose that we define the P Cygni type stars as follows:

"<u>P Cygni type stars are luminous supergiants (with $M_V \lesssim -7$) of spectral types O, B and A, which in their visual spectrum show or have shown P Cygni profiles (emission components with blue shifted absorption components) of not only the strongest Balmer and He I lines (Hα, Hβ, He I 6678, He I 5875) but also of other lines, such as lines of higher Balmer numbers and/or lines of other ions.</u>"

Since the phenomenon which produces the P Cygni type characteristic (severe mass loss) seems to be variable, a P Cygni type star may not show this characteristic at all times. The star HD 269006 is a good example. It showed the P Cygni type spectrum in 1973 during maximum brightness, but during minimum brightness in 1981 it had an absorption line spectrum in the visual with [Fe II] emission lines (Wolf et al., 1981b).

With the definition given above we have restricted ourselves to a group of stars in a certain part of the HR diagram which have similar characteristics. It is possible, but not proven, that these stars are in the same evolutionary stage. If that is the case, they will really constitute a specific class of stars.

The class of P Cygni stars is not an isolated class. There seems to be a general trend going from A and B supergiants, which may show P Cygni profile in Hα only, through the A and B hypergiants or superluminous supergiants of class Ia$^+$ or Ia-0, which show P Cygni profiles in Hα, Hβ, He I 6678 and He I 5875 (Underhill, 1982, p. 131), to the P Cygni stars.

P CYGNI TYPE STARS

I propose to indicate the class of P Cygni type stars by: PCT. (Notice that the sympol PC was used by van Genderen et al. (1983) to indicate that the star had P Cygni characteristics in their visual spectrum, irrespective of the number or type of lines. Some of their PC stars may not meet our definition of the PCT stars.)

2. THE RELATION BETWEEN P CYGNI TYPE STARS AND S DOR OR HUBBLE-SANDAGE VARIABLES

Using the definition of the PCT stars in the previous paragraph, we can identify the PCT stars in our galaxy, the LMC, M31 and M33. There are no known PCT stars in the SMC. They are listed in Table I. The PCT stars in the LMC were selected on the basis of the description of their spectru* in the classical paper of the brightest stars in the LMC and SMC by Feast et al. (1960). For several of the stars we listed a range in spectral types because most of them show drastic variations in brightness as well as in color or spectral type.

The reader will notice that several of these P Cygni type stars were also discussed by Wolf in his review of the S Doradus and Hubble-Sandage variables in this symposium. The reason is obvious: The S Dor or HS-variables are identified on the basis of their spectral

Table I. P Cygni type stars in the galaxy, LMC, M31 and M33.

System	HD	Name	Type	Class	Ref.
GAL	193237	P Cyg	B1 Ia$^+$	PCT	1
	94910	AG Car	B0 I-A1 I	PCT, S Dor	2
	90177	HR Car	B2 eq	PCT, S Dor	3
LMC	268835	R66	B7, Aeq	PCT	4
	269006	R71	B2.5 Iep-A1 Ieq	PCT, S Dor	5
	269128	R81	B2.5 eq	PCT	6
	35343	R88=S Dor	A2-5 eq	PCT, S Dor	7
	269859	R127	0Ia fpe	PCT, S Dor	8
M31		AE And	---	PCT, HS	9
		AF And	---	PCT, HS	9
		Var A-1	---	PCT, HS	9
M33		Var B	---	PCT, HS	9
		Var C	---	PCT, HS	9

References: 1 - de Groot (1969); 2 - Wolf and Stahl (1982); 3 - Bond and Landolt (1970); 4 - Stahl et al. (1983b); 5 - Wolf et al. (1981b); 6 - Wolf et al. (1981a); 7 - Wolf et al. (1980); 8 - Stahl et al. (1983a); 9 - Kenyon and Gallagher (1985).

type and variability. The PCT stars are identified on the basis of
their visual line spectrum. Clearly, many S Dor variables have a PCT
spectrum and many PCT stars show S Dor type variations. The interesting question is: are both groups identical?

Consider the stars in Table I. Only three of them, P Cygni, R66
and R81, are not classified as S Dor variables because they have not
shown the characteristic brightness variations during the last decades.
We know, however, that P Cygni had an outburst in AD 1600 (see Sec.
5.1) and on the basis of this variation P Cygni should be called an
S Dor variable. This leaves R66 and R81 as the only PCT stars which
are not classified as S Dor variables. We may speculate that these
stars are also S Dor variables but that their variations have not been
detected yet. This suggests that all P Cygni type stars may be S Dor
variables, which are the same as Hubble-Sandage variables (see Wolf's
review and Humphreys et al., 1984).

Is the reverse also true? To answer this question we have to
look at visual spectra of the S Dor or H-S variables. All the S Dor
or H-S variables listed in Table I are PCT stars. I am not aware of
other variables of this nature that have been observed at sufficient
resolution to check whether they are PCT stars. The only exception
is η Car, which has an emission line spectrum in its present phase.
We may conclude that all S Dor and Hubble-Sandage variables may show
PCT spectra during some phases, and therefore they are PCT stars. So
the classes of PCT stars and the S Dor or H-S variables are possibly
identical.

3. THE BASIC PARAMETERS OF THE P CYGNI TYPE STARS AND THEIR LOCATION IN THE HR DIAGRAM

The characteristics of the PCT stars are listed in Table II. Since
most of the stars are variable, I have selected the data in such a way
that they apply as much as possible to the phase when the star showed
the PCT characteristics. For instance, for R71 this is at maximum
brightness, whereas for R127, I have selected the phase just after the
onset of the brightening in Jan 1982. Since the time coverage of the
spectra is poor, this selection could not be applied strictly. The
phase is given at the end of the table, together with the references.
For the stars in M31 and M33 the characteristics were determined at
one epoch only. We do know if the stars were at minimum or maximum
brightness at that time. The temperatures for these stars were determined by means of a blackbody fit to the UV and visual energy distribution (Humphreys et al., 1984). The temperature of the star AG Car
is extremely uncertain. The value of T_{eff} = 9000 K was adopted by
Wolf and Stahl (1982) on the basis of the Hβ profile which is formed
in the wind and may give rise to a serious underestimate. Therefore,
M_{bol} of AG Car should be considered as a lower limit.

I have estimated the masses of the stars on the basis of their
luminosity by using the evolutionary tracks of Doom (1982), in which
mass loss and convective overshooting were taken into account. An
upper limit to the mass, M_u, can be derived from the relation between

Table II. The characteristics of the P Cygni type stars.

Star	M_V	M_{bol}	T_{eff} (K)	$R_\star$ ($R_\odot$)	M_u ($M_\odot$)	M_ℓ ($M_\odot$)	$M_\star$ ($M_\odot$)	log g (cm/s^2)	Γ	log g_{eff} (cm/s^2)	V_{esc} (km/s)	Phase	Ref.
P Cygni	-8.3	-9.9	19,300	76	45	35	40	2.29	0.48	2.01	330	quiescent	1,2
AG Car	-7.6	-7.8	9,000	130	--	--	16:	1.50	0.14	1.43	220	max	3
HR Car	-8.0	--	--	--	--	--	--	--	--	--	--		4,5
R66	-8.4	-8.9	12,000	125	23	19	21	1.57	0.37	1.36	200	quiescent	6
R71	-9.4	-10.6	10,000:	390	68	55	61	1.04	0.60	0.64	150	max	7,8
R81	-8.2	-10.0	20,000	75	48	37	42	2.31	0.50	2.01	330	quiescent	9
S Dor	-9.3	-10.6	8,500	540	68	55	61	0.76	0.60	0.36	130	max	10,14
R127	-8.7	-10.6	17,000	135	68	55	61	1.97	0.60	1.57	260	brightening	11
AF And	-8.8	-10.9	25,000:	72	81	67	74	2.60	0.65	2.14	370	unknown	12,13
AE And	-7.0	-8.3	15,000:	60	--	--	19:	2.16	0.23	2.05	310	unknown	12,13
Var A-1	-8.8	-11.0	27,000:	65	85	72	78	2.71	0.68	2.22	390	unknown	12,13
Var B	-8.4	-10.3	22,900:	65	57	44	50	2.51	0.56	2.15	360	unknown	12,13
Var C	-7.5	--	--	--	--	--	--	--	--	--	--	unknown	12,13

References: 1 - Lamers et al. (1983); 2 - Cassatella et al. (1979); 3 - Wolf and Stahl (1982); 4 - Bond and Landolt (1970); 5 - Viotti (1971); 6 - Stahl et al. (1983b); 7 - Wolf et al. (1981b); 8 - Thackeray (1974); 9 - Wolf et al. (1981a); 10 - Wolf et al. (1980); 11 - Stahl et al. (1983a); 12 - Humphreys et al. (1984); 13 - Kenyon and Gallagher (1985); 14 - Leitherer et al. (1985).

L and the mass at the end of the core-H burning. This assumes that the PCT stars are beyond the core-H burning phase. A lower limit to the mass, M_ℓ, can be derived by adopting the relation between L and the mass for which the surface composition of H has dropped to X = 0.40. If the mass were lower, X would be smaller and the star would be a WR star. I adopted the mean value M_* between this upper and lower limit. Since the difference between the upper and lower limit is small, this should give a reasonable guess of the actual stellar mass. For the stars AG Car and AE And, which are outside the range of initial masses calculated by Doom, I adopted a mass of 0.75 times the mass that a star with that L would have when it is at the end of main sequence according to the tracks by de Loore et al. (1978). The data in Table II also give the resulting values of log g, log g_{eff} and v_{esc}, where $g_{eff} = g(1-\Gamma)$ and $\Gamma = 2.66 \times 10^{-5} (M/M_\odot)(L/L_\odot)^{-1}$ is the correction due to radiation pressure by electron scattering. For this I assumed that H is ionized in the atmospheres of the PCT stars.

Figure 1 shows the HR diagram of the stars on the P Cygni phase. Different symbols are used to indicate whether the star is in a

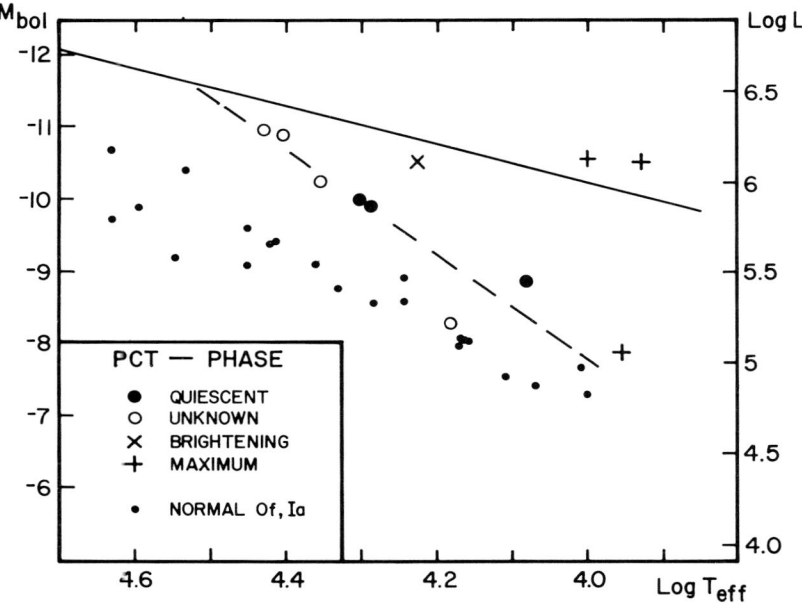

Fig. 1. The HR diagram of the PCT stars. Different symbols are used to indicate the phase in which stars showed their P Cygni type characteristics. The data for normal Ia supergiants are shown for comparison. The full line indicates the location of the Humphreys-Davidson luminosity upper limit. The broken line gives the mean relation for the quiescent phase of the PCT stars.

quiescent phase or at maximum brightness. For comparison I have also plotted the location of a number of well-studied bright supergiants of class Ia with types between O9 Ia and B9 Ia and the Of stars, from Lamers (1981). If we ignore the stars at maximum brightness, there is a suggestion of a trend between M_{bol} and T_{eff} (dashed line) for the PCT stars which goes approximately as

$$\log L \simeq 5.0 + 3 \log(T_{eff}/10^4) \quad . \tag{1}$$

This relation is very uncertain since it depends on a small sample of stars only, but the trend is suggestive, especially if we take into account the fact that the stars at maximum brightness have moved horizontally to the right in the HRD (see Wolf's review of S Dor variables). It is interesting to notice that this relation is almost parallel to the relation determined by the Of and Ia supergiants, except that it is about 1 to 1.5 magnitudes brighter. The star AE And in M31 is an exception as it is as bright as the normal Ia supergiants. It has the lowest extinction, A_v = 0.70, compared to $A_v \simeq$ 1.1 to 1.7 for the other stars in M31 studied by Humphreys et al. (1984).

We can compare relation (1) with the one shown by van Genderen et al. (1983) in their Figure 4 for LMC stars with P Cygni profiles (but not necessarily PCT stars!). Their mean relation through 15 stars goes as

$$\log L \simeq 5.6 + 1.7 \log(T_{eff}/10^4) \quad . \tag{2}$$

The difference in the slope between equations (1) and (2) might be due to the fact that our sample is too small or the sample used by van Genderen et al. may contain stars during their maximum.

In Figure 1, I also show the luminosity upper limit from Humphreys and Davidson (1979) (solid line). We see that the PCT stars are below this upper limit, and only those at maximum reach the limit. This suggests that the actual upper limit for stars during their quiescence or minimum (which is the limit to be compared with evolutionary calculations!) is lower and steeper than the one adopted by Humphreys and Davidson.

In Figure 2A I have plotted the adopted masses of the PCT stars and their effective gravities as a function of L. Since the derived masses depend only on the luminosity and not on T_{eff}, the mass is independent of the phase (maximum or minimum brightness) of the star because the light variations occur at almost constant L. The error bars indicate the upper and lower limits of the mass. The mean relation through the data is

$$\log(M/M_\odot) = 1.68 + 0.67 \log(L/10^6) \tag{3}$$

for $L \gtrsim 5.5$. The tightness of the relation shown in Fig. 2A is due to the fact that the masses were derived from evolutionary tracks by assuming a unique relation between M and L. The normal Ia supergiants (not shown in this graph) follow the relation of M_u, i.e. over the top of the error bars. Figure 2B shows the values of g_{eff} as a function

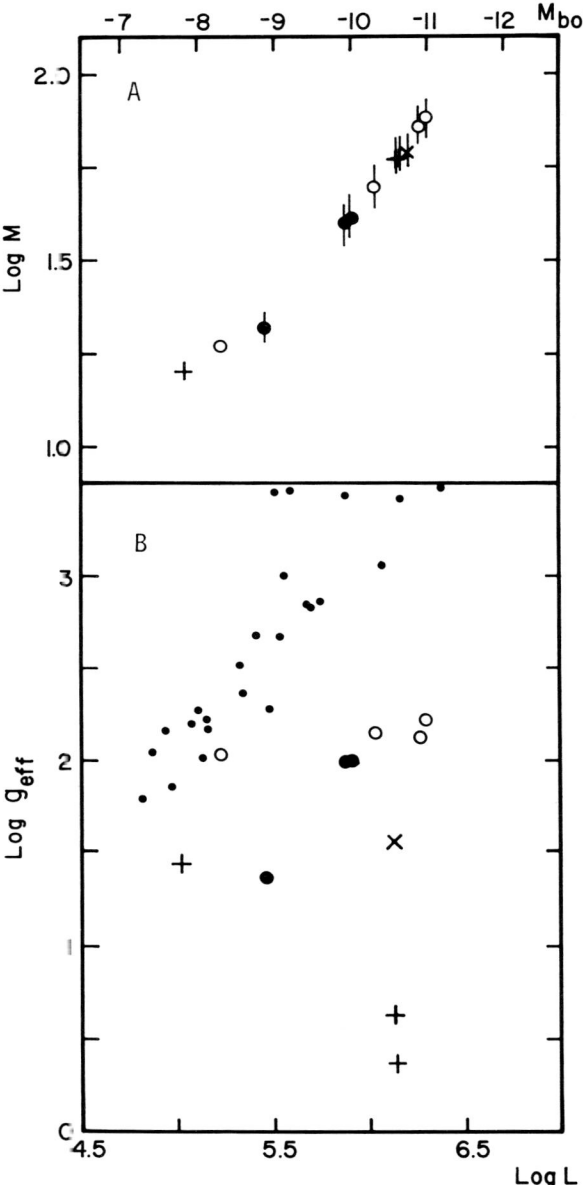

Fig. 2. A: The relation between the mass and the luminosity, derived from the evolutionary calculations. B: The effective gravities of the PCT stars, compared to those of Ia supergiants. The symbols are the same as in Fig. 1. The PCT stars during quiescence have a 5 to 10 times lower gravity than normal supergiants.

of luminosity for the PCT stars as well as for the Ia and Of supergiants of Figure 1. We see that g_{eff} of the PCT stars is about a factor 5 to 10 smaller than in Ia and Of stars. The two stars R71 and R88 which show a PCT spectrum during maximum have an even lower value of $\log g_{eff} \simeq 0.5$ during maximum.

It is likely that this low gravity plays an important role in the fact that the stars are PCT stars. One might speculate that the low gravity is in some way responsible for a high mass loss rate which in turn produces the PCT characteristic of the spectrum.

4. MASS LOSS FROM P CYGNI TYPE STARS

The PCT characteristic shows that the stars are suffering a high mass loss rate. The mass loss rates have been estimated for most of the known PCT stars. They are listed in Table III together with some of the relevant basic data. The phase and the references are the same as those of Table II. In Figure 3A the mass loss rate is plotted versus the luminosity and compared with that of the normal Ia and Of stars from Lamers (1981). We see that the mass loss rates of the PCT stars are a factor 3 to 10 higher than those of normal stars, even during the quiescent phase. Var A-1 is an exception, as it has a normal mass loss rate. In Figure 3B I have plotted the terminal velocities v_∞ as a function of T_{eff} for the PCT stars and the normal Ia and Of stars. The values of v_∞ for the normal stars are from the references given by Lamers (1981). We see that the terminal velocity of the PCT stars is

Table III. Mass loss from P Cygni type stars.

Star	$\log L$	$\log g_{eff}$	V_{esc} (km/s)	V_∞ (km/s)	$\dot{M}$ ($M_\odot$/yr)	V_∞/V_{esc}	$\log n_H$ (2 R_*)
P Cygni	5.86	2.01	330	300	2×10^{-5}	0.91	10.92
AG Car	5.02	1.43	220	166	3×10^{-5}	0.75	10.88
HR Car	--	--	--	--	--	--	--
R66	5.46	1.36	200	307	3×10^{-5}	1.54	10.65
R71	6.14	0.64	150	127	5×10^{-5}	0.85	10.27
R81	5.90	2.01	330	250	3×10^{-5}	0.76	11.18
S Dor	6.14	0.36	130	130	7×10^{-5}	1.00	10.12
R127	6.14	1.57	260	199	6×10^{-5}	0.77	11.07
AF And	6.26	2.14	370	350	5×10^{-5}	0.95	11.29
AE And	5.22	2.05	310	50	--	0.16	--
Var A-1	6.30	2.22	390	200	1.5×10^{-5}	0.52	11.10
Var B	6.02	2.15	360	350	--	0.97	--
Var C	--	--	--	200	--	--	--

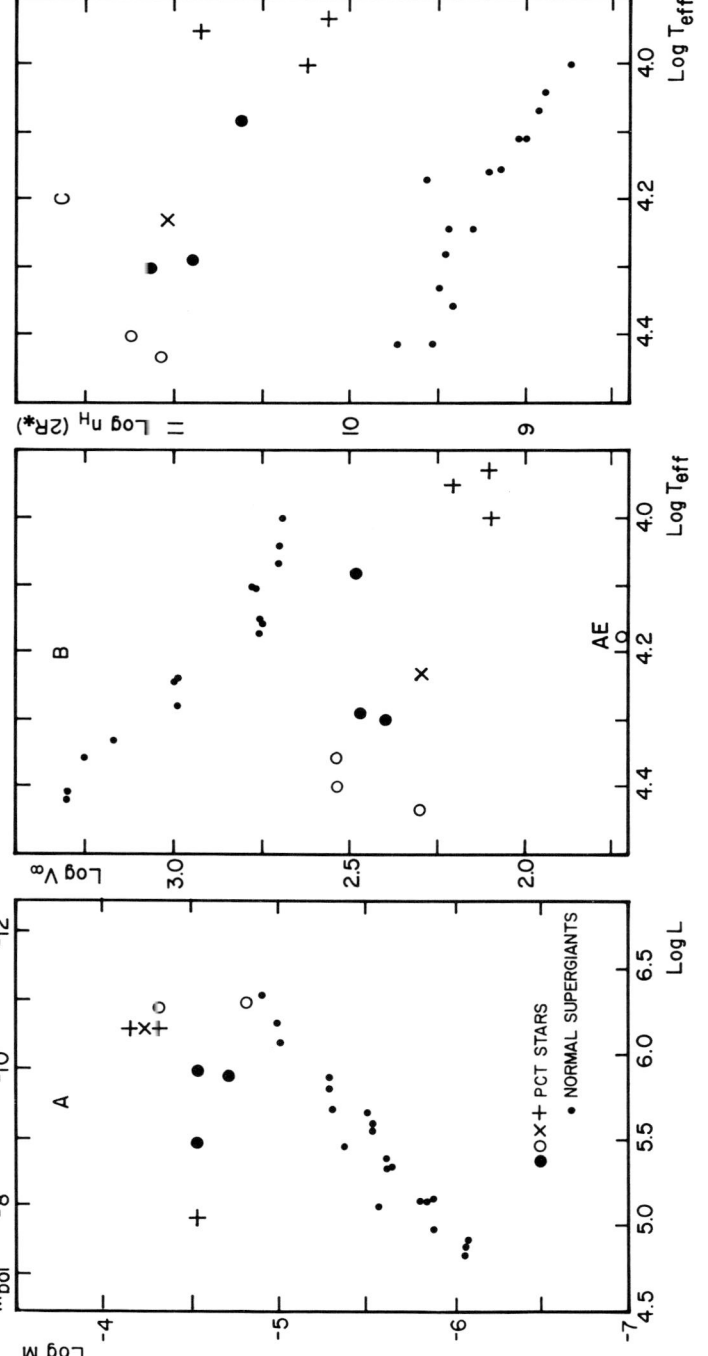

Fig. 3. The mass loss characteristics of the PCT stars are compared with those of normal supergiants. A: The mass loss rate is 3 to 10 times higher than in normal supergiants of the same luminosity. B: The terminal velocity of the winds of PCT stars is of the order of 100 to 300 km/s, and 3 to 10 times lower than in normal supergiants. C: The density of the wind at 2 R_* in PCT stars is 30 times higher than in normal supergiants.

of the order of 100 to 400 km/s, whereas the normal supergiants have a range of $v_\infty \simeq$ 500 km/s for $T_{eff} \simeq$ 10,000 K to 2000 km/s for 25,000 K. So the values of v_∞ for the PCT stars is about 0.1 to 0.5 times as small as that of normal supergiants. Even if we compare the ratio v_∞/v_{esc} of the PCT stars with those of the normal supergiants we find that the PCT stars have a lower velocity than the normal stars. The PCT stars have $v_\infty/v_{esc} \simeq$ 0.2 to 1.5, whereas the supergiants have $v_\infty/v_{esc} \simeq$ 3.5 at 25,000 K to 1.5 at 10,000 K. If we concentrate on the PCT stars in Figure 3B and ignore the discrepant value of AE And we find an indication of a trend which shows that v_∞ decreases with decreasing T_{eff}, similar to the case of the normal supergiants but about a factor 0.25 lower. Such a relation implies that the terminal velocity will decrease when the PCT star goes into a brighter state, since the temperature decreases with increasing visual brightness. The stars which are in their maximum/brightness state have indeed the lowest velocity in Figure 3B.

The high mass loss rate and the low velocity of the PCT stars imply that the density in the wind is much higher than in normal supergiants. The density in a stationary spherically symmetric flow is

$$\rho(r) = \dot{M}/4\pi\, r^2 v(r) \quad . \tag{4}$$

Not only is $\dot{M}$ higher and v_∞ smaller in PCT stars than in supergiants, but also the velocity law, i.e. the increase of $v(r)/v_\infty$ with r/R_*, is slower in the PCT stars. The velocity law in the wind of P Cygni was determined from an accurate analysis of the Balmer profiles by Van Blerkom (1978) and from the IR excess by Waters and Wesselius (1985). They found that the velocity increases about linearly with distance from 0.1 v_∞ at the photosphere to v_∞ at 15 R_*. A similar velocity law was derived for S Dor from the IR excess by Leitherer et al. (1985). This implies that at a distance of 2 R_* the velocity is only 0.16 v_∞. In a normal supergiant, the velocity law goes as $v_\infty[1-(R_*/r)]^\beta$ with $\beta \simeq 1$ (Castor and Lamers, 1979), so $v \simeq 0.5\ v_\infty$ at 2 R_*. In Table III I have listed the resulting densities at 2 R_* from the stellar center, assuming that all PCT stars have the same velocity law as P Cygni. The density is expressed in the hydrogen density, for which I assumed a ratio of $n_H/\rho = 4.42 \times 10^{23}$. The densities are plotted versus T_{eff} in Figure 3B. We see that the density in the PCT stars follows about the same trend of decreasing n_H with decreasing T_{eff} as the supergiants, but that the density in the winds of the PCT stars is about a factor 30 higher! <u>It is this high density which is responsible for the PCT spectrum.</u>

We conclude that the presence of the P Cygni profiles in the spectrum of the PCT stars is due to high density in the wind. This high density is a result of three effects:

    a.   The mass loss rate is about 3 to 10 times higher than in normal supergiants (Fig. 3A).

    b.   The terminal velocity of the wind is 0.1 to 0.5 times smaller than in normal supergiants (Fig. 3B).

    c.   The velocity in the wind increases much more slowly than in normal supergiants.

The key question is: is this only due to high luminosity and the low effective gravity of the PCT stars (Figs. 1 and 2)? In order to answer that question we will consider one star in detail.

5. THE STAR P CYGNI: A CASE STUDY

5.1. The History of P Cygni

The star P Cygni was discovered on August 18 in 1600 by the famous Dutch chartmaker Blaeu when it suddenly brightened as a nova to the third magnitude. Between 1606 and 1626 it faded until it became invisible. Then in 1654 it brightened again to third magnitude and it remained that bright until 1659. From 1660 to 1683 it was faint and strongly variable with occasional drops to invisibility. The observers in the seventeenth century noticed that P Cygni was a "red" star. Between 1683 and 1780 the star gradually brightened to magnitude 5.2. Between 1781 and 1786 P Cygni seems to have been 0.4 magnitude fainter than before, but after that it increased in brightness again between 1786 to 1870 to its present magnitude of 4.9 in the visual. Since that time, the brightness is about constant, apart from some typical irregular variability of about $0^m.2$. (For the light history of P Cygni see Zinner, 1952; de Groot, 1969; van Gent and Lamers, 1985.) The history is sketched schematically in Figure 4.

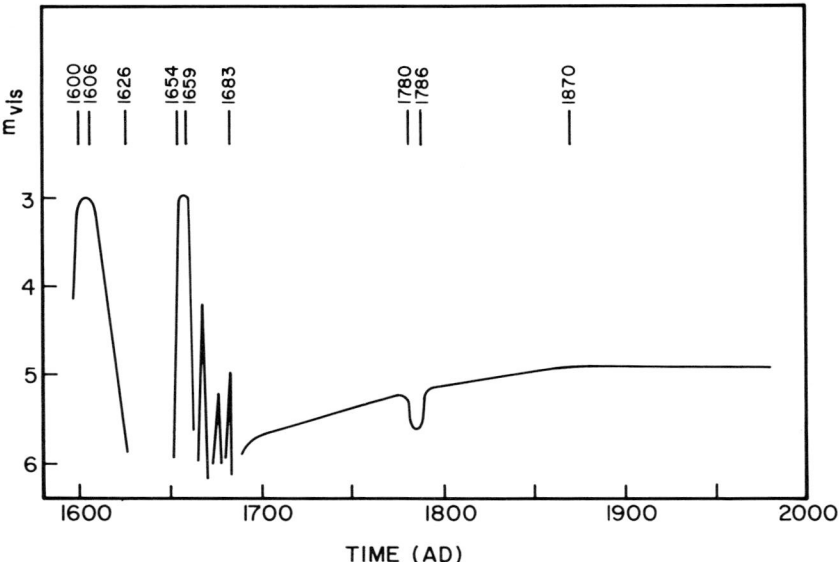

Fig. 4. A sketch of the visual light history of P Cygni.

It is very likely that the sudden visual brightness changes of P Cygni were due to mass ejection at almost constant luminosity, as is the case for the other S Dor variables. The irregular fadings to below the limit of visibility might be due to the formation of dust during the activity in the seventeenth century. The red color which was observed when the star was bright corroborates this.

Did the outbursts in the seventeenth century leave any remnant? Wendker (1982) has found an arc-like structure around P Cygni at 6 cm with the "head" of the arc pointing in the direction of the proper motion of P Cygni. The diameter of the arc is about 1.2 arc min which corresponds to a diameter of 0.62 pc at $d = 1.8$ kpc. If this is a result of the ejection of AD 1600, the ejection velocity must have been 800 km/s which is very high for the ejection velocity of S Dor outbursts. Their velocities are typically of the order of 50 to 200 km/s. If the ejection velocity is between 50 and 200 km/s, the outburst which might be responsible for the radio arc, should have occurred 1500 to 6000 years ago.

Very recently, van Gent (1985, private communications) has found a letter written by a French astronomer Boulliau to Huygens in 1661 in which he mentions that he has seen P Cygni through the telescope of Hevelius and that P Cygni looks similar to the Orion nebula! If this observation is correct, it might imply that a remnant of the ejection was actually observed about 60 years after the outburst. A drawing of the Orion nebula as seen by Huygens in 1694 through a telescope of 13.4 m focal length (comparable to the one of Hevelius) shows a size of 130" × 230". If a nebulosity around P Cygni had a similar size of, say, 100" in 1661, the expansion velocity would have been 7000 km/s. This velocity is about a factor 10 higher than the one derived from the distance of the present radio arc. This might indicate that the velocity of expansion has slowed down considerably. A historical verification of the observations of P Cygni made in Huygens' time would be extremely interesting!

5.2. Photometric Variations of P Cygni

Apart from the great activity in the seventeenth century, P Cygni is presently an irregular variable with an amplitude of $\Delta V \simeq 0.2$ mag. The photometric periodicities quoted in the literature range from 0.5 days to 18 years! The existence of a true periodicity or the determination of a timescale if the variations are quasi-periodic can give insight into the structure of the star. In a recent paper van Gent and Lamers (1985) have reviewed all the evidence for periodicities in the brightness variations, the radial velocity variations and the polarimetric variations. They conclude that the photometric variations are not periodic but occur on a timescale, τ, between 12 days and 125 days, with a mean characteristic timescale of about 25 days. There is some evidence that the characteristic timescale varies with time; sometimes the Fourier periodograms show peaks near $\tau \simeq 12$ days, at other times the peaks occur near 30 days. The polarimetric observations by Hayes (1985) during 1978/1979 also show evidence of a periodicity with $\tau \simeq 12$ days. We can compare this with the expected timescale for non-radial pulsations:

$$P = Q(\bar{\rho}/\bar{\rho}_\odot)^{-0.5} \tag{5}$$

where the pulsational constant Q is expressed in days. Adopting the parameters of P Cygni from Table II, we find $(\bar{\rho}/\rho_\odot)^{-0.5} \simeq 105$. The empirical value of Q can be derived from the study of Maeder (1980) who found $Q \simeq 0.16$ for B1 to B3 supergiants. If we take into account the fact that $Q \sim L^{0.27}$ and that P Cygni is about 1.5 mag brighter than normal supergiants, we expect $Q \simeq 0.23$ and $P \simeq 24$ days. The good agreement between the observed characteristic timescale and the one predicted for non-radial pulsations suggests that the photometric variations of P Cygni are due to non-radial pulsations, similar to those in other supergiants.

5.3. Variations in the Mass Loss and Shell Ejections

In addition to the quiescent mass loss rate of $\dot{M} = 2 \times 10^{-5}$ $M_\odot$/yr derived from the IR and radio excess and from the Balmer profiles, P Cygni seems to eject shells on timescales of months. This shell-ejection can most easily be observed in the Balmer lines, which show variable absorption components. The first extensive analysis of these shells was made by de Groot (1969) who found three absorption components near -110, -160 and -215 km/s. He suggested that the last component was showing periodic radial velocity variations with a semi-amplitude of 30 km/s and a period of 114 days. Luud et al. (1975) argued that the first component (-110 km/s) was also variable with a period of 57 days, i.e. half the period found by de Groot. A re-analysis of these data by van Gent and Lamers (1985) showed that both periods are spurious.

A very extensive study of the Balmer lines by Markova and Kolka (1984) during 1981 has shown that the absorption components are indeed variable, but not in the way described by de Groot and Luud. They find that at any epoch four absorption components are present in the velocity range of -100 to -250 km/s. (No components are found closer to the line center because of confusion with the core of the photospheric profile.) Each component travels through this velocity range from -100 to about -220 km/s with a mean acceleration of the order of 0.5 km s^{-1}/day or 0.6 cm/s^2. The shells are ejected at a rate of about six shells per year (see Fig. 5).

If we assume that the shells leave the star with an initial velocity of 30 km/s (which is the velocity at the base of the wind of P Cygni) and that the acceleration is constant and 0.6 cm/s^2, we find that they reach their terminal velocity of 220 km/s about 400 days after they were ejected. During that time the shells have reached a distance of $r_{shell} \simeq 90$ R_*. So these shells can be observed in the Balmer components up to a distance of about 10^2 R_*.

Similar shell components have been observed in the UV spectrum of P Cygni by Lamers et al. (1985). They found that the lines of Fe II, Ni II, Cr II etc. have a stationary absorption component at

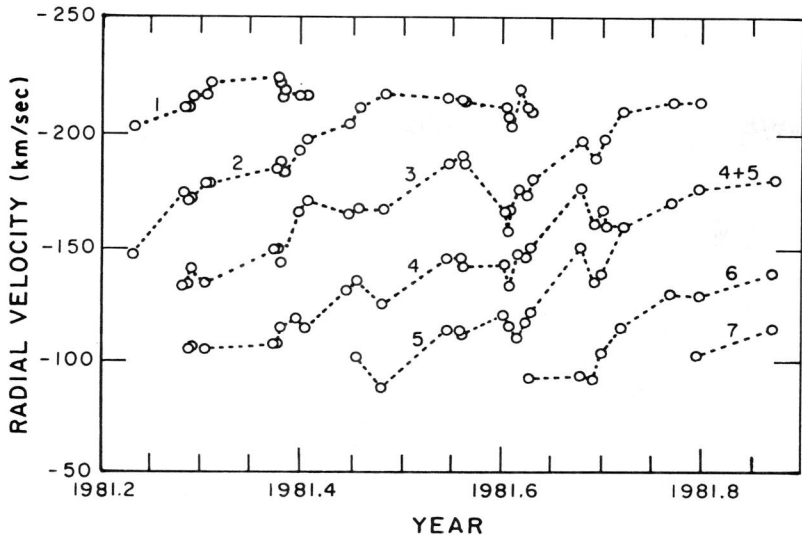

Fig. 5. The velocities of the shell components, observed in the Balmer lines of P Cygni by Markova and Kolka (1984). The average rate of shell ejection is six per year. The mean acceleration of the shells is 0.6 cm/s^2.

v = −206 km/s, which is constant in depth and velocity over a period of at least five years. This component could be due to a massive shell at a distance of at least 1100 R_* with a mass $M_{sh} > 6 \times 10^{-5}$ $M_\odot$. Alternatively, the component at −206 km/s might be due to material in the quiescent wind of P Cygni at a very large distance r ≳ 3000 R_* which has recombined from Fe III to Fe II. In addition to this stable component, variable components have been observed in the lines of once and twice ionized metals. These variable components increase in radial velocity from −100 to −200 km/s with a mean acceleration of 0.17 cm/s^2. Such variable Fe II and Fe III components are ejected about once per year with a mass of the order of 10^{-5} $M_\odot$.

The shells observed in the UV and in the Balmer lines show the same characteristic of a velocity increase to 200 km/s. The difference is in the recurrence of the shells (about six per year for the Balmer lines and one per year in the metal lines) and the acceleration (0.6 cm/s^2 for the Balmer lines and 0.17 cm/s^2 for the metal lines). I would argue that this is due to the amount of mass in the shell. It is likely that P Cygni ejects shells continuously with different masses. The less massive shells, which can be detected in the Balmer lines only, are accelerated faster than the less frequently ejected, but more massive shells, which can be observed in the metal lines. Until now, no simultaneous study of the shells in Balmer and metal lines has been performed.

Independent evidence for the ejection of massive shells by
P Cygni comes from the variation in the radio flux. Van den Oord
et al. (1985) observed a sudden drop in the radio flux of P Cygni by
a factor 4 within one month. This timescale is much faster than the
timescale in which the extended radio emitting region ($\Delta r \simeq 300\ R_*$)
can change its density. The timescale for changes in the radio flux
due to density changes in the emitting region is $\tau \simeq \Delta r/v \simeq 300\ R_*/$
200 km/s $\simeq$ 2.5 yr. So the observed much faster drop in the radio flux
can only be explained by a sudden recombination of the radio emitting
region. The recombination timescale in the region where the radio flux
is emitted is $\tau_{rec} \simeq (\alpha \cdot n_e)^{-1} \simeq 1$ month (van den Oord et al., 1985).
Why would the wind of P Cygni suddenly recombine? This can only be due
to the ejection of a thick shell which temporarily blocks the ionizing
flux from the star. I expect that such an ejection would also result
in a temporary change in B-V with a duration of about a week.

It is important to monitor photometrically P Cygni with a small
telescope with a timescale of days in order to determine the time of
the ejections of the shells which are later observed in the Balmer and
UV lines or in the radio flux.

The recurrence of the shell ejections observed in the Balmer
lines with a characteristic timescale of about 2 months is about twice
as long as the timescale of 1 month for the photometric variations.
If the star also ejects more shells of lower mass, which may be undetected, the ejection timescale may be shorter and about the same as
the photometric timescale, which is probably due to nonradial pulsations. This would suggest that the ejection of shells by P Cygni may
be due to nonradial pulsations.

5.4. The Acceleration of the Wind of P Cygni and the Origin of its Mass Loss

The velocity law in the quiescent wind of P Cygni has been determined
from Balmer profiles and the IR excess by Van Blerkom (1978) and
Waters and Wesselius (1985). They find that the velocity law is
linear and goes as

$$v(r) = 0.10\ v_\infty + 0.057\ v_\infty \{(r/R_*) - 1\} \quad \text{for} \quad r \leq 15\ R_* \quad (6)$$

with $v_\infty \simeq 300$ km/s. From this velocity law we can derive the net outward acceleration which turns out to be only 1 to 3 cm/s^2 at $1 < r <$
4 R_*. If this is compared with the acceleration due to the gravity,
which is $100(R_*/r)^2$ cm/s^2 (Table II), we come to the surprising conclusion that the acceleration in the wind of P Cygni must be due to
some force which barely overcomes the gravity and which varies approximately as $(r/R_*)^{-2}$. Lamers (1985) has studied the various processes which might produce such a force. If the force were due to
turbulent pressure, which was proposed by de Jager (1984) to explain
the mass loss of the hypergiants, the turbulent velocity has to be of
the order of 100 km/s. Such a high turbulent velocity is unlikely, as
it would exceed the sound velocity by a factor 10. On the other hand,
the acceleration can be produced by radiation pressure due to a large

number (10^3) of weak metal lines in the Balmer continuum. These lines
have actually been observed in the IUE spectra of P Cygni (Cassatella
et al., 1979; Luud and Sapar, 1980). They produce a blocking of about
40% of the continuum flux in the Balmer continuum.

This strongly suggests that the mass loss from P Cygni is due to
radiation pressure by a large number of optically thin lines in the
Balmer continuum. Such a mechanism will be effective when a luminous
star evolves away from the main sequence and reaches a value of $T_{eff} \lesssim$
25,000 K, where most of its energy is emitted in the Balmer continuum
and the many lines from once or twice ionized metals become effective
absorbers. Since the effective gravity decreases with increasing
radius as the star evolves away from the main sequence, the mass loss
rate is expected to increase very drastically when a very luminous
star passes the limit of $T_{eff} \lesssim$ 25,000 K.

5.5. A Summary of the Processes in P Cygni

P Cygni has suffered large outbursts in the seventeenth century.
After that time the star has gradually stabilized and is now a B1 Ia$^+$
PCT star. The star shows irregular brightness variations with $\Delta V \simeq$
0.2 mag. These seem to occur on different timescales between 12 and
125 days with a mean value of about 25 days. This suggests that the
star is nonradially pulsating. In addition to the steady mass loss,
P Cygni ejects shells at intervals of about 2 months or less. The
most massive shells are observed as absorption components in many UV
metal lines, the less massive shells are observed in the Balmer lines
only. It is possible that the shell ejection is related to the photometric variability and thus to the nonradial pulsations. There is a
correlation between the mass of the shells and their acceleration:
the most massive shells (observed in the UV) have the lowest acceleration, the less massive shells (observed in the Balmer lines) have a
higher acceleration. The quiescent wind is accelerated faster than
the shells, but considerably slower than the less dense winds of
normal supergiants.

The acceleration of the wind can be explained by radiation pressure
due to metal lines in the Balmer continuum. It is likely that this same
mechanism is responsible for the high mass loss rate in PCT stars.

6. THE NATURE AND ORIGIN OF THE P CYGNI TYPE PHENOMENON

After having discussed the general properties of the PCT stars and the
physical processes in one of them we may combine this information and
try to form a coherent picture.

The presence of the P Cygni profiles in the visual spectrum implies that the high mass loss rate and a rather low wind velocity. In
comparison with normal supergiants of type Ia the mass loss rate of
the PCT stars about 3 to 10 times higher for their luminosity, the
wind velocity is about 0.10 times lower, and the density at 2 R_* is 30
times higher (Fig. 3). This high density is responsible for the occurrence of the P Cygni lines in the visual spectrum.

Which mechanism produces the high mass loss rate and the low velocity, which together give rise to the high density? I suspect that this is due to the combination of two effects: the low gravity and the radiation pressure due to metal lines in the Balmer continuum. If a massive star evolves beyond the hydrogen core-burning phase it will expand. The surface gravity will decrease as R_*^{-2}, but since the luminosity remains about constant, the effective gravity decreases faster than R_*^{-2}. This by itself would produce an <u>extended atmosphere</u>. Other effects, such as turbulent pressure (proposed by de Jager, 1984) or rotation might make the atmosphere even more extended, although this may not be a necessary condition. At the same time the expansion of the star will result in a decrease of T_{eff}. When T_{eff} reaches a value below about 25,000 K, the radiation pressure will increase due to the fact that the many metal lines in the Balmer continuum (where most of the stellar flux is emitted) become efficient absorbers. This will result in an increase in the mass loss rate and a decrease of the wind velocity, since $\dot{M} v_\infty$ is approximately proportional to L/c and L remains constant. The net effect is a drastic increase in the density of the wind. I expect that this will develop into a new equilibrium for a stationary wind, where the mass loss and the acceleration are both due to radiation pressure in the Balmer continuum. The data in Table II suggest that this mechanism will work for stars which have $g_{eff} \lesssim 10^2$ cm/s^2 and $L \gtrsim 10^5 L_\odot$.

This mechanism will be operating only in the luminous stars. If $L \sim M^2 \sim R^2 T_{eff}^4$ and the stars evolve at about constant L, the gravity of a star will vary as $g \sim T_{eff}^4/M$ and the correction due to the electron scattering will vary as $1-\Gamma$ with $\Gamma \sim L/M \sim M$. So for a given value of T_{eff} the effective gravity decreases faster than M^{-1}.

The mass loss mechanism which I have proposed here will result in a high but steady mass loss, as observed, e.g., in the quiescent phase of P Cygni. However, it cannot be stable over a long timescale. As the star keeps expanding, its g_{eff} will rapidly go to zero and may even become negative. This would result in a sudden ejection of its complete atmosphere. Maeder (1983) and Humphreys and Davidson (1984) have proposed that such a sudden mass loss at the instability limit may result in the ejection of a considerable amount of mass, so that the star "recoils" from this limit and quietly evolves into this limit again. This might explain the multiple outbursts of the PCT stars.

If the recoil interpretation is correct, we would expect that after a more violent outburst (larger amount of mass ejected) the star will remain quiet for a longer time. The long period of quiescence of P Cygni (apart from the recurrent shell ejections) after the violent outbursts in the seventeenth century may be indicative of such a correlation.

It is interesting to realize that the mechanism for the mass loss from P Cygni stars, i.e. radiation pressure due to metal lines, proposed by Lamers (1985), is the same as proposed independently by Appenzeller (these proceedings) to explain the variability in the S Dor stars.

7. THE EVOLUTION OF THE P CYGNI TYPE STARS

The PCT stars are luminous stars, with $-7.8 \lesssim M_{bol} \lesssim -11.0$. The luminosities of the two faintest stars in Table II are very uncertain and may have been underestimated. For AG Car the temperature was probably underestimated and for AE And the extinction is uncertain and lower than that of the other PCT stars in M31. If we omit these two stars, we find $-8.9 \lesssim M_{bol} \lesssim -11.0$ for the PCT stars. This lower limit agrees very well with the lower limit of $M_{bol} = -8.9$ for the emission line stars in the LMC (van Genderen et al., 1983).

The masses of the PCT stars are estimated to be in the range of $21 < M < 78 \ M_\odot$, if we ignore the two lowest luminosity stars. These are present masses. These masses were derived from the evolutionary tracks of Doom (1982). The initial masses of the PCT stars are related to the adopted masses as

$$M_{initial} \simeq 1.32 \ M_{adopted}$$

so the initial masses of the PCT stars are $28 \lesssim M_{init} \lesssim 100 \ M_\odot$.

The PCT stars are clearly in their post-main-sequence phase. The most compelling evidence of this is the indication that several of these objects have an overabundance of N and an underabundance of C and O. Examples are P Cygni (de Groot, 1969), AG Car (Viotti et al., 1984). S Dor (Leitherer et al., 1985) and R127 which has also been classified at one phase as WN9-10 (Walborn, 1982). Such an abundance pattern is expected to occur when the products of the CNO-cycle of hydrogen burning reach the stellar surface. This implies that the PCT stars are intermediate between the main-sequence and the Wolf-Rayet phase.

Since the PCT stars are near the Humphreys-Davidson upper limit for the luminous stars, they must be located in he HRD at the region where the post-main-sequence evolutionary tracks stop their redward motion and return to the left to become Wolf-Rayet stars (Humphreys and Davidson, 1979; Maeder, 1983). A massive star can become a WR star if it has got rid of 30% of its initial mass (Doom, 1982). During the hydrogen core burning phase the star may lose about 15% cent of its initial mass (Lamers, 1981) so about 15% has to be lost during the PCT phase.

The duration of the PCT phase can be estimated as the time needed to lose 15% of the initial mass

$$\tau_{PCT} \simeq 0.15 \ M_{init} / \langle \dot{M} \rangle_{PCT} \ . \tag{7}$$

What is the average mass loss rate in the PCT phase? Let us take the star P Cygni as an example. It loses mass on three different timescales, given below.

Process	Mass Loss ($M_\odot$)	Recurrence (yr)	dM/dt ($M_\odot$/yr)
Violent Outbursts	$10^{-3}-10^{-1}$	10^2-10^3	$10^{-6}-10^{-3}$
Shell Ejections	$10^{-6}-10^{-5}$	0.2	$5 \times 10^{-6} - 5 \times 10^{-5}$
Quiescent Mass Loss	---	---	2×10^{-5}

$$\langle \dot{M} \rangle = 3 \times 10^{-5} - 10^{-3}$$

These values are very uncertain, especially those of the violent outbursts. Taking a mean of 10^{-3} to 10^{-4} $M_\odot$/yr and M_{init} = 53 $M_\odot$ we find $\tau_{PCT} \simeq 10^4-10^5$ yr. This is about 0.3 to 3% of the main-sequence lifetime.

8. UNRESOLVED PROBLEMS AND FUTURE WORK

The most important problem to be solved is the exact location of the PCT stars in the HR diagram (accurate T_{eff} and L determinations). The data in Figure 1 and in van Genderen et al. (1983) suggest that they are near the Humphreys-Davidson limit. However, they do not seem to be the only stars in that region (see e.g. the preliminary study by Bohannan in these proceedings). What is the difference between the PCT stars and the more normal super- or hypergiants in the same region of the HR diagram? Are the hypergiants, such as ζ^1Sco and Cyg OB2 Nr. 12 with $M_{bol} \leq -10$, also PCT stars which are presently in a long quiescent phase or is there an evolutionary difference, in the sense that the PCT stars are more evolved? Since the luminous stars may spend about 10^4 to 10^5 yr in the vicinity of the instability limit, the region near this limit may contain "freshmen," which are just entering the PCT phase and show little or no PCT characteristics, and "veterans" showing well-developed PCT characteristics (because of their lower g_{eff}) and the scars of their mass loss (increased N-abundance at the surface).

Abundance studies of PCT stars and their ejecta, such as the ring nebula around AG Car (Thackeray, 1974; Viotti et al., 1984), are needed to give critical tests of the evolution theory. Is the CNO abundance in PCT stars different from those of other stars in the same region of the HRD? This would indicate that the PCT stars are more evolved.

How much mass does a PCT star or S Dor star lose during its outbursts? This is important for estimating the lifetimes of the PCT stars. In addition we should estimate the lifetime on the basis of stellar statistics. The LMC seems to be the most suitable system for this.

What is the lower limit for the luminosity or mass of the PCT stars? If the PCT stars occur at $M_{bol} \geq -9.5$ (the upper limit for the red supergiants), their outbursts are not sufficient to prevent the star from becoming a red supergiant, unless these PCT stars are different from other stars at that luminosity. Rotation might play an important role in this evolution (Sreenivasan and Wilson, 1982). What is the minimum mass loss rate in the post main-sequence phase which prevents a star from becoming a red supergiant? Are the observed mass loss rates for stars with $M_{init} \geq 50\ M_\odot$ sufficiently large?

What is the reason for the high mass loss rate and the instability near the Humphreys-Davidson limit? Turbulent pressure was suggested by de Jager (1984) and radiation pressure on metal lines in the Balmer continuum was proposed to explain the quiescent high mass loss (Lamers, 1985) and the S Dor type outbursts (Appenzeller, these proceedings). Stothers and Chin (1983) have investigated various other mechanisms.

What is the reason for the shell-ejections in P Cygni and how much mass is ejected? Is this related to nonradial pulsations of the star? To answer these questions a spectroscopic and photometric monitoring of a few bright PCT stars on a timescale of a few days is urgently needed.

ACKNOWLEDGMENTS

This research was partially supported by the National Science Foundation under a grant to Dr. P. S. Conti. The author acknowledges the staff of JILA for their hospitality during the summer of 1985 and Mrs. L. Volsky for efficient help in the preparation of the manuscript.

REFERENCES

Beals, C.S.: 1950, Publ. Dom. Astrophys. Obs. Victoria **9**, 1.
Bond, H.E., Landolt, A.U.: 1970, Publ. Astron. Soc. Pac. **82**, 313.
Cassatella, A., Beeckmans, F., Benvenuti, P., Clavel, J., Heck, A., Lamers, H.J.G.L.M., Macchetto, F., Penston, M., Salvelli, P.L., Stickland, D.: 1979, Astron. Astrophys. **79**, 223.
Castor, J.I., Lamers, H.J.G.L.M.: 1979, Astrophys. J. Suppl. **39**, 481.
de Groot, M.: 1969, Bull. Astron. Inst. Netherlands **20**, 235.
de Jager, C.: 1980, The Brightest Stars (Dordrecht: Reidel).
de Jager, C.: 1984, Astron. Astrophys. **138**, 246.
de Loore, C., de Greve, J.P., Vanbeveren, D.: 1978, Astron. Astrophys. Suppl. **34**, 363.
Doom, C.: 1982, Astron. Astrophys. **116**, 303.
Feast, M.W., Thackeray, A.D., Wesselink, A.J.: 1960, Mon. Not. R. Astron. Soc. **121**, 337.
Hayes, D.P.: 1985, Astrophys. J. **289**, 726.
Humphreys, R.M., Blaha, C., D'Odorico, S., Gull, T.R., Benvenuti, P.: 1984, Astrophys. J. **278**, 124.
Humphreys, R.M., Davidson, K.: 1979, Astrophys. J. **232**, 409.
Humphreys, R.M., Davidson, K.: 1984, Science **223**, 243.
Kenyon, S.J., Gallagher, J.S.: 1985, Astrophys. J. **290**, 542.

Lamers, H.J.G.L.M.: 1981, Astrophys. J. **245**, 593.
Lamers, H.J.G.L.M.: 1985, Astron. Astrophys. (in press).
Lamers, H.J.G.L.M., de Groot, M., Cassatella, A.: 1983, Astron. Astrophys. **128**, 299.
Lamers, H.J.G.L.M., Korevaar, P., Cassatella, A.: 1985, Astron. Astrophys. (in press).
Leitherer, C., Appenzeller, I., Klare, G., Stahl, O., Waters, L.B.F.M., Wolf, B.: 1985, Astron. Astrophys. (in press).
Luud, L., Gollandsky, O., Yarygina, T.: 1975, Publ. Tartu Astrofiz. Obs. **43**, 250.
Luud, L., Sapar, A.: 1980, Tartu Observatory Preprint A-1.
Maeder, A.: 1980, Astron. Astrophys. **90**, 311.
Maeder, A.: 1983, Astron. Astrophys. **120**, 113.
Markova, N., Kolka, I.: 1984, Astrofiz. **20**, 465.
Sreenivasan, S.E., Wilson, W.J.F.: 1982, Astrophys. J. **254**, 287.
Stahl, O., Wolf, B., Klare, G., Cassatella, A., Krautter, J., Persi, P., Ferrari-Tonioli, M.: 1983a, Astron. Astrophys. **127**, 49.
Stahl, O., Wolf, B., Zickgraf, F.J., Bastian, U., de Groot, M.J.H., Leitherer, C.: 1983b, Astron. Astrophys. **120**, 287.
Stothers, R., Chin, C.W.: 1983, Astrophys. J. **264**, 583.
Thackeray, A.D.: 1974, Mon. Not. R. Astron. Soc. **168**, 221.
Underhill, A.B.: 1960, in Stellar Atmospheres, ed. J.L. Greenstein (Chicago: Univ. of Chicago Press), p. 423.
Underhill, A.B.: 1982, in B-Stars With and Without Emission Lines, eds. A. B. Underhill and V. Doazan, NASA SP-456.
Van Blerkom, D.: 1978, Astrophys. J. **221**, 186.
van den Oord, G.H.J., Waters, L.B.F.M., Lamers, H.J.G.L.M., Abbott, D.C., Bieging, J.H., Churchwell, E.: 1985, Proceedings of Workshop on Radio Stars, Boulder
van Genderen, A.M., Groot, M., Thé, P.S.: 1983, Astron. Astrophys. **117**, 53.
van Gent, R.H.: 1985, private communication.
van Gent, R.H., Lamers, H.J.G.L.M.: 1985, Astron. Astrophys. (in press).
Viotti, R.: 1971, Publ. Astron. Soc. Pacific **83**, 170.
Viotti, R., Altamore, A., Barylak, M., Cassatella, A., Gilmozzi, R., Rossi, C.: 1984, in Future of Ultraviolet Astronomy Based on Six Years of IUE Research, NASA Conf. Publ. 2349, p. 231.
Walborn, N.R.: 1982, Astrophys. J. **256**, 452.
Waters, L.B.F.M., Wesselius, P.P.: 1985, Astron. Astrophys. (in press).
Wendker, H.J.: 1982, Astron. Astrophys. **116**, L5.
Wolf, B., Appenzeller, I., Cassatella, A.: 1980, Astron. Astrophys. **88**, 15.
Wolf, B., Appenzeller, I., Stahl, O.: 1981b, Astron. Astrophys. **103**, 94.
Wolf, B., Stahl, O.: 1982, Astron. Astrophys. **112**, 111.
Wolf, B., Stahl, O., de Groot, M.J.H., Sterken, C.: 1981a, Astron. Astrophys. **99**, 351.
Zinner, E.: 1952, Kleine Veröff. Remeis-Sternw Bamberg **1**, nr. 7.

Discussion : LAMERS.

DE JAGER :
In your talk you ascribed the cause of the wind of P Cyg to radiation pressure by the many UV absorption lines. I would rather believe that these lines, visible in the wind spectrum, cause the acceleration, as you actually suggested yourself. I believe there are strong arguments that the cause of the wind is seated in the photosphere and related to shock-driven mass loss initiated by the fact that $g_{eff} < 0$. As shown in Nieuwenhuijzen's and my contribution to these Proceedings : the assumption $g_{eff} = 0$ allows one to calculate the necessary mechanical flux; this appears first to be sufficient to cause photospheric supersonic microturbulence, and secondly, if one assumes that the energy of the mechanical flux is transformed into stellar wind energy one finds the observed rate of mass loss. Radiation acceleration by lines, on the other hand, may not work in the photosphere, which is optically thick in the continuum.

LAMERS :
It is true that my analysis of P Cygni only gives information about the acceleration in the wind. However, if the multitude of faint lines in the Balmer continuum act as a "pseudo-continuum in the wind", they might do the same in the photosphere. In that case the radiation pressure will produce a low effective gravity in the photosphere ($g_{eff} \geq 0$). This, together with the larger radiation pressure in the layers above the photosphere (because the degree of ionization is decreasing outwards) will initiate and accelerate the mass loss.
If the turbulent pressure gradient is large, this may certainly help the initiation of the wind, but whether turbulent pressure alone is responsible for the initiatiion of the wind remains to be seen.

WOLF :
You said P Cygni stars are B or A supergiants. The only "A star" in your list, however, is R66. By fitting the continuum from the satellite UV to the IR, however, we could show that this star is of type B as well. The A type spectrum is the equivalent spectral type of the cooler envelope (see also the contribution by Zickgraf et al., this volume).

LAMERS :
So it may be that the P Cygni stars are always of spectral type B. In that case, most of them, or possibly all of them, are of spectral type early-B. If this is true, it suggests that the mechanism which produces the large mass loss rates, characteristic for the P Cygni stars, is effective in a small temperature regime around $T_{eff} \sim 20,000$ K. The proposed mechanism of pseudo-continuum radiation pressure might explain this.

MELNICK :

Your interpretation of the Humphreys-Davidson-de Jager limit predicts that very low metallicity galaxies should contain no P Cygni S Dor or in general, Hubble Sandage variables. Is that the case for the SMC?

LAMERS :

There are only three known HS variables in the LMC and none in the SMC.

VIOTTI :

FeII is indeed the most important ion for line opacity in the UV of P Cyg and in Hubble-Sandage variables in other galaxies. The contribution is however difficult to evaluate because most lines are saturated and weak. High excitation energy lines might give an important contribution. One should try to compute line formation in stellar winds, and derive synthetic spectra. We indeed find a strong UV opacity from FeII in many galactic and MC stars. Concerning the light history of P Cygni, if you assume for the 1600 maximum a bolometric correction near zero - i.e. a cooler blackbody distribution - you get nearly the same M_{bol} as at the present times. This again pushes in favour of a variability at constant bolometric luminosity over a period of 300 years, without "outburst" events.

LAMERS :

I agree with your suggestion that the luminosity of P Cyg may have been the same during the AD 1600 mass ejection as it is now. The early observers of the seventeenth century call P Cygni a "red" star.

As far as the radiation pressure due to FeII lines is concerned, I have estimated this effect in P Cygni by measuring the observed lines. I agree that a theoretical study of this nature is needed, but difficult.

STALIO :

How reliable is the estimate of g, and what could be that effect of 10 - 20% error in g on your model of the mass loss mechanism?

LAMERS :

The gravity is indeed uncertain, since we have adopted the mass of P Cygni based on evolutionary tracks. This gives an uncertainty of about a factor two in mass and gravity. But even with this uncertainty, the effective gravity of P Cygni is about a factor 10 to 30 smaller than of normal B supergiants. Whether the uncertainty of a factor two will make the proposed origin of the mass ejection impossible is hard to say

KONTIZAS :

How could we explain the high number of narrow UV components in P Cygni stars compared to the Be stars?

LAMERS :

I think that this is due to the small acceleration in the P Cygni stars. If a B supergiant would eject a shell every two months, each one would have become invisible because of its large distance to the star. In a P Cygni star, however, the velocity and the acceleration is so small that the shells travel a much smaller distance in 2 months. So one will be able to see more of the previous shells when new ones are ejected in P Cygni. In addition to this, the shells ejected by P Cygni may be more massive than those in other B supergiants, but this is not very well known at present.

KUDRITZKI :

You know that in my talk I claimed that the improved radiation wind theory can explain the mass loss rate and v_{inf} of P Cygni. What is your opinion of this?

LAMERS :

It is true that a reduction of the gravity increases the mass loss rate and decreases v_{inf} in the framework of the radiation driven wind theory of Castor et al. (1975) and Abbott (1982). However, in P Cygni another effect comes into view : if the mass loss rate is large, the ionization in the wind decreases to FeII, NiII etc, which can absorb very efficiently in the Balmer continuum. For this reason, I think that the effect of lowering the gravity is even more severe than in your calculations. You predicted v_{inf} = 600 km/s, whereas the observed value is between 200 and 300 km/s.

MAEDER :

There is some empirical relation between the characteristic time of supergiant pulsation, luminosity and colour (P-L-C relation). Do you have anything similar for the P Cygni stars? I especially think of the timescale between two successive shell ejections.

DE GROOT :

I want to make two remarks. The first one relates to Maeder's question. Taking a 114-day time scale, I found that this fits the P-L-C relation for early-type supergiants quite well. Lamers has just reduced this time scale to 60 days (six shells per year). In view of the uncertainties involved, this is still a reasonable agreement.

My second remark concerns the time scale of the P Cygni phenomenon. The P Cygni phase depends on the presence of hydrogen in the star's atmosphere. For a P Cygni-type star this is between 20 and 60 $M_\odot$. Since mass loss including major outbursts every 100 to 1000 years, is about 10^{-3} $M_\odot$/yr, the P Cygni phase lasts between 2×10^4 and 6×10^4 years. Furthermore, in the LMC the P Cygni stars number about 20% of all B-type supergiants. Thus, a B-type supergiant will spend between 10^5 and 3×10^5 years off the main-sequence. This figure agrees reasonably well with Maeder's (1983) evolutionary calculations.

MOFFAT :

Since about half of all stars are in binaries and the fact that no P Cyg star shows strict periodic variations could lead one to conclude that P Cyg stars are preferentially single stars.

LAMERS :

I agree.

APPENZELLER :

I would like to comment on the range of luminosities at which P Cyg stars seems to be observed. Regardless of whether the mass loss is driven by turbulent pressure or radiation pressure, the acceleration will be influenced by the chemical composition. Hence, the mass loss rate should depend on L, Teff, and Z, which perhaps explains the scatter of the location of P Cyg stars in the HR diagram.

LAMERS :

I agree with you. However, what worries me is the fact that Bohannan et al. (these proceedings) from spectroscopic observations find a large scatter in L of the LMC P Cygni stars, whereas van Genderen (1983) from Walraven photometric observations found that the P Cygni stars in the LMC show a clear relation between L and Teff, and follow the Humphreys-Davidson limit quite nicely.

BOHANNAN :

Henny, I would not lose any sleep over the difference between my H-R diagram for the LMC P Cygni-like stars and that of van Genderen. The difference probably lies in making the transformation between the domain of observation and of theory, the calibration of temperature and bolometric correction with spectral type. Until we do profile analyses to measure Teff and log g, there will be significant uncertainty in locating these stars in the M_{bol}-Teff diagram. The uncertainty is much larger if the temperature is derived from photometry.

SESSION 4.

OTHER LUMINOUS STARS WITH EMISSION LINES.

Chairman : V. NIEMELA.

1. N.R.WALBORN: Optical and UV Spectral Morphology of Luminous OB Stars in the Galaxy and the Magellanic Clouds.

2. P.S.CONTI: Parameters of Wolf-Rayet Stars.

3. P.MASSEY: Wolf-Rayet Stars in Nearby Galaxies.

OPTICAL AND ULTRAVIOLET SPECTRAL MORPHOLOGY OF LUMINOUS OB STARS
IN THE GALAXY AND THE MAGELLANIC CLOUDS

Nolan R. Walborn[*]
Space Telescope Science Institute
3700 San Martin Drive
Baltimore, MD 21218, U.S.A.

ABSTRACT

Three areas of current progress relevant to the theme of this Symposium will be discussed. (1) New spectroscopic observations of the 30 Doradus central cluster, obtained independently by the author and by J. Melnick, confirm the presence of numerous very early O-type members, including several of type O3. In combination with sophisticated new direct imagery of the luminous central object R136 by A. Walker and by G. Weigelt, these results have evident implications for understanding the ionization of the supergiant H II region, as well as for the interpretation of R136 itself and of the apparently similar regions seen in more distant galaxies. In particular, no evidence remains for a supermassive object in 30 Doradus, but its central cluster is revealed as a spectacular grouping of very massive hot stars. (2) A further member of the Ofpe/WN9 category in the LMC has been identified, bringing their number to seven, with no exact spectroscopic counterparts yet known in the Galaxy. One of these objects is currently in a state of outburst and has been interpreted by O. Stahl et al. as the hottest known Hubble-Sandage variable. (3) An extensive survey of IUE high-resolution data has revealed a strong correlation between the ultraviolet stellar wind features and the optical spectral classifications for the majority of normal O stars. These results are relevant to future studies with the High Resolution Spectrograph on the Hubble Space Telescope, which may observe restricted UV wavelength ranges in faint extragalactic OB stars lacking optical data of comparable quality.

I.   THE STELLAR CONTENT OF 30 DORADUS

There has been substantial progress toward an understanding of the 30 Doradus cluster and its luminous central object R136 since this

---

[*]Visiting astronomer, Cerro Tololo Inter-American Observatory, which is supported by the National Science Foundation under contract No. AST 78-27879.

subject was reviewed at IAU Symposium 108 (Walborn 1984). Indeed, it is not too strong to say that the situation has developed from one of predominantly indirect, hypothetical, and plausibility arguments at that time, to one dominated by direct evidence now. This evidence shows conclusively that R136 is a complex multiple system, which constitutes the core of a spectacular cluster rich in the most massive hot stars. Important new data have been contributed by Chu, Cassinelli, and Wolfire (1984), Chu and Daod (1984), and Moffat, Seggewiss, and Shara (1985). Qualitatively new information has been provided by two other approaches: sophisticated direct imagery of R136 itself by Walker and O'Donoghue (1984) and by Weigelt and Baier (1985); and detailed spectroscopy of the cluster stars surrounding R136 by Melnick (1985) and by this author (to be presented here).

Walker and O'Donoghue (1984) have analyzed CCD images of R136 by maximum entropy methods, achieving a resolution of $0''.4$ and isolating 22 components with V magnitudes between 11.4 and 15.7 within $4''.6$ of R136a_1, including the $0''.5$ component of a as well as b and c. Their outermost faint components are visible in the best direct photographs, such as Figure 3 below. Weigelt and Baier (1985), using the remarkable technique of holographic speckle interferometry, have determined the structure within R136a at a resolution of $0''.09$, discovering 8 components within $0''.70$. Two of these components are only 0.3 mag fainter than a_1 in the red (note that Weigelt and Baier have reversed the definition of a_2 and a_3 with respect to usage in the previous literature), three are 2 mag fainter, and two more are 2-3 mag fainter. On the assumptions that the visual magnitude differences are the same, that V = 10.8 for all of R136a (Schmidt-Kaler and Feitzinger 1981), that A_V = 1.2 mag (Panagia, Tanzi, and Tarenghi 1983; Savage et al. 1983; Fitzpatrick and Savage 1984), and that $V_o - M_V$ = 18.6, it follows that M_V = -7.7 for R136a_1, -7.4 for a_2 and a_3, -5.7 for a_4-a_6, and -5.2 for a_7-a_8. R136b has M_V = -5.2 and c has -5.7. It will be shown below that the same range of values is found for stars in the surrounding cluster. Thus we now have 28 resolved components in R136 within a radius of $4''.6$, all of which are included in the IUE large aperture as well as in the original optical photometry which was interpreted as due to a single object.

Melnick (1985) has performed an extensive investigation of the spectral types of the 30 Doradus cluster stars surrounding R136. Independently, I have observed about 40 of them with the CTIO 4-meter SIT vidicon system at 1.5 Å resolution, during four nights in Dec. 1984/Jan. 1985. These studies have revealed the most spectacular clustering of massive hot stars for which such detailed information is available. As already indicated by the preliminary results of Melnick (1983), it contains several examples of what may be considered a new class of hot stars, whose spectra show the strong absorption lines of He II and N V and the narrow N IV λ4058 emission characteristic of type O3 If*, along with the broad He II λ4686 emission typical of WN-A spectra. I have classified such spectra as intermediate O3 If*/ WN-A. The first example to be described was Sanduleak -67°22, also in

OPTICAL AND ULTRAVIOLET SPECTRAL MORPHOLOGY OF LUMINOUS OB STARS

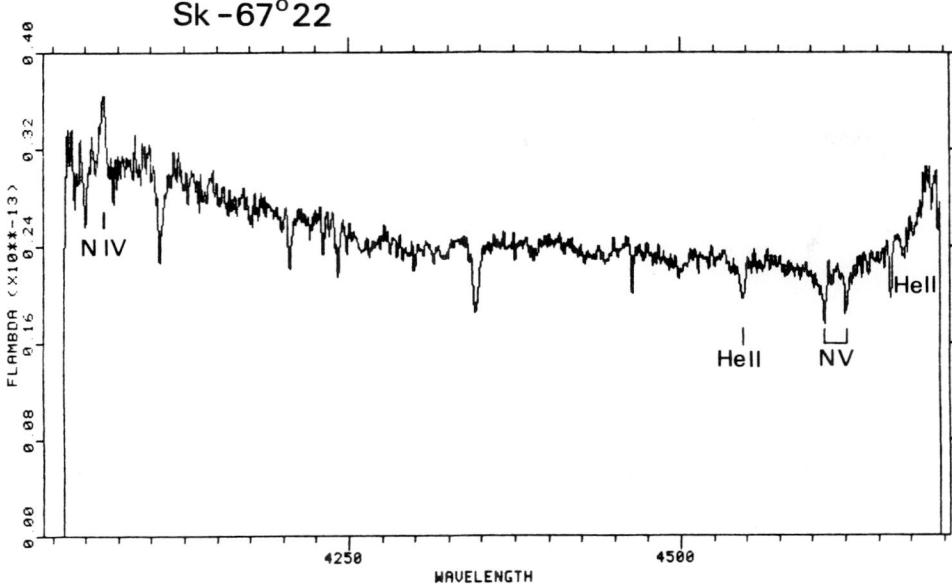

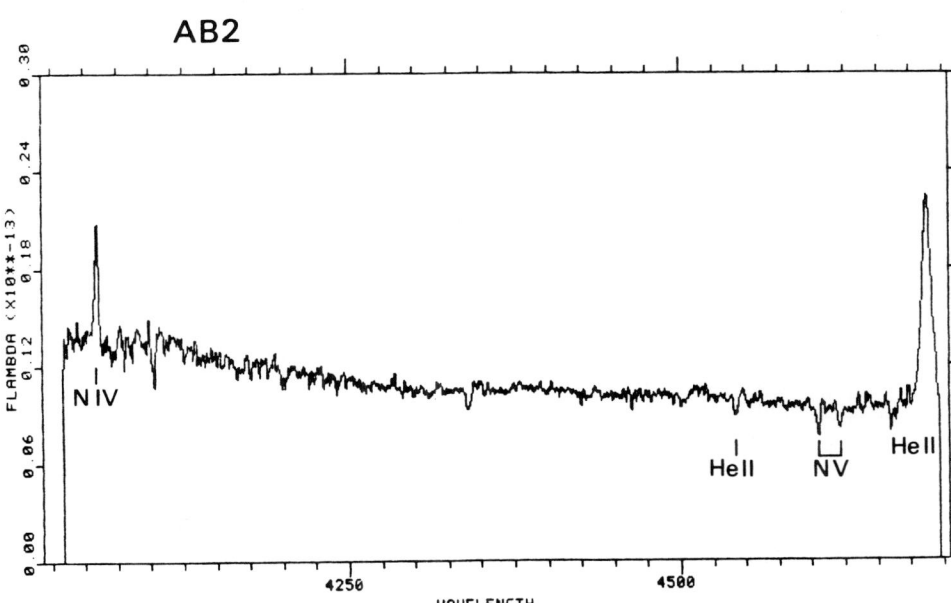

Fig. 1 — Vidicon spectrograms of the O3If*/WN6-A stars Sk-67°22 (M_v = -5.4) in the LMC and AB2 (M_v = -5.2) in the SMC. The flux scales are correct as given. The spectral features identified are N IV $\lambda 4058$ emission, He II $\lambda 4541$ and N V $\lambda\lambda 4604$-4620 absorption, and He II $\lambda 4686$ emission.

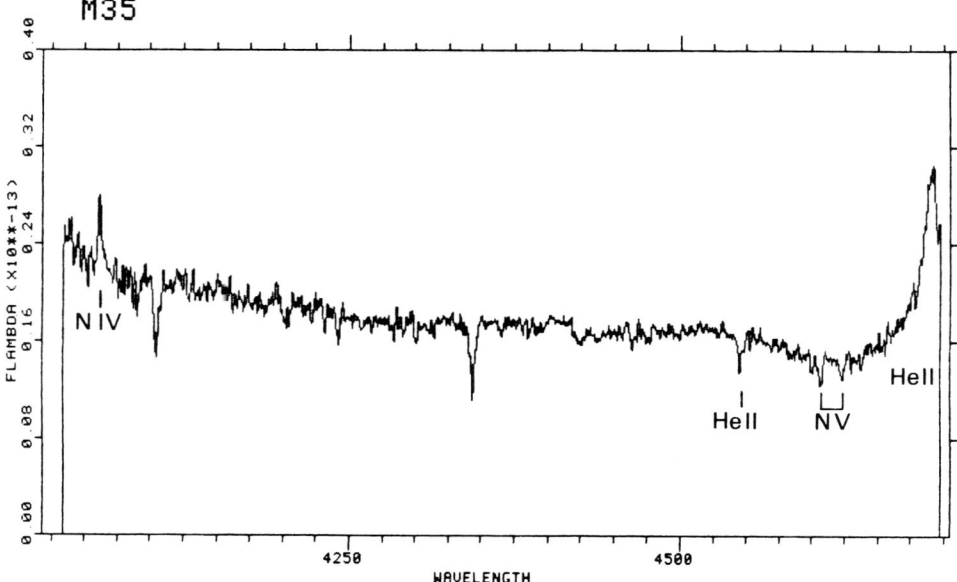

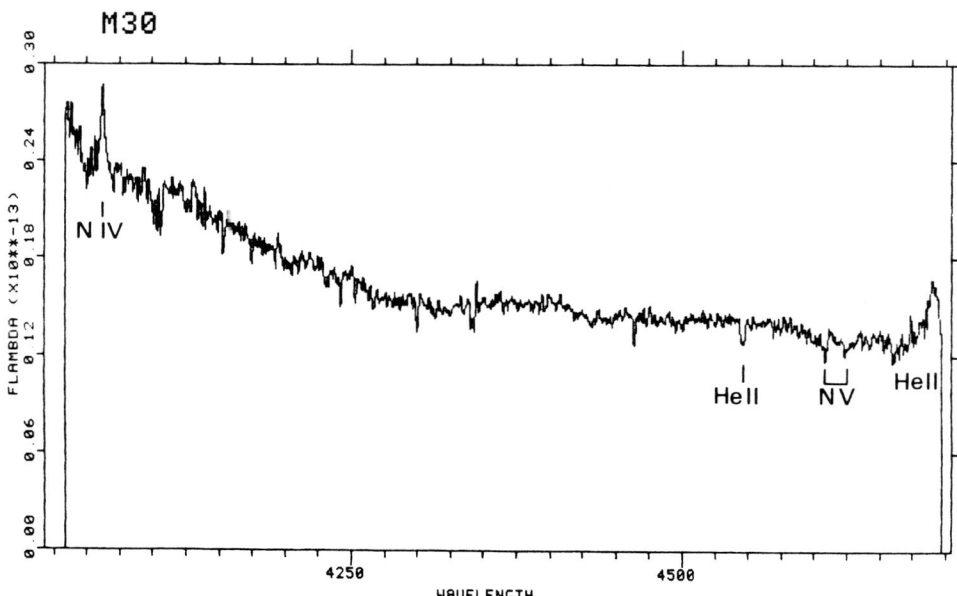

Fig. 2 - Vidicon spectrograms of the O3If*/WN6-A stars Melnick 30 and 35 in 30 Doradus. The flux scale for M30 should be brighter by a factor of 1.9 at 4330 Å and that for M35 by 1.45. The spectral features identified are as in Figure 1.

the LMC (Walborn 1982a); recently another has been found in the SMC, namely AB2 (Azzopardi and Breysacher 1979a; Garmany and Massey 1984). The vidicon spectrograms of these two stars are shown in Figure 1, and those of the similar stars Melnick 30 and 35 in 30 Doradus are shown in Figure 2.

Table 1 gives spectral classifications for 56 stars in the 30 Doradus cluster exclusive of R136, from either my results (W) or Melnick's (M); there is excellent agreement in detail between the two studies. Many of my classifications are preliminary, because the intended tailored nebular emission-line subtractions have not yet been accomplished; in these cases I have given spectral-type ranges which are unlikely to be exceeded. The table contains 15 definite or potential O3 stars, whereas only 12 members of this class were previously known (the 10 discussed by Walborn 1982a plus R136a and AB2). None of the spectral types earlier than O6 was known before these investigations, confirming the suggestion of Melnick (1983) and Walborn (1984) that the principal sources of ionization in 30 Doradus other than R136 had not yet been identified. Table 1 also lists blue magnitudes derived from the vidicon fluxes; these have had to be corrected, with reference to the photographic photometry of Westerlund (private communication), for an improper positioning of a guide probe into the telescope beam affecting three of the four nights, and they are accurate to a few tenths of a magnitude. Accurate photometry of the 30 Doradus cluster is in preparation by Melnick and by Chu and Seitzer. The absolute visual magnitudes in Table 1 are approximate

Table 1. The Stellar Content of 30 Doradus (Exclusive of R136)

Melnick/ Radcliffe	Spectral Type	Source	m_B	M_V	Remarks
30	O3 If*/WN6-A	W	13.3	-6.6	
35	O3 If*/WN6-A	W	13.4	-6.5	
39	O3 If*/WN6-A	W	12.6	-7.3	
42	O3 If*/WN6-A	W	12.6	-7.3	
51	O3 If*/WN7-A	W	14.0	-5.9	Moffat, Seggewiss, and Shara (1985)
13	O3-4	W	13.9	-6.0	
23	O3-4 (f?)	W	14.3	-5.6	
24	O3-4 V	W	14.0	-5.9	
25	O3-4 V	W	12.9	-7.0	
26	O3-4 V	W	13.8	-6.1	
36	O3-4 V	W	14.2	-5.7	
10	O3-6 V	W	14.2	-5.7	
14	O3-6 V	W	14.5	-5.4	
15S	O3-6 V	W	14.3	-5.6	
55	O3-6 V	W	14.5	-5.4	

Table 1. The Stellar Content of 30 Doradus (continued)

Melnick/Radcliffe	Spectral Type	Source	m_B	M_V	Remarks
28	O4 V	M			
47	O4-6 (f)	W	13.9	-6.0	
4	O5-6 V:	W	14.4	-5.5	
8	O5-6	W	14.5	-5.4	
35S	O5-6 V	W	13.8	-6.1	
35N	O5-7:	W	14.0	-5.9	
48	O5-7	W	14.4	-5.5	
R139	O6-7 Iaf	W	12.1	-7.8	
7	O(7) V	M			
15	O7 V	W	13.5	-6.4	
21	O7 V	M	14.2	-5.7	
R133	O7-8 II	W	12.6	-7.3	
32	O8 II	W	13.1	-6.8	
6	O8-9	W	14.6	-5.3	
58	O8-9	W	13.9	-6.0	
59	O8-9	W	14.4	-5.5	
1S	O9-9.5	W	14.7	-5.2	90" south of Melnick 1
38	O9.7 Iab	W	13.6	-6.3	Could be OC
14N	O V	M			
22AB	O	M	13.8	-6.1	
33	WC5+O4	M			
R140N	WC5+WN4	M			
34	WN4.5	M			
37	WN7-A	W	12.7	-7.2	Strong, symmetrical Hγ absorption
49	WN7	M			
53	WN8	M			Azzopardi and Breysacher (1979b)
R134	WN7	M			
R135	WN6	M			
R140S	WN4.5	M			
R145	WN6	M			
27	B0 I	W	13.5	-6.4	
50	B0: I:	W	14.3	-5.6	
11	B0-0.5 Ia	W	13.2	-6.7	
12	B0-0.5 Ia	W	12.4	-7.5	
54	B0.5 Ia	W	13.0	-6.9	
R141	B0.5 I	W	12.4	-7.5	
R142	B0.5-0.7 I	W	11.8	-8.1	
R137	B0.7-1.5 I	W	12.0	-7.9	
52	B1 Ia	W	13.8	-6.1	
5	B2: Ip?	W	14.3	-5.6	Hγ P Cyg profiles?
R138	A0 I	W			

and have been derived with the assumptions of B - V = 0.1, A_V = 1.2, and $V_o - M_V$ = 18.6 for all stars. The absolute magnitudes of Melnick 39 and 42 are essentially the same as those of $R136a_1-a_3$.

The distribution of the (abbreviated) spectral types is shown on photographs of the 30 Doradus cluster in Figure 3 for types O3-O6 and WR, and in Figure 4 for the later types. Note that four of the O3 If*/WN-A objects immediately surround R136. On the other hand, the later-type stars show rather less of a concentration toward R136. Whether the several late O and early B supergiants present belong to the 30 Dor cluster or are related to the general field in which the nebula is situated will have to be determined by comparably detailed spectroscopy of the latter.

The brightest unresolved component in R136, namely a_1, is about 1 mag brighter in M_V than HD 93129A; following the discussion of Walborn (1984), the upper limit to its mass is therefore 250 $M_\odot$. The entire 30 Doradus cluster, including R136, may contain 15-20 stars with masses of 100-200 $M_\odot$, and there is no evidence for the presence of substantially greater masses than these. The nebular ionization balance has been recalculated with the new spectral types by Melnick (1985), with the result that R136a contributes about one-third of the ionization and the other cluster stars provide the remainder.

As a footnote to this discussion, I would like to mention the interesting VLA study of the optically obscured, galactic supergiant H II region W49 by Dreher et al. (1984). They have discovered a very luminous, multiple stellar system therein, which may well represent an earlier evolutionary state of those in 30 Doradus and NGC 3603 (Walborn 1973), since it is associated with masers and compact H II regions.

II. A NEW Ofpe/WN9 STAR IN THE LMC

A large number of OB supergiants with P-Cygni and other emission-line characteristics have been known in the LMC since the pioneering work of Henize (1956) and Feast, Thackeray, and Wesselink (1960). Recently a considerable wealth of information about them at both optical and ultraviolet wavelengths has become available through the work of Shore and Sanduleak (1984) and Stahl et al. (1985 and references therein; see also the review by B. Wolf in this volume). I have been interested in the hottest subgroup of these objects, which display peculiar Of-like spectra (Walborn 1977, 1982b). They may be further subdivided into three categories: (1) HDE 269858 and Sk-67°266, which have exceptionally narrow Si IV λ4089, N III λ4097 absorption lines and expanding nebular shells; (2) HDE 269227, HDE 269927C, and Bohannan-Epps 381, which appear most strongly related to the narrow-line WN-A sequence; and (3) HDE 269445, a unique superluminous object with a nearly pure emission-line spectrum in the blue-violet. HDE 269858 (=R127) is currently in a remarkable state of

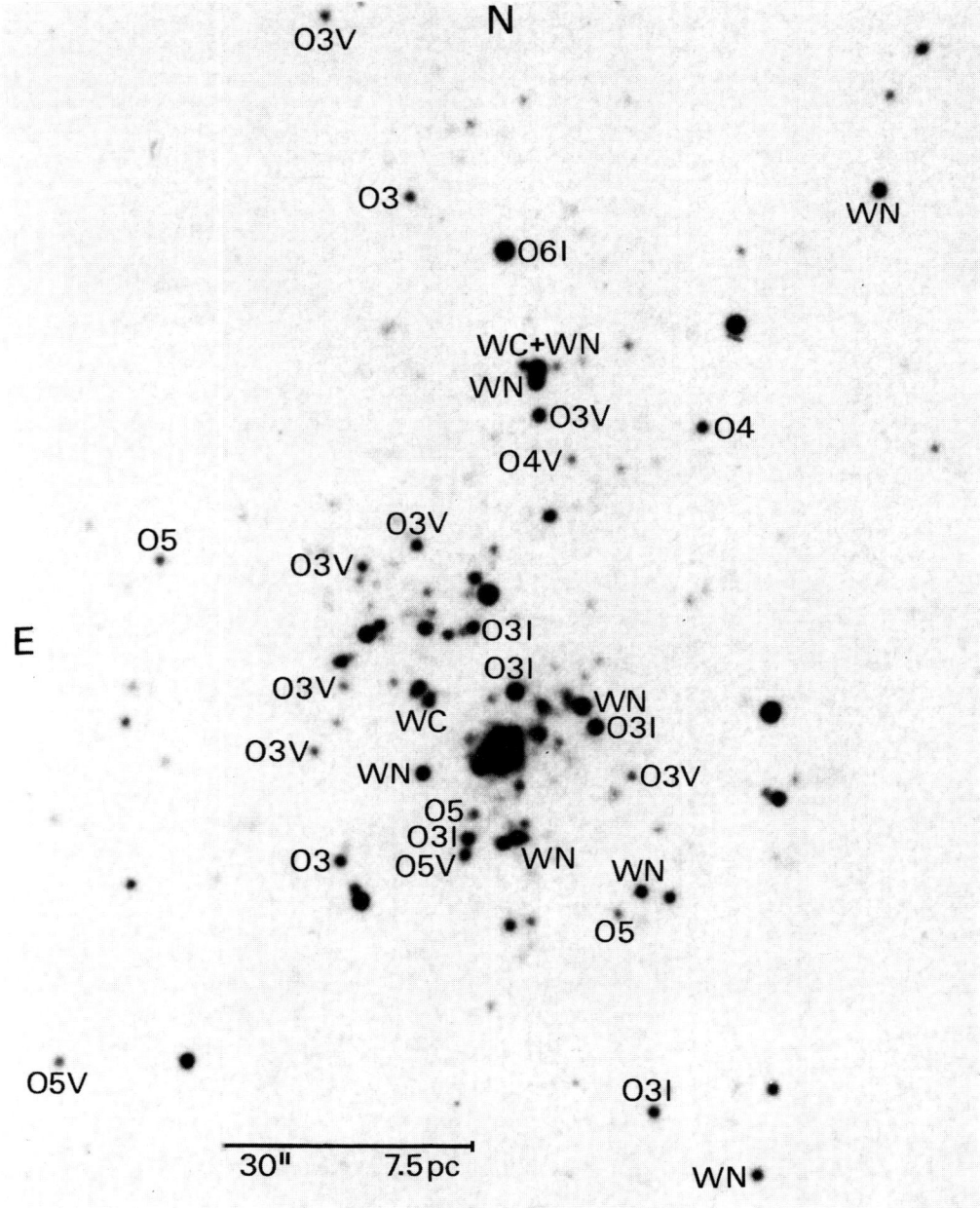

Fig. 3 - The distribution of the O3-O6 and WR stars (exclusive of R136) in the 30 Doradus cluster. This excellent photograph was obtained by Dr. Y.-H. Chu through a blue-continuum (λ4765) interference filter at the CTIO 4-meter prime focus.

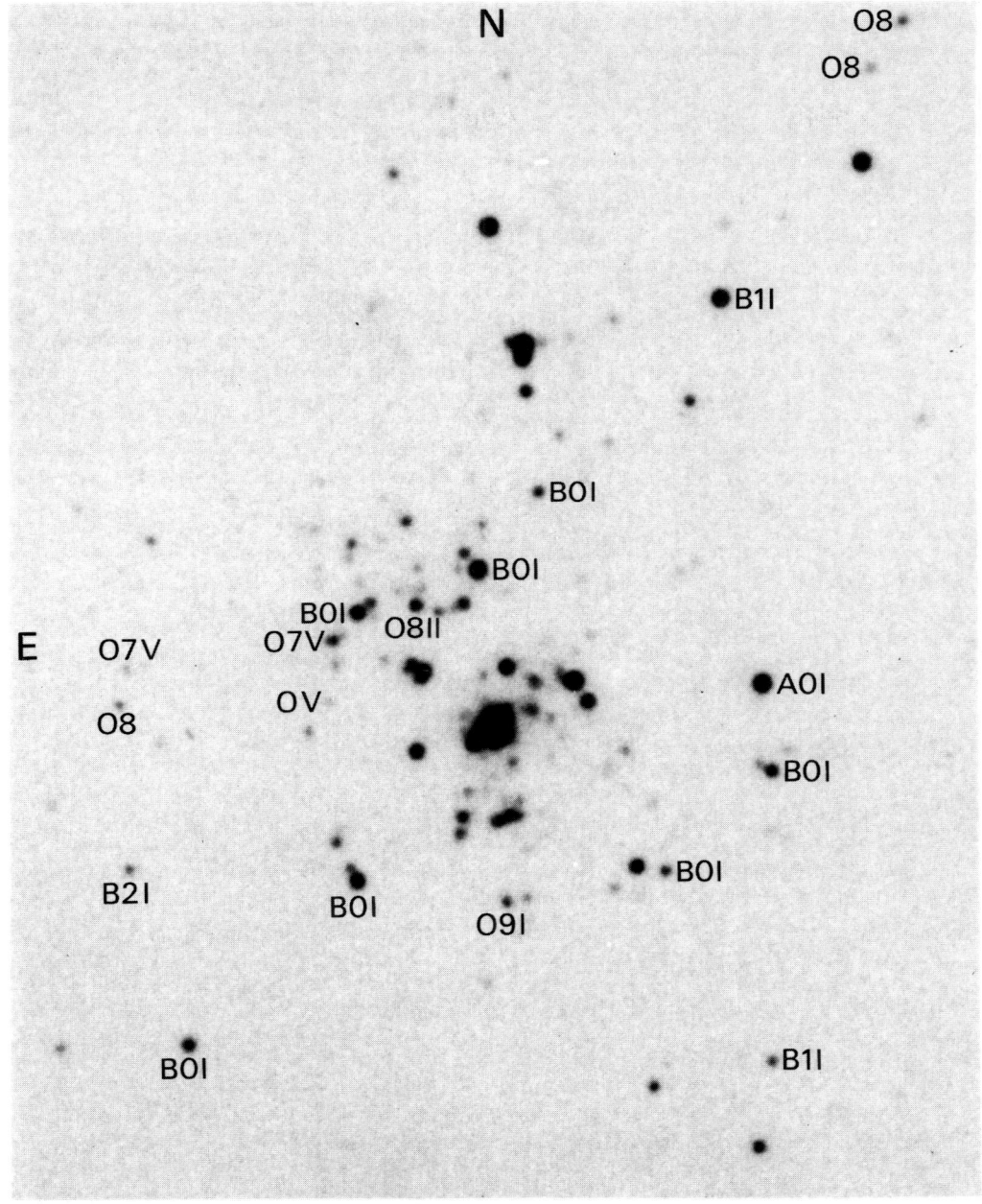

Fig. 4 - The distribution of the later-type stars on an He II λ4686 interference-filter photograph of 30 Doradus (which enhances the WR images), also obtained by Dr. Chu at the CTIO 4-meter.

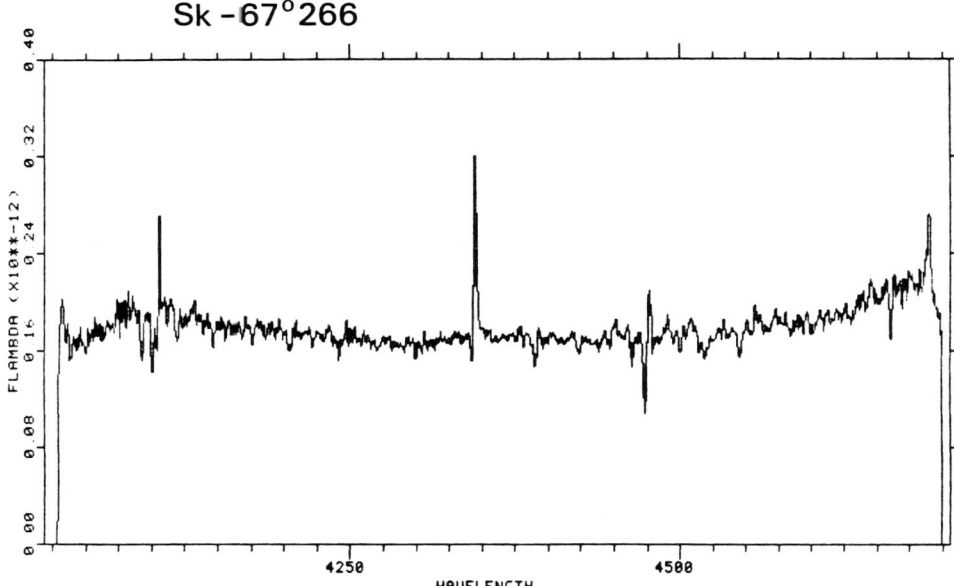

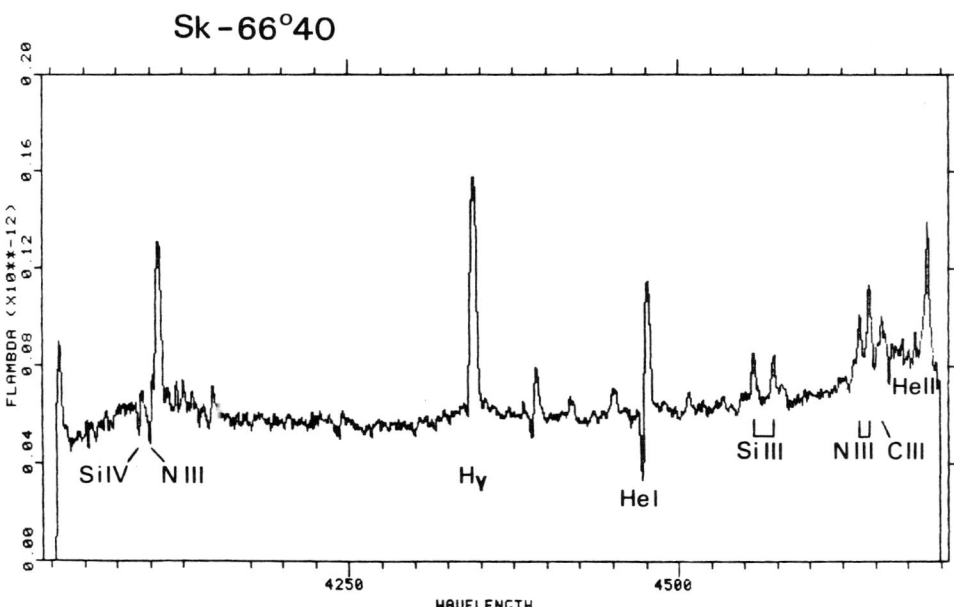

Fig. 5 - Vidicon spectrograms of the Ofpe stars Sk-66°40 and -67°266 in the LMC. Both flux scales should be fainter by a factor of 1.7 at 4330 Å. The spectral features identified are Si IV λ4089 and N III λ4097 absorptions, Hγ emission, He I λ4471 P Cygni profile, and Si III λλ4552-4568, N III λλ4634-4641, C III λ4650, and HeII λ4686 emissions.

outburst and has been shown to be the hottest known Hubble-Sandage variable by Stahl et al. (1983) and Stahl and Wolf (1985); also its nebular shell has apparently been directly resolved by Stahl (1985). The purpose of this note is to establish that the star Sk-66°40 is a further member of the HDE 269858/Sk-67°266 subclass. A SIT vidicon observation of its spectrum is shown along with one of -67°266 in Figure 5. The narrow Si IV and N III absorptions are easily seen; the strength of the Si III and C III emission lines in -66°40 is also noteworthy. Thus the Ofpe/WN9 category in the LMC now contains seven members, while no spectroscopically identical counterparts have yet been identified in the Galaxy.

III. THE ULTRAVIOLET SPECTRA OF THE O STARS

Recently an extensive survey of the 1200-1900 Å region of O-type spectra has been completed, based upon the unprecedented sample of homogeneous, high-resolution data contained in the International Ultraviolet Explorer archives. The primary result of this survey, comprising about 120 different objects, is conclusive evidence for a high degree of detailed correlation between the UV spectra and the optical spectral classifications, as well as between the photospheric and stellar-wind features, for the great majority of the O stars. The pronounced luminosity dependence of the Si IV stellar-wind effect, the main-sequence phenomena, and the ON/OC spectra have been discussed by Walborn and Panek (1984a, b; 1985), respectively. The Of supergiant and WN-A sequences provide further striking examples of detailed correspondences between stellar-wind features and the optical spectral types. For instance, the $\lambda\lambda 1300$-1600 Å region provides a unique signature of an O3 If* spectrum, with a strong O V $\lambda 1371$ wind profile, no Si IV $\lambda\lambda 1394$-1403, and very strong C IV $\lambda\lambda 1548$-1551. Similarly, the O4 If spectra, with no O V, intermediate Si IV, and strong C IV can be readily discriminated from both earlier and later types. Such effects assume considerable significance in the context of future programs with the Hubble Space Telescope High Resolution Spectrograph, since they provide a framework relative to which one can interpret observations of restricted UV wavelength ranges in faint extragalactic OB spectra lacking high-quality optical data. Unfortunately, space does not permit these spectral sequences to be reproduced here, but an extensive atlas displaying montages of the 1200-1900 Å range in about 100 objects is currently in press at NASA for wide distribution (Walborn, Heckathorn, and Panek 1985).

REFERENCES

Azzopardi, M. and Breysacher, J. 1979a, Astron. Astrophys., **75**, 120.
―――――. 1979b, Astron. Astrophys., **75**, 243.
Chu, Y.-H., Cassinelli, J. P., and Wolfire, M. G. 1984, Ap. J., **283**, 560.
Chu, Y.-H. and Daod, N. A. 1984, P.A.S.P., **96**, 999.

Dreher, J. W., Johnston, K. J., Welch, W. J., and Walker, R. C. 1984, Ap. J., **283**, 632.
Feast, M. W., Thackeray, A. D., and Wesselink, A. J. 1960, M.N.R.A.S., **121**, 337.
Fitzpatrick, E. I. and Savage, B. D. 1984, Ap. J., **279**, 578.
Garmany, C. D. and Massey, P. 1984, B.A.A.S., **16**, 508.
Henize, K. G. 1956, Ap. J. Suppl., **2**, 315.
Melnick, J. 1983, The Messenger, ESO, No. 32, p. 11.
──────. 1985, Astron. Astrophys., in press.
Moffat, A. F. J., Seggewiss, W., and Shara, M. M. 1985, Ap. J., in press.
Panagia, N., Tanzi, E. G., and Tarenghi, M. 1983, Ap. J., **272**, 123.
Savage, B. D., Fitzpatrick, E. L., Cassinelli, J. P., and Ebbets, D. C. 1983, Ap. J., **273**, 597.
Schmidt-Kaler, Th. and Feitzinger, J. V. 1981, in The Most Massive Stars, ed. S. D'Odorico, D. Baade, and K. Kjär (Garching: ESO), p. 105.
Shore, S. N. and Sanduleak, N. 1984, Ap. J. Suppl., **55**, 1.
Stahl, O. 1985, The Messenger, ESO, No. 39, p. 13.
Stahl, O. and Wolf, B. 1985, Astron. Astrophys., in press.
Stahl, O., Wolf, B., de Groot, M., and Leitherer 1985, Astron. Astrophys. Suppl., in press.
Stahl, O., Wolf, B., Klare, G., Cassatella, A., Krautter, J., Persi, P., and Ferrari-Toniolo, M. 1983, Astron. Astrophys., **127**, 49.
Walborn, N. R. 1973, Ap. J. Letters, **182**, L21.
──────. 1977, Ap. J., **215**, 53.
──────. 1982a, Ap. J. Letters, **254**, L15.
──────. 1982b, Ap. J., **256**, 452.
──────. 1984, in IAU Symp. 108, Structure and Evolution of the Magellanic Clouds, ed. S. van den Bergh and K. S. de Boer (Dordrecht: Reidel), p. 243.
Walborn, N. R., Heckathorn, J. N., and Panek, R. J. 1985, International Ultraviolet Explorer Atlas of O-Type Spectra from 1200 to 1900 Å (NASA), in press.
Walborn, N. R. and Panek, R. J. 1984a, Ap. J. Letters, **280**, L27.
──────. 1984b, Ap. J., **286**, 718.
──────. 1985, Ap. J., **291**, 806.
Walker, A. R. and O'Donoghue, D. E. 1984, Astron. Express, **1**, 45.
Weigelt, G. and Baier, G. 1985, Astron. Astrophys., in press.

PARAMETERS OF WOLF-RAYET STARS

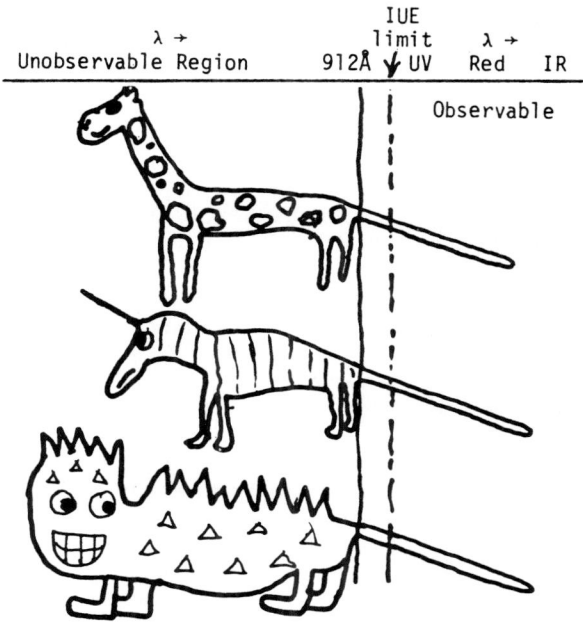

Fig. 2. Cartoon showing difficulty in discerning the nature of the "beast" existing below λ912 Å, the Lyman limit, if one only has the "tail" of the distribution to observe in the UV, optical, and IR wavelengths.

the nature of the problem with the data: we are in a situation, even with IUE observations, in which we only measure the "tail" of the distribution of energy coming from a hot star. The actual "beast" that lurks shortward of the Lyman limit is only dimly perceived through theoretical modeling of a moving atmosphere, which is an incompletely solved problem. In particular, the backscattering of photons from the wind may affect the emergent continuum of the underlying stars (Hummer 1982). This is only now being incorporated in models for O stars (Abbott and Hummer 1985), and will play a critical role for W-R stars. My caution here is to realize the models of W-R stars are very inadequate and that conclusions which rest heavily upon them are subject to considerable revision.

It should also be kept in mind that the observed data need to be corrected for the interstellar absorption. This is by no means a straightforward procedure as the extinction "law" is not universal in the UV regions (Massa et al. 1983; Garmany et al. 1984).

What then can be done about effective temperatures for W-R stars? There are a very few eclipsing binaries which do lend themselves to determinations of this quantity. Very recently a detailed discussion of the light curves of V444 Cyg = HD193576 has been carried out by Cherepashchuk et al. (1984). Using data extending over the UV to near IR regions, they find the eclipse durations are a function of wavelength (as would be expected if the opacity depends strongly on this quantity). They are thus able to model the eclipses and find a relatively small core radius for HD 193576 of some 2.9 $R_\odot$, implying a large effective temperature of 90,000°K for this WN5 star. Pauldrach et al. (1985) have reported an improved code for calculating radiatively driven wind models. They were able to duplicate the inferred density, radius, and velocity relationships for the V444 Cyg found empirically by Cherepashchuk et al. with their high effective temperature.

Hillier (1983) found that an effective temperature of 60,000°K fit the emission line measures of HD 50896, another WN5 star, but pointed out his data did not yield a sensitive determination of this quantity. Cherepashchuk (1982) reviews data on three other W-R eclipsing binary systems which might be used to determine radii and thus empirical effective temperatures. The relatively small core radii found suggest relatively high values for the temperature, corresponding more closely to the excitation/ionization values than to those implied by the emergent continua.

My conclusion is that at present the effective temperature scale for W-R stars is not well known. The earliest types may have values near 100,000°K and the latest near 30,000°K but the relationship with the spectral types is not at all certain. Presumably, the effective temperature will scale with the excitation/ionization sequence but I would not be completely surprised if it eventually turned out that two, say, WN5 stars ended up with different effective temperatures. I would again like to stress that the spectrum of the W-R stars is one of the wind and this may be only loosely coupled to the intrinsic parameters of the star itself.

MASSES

There is not too much new to add to the review of this problem given by Massey (1981). There are values for some dozen or so W-R binaries, either members of the eclipsing systems or those for which the spectrum of the O-type companion is estimated well enough that its mass can be found. The values for W-R stars range from some 10 to 40 $M_\odot$; the extremes are for two stars of the WN6 class so there is no simple dependence on spectral subtype. Although one might expect that WC stars are more evolved than WN stars, given their compositions, the former are not less massive than the latter. Thus with the data presently in hand it is not possible to conclude that all WN stars become WC, nor that all WC were once WN. Massey suggests that the WC stars might have evolved from the more massive WN type but we cannot go much beyond this at the present time, at least in my opinion.

COMPOSITION

I have already alluded to the compositions of W-R stars. The pertinent references are Willis and Wilson (1978), Smith and Willis (1982), Nugis (1975, 1982), Conti et al. (1983a), Torres et al. (1985). These results suggest that the WN stars have greatly enhanced helium and nitrogen and little or no hydrogen present. Carbon and oxygen are found in WN stars in more or less normal abundance. In WC stars helium, carbon and oxygen are found to be in great abundance and there is no evidence for hydrogen or nitrogen. These data are qualitatively in agreement with the predictions of massive star evolution with mass loss and mixing revealing the underlying nuclear reaction core (Noels and Gabriel 1981; Maeder 1983). The quantitative agreement is a factor of a "few" but this may reflect problems in the evolution of the models or in the absolute abundance or some combination of these features.

MASS LOSS RATES

Abbott et al. (1985) have completed a detailed study of detections of thermal free-free emission from the winds of O and W-R stars. These are mostly at 6 cm wavelength and all with the VLA. This emission is proportional to the mass loss rate of the star, with practically no model dependence. The $\dot{M}$ can be found if the distance of the star is known (the uncertainty here scales as the 3/2 power) and if the terminal velocity of the wind can be obtained (typically with IUE measures). A presumption is that the terminal velocity measured a few tens of stellar radii away is unaffected out to a few hundred radii where the free-free emission is measured. In spherical flow models this should be readily satisified.

Recently it has been recognized that some of these hot luminous stars have a nonthermal component to the radio emission measures. Abbott et al. (1985) are able to show that for most W-R stars, the nonthermal contribution is small since the wavelength dependence of the flux follows a power law with exponent −0.6 they list 7 W-R stars as definite thermal emitters and another 17 as probable thermal sources, thus giving 24 stars with well-determined mass loss rates. I have appended their data in Table 2. Is there any dependence of $\dot{M}$ on the W-R subtypes? Figure 3, adapted from Abbott et al., shows there is not. The WN and WC subtypes appear to have similar $\dot{M}$ and there is no dependence on excitation class. There is a certain spread in the values of $\dot{M}$ at a given subtype, a point to which I will return shortly.

Abbott et al. (1985) have determined $\dot{M}$ for five binary systems for which reasonable mass estimates are in hand. Figure 4, adapted from their paper, shows that as far as these data are concerned, there is a relation between the $\dot{M}$ and the stellar masses, the mass loss rates being larger in the larger mass stars. One would certainly like to have more data points, but Figure 4 looks pretty convincing. Why does the $\dot{M}$ depend on the stellar mass? Abbott et al. suggest this is

Table 2. Mass Loss Rates for W-R Stars[a]

WR	Spectral type	log $\dot{M}$ ($M_\odot$ yr^{-1})	WR	Spectral type	log ($M_\odot$ yr^{-1})
142	WO2	<-4.7	2	WC9	<-4.6
111	WC5	-4.8	133	WN4.5+O9I	≤-5.0
114	WC5	<-4.8	1	WN5	-4.5
143	WC5	<-5.1	6	WN5	-4.5
92	WC6+Abs	-4.5	138	WN5+Abs	-4.9
79	WC7+O6	-4.3	139	WN5+O6	-5.0
86	WC7+Abs	-4.6	110	WN6	-4.3
90	WC7	<-4.8	115	WN6	-5.0
137	WC7+Abs	-4.7	134	WN6	-4.5
11	WC8+O9I	-4.1	136	WN6	-4.5
113	WC8+O8V	-4.5	141	WN6	-4.9
135	WC8	-4.6	78	WN7	-4.4
81	WC9	-4.9	87	WN7	<-4.7
95	WC9	<-4.8	89	WN7	-4.4
103	WC9	<-4.7	130	WN8	<-4.8
104	WC9	<-4.7	145	WN/C	-4.7
121	WC9	<-4.8			

[a]From Abbott et al. 1985.

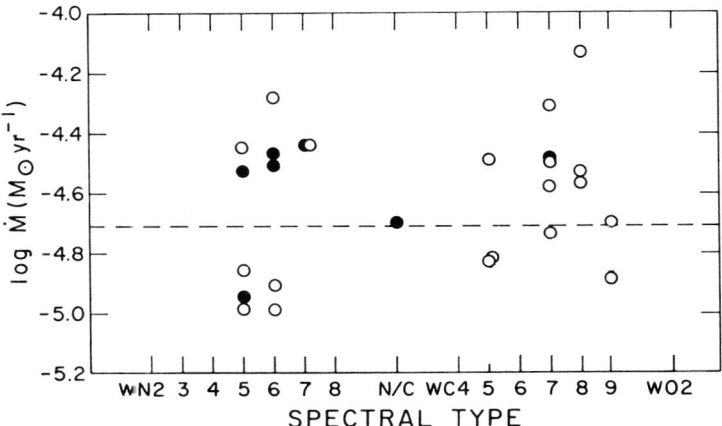

Fig. 3. Mass loss rates versus W-R subclasses (from Abbott et al. 1985). The filled symbols represent definite thermal sources; the open symbols the probable thermal sources.

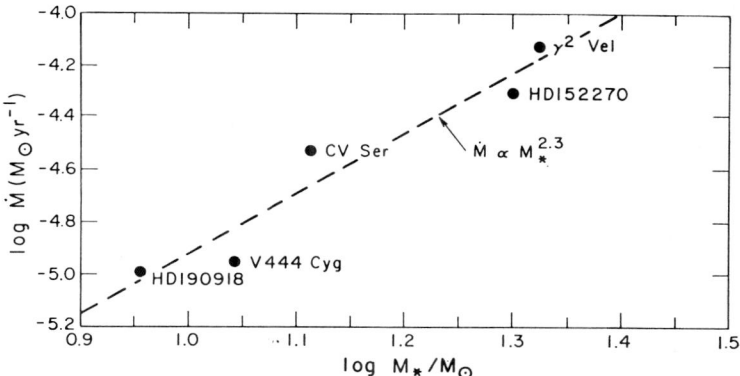

Fig. 4. Mass loss rates versus masses for five binary systems with well determined parameters (from Abbott et al. 1985).

because the more massive stars are the more luminous. As I have discussed previously, the bolometric luminosities of W-R stars are not well estimated. However, Abbott et al. were able to skirt this problem by taking advantage of an artifact. They took a mean M_v for each spectral subtype and then a deviation from the mean for each W-R star with a measured $\dot{M}$. The correlation of this artificial luminosity with $\dot{M}$ is shown in Figure 5. Here there are more data points and there is a tendency for the stars with the higher $\dot{M}$ to be brighter (in M_v within their subtypes). This is to be expected if the winds in W-R stars are radiatively driven. Figure 5 is currently the best indica

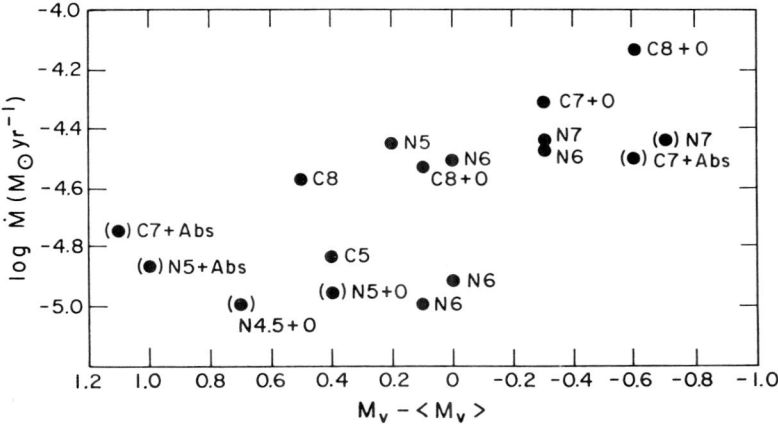

Fig. 5. Mass loss rates versus artificial luminosities (M_v minus means M_v) adapted from Abbott et al. (1985). The symbols with parentheses have less certain M_v.

tion of mass loss rate depending on the luminosity for W-R stars. I will also recall the recent theoretical model of Pauldrach et al. (1985) in which they are able to match the wind law of V444 Cyg found by Cherepashchuk et al. (1984) with a radiatively driven wind from a core of 90,000°K effective temperature.

The substartial mass loss rates found for W-R stars by Abbott and his associates are significant on the evolution time scale for these objects. These values will peel down a considerable fraction of the remaining mass in the helium burning lifetime. As the star becomes less massive, its luminosity also may decrease and the loss may taper off. The final state of a W-R star is by no means clear, whether they just "poop out" or violently explode is an interesting question.

DISTRIBUTION OF W-R STARS IN OUR GALAXY

I should now like to address the question of where W-R stars are found in our galaxy. Garmany (1986) has given a plot of the longitude distribution of OB stars more massive than 40 $M_\odot$ projected onto the plane of the galaxy within 3 kpc. In Figure 6, I show the analogous distribution of W-R stars making use of the M_v calibration of Table 1 and the available photometry (e.g. van der Hucht et al. 1981). We see considerable numbers of these objects in the inner Scorpio-Saggitarius arm, and in the Cygnus-Carina arm but very few outward from the Sun in

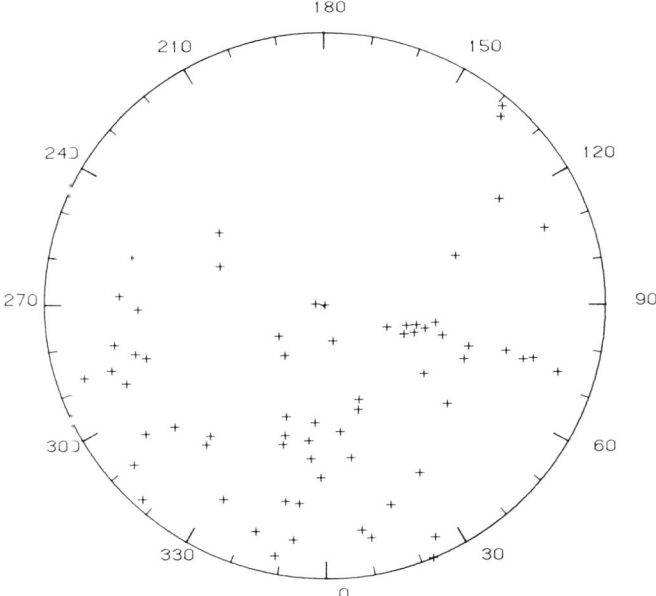

Fig. 6. W-R stars within 3 kpc, plotted on the galactic plane. This figure can be compared with that shown by Garmany (1986).

the Perseus arm. The distribution is similar to that of the massive O stars, suggesting that the W-R stars are decendents of these objects (Conti et al. 1983b). There are 245 progenitor massive OB stars in Garmany's 3 kpc volume; this is to be compared with 63 W-R descendants in Figure 6. Thus if all stars more massive than 40 $M_\odot$ become W-R stars, the lifetimes are in the ratio of about 4. I will return to this issue presently.

The galactic latitude, or better, the vertical Z distribution of massive stars in our galaxy can be found from the distances and the coordinates. That for the massive OB stars within 3 kpc is given by Garmany (in this volume). The OB stars are indeed closely confined to the galactic plane as would be expected from an extreme Population I sample. The vertical extent of the extreme Population I stars seems to be some ±100 pc. A very few OB stars are found substantially further from this range and would be described as "runaway" stars. Gies (1984) has recently considered the nature of such objects and concludes they have escaped from close star-star interactions in the cores of young clusters and associations.

Figure 7 shows the galactic latitude (in terms of Z) distribution of the W-R stars within 3 kpc. These stars are similarly closely coupled to the massive O-star distribution. The vertical extent is like that of the parent population. We see two stars appreciably far from the plane, presumably decendents of the "runaway" OB stars found there. There is no need to suggest a different Z dependence for W-R stars compared to the massive OB stars, as has sometimes been done in the literature.

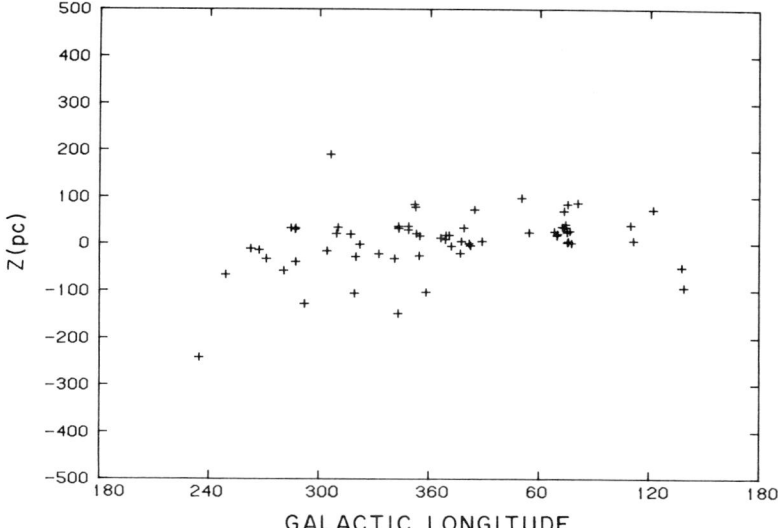

Fig. 7. W-R stars distribution in "Z" distance from galactic plane, within 3 kpc.

CONCLUSIONS

Let me first raise some questions concerning W-R stars. We see that they all have substantial mass loss rates. Is it possible that these large values are a result of the high effective temperatures, thus very large luminosities, and the winds are driven by radiative forces? I think this is a plausible conclusion but confirmation will await further work. The wind law (that is the run of opacity and velocity with radius) in W-R stars seems to be rather "soft" compared to OB stars. By this I mean the velocity increases rather slowly with radius. What physically causes this? Here I don't really have any good suggestions except to note that the hydrogen/helium ratio, which is very low in most W-R stars may play a role in the structure of the wind. Why do different subclasses of WN and WC stars exist? Is this merely because of the different effective temperatures? Even more importantly, what is the t_{eff} scale for the WN and WC subtypes? Is there an evolutionary connection between the WN and WC subclasses; for example, do late WN evolve to early WC? Or early WN evolve to late WC? At present there is no real understanding of the W-R subtypes and their evolutionary connection.

The number ratio of massive OB stars to W-R stars in the solar vicinity, presumably a reasonably complete sample of each to 3 kpc, has the value 4. Thus if all OB stars more massive than 40 $M_\odot$ become W-R stars, the lifetime of the latter phase is some 25% of the main sequence phase. Traditionally, the helium burning lifetime of a star is some 10% of the hydrogen burning time so the observed value seems a little large. However, on the bus ride to this meeting, Maeder told me that the new rates for carbon-alpha particle reactions will lengthen the helium burning lifetime of massive stars by about a factor of two. Does this take care of the discrepancy? (As an aside, later during the Symposium both Chiosi and de Loore asserted the new rate would not affect the ratio; Maeder still concluded it did.) Possibly a few close binaries, with masses less than 40 $M_\odot$, could become W-R stars because of interactions with their companions, or perhaps, as Maeder has stressed, the metal abundance of the progenitor star will affect its ease of evolving to the W-R stage.

What evidence supports the general scenario? We see that the massive OB stars and the W-R stars share a similar galactic longitude and latitude distribution. In particular there are fewer of both objects toward the anti-center region. Humphreys et al. (1985) have pointed out that in associations in our galaxy and in M33, the W-R stars and the red supergiants are anti-correlated, that is they are not found together. Humphreys and Davidson (1979) had previously noted that in the galaxies of the Local Group, red supergiants are never found brighter than M_v brighter than -8.0 magnitude, corresponding to an initial mass between 30-50 $M_\odot$, depending on whose evolution tracks are used. It seems clear that above some initial mass, which is of the order of 40 $M_\odot$, stars do not become red supergiants when they begin to burn helium in the core. But helium burning must go on someplace, which one can readily identify with the W-R stars. We have already noted the evidence that these objects are, in fact, helium rich. They must also be helium burning for the most part since the mass/luminosity ratio of those in binaries indicates they are over-luminous for their mass (e.g. Paczynski 1973).

So, we see solid and consistent evidence that the W-R stars are the helium burning descendants of massive stars. We don't understand all the details of the evolution but we can begin to use these objects as probes of the luminous stellar population in other galaxies, the subject of the next talk.

This work has been supported by the National Science Foundation. I am indebted to my colleagues Drs. Garmany and Massey for their intimate involvement with much of the work reported here, to Dr. Abbott for an advance look at his newest results on the radio observations and Mike van Steenburg for construction of several of the figures.

REFERENCES

Abbott, D. C., Bieging, J., Churchwell, E., and Torres, A. V. 1985, Astrophys. J., submitted.
Abbott, D. C. and Hummer, D. G. 1985, Astrophys. J., in press.
Cherepashchuk, A. M. 1982, Astrophys. Space Sci. **86**, 299.
Cherepashchuk, A. M., Eaton, J. A., and Khaliullin, Kh. F. 1984, Astrophys. J. **281**, 774.
Conti, P. S., Leep, E. M., and Perry, D. N. 1983a, Astrophys. J. **268**, 228.
Conti, P. S., Garmany, C. D., de Loore, C., and Vanbeveren, D. 1983b, Astrophys. J. **274**, 302.
Conti, P. S. and Underhill, A. B. 1986, in O and W-R Stars, NASA Monograph (in press).
Garmany, C. D. 1986, in this volume.
Garmany, C. D., Massey, P., and Conti, P. S. 1984, Astrophys. J. **278**, 233.
Gies, D. 1984, Thesis, University of Toronto.
Hillier, J. D. 1983, Thesis, University of Canberra.
Hummer, D. G. 1982, Astrophys. J. **257**, 724.
Humphreys, R. M. and Davidson, K. 1979, Astrophys. J. **232**, 409.
Humphreys, R. M., Nichols, M. and Massey, P. 1985, Astron. J. **90**, 101.
Kudritzki, R. P. and Hummer, D. G. 1986, in this volume.
Lundström, I. and Stenholm, B. 1984, Astron. Astrophys. Suppl. **58**, 163.
Maeder, A. 1983, Astron. Astrophys. **120**, 113.
Massa, D., Savage, B. D., and Fitzpatrick, E. L. 1983, Astrophys. J. **266**, 622.
Massey, P. 1981, Astrophys. J. **276**, 153.
Noels, A. and Gabriel, M. 1981, Astron. Astrophys. **101**, 215.
Nugis, T. 1975, in Proceedings of IAU Symposium 67 (Dordrecht: Reidel), p. 291.
Nugis, T. 1982, in Proceedings of IAU Symposium 99 (Dordrecht: Reidel), p. 131.

Paczynski, B. 1973, in Proceedings of IAU Symposium No. 49
   (Dordrecht: Reidel), p. 143.
Pauldrach, A., Puls, J., Hummer, D. G., and Kudritzki, R. P. 1985,
   Astron. Astrophys. Lett., in press.
Smith, L. E. 1968, Monthly Notices Roy. Astron. Soc. **138**, 109.
Smith, L. J. and Willis, A. J. 1982, Monthly Notices Roy. Astron.
   Soc. **201**, 451.
Torres, A. V., Conti, P. S., and Massey, P. 1985, Astrophys. J.,
   submitted.
van der Hucht, K. A., Conti, P. S., Lundström, I., and Stenholm, B.
   1981, Space Sci. Rev. **28**, 227.
Willis, A. J. 1982, in Proceedings of IAU Symposium 99 (Dordrecht:
   Reidel), p. 87.
Willis, A. J. and Wilson, R. 1978, Monthly Notices Roy. Astron. Soc.
   182, 559.

Discussion   :   CONTI.

APPENZELLER :

You mentioned the great range of mass values observed for WR stars. Can the lowest mass WR's form according to the same mechanism as the high mass ones?

CONTI :

We don't really know. Presumably higher mass W-R stars come from higher mass progenitors by mass loss and mixing processes.

MELNICK :

The most massive stars in Irr galaxies are located in giant HII regions where crowding and differential extinction exclude them from any magnitude limited survey. Thus the luminosity function of any galaxy that has giant HII regions will be severely incomplete at the high as well as the low luminosity ends. This applies both to your work on the LMC as well as to that of Freedman on M33. On the other hand, since bursts of star formation are stochastic, if you look at a galaxy that has no giant HII regions you will still find a luminosity function that has too few massive stars. That does not mean that the IMF of that galaxy is abnormal but instead that at the particular time of your observation there are no massive O stars alive! This stochastic limitation is a fundamental one and in my opinion completely precludes any effort to determine the IMF in late type galaxies.

CONTI :

   I have mentioned that even with a substantial incomplete search the LMC has appreciably more O stars than would be expected compared to the SMC.

LORTET :

   Should we not systematically reintroduce the subdivision into narrow (A) and broad lines (B) for WN stars and especially for WN6 stars, even if intermediate cases (A(B) or (A)B) also exist. There is increasing evidence that WNA, either in binaries or clusters, are associated with the hottest O stars (e.g. core of 30 Dor region) while WNB are associated with evolved late O and B supergiants : this is striking when reading the Schild and Maeder paper (1984, Astron. Astrophys., **136**, 237). A new example is Brey 72 in LMC which has been found to be a binary WN6 with very broad lines +BlIa (Moffat, private communication).

CONTI :

   I think the "A" and "B" designations are useful for the extreme cases but it must be remembered that line widths in WR stars show a continuous range. This is shown in studies of some 60 WN stars (Conti et al. Ap.J. 1984) and so is the case for WC stars (Torres et al. Ap.J. 1985).

GRAHAM :

   Do you ever find WR stars which happen to be unusually heavily reddened?

CONTI :

   It is not at all uncommon to find substantial reddening ($E_{B-V} > 1.0$) for W-R stars within 3 Kpc.

NIEMELA :

   Do you include those WN stars with H in their spectra also in the He-burning phase? Were these included in your statistics?

CONTI :

   I would make a restriction for the Carina WN stars which are H-burning stars, that is 3 stars. The other WN stars with hydrogen may also have been counted as He burning stars in WR statistics, so maybe this is not very consistent. It would bring down the WR/O number by a few percent in the totals.

VIOTTI :

Concerning the problem of the effective temperature of WR stars, I would like to highlight the case of HD 93162 and 193793, whose high-resolution UV spectra display, according to Fitzpatrick (Ap.J. 261, L91), narrow absorption lines of FeIV and FeV probably formed in the WR photospheres. Certainly, the identification of these lines will narrow the range of possible T_{eff} in WR stars.

MAEDER :

Classically, the ratio t_{He}/t_H of the helium to hydrogen lifetimes is 10%. With the new $^{12}C(\alpha,\gamma)^{16}O$ cross-sections by Kettner and colleagues, this ratio is about 20% which is in agreement with the observations by Conti et al.

CONTI :

I'm pleased to hear this, since the observed WR/O ratio is also near this value.

WOLF-RAYET STARS IN NEARBY GALAXIES

P. Massey, Kitt Peak National Observatory, NOAO

I have recently reviewed the reasons for studying the Wolf-Rayet (WR) content of nearby galaxies, and what these findings seem to be telling us (Massey 1985). The only data which has become available since that review is discussed elsewhere in this volume by Armandroff et al. Here I will only outline the salient points.

Unless metallicity plays a dominant role in the evolution of massive stars (and it may), the number density of WR stars in a galaxy should be proportional to the number of massive star progenitors. In any event, the relative numbers of WR stars in galaxies of similar metallicity should provide a direct comparison of the relative numbers of massive stars. Detecting WR stars is far more reliable than attempting to determine the unevolved massive star population, given the degeneracy in broad-band colors for very hot stars. By using CCD photometry with "optimized" WR filters, and crowded field photometry algorithms, reasonable completeness can be achieved.

So far, this is what is known:

(1) The LMC and SMC differ significantly in the number density of WR stars, with the LMC richer than the SMC by a factor of three or four over that expected on the basis of either mass or surveyed area (Azzopardi and Breysacher 1979). The LMC and SMC both have a considerably higher WN:WC ratio (5 or 6 : 1) than that found within a few kiloparsecs of the sun. Most of the WR stars in the Clouds are of high excitation type (WN3, WN4, WC4), although a few lower-excitation WN types are found in the 30 Doradus region of the LMC. In the vicinity of the sun, roughly equal numbers of high and low excitation stars are found.

(2) The Sc galaxy M33 has a number density of WR stars in its inner region considerably greater than that found in the solar neighborhood. The number density falls off with increasing radius; presumably this principally reflects the general decline of the star formation rate with galactocentric distance. In addition, the there is a conspicuous absence of WR stars in the northern arm, in accord with the study of Boulesteix et al. (1981), who suggested a lack of very massive stars in that region, based upon the scarcity of HII regions there. Most of the WR stars are found within the OB associations catalogued by Humphreys and Sandage (1980), although a few fall well away from obvious star-forming complexes. The ratio of WC to WN types is a strong function of galactocentric distance, falling from 2:1 near the center to 1:5 at the Holmberg radius. However, at the radius at which the oxygen abundance is similar to that of the LMC, the ratio of WC to WN types is roughly a factor of 5 higher than that found in the LMC. This suggests that metallicity alone does not control the relative numbers of WC and WN types, as had been inferred from comparing the Magellanic Clouds with the solar neighborhood. The details of the M33 study, including references to earlier work, can be found in Massey and Conti (1983). Further analysis by Humphreys et al. (1985) found that WR stars and red supergiants (RSG) were usually not found in the

same OB associations in M33, which is to be expected if most RSG's come from lower mass progenitors than do WR stars.

(3) Armandroff and Massey (1985) have recently completed a study of the Magellanic-type irregulars IC 1613 and NGC 6822. They found that the surface density of WR stars in NGC 6822 was low, similar to that of the SMC, while the number density of WR stars in IC 1613 was as high as that of the LMC. Since the metallicity of NGC 6822 is like that of the LMC, while that of IC 1613 is like that of the SMC, this **must** mean that NGC 6822 is poorer in massive stars than the LMC, while IC 1613 is considerably richer in massive stars than the SMC, no matter what role metallicity plays in the evolution of massive stars. Furthermore, all of the WR stars in NGC 6822 are of WN type, which again suggests that this ratio is not simply correlated with galaxian metallicity. The authors speculate that the ratio of WC to WN type may reflect differences in the IMF of recently formed massive stars, but the effects of small numbers could easily dominate the statistics at this point.

(4) In this symposium, Armandroff et al. present the results of a survey of several OB associations in M31. In two of the regions, NGC206, and OB48, the number density of WR stars is as large, or larger, than that of any area of M33. Thus there are regions in M31 which are extremely active in forming massive stars. Their data are too scarce, as yet, to comment on the global ratio of WC to WN types, or to search for gradients, but in the four well-studied regions, WN types clearly dominate.

REFERENCES

Armandroff, T. E., and Massey, P.: 1985, Astrophys. J., 291, 685.
Azzopardi, M., and Breysacher, J.: 1979, Astron. Astrophys., 75, 120.
Boulesteix, J., Dubout-Crillon, R., and Monnet, G.: 1981, Astron. Astrophys., 104, 15.
Humphreys, R. M., and Sandage, A.: 1980, Astrophys. J. Suppl., 44, 319.
Humphreys, R. M., Nichols, M., and Massey, P.: 1984, Astron. J. 90, 101.
Massey, P.: 1985, Publ. Astron. Society of the Pacific, 95, 5.
Massey, P., and Conti, P. S.: 1983, Astrophys. J., 273, 576.

Discussion   : MASSEY

MELNICK :

I think you should be very cautious about quoting WN/WC ratios until you have slit spectra of all your candidates. Also the WN/WC ratio in 30 Dor is different from that of the LMC as a whole so when you look at M31 you may be seing local "snapshots" of the real M31 ratio.

MAEDER :

The differences in the population of massive stars in galaxies may result from several facts : star formation rate (SFR) - initial mass function (IMF) - effects of metallicity Z on stellar structure - effects of Z on mass loss rates, etc. It is thus very difficult to exclude metallicity effects, since the other causes should be kept constant. It seems that one of the best ways to see whether Z-effects are present is to compare the galactic gradient of WR stars with the galactic gradient of their O-type precursors of initial mass larger than 40 Mo. As mentionned by Peter Conti, the gradient of WR stars is steeper than the gradient of O-type stars initially more massive than 40 Mo. Since neither the SFR nor the IMF can be advocated here, metallicity effects are a likely candidate for explaining the differences of these galactic distributions.

SHARA :

Though you didn't state it explicitely in your talk, your results imply that the WN/WC number ratio in M31 (or at least in its OB associations) is about 5. It is crucial that you obtain and publish spectra of your candidate stars to support or disprove this high WN/WC ratio. The fact that you are finding few new WC stars implies that our global survey of M31 is quite complete for WC's, and that there are only 30-40 WC's in all of M31. Even if WN/WC is about 5 in stars in total in M31 the Galaxy probably contains > 1,000 WR stars, and has only half the mass of M31. Thus M31 is much quieter at present, than the Galaxy. Your abstract rejected this "conventional wisdom", could you explain why?

MASSEY :

We have surveyed only a few regions in M31, and we selected the ones likely to be the most active. Interestingly, we found that these regions were very active in forming massive stars. Your statement that the total WC population in M31 has been identified is probably wrong by factors of 10-1000, as you have ignored (Moffat and Shara Ap.J. 273, 533 and your poster paper) differential reddening. While there may be some controversy as to whether M33 has different reddening from one region to another, nobody else here is likely to claim that for M31. We selected regions likely to be unreddened; even so, the B-V color of the blue supergiant plume in OB48 is .2 mag redder than in NGC 206. Furthermore, we found WC stars which you missed.

ROSA :

a) by going back to HII regions like NGC 604, do you detect the WR stars spectroscopically with your technique?
b) you might be missing a large fraction of the WR population in your statistics.

MASSEY :

We found a number of the Conti and Massey objects in NGC 588 (Ap.J. 249,471) in our "test" M33 field (Armandroff and Massey, Ap.J. 291, 685). We didn't try NGC 604, but we would probably have found them, given their high luminosity (good S/N). The HII regions matter only a little since we do local sky determinations. However, it is not the optimum way of detecting them.

ROSA :

a) can you find with your technique the W/C stars in giant HII regions - i.e. have you tried to reobserve NGC 604 to find the ones known already? b) the point of my question is, that you may be missing a substantial fraction of a galaxy's WR population. c) with an EQW of only 5-20 for the blue WR band the detection is very much hindered in such regions - may it be due to the dilution by OB star continua or the intrinsic low EQW of peculiar objects.

ZINNECKER :

You want to extend your WR star search. What is your strategy, what questions do you want to answer? It would perhaps be interesting to find WR stars outside OB associations. Whether very massive stars form in OB associations only, or in the field as well, would be an important question to answer.

MASSEY :

We answered that, I think, in our study of M33 (Massey and Conti : Ap.J. 273, 576) : while most WR stars were found in OB associations, a few of them are out in the boon-docks, and must have been formed there.

POSTER PAPERS 2.

Chairman : J. GRAHAM

1. P.HELLINGS: Bolometric Corrections and Magnitudes of WO+O Stars.

2. T.NUGIS: Two Possible Wind Models for Wolf-Rayet stars.

3. M.AZZOPARDI and N.MEYSSONIER: A New Survey for H alpha Emission-Line Stars in the SMC.

4. B.BOHANNAN: A Large Sample of Emission-Line Stars in the LMC:Their Location in the H-R diagram.

5. D.J.STICKLAND, C.LLOYD and A.J.WILLIS:
   Is AS 431 a Superluminous WR Star?

6. M.SHARA and A.MOFFAT: Wolf-Rayet Stars in M31.

7. A.Moffat, M.Shara and W.Seggewis:
   The Luminous Stellar Content of 30 Doradus and NGC 3603 - The Nearest Visible Giant HII Regions.

8. C.LEITHERER, I.APPENZELLER, G.KLARE, H.J.G.L.M.LAMERS, O.STAHL, L.B.F.M.WATERS and B.WOLF:
   The Massive Stellar Wind of the Hubble-Sandage Variable S Doradus.

9. D.VANBEVEREN: The WR/OB Number Ratio within 2.5 kpc from the Sun.

10. T.E.ARMANDROFF, P.MASSEY and P.S.CONTI:
    Wolf-Rayet Stars in M31's OB Associations.

11. K.A.VAN DER HUCHT, T.A. JURRIENS, F.M. OLNON, P.S.THE, P.R. WESSELIUS, P.M. WILLIAMS:
    IRAS Observations of Wolf-Rayet Stars.

12. P.J.McGREGOR, A.R.HYLAND and J.D.HILLIER:
    Infrared Spectroscopy of Southern P Cygni Stars.

13. O.STAHL, B.WOLF, M.DE GROOT and C. LEITHERER:
    High-Dispersion Spectroscopy of the Brightest Emission-Line Stars of the Magellanic Clouds.

14. R.VIOTTI, L.ROSSI, A.ALTAMORE, A.CASSATELLA:
    New Results on Eta Carinae.
    Evidence for an Asymmetric Inhomogeneous Wind.

15. B.WOLF and O.STAHL: MWC 300: A Runaway Hypergiant.

16. C.de JAGER and H.NIEUWENHUIJZEN:
    Stellar Atmospheric Instability in the Upper Part of the H-R Diagram.

17. J.P.DE GREVE, P. HELLINGS and E.P.J.VAN DEN HEUVEL:
    On the Occurrence of WR+O Binaries.

18. A.J.WILLIS, P.S.CONTI, C.D.GARMANY and I.D.HOWARTH:
    Rapid Ultraviolet Spectral Variations in HD 50896 (WN + ?)

19. J.HILLIER: The Formation of Nitrogen and Carbon Lines in HD 50896 (WN5).

20. F.J.ZICKGRAF, B.WOLF, O.STAHL and C.LEITHERER:
    B(e)-Stars in the Magellanic Clouds

21. A.J.WILLIS, I.D. HOWARTH, K.NANDY and D.H. MORGAN:
    The Mass Loss Rate of Sk 80 (O7iaf) in the Small Magellanic Cloud.

22. H.ZINNECKER: How to Form a 200 M_o Star.

23. G.MURATORIO, M.FRIEDJUNG and R.VIOTTI:
    FeII in the UV Spectrum of Luminous Emission Line Stars.

BOLOMETRIC CORRECTIONS AND MAGNITUDES OF WR+O STARS

Paul Hellings
Astrophysical Institute, Free University of Brussels,
Pleinlaan 2, B-1050 Brussel, Belgium

Bolometric corrections and absolute magnitudes for WR stars have been evaluated by various authors. A review of the results may be found in De Jager (1980). The calibrations, relating BC or Mv with the spectral subtype, however strongly differ from author to author. Moreover, even with the same observational method a large scatter is found for the same subtype, as was recognised by Conti et al. (1983). The only trend found is the increase of Mv with the subtype from WN3 to WN7, and for the WC subtypes. In this study BC and Mv are computed for a number of WR+O binaries.

The results are based on model calculations for double lined spectroscopic WR+O binaries, for which mass estimates have been derived (Hellings, 1984). This was performed under the assumption that these binaries are formed by mass transfer in massive close binaries. The bolometric magnitude Mb of the WR star is obtained with the mass-luminosity relation for RLOF remnants, stars that may be identified with WR components in WR+O binaries. The Mb value of the O star is obtaind from the model calculations described by Hellings (ibid.). For these models a complete stellar structure picture is available. The visual absolute magnitude Mv of the O star is obtained from the BC scale of Underhill et al. (1979). If the difference between the magnitudes of the two components is known, Mv of the WR star is easily computed. If not, we apply the Mv scale of Conti et al. (1983), relating Mv with the subtype of the WR star. The error on the magnitudes of the O conponents is +/-0.3 mag. For the WR stars the error on Mb is +/-0.6, which is due to the width of the mass-luminosity relation, and the uncertainty of M(WR). Finally we conpute the distances of these systems with the results of Mv and with the photometric data ($b-v, b-v_o, m_v$) listed by Hidayat et al. (1982, 1984) and Lundstrom and Stenholm (1984). Possible membership of clusters and associations is discussed.

The results show no obvious relation between any of the variables Mb, Mv or BC and the subtype, which is hardly to expect with such a limited sample. Only for the WN stars, a fair correlation between Mv and the subtype is possible. The Mb values are in the range -7.6 to -8.9. This corresponds to a factor three in luminosity. The visual magnitudes are in the same range as the calibrations found in the literature : -3.5

WR	Spc	Mb W	Mb O	Mv O	dMv	mv	Av	Mv W	BC W	y_o	y_o AC	AC
21	WN4	-8.5	-8.0	-4.5		9.80	2.58	-4.0	-4.5	12.2	12.1	Car OB1
31	WN4	-8.0	-8.5	-5.1		10.69	2.50	-4.0	-4.0	13.6		
42	WC7	-8.7	-8.1	-4.7	0.20	8.25	1.19	-4.5	-4.2	12.5	12.1	Car OB1
79	WC7	-8.3	-8.7	-5.2	0.75	6.95	1.39	-4.6	-3.7	11.3	11.5	NGC 6231
97	WN3	-7.6	-8.2	-4.9		11.15	3.65	-3.6	-4.0	12.7		
113	WC8	-8.8	-8.3	-5.0	0.00	9.43	3.36	-5.0	-3.8	11.8	11.5	Ser OB1
127	WN4	-8.1	-7.5	-4.6	1.30	10.36	2.21	-3.3	-4.8	13.1	13.2	Vul OB2
139	WN5	-8.2	-8.7	-5.3	1.50	8.27	2.91	-3.8	-4.4	10.9	11.2	Be 86
151	WN5	-7.8	-8.0	-4.7	0.00	12.40	4.22	-4.7	-3.1	13.7		

Table 1 : Photometric data for the sample : WR number, Spectral type, Mb for the WR and the O components, Mv for the O star, difference in Mv, apparent magnitude of the system, interstellar extinction correction, Mv of the WR star, Bc of the WR star, distance moduli of the system and of the cluster or association (if present), name of cluster or association.

to 4.5. BC values for the WR components are between -4 and -5, which is about one magnitude more than for O stars. For the WC stars, we find BC around -4, which is slightly less than for WN. On the other hand Mv for the WC stars is on the average 0.8 mag. higher. We stress that these results have only individual meaning. The sample is too small for drawing general calibrations. The Mv values of the O stars are scattered wihtin the error bars of the calibration for O stars of Conti et al. (1983).

The distances of these systems may then be computed with our magnitudes based on the model computations of the individual systems, and with the observational data. Six of the systems considered here are observed in the field of a cluster or an association. The conclusions on possible membership by Lundstrom and Stenholm (1984) are confirmed for WR 42, 79, 113, 127 and 133, taking into account the error bars. Lundstrom and Stenholm classify WR21 as a probable background object. We find its distance modulus to differ only 0.1 from the distance modulus of Car OB1, suggesting a probable membership. This difference is totally due to the magnitudes adopted for the O star, since we both used the Mv scale of Conti et al. (1983) for the WR component of this binary.

References :
De Jager C. (1980) "The brightest stars", D. Reidel, Dordrecht
Conti P.S., Garmany C.D., de Loore C., Vanbeveren D. (1983) Astrophys. Journ. 274, 302
Hellings P. (1984) Astrophys.Space Sc. 104, 83
Hidayat B., Supelli K., Van der Hucht K.A. (1982) "Wolf Rayet stars : observations, physics and evolution", C. de Loore, A. Willis (eds), IAU symp. 99, p27, Reidel, Dordrecht.
Hidayat B., Admiranto A.G., Van der Hucht K.A. (1984) Astrophys.Space Sc. 99, 175
Lundstrom I., Stenholm A.G. (1984) Astronom.Astrophys.Suppl 58, 163
Underhill A.B., Divan L., Prévot-Burnichon M.L., Doazan V. (1979) Mont.Not.R.A.S. 189, 601

TWO POSSIBLE WIND MODELS FOR WOLF-RAYET STARS

T. Nugis
W. Struve Astrophysical Observatory of Tartu
Estonia, USSR

Frequently the models proposed to explain some observational data are in conflict with the other observational data. Our purpose was to find out the empirical models of the envelopes of WN5 and WN6 stars which are not in conflict with the continuum runs in the IR spectral region and with the emission line intensities and profiles. In our recent study (Nugis, 1984) we proposed for Wolf-Rayet stars the spherically symmetric outflow model having an increasing velocity zone (with some coronal heating near the stellar surface) which is followed by the decreasing velocity zone extending up to the distance where the matter starts flowing with constant velocity. Let us call this model the AD-model (Acceleration-Deceleration).

The analysis of theoretical line intensities of HeI and HeII found by us by solving the statistical equilibrium equations for level populations at different distances from the stellar surface (by using the Sobolev escape probability method (Sobolev, 1960)), has confirmed that from spherically symmetric outflow models only the AD-model agrees with the observed line spectra of WN5 and WN6 stars. The intensity ratios suggest that HeI lines are effectively formed at distances $R > 4R_*$ (R_* is the stellar radius). In the case of monotonically increasing or constant velocity flows the optical depths of HeI lines are small at such distances and therefore no strong absorption components can be obtained in calculations. The agreement with the observed HeI spectrum can be obtained only in the case of the AD-type models (from spherically symmetric flows).

The specific feature of HeII line spectra of WN5 and WN6 stars is that lower members of the Pickering series have no apparent absorption components although from the line intensity ratios it can be concluded that the line optical depths for them must be greater than unity, provided the level populations have a normal run in the sense that the Menzel coefficients b_n are approaching unity from above ($b_k/b_i < 1$ if $i < k$ and $i < n_o$, where n_o is the principal quantum number beginning from which $b_n \approx 1$). Usually Menzel coefficients have a normal run. The only possibility to get agreement with observations is the case when $S_L > I_\nu^*$ at high expansion velocities (in the case of Planckian approximation to the stellar radiation: $S_L/I_\nu^* = (\exp(h\nu/kT_*)-1)/(b_i/b_k\exp(h\nu/kT_e)-1))$. This condition cannot be fulfilled as it appeared from our present calculations for the models having comparatively low T_* values ($T_* \approx 30000 - 50000$), even if $T_e > T_*$. This condition was not fulfilled also in the calculations by Castor and Van Blerkom (1970) for $T_e > T_*$ models. Our calculations showed that the condition $S_L > I_\nu^*$ can be fulfilled for the Pickering lines in the HeIII zone of WN5 and WN6 stars if we assume that the coronal heating takes place only in the thin region close to the stellar surface and that due to this heating the radiation power of the core in the first series continuum increases.

If we want to have a model without decelerating flow region, then we must give up the assumption of spherical symmetry. Let us consider the flow, which is on a large scale average, spherically symmetric. In our "alternative" model the radial matter streams from different spherical zones on the stellar surface are moving outward at periodically different speeds ($v(r,\varphi) = v(r,\varphi+n\varphi_o)$, where φ is the latitudinal angle of the spherical zone on the stellar surface). If φ_o is small, then the line-of-sight velocities above the stellar disk may be very different even at a comparatively large distance from the star. To get the agreement with the observed line spectrum of HeI it is yet necessary that the matter streams from different spherical zones on the stellar surface (with $\Delta\varphi < \varphi_o$) would be compressed into a smaller opening angle beginning at some distance from the star. The proposed "alternative" model is only a schematic empirical model. The estimates show that the mass loss rates and the optical depths of the envelopes in electron scattering of this model are smaller as compared to the case of the spherically symmetric AD-model. This "alternative" model belongs to the category of alternative models proposed for the atmospheres of Wolf-Rayet and O stars by Underhill (1983).

The crucial test for our models ought to be far IR spectrophotometry. The presence in the far IR spectral region of the lines of high ionization potential ions, such as of NV, favours the second model ("alternative" model) because in the case of a spherically symmetric model one cannot see at long wavelengths the regions of the envelope which are situated close to the stellar surface.

REFERENCES

Castor, J.I., and Van Blerkom, D.: 1970, Astrophys. J. 161, 485.
Nugis, T.: 1984, Tartu Astrofüüs. Obs. Publ. 50, 101.
Sobolev, V.V.: 1960, The Moving Envelopes of Stars. Cambridge, Harvard University Press.
Underhill, A.B.: 1983, Astrophys. J. 265, 933.

A NEW SURVEY FOR Hα-EMISSION-LINE STARS IN THE SMALL MAGELLANIC CLOUD

Marc Azzopardi
European Southern Observatory
Karl-Schwarzschild-Str. 2, D-8046 Garching bei München, F.R.G.

Nicole Meyssonnier
Observatoire de Marseille
2, Place le Verrier, F-13248 Marseille Cedex 4, France

A number of surveys for emission-line objects have been made in the Magellanic Clouds (see Westerlund, 1983). Hα-emission-line objects in the Small Magellanic Cloud (SMC) have been identified mainly by Henize (1956) and Lindsay (1961). Since this pioneering work took place, no other extensive survey for this kind of object - such as the one by Bohannan and Epps (1974) in the Large Cloud - has been carried out in the SMC. Consequently the detection of Hα-emission-line objects in the SMC is still rather incomplete.

Recent observations with the Curtis Schmidt telescope at CTIO allowed one of us (MA) to secure a set of very good SMC plates with the 10° objective-prism (420 Å mm^{-1} dispersion at Hα) through a 120 Å bandwidth Hα interference filter (λ_o = 6565 Å). Exposures of $\frac{1}{2}$, 1, 2 and 4 hours on hypersensitized IIIa-F plates permitted us to survey a field of 3.5 × 3.5 sq. degrees fully covering the central regions of the SMC.

Many new Hα-emission-line objects have been detected, especially in the very crowded fields of the SMC Bar. Since our spectra are very short, the number of overlaps is greatly reduced. In addition, the use of interference filters cuts down the sky background, allowing longer exposures and hence the possibility of reaching fainter stars (Azzopardi, 1983). For instance, our 4 hour exposure plate allowed us to reach the continuum of Be stars 2 to 3 magnitudes fainter than those found by previous Hα-objective-prism surveys.

This new SMC survey - which is in an early stage - resulted in the visual identification of about 3000 Hα-emission-line objects, quintupling the number of these objects found during the previous most complete survey (593 objects) by Lindsay (1961). Work is in progress. The 4 hour exposure objective-prism plate has been fully scanned with the PDS microdensitometer of the "Laboratoire d'Astronomie Spatiale" of Marseille in order to detect the continuum, if any, of the very faint Hα-emission-line objects, to compute accurate positions and also to determine the shape and intensity of Hα-emission-lines.

The main goal of this new survey is the most complete detection possible of Be stars in order to determine more accurately their distribution throughout the large SMC stellar complexes and to understand better their role in stellar evolution. For example, using our technique, 14 Hα-emission-line objects have been identified in the region (5 sq. arc min.) centered on NGC 330 where just 3 (Nos. 298, 303 and 305) were previously known (Lindsay, 1961). All of these are Be stars except L 305 which is a planetary nebula. Its objective-prism spectrum clearly shows the λ6584 [N II] nebular excitation line previously detected by Webster (1976).

An interesting by-product of this survey is the discovery of 15 new faint nebulae - probably planetary nebula candidates - showing λ6584 [N II] lines. For two of them, these spectral features have already been confirmed on Boller & Chivens spectra (8 Å FWHM) obtained at the ESO 3.6 m telescope. If the λ6584 [N II] lines are also confirmed in the remaining objects, they will allow us to investigate whether nitrogen is really deficient in the SMC as suggested by Sanduleak and his associates (1972, 1978).

REFERENCES

Azzopardi, M.: 1983, I.A.U. Coll. 78, 351.
Bohannan, B., Epps, H.W.: 1974, Astron. Astrophys. Suppl. **18**, 47.
Henize, K.G.: 1956, Astrophys. J. Suppl. **2**, 315.
Lindsay, E.M.: 1961, Astron. J. **66**, 169.
Sanduleak, N., MacConnell, D.J., Hoover, P.J.: 1972, Nature **237**, 28.
Sanduleak, N., MacConnell, D.J., Philip, A.G.D.: 1978, Publ. Astron. Soc. Pacific **90**, 621.
Webster, B.L.: 1976, Monthly Notices Roy. Astron. Soc. **174**, 513.
Westerlund, B.E.: 1983, I.A.U. Coll. 78, 333.

A LARGE SAMPLE OF EMISSION-LINE STARS IN THE LARGE MAGELLANIC CLOUD: THEIR LOCATION IN THE HERTZSPRUNG-RUSSELL DIAGRAM

Bruce Bohannan
Astrophysical, Planetary and Atmospheric Sciences
University of Colorado Box 391
Boulder, CO 80309 U.S.A.

The paths that massive stars follow in their evolution can potentially be determined by studying a homogeneous collection of luminous stars. In this investigation, mass-losing stars in a galaxy are studied through a sample of 59 H alpha emission-line stars in the Large Magellanic Cloud (LMC). The most luminous, most massive stars in the LMC, like their counterparts in the Galaxy, are losing mass at a rate that significantly alters their spectroscopic appearance and that affects their evolution.

What do emission-line stars tell us about the evolution of massive stars? Since the Humphreys-Davidson limit (Humphreys and Davidson 1979) cuts through the upper Hertzsprung-Russell (H-R) diagram at an angle, different mass stars will become unstable at different temperatures. One would then expect that there should be certain zones in the H-R diagram where stars of similar peculiar spectroscopic appearance are found.

Eighty-two LMC emission-line stars in the survey by Bohannan and Epps (1974) are brighter than V=12.6 (M_V=-6). Only 28 of the 82 had well established spectral types. New spectral classifications were made from image-tube spectra (43 A/mm) obtained at Cerro Tololo InterAmerican Observatory of 50 of these stars, while published types were used for 12 stars. Here, I consider 59 stars with published UBV photometry. The 13 Wolf-Rayet stars in this magnitude range were not re-observed.

I found four distinct groups of LMC emission-line stars with similar spectroscopic appearance.

- O, B and A supergiants with to varying degrees relatively normal spectroscopic appearance (51% of the sample).

- O and B stars with strong emission features (20%): ranging from stars with P Cygni-like hydrogen Balmer lines, but otherwise normal absorption lines, to stars with HeI 4471A in emission and no absorption features present in the blue.

- S Doradus-like stars with FeII and [FeII] emission (12%).

- Wolf-Rayet stars (17%).

The sample of emission-line stars are plotted in the bolometric magnitude -- effective temperature diagram (Figure 1) for comparison with tracks of theoretical stellar evolution by Bressen, Bertelli and Chiosi (1981) that include convective overshooting and mass loss by radiation pressure. The Wolf-Rayet stars are not plotted.

No clear separation of spectroscopic groups into different regions of the H-R diagram is evident. Indeed, in any small range of luminosity and temperature, there are examples of several different spectroscopic groups. In general, the emission-line stars are found at temperatures cooler than the band of core hydrogen-burning models. In addition, all of the groups are found over a wide range of luminosities and extend to relatively lower masses than one would expect. For example, the S Doradus-like stars span a range of over 4 magnitudes in bolometric magnitude, from nearly -11 for S Doradus itself to -6.2 for BE 436 (HD 38489). Surprisingly, two apparently normal supergiants are above the Humphreys-Davidson limit.

Since there are large uncertainties in the derived effective temperatures, especially for stars with strong emission lines, these conclusions must be considered as extremely tentative and subject to re-evaluation when more direct measures of temperature are made.

REFERENCES:

Bressen, A.G., Bertelli, G., and Chiosi, C. 1981, Astron. Astrophys. 102, 25.
Bohannan, B. and Epps, H. W. 1974, Astron. Astrophys. Suppl. 18, 47.
Humphreys, R. M. and Davidson, K. 1979, Astrophys. J. 232, 409.

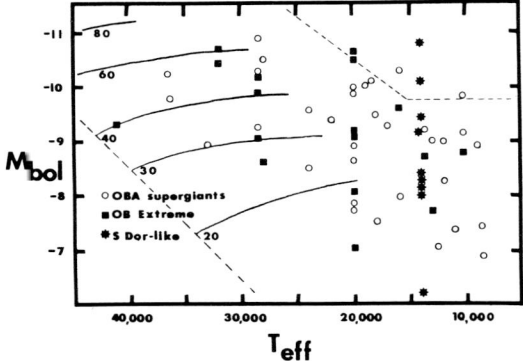

Figure 1: M_{bol}-T_{eff} diagram for 59 emission-line stars in the LMC with theoretical mass tracks of Bressen, Bertelli and Chiosi (1981). The Humphreys-Davidson limit is the dashed line in the upper right.

IS AS 431 A SUPERLUMINOUS WR STAR?

D. J. Stickland and C. Lloyd
Rutherford Appleton Laboratory, Oxfordshire, U.K.

A. J. Willis
University College, London, U.K.

The recent study by Caillault et al. (1985) has shown the emission-line star AS 431 (= WR 147) to be a strong X-ray source and moderately strong radio emitter. Combining optical, near-infrared and radio data, they deduced that its mass loss rate was ~ 4×10^{-4} $M_\odot yr^{-1}$, an order of magnitude greater than is normal for WR stars. They also suggested that it would show up in the IRAS survey and that such data would help to elucidate whether its extreme reddening had a significant circumstellar component. To investigate this possibility and to study the general properties of the object, we have raided the IRAS Point Source Catalogue.

Our reward was fluxes at 12 and 25 microns of 5.26 and 2.57 Jy respectively after colour correction; the background in the Galactic plane is too bright to permit 60 and 100 micron measurement. These data have been added to the ground-based (UBVRIJHK) photometry of Caillault et al., the narrow-band b, v and r measures of Massey (1984) and the radio and X-ray fluxes also presented by Caillault et al. The only other information to hand is the spectral type - given as WN8 by Caillault et al. and as WN7 by Massey. We have adopted the latter.

We have explored two methods of correcting for interstellar extinction. The first is based upon the assumption that AS 431 is a normal WN7 star and that a colour excess can be obtained from the observed b-v and the intrinsic index given by van der Hucht et al. (1981). The result, corresponding to A_v = 9.44 mag., led to an energy distribution quite unlike any single WR star and implies that the intrinsic index is, for some reason, inappropriate.

The other method, recognising that there may be free-free contamination even into the near infrared, relies upon a least-squares fitting of all the UBVR and bvr data to a 30000 K black-body energy distribution with the reddening, represented by the Savage and Mathis (1979) law, as the variable. A value for A_v of 12.0 mag. was found.

confirmed WR stars found to date, with the addition of two new, high pricrity candidates; (2) as in Fig.1, but for the 60 μm IRAS map of M31 (Habing et al. 1984, Ap.J.(Lett.), 278, L59); and (3) the distribution of various WR subtypes as a function of galactocentric radius in M31. The symbols are the same in all three Figures.

(1)
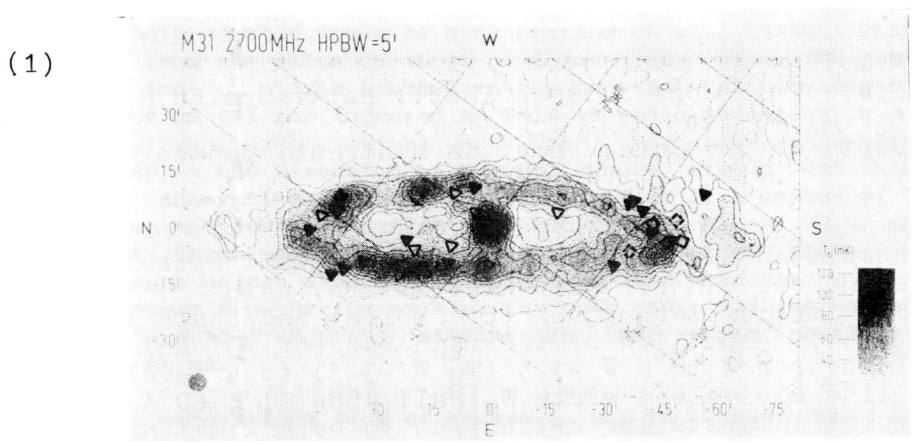

(2)
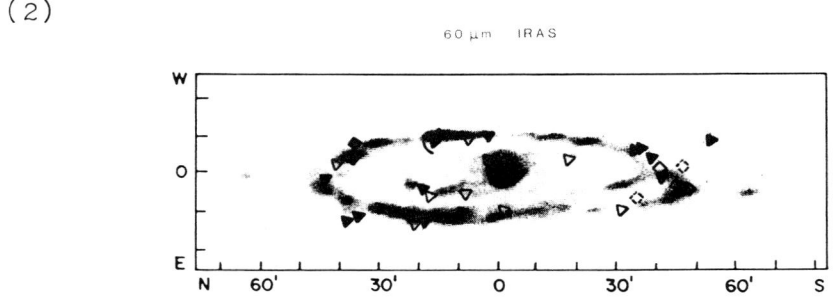

(3)
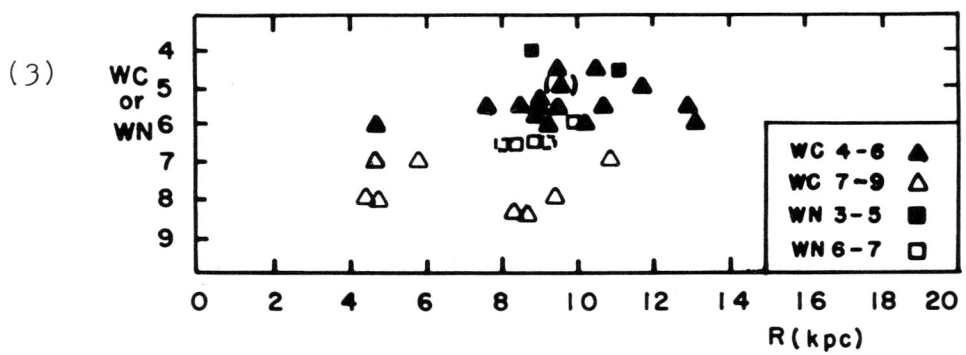

THE LUMINOUS STELLAR CONTENT OF 30 DORADUS AND NGC 3603 - THE NEAREST VISIBLE GIANT HII REGIONS

A.F.J. Moffat
Dépt. de physique, Univ. de Montréal, Canada
M.M. Shara
Space Telescope Science Inst., Baltimore, U.S.A.
W. Seggewiss
Univ.-Sternwarte Bonn, F.R. Germany

30 Dor in the LMC and NGC 3603 in the Galaxy are the nearest visible examples of similar giant (or even more massive supergiant) HII regions being studied in other more distant galaxies, where spatial resolution is a much more serious problem. Hence, understanding 30 Dor and NGC 3603 may provide important clues to understanding extragalactic giant HII regions in general.

On an angular scale, the bright stellar core regions of 30 Dor and NGC 3603 are very similar. On the basis of CCD imagery, Fig.1 shows that the central radial light distribution of each region satisfies a semi-empirical King profile for spherically symmetric, isothermal stellar systems, over a span of two orders of magnitude in radius. Allowing for background starlight (nebular light is almost negligible), the partially resolved, luminous central objects R136 (which emits $\sim$ 20% of the total starlight in 30 Dor) and HD 97950, respectively, do not contain stars that are significantly brighter than adjacent, isolated stars of high but normal luminosity (and mass), i.e. we find no evidence for supermassive stars.

Wolf-Rayet stars were detected using CCD image subtraction (narrow band minus broad band filter centered on the intense 4650-4686 Å emission feature). We confirm that the number ratio of WR to O stars in 30 Dor and NGC 3603 is normal (~ 0.05) compared to other less exotic regions. This may mean that star formation bursts do not necessarily lead to high WR/O ratios. With the nearer distance modulus for the LMC of 18.2 (Schommer et al. 1984, Ap.J.(Lett.),$\underline{285}$, L53), the WN6/7 and WN+O/Of stars in 30 Dor have identical mean absolute magnitudes $M_V \cong -6.7 \pm 0.6$ (σ), similar to their Galactic counterparts (Moffat et al. 1985, in prep.)

On an absolute scale, the stellar core of NGC 3603 is much more compact, with central density ~ two orders of magnitude greater than that of the 7 x more distant core of 30 Dor. The latter is similar to Galactic globular clusters. This may have some interesting dynamical consequences and may reflect a fundamental difference between massive star clusters in the Galaxy and in the LMC.
More details will be published in the Astrophysical Journal.

Fig.1: Accumulative magnitude versus radius from the center of 30 Dor and NGC 3603, relative to single stars on the same image, based on broad band CCD images at 4700 A. True distance moduli of 18.6 and 14.2, respectively, are assumed. Two extreme limits (a and b) for background subtraction in 30 Dor are indicated.

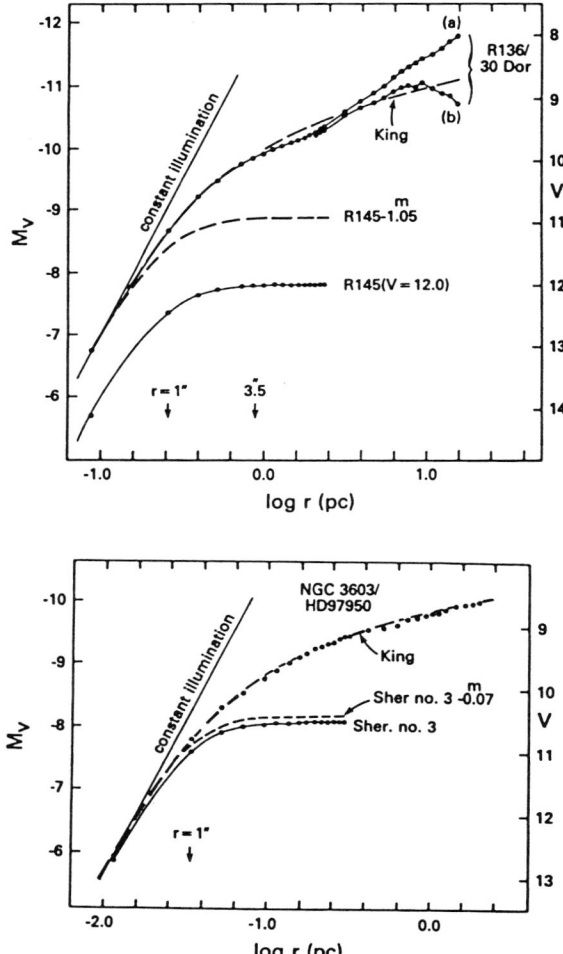

THE MASSIVE STELLAR WIND OF THE HUBBLE-SANDAGE VARIABLE S DORADUS

C. Leitherer[1], I. Appenzeller[1], G. Klare[1],
H.J.G.L.M. Lamers[2], O. Stahl[1], L.B.F.M. Waters[2],
B. Wolf[1]
1 - Landessternwarte Königstuhl, Heidelberg, Germany
2 - Space Research Laboratory, Utrecht, The Netherlands

We present coordinated spectroscopic and photometric observations obtained during the present bright phase of the luminous variable star S Dor in the LMC. High resolution spectrograms in the satellite UV and in the visual range were obtained with IUE and with CASPEC (attached to the ESO 3.6-m telescope), respectively. Moreover, photometric UBVRIJHKL observations were carried out.

Our new observations support earlier suggestions that S Dor consists of a hot stellar core which during the maximum state is surrounded by a very extended (R ≃ 300 $R_\odot$, T_{eff} ≃ 8000 K) optically thick gaseous envelope. During the maximum state the surface of this optically thick envelope is observed as the star's pseudo-photosphere. From the line-intensity ratio of temperature sensitive absorption lines (Mg II 4481, He I 4471, etc.), we find physical conditions typical of an early A star in July 1983. In August 1984, when S Dor was visually fainter, the spectrum corresponds to a late B star. The radial velocity of the absorption lines varies around the systemic velocity with an amplitude of about 10 km s^{-1} and on time scales of months indicating pulsation-like motions of the pseudo-photosphere. Apart from these pulsating motions, a systematic, directed outflow of the pseudo-photosphere layers is not measurable. Only above the pseudo-photosphere (where the stellar wind flows) the outflow velocity increases.

The LWR-IUE spectrum is dominated by blue-shifted singly ionized metal lines arising in the stellar wind. The edge velocity v_{edge} of these lines increases with decreasing excitation potential. This relation can be understood in terms of a wind velocity field with outward increasing velocity.

The velocity field as derived from the UV absorption lines is consistent with the results obtained from the forbidden lines. The maximum wind velocity derived from the FWZI of the [Fe II]-lines in the visual is about 130 km s^{-1}.

Further support for an outward increasing velocity field provide the asymmetric line profiles of the forbidden [Fe II] lines: They are in striking agreement with those predicted theoretically for optically thin lines originating in expanding circumstellar envelopes with outward increasing velocity fields.

Our spectra also contain weak but reliably detected emission lines of [N II]. The [N II] $\lambda\lambda$6548, 6583 lines show approximately flat-topped and slightly double-peaked profiles, as predicted for spherically symmetric constant-velocity flows. As these lines (which have much lower critical densities than the [Fe II] lines) can only form far from the stellar surface, constant-velocity profiles are not unexpected. However, interestingly, the profiles of these lines have FWZIs of only 117 km s^{-1}, corresponding to flow velocities of <60 km s^{-1}. Hence, at the distance where these lines form ($10^3 - 10^4$ R_*) the flow velocity is significantly lower than the maximum velocity in the inner parts of the envelope. With a velocity field consistent with the other observational data a ballistic deceleration to such a low end velocity would require an unreasonably high mass of S Dor. Hence it seems more likely that the observed deceleration is due to an interaction between the wind and ambient interstellar matter. The density field and (in combination with the velocity field) the mass-loss rate $\dot{M}$ of S Dor can be derived from the IR emission and the Balmer emission lines. We find that a slow increase of the wind velocity with the terminal velocity being reached only at large distances (>10 R_*), and a mass-loss rate of $5 \cdot 10^{-5}$ $M_\odot yr^{-1}$ < $\dot{M}$ < $1 \cdot 10^{-4}$ $M_\odot yr^{-1}$ can account for the observed infrared excess and Balmer-line profiles. This velocity law provides also an explanation for the χ-dependence of the edge velocity of the UV lines.

THE WR/OB NUMBER RATIO WITHIN 2.5 KPC FROM THE SUN.

D. VANBEVEREN
K.I.H. Group T   and   Vrije Universiteit Brussel
Campus Blauwput   Dept. of Physics
Vuurkruisenlaan 4   Pleinlaan 2
3000 Leuven   1050 Brussels
Belgium   Belgium

ABSTRACT. Comparing the observed and theoretical HR diagram for massive stars, it is concluded that <u>at least</u> 34 % of the stars within 2.5 kpc from the sun with initial ZAMS mass larger than 35 M_o have not yet been detected. It is argued that the number of <u>core helium burning pop I WR stars</u> within 2.5 kpc from the sun may be as low as 26. If the progenitors of WR stars have initial ZAMS masses larger than 35 M_o, one then concludes that the observed number ratio (core helium burning pop I WR stars)/(WR progenitors) within 2.5 kpc from the sun ranges between 0.05 and 0.09.

1. THE NUMBER OF OB STARS WITH $M \geqslant 35\ M_o$ WITHIN 2.5 KPC FROM THE SUN.

If the observed HR diagram for massive stars within 2.5 kpc from the sun is compared with the theoretical HR diagram (evolutionary tracks without, respectively with overshooting ; the T_{eff}-spectral type relation given by Humphreys and McElroy (1985, Ap. J. in press) is used), one finds the surprising result that ∼ 90 % of the observed stars with initial ZAMS mass $M_i \geqslant 35\ M_o$ are <u>older</u> than 2.10^6 yrs (resp. $2.5\ 10^6$ yrs). This means that if stellar evolution represents reality (at least for luminosity class V and IV stars), within 2.5 kpc from the sun we are missing a large number of stars with $M_i \geqslant 35\ M_o$ <u>younger</u> than 2.10^6 yrs (resp. $2.5\ 10^6$ yrs). This number can be estimated as follows. The average core hydrogen burning lifetime for stars with $M_i \geqslant 35\ M_o \approx 4.10^6$ yrs (resp. 5.10^6 yrs). Within 2.5 kpc from the sun there are 260 stars (resp. 220) with $M_i \geqslant 35\ M_o$ older than 2.10^6 yrs (resp. $2.5\ 10^6$ yrs). We then obviously expect to see ∼ 260 (resp. 200) younger than 2.10^6 yrs (resp. $2.5\ 10^6$ yrs) and we only observe 60 (resp. 50) stars. I then conclude that there are 520 (resp. 440) OB type stars within 2.5 kpc from the sun with $M_i \geqslant 35\ M_o$; 200 (resp. 150) have not yet been observed (some if not most of them may still be hidden in clouds, Mezger, 1976, Proc. 3rd Eur. Astr. Meeting).

2. THE NUMBER OF WR STARS WITHIN 2.5 KPC FROM THE SUN BURNING HELIUM IN THEIR CORE.

I will use the WR stars within 2.5 kpc from the sun compiled by Conti, Garmany, de Loore, Vanbeveren (1984, Ap. J. 274, 302). Stellar aggregate membership is taken from Lundström and Stenholm (1985, A.A. in press). There are 34 WR stars within 2.5 kpc from the sun which are certainly burning helium in their core (i.e. WNE and WC types). There are 9 WNL stars, however I propose that at least half of them are core hydrogen burning stars. The number 34 then is obtained by assuming that all WR stars which are observed are pop.I WR stars, i.e. we can apply the magnitude calibration obtained from the aggregate members, to all field WR stars. This method however may overestimate the number of pop I WC field stars if a number of them are in reality pop II. In order to have an estimate of this number, let us proceed as follows. Accounting for the fact that the IMF for massive field stars (index f) is steeper than the IMF of aggregate stars (index a), whereas the minimum progenitor mass for the formation of WC stars may be larger than for WNE stars, it follows from evolution that for the number ratio WNE/WC the following theoretical relation should hold :

$$(WNE/WC)_a \leqslant (WNE/WC)_f \quad (1)$$

If for obvious reasons we only consider single WR stars, the observations tell us that $(WNE/WC)_a \approx 0.5-0.6$ and $(WNE/WC)_f \approx 0-0.1$ (the intervals for each ratio reflect the fact whether or not possible aggregate members in the study of Lundström and Stenholm are considered as real members). It is clear that relation (1) is not fulfilled. There are two possible ways out for this discrepancy, i.e. or at least 7 out of the 12 WC field stars have not yet been recognised as aggregate members, or at least 8 WC field stars which are considered as pop I WR stars are in reality pop II WR stars. In the latter case obviously we only have to consider 26 WR stars within 2.5 kpc from the sun which are pop I core helium burning stars.

3. THE WR/OB NUMBER RATIO WITHIN 2.5 KPC FROM THE SUN.

It has frequently been argued in litterature that WR stars are descendants from OB type stars with $M_i \geqslant 35 M_\odot$. Using direct number counts, the observed WR/OB ($M_i \geqslant 35 M_\odot$) number ratio within 2.5 kpc from the sun ≈ 0.13 (resp. 0.17 when overshooting models are adopted). Using the results of section 1, this ratio decreases down to 0.08 (resp. 0.1). Accounting finally for the results of section 2, i.e. the number of core helium burning pop I WR stars ranges between 26 and 38, I conclude that the observed (core He burning pop I WR stars)/(WR progenitors) number ratio within 2.5 kpc from the sun $\approx 0.05-0.07$ (resp. 0.06-0.09).

WOLF-RAYET STARS IN M31'S OB ASSOCIATIONS

T.E. Armandroff[1], P. Massey[2], and P.S. Conti[3]

(1) Yale University Observatory
(2) Kitt Peak National Observatory, NOAO
(3) Joint Institute for Laboratory Astrophysics

It is well-known that the massive star population exerts a disproportionately great influence on the environment within a galaxy, especially through the ejection of chemically enriched material and the input of energy into the interstellar medium. In order to more fully understand the structure and evolution of galaxies, it is important to determine how the massive star content of a galaxy changes with Hubble type, galaxian luminosity and metallicity. Conversely, if our goal is to understand the evolution of massive stars, it is helpful to observe them in a variety of environments, as stellar wind properties should depend on chemical abundances. We are involved in a long-term project to survey the massive star population in a diverse sample of nearby galaxies. Unfortunately, it is extremely difficult to isolate a sample of relatively unevolved massive stars in an external galaxy, since the colors and absolute visual magnitudes of a 20 M_0 and 100 M_0 star are nearly identical, and can be further confused by differential internal extinction. Wolf-Rayet (WR) stars, however, are relatively easy to detect due to their strong emission-line spectra and high luminosities. Since WR stars are known to have evolved from O stars and because their space distribution in the solar neighborhood is identical to that of the most massive O stars, we have decided to use WR stars as tracers of the massive star population in nearby galaxies.

When we began these studies, it was accepted that the surface density of WR stars decreased substantially from the Galaxy, to the LMC, to the SMC. Similarly, the ratio of WC to WN types fell fom 1:1 in the solar neighborhood, to 1:4.5 in the LMC, to 1:7 in the SMC. This behavior was widely attributed to the metallicity differences known to exist between these galaxies. Subsequently, Massey and Conti (1983) surveyed M33 and found that the WC to WN ratio in its central regions is about five times higher than that of the LMC, despite their similarity in chemical abundances. This pointed to some factor in addition to metallicity, the IMF for example, which controls the WR content. Furthermore, Armandroff and Massey's (1985) survey of the Magellanic-type Irregulars NGC 6822 and IC 1613 revealed that 6822 resembles the SMC in its WR surface density and WC to WN ratio, while 1613 is like the LMC. Since 6822 and the LMC appear to have similar abundances, while 1613 and the SMC are both more metal-poor, metallicity variations cannot account for the differences in WR content among these galaxies.

M31's prominence as the most massive and most metal-rich galaxy in the Local Group, and its morphological similarity to our galaxy made its inclusion in this survey essential. Unfortunately, M31's large inclination and substantial internal extinction have discouraged large-scale surveys, and its massive star content remains relatively unexplored. Recognizing these problems, we have decided to concentrate on individual OB associations (van den Bergh 1964), selecting the ones with the least amount of reddening and the ones apparently richest in massive stars, as evidenced by their Hα nebulosity. This procedure should not bias our conclusions since most of the WR stars identified in M33, NGC 6822 and IC 1613 are located within the boundaries of previously catalogued associations.

The observational technique used to survey M31 is nearly identical to that employed by Armandroff and Massey (1985) in their study of NGC 6822 and IC 1613. KPNO 4-m PFCCD exposures have been obtained of selected OB associations through UBV filters and three narrow-band interference filters. The interference filters were designed for the optimal detection of WR stars, and the unambiguous separation of these into WC and WN types. Consequently, two of the filters include strong WR emission features, while the third passes only continuum. WR stars can be detected by blinking either of the emission-line exposures with the continuum exposure, and stars can be classified as WC or WN by comparing their brightness in all three exposures. However, in order to be complete to small magnitude differences (e.g. weak-lined WN's) and in very crowded regions, it is necessary to go a step beyond blinking. Accordingly, a magnitude for every image on a set of exposures is determined using Peter Stetson's crowded-field photometry program, DAOPHOT. Magnitude differences between the emission-line and continuum frames are then calculated for each image, significant differences ($>2.5\sigma$) indicating a WR candidate. The UBV exposures are also reduced using DAOPHOT, and yield a color-magnitude diagram for each association.

Poor weather last Fall allowed us to search only nine associations (NGC 206, OB 8, 9, 10, 48, 54, 102, 136, 139) and limited us to UBV photometry of six (NGC 206, OB 8, 9, 10, 48, 102). Although we plan to obtain more observations this year, the data in hand are very suggestive. 30 WR candidates were detected in the regions surveyed, of which 17 are extremely probable ($>3.2\sigma$) WR stars. Of this extremely probable group, 11 are of WN type, yielding a WC to WN ratio of about 1:2.

The results outlined above are in sharp contradistinction with the photographic WR survey of Moffat and Shara (1983). In a survey covering two-thirds of M31's surface area, they identified 21 WR candidates. Subsequent spectroscopy indicated that 14 of these were WC stars, 3 were WN stars and the remaining 4 were non-WR, resulting in a WC to WN ratio of about 5:1. There are several reasons to suspect that the Moffat and Shara study suffers from substantial incompleteness. The root of the problem is that WC stars, on average, have much stronger lines than WN stars. The strongest line in WC stars, CIII $\lambda 4650$, typically exceeds 1000A in equivalent width, while the strongest line in WN stars, HeII λ4686, ranges from 30A to 200A in equivalent width. Imaging surveys are complete to a given magnitude difference, which translates directly to a limiting equivalent width. If this completeness limit exceeds the typical equivalent width for WN stars, then primarily WC stars will be found. One of the OB associations which we surveyed falls within the search area of Moffat and Shara (1983). We detect both of their spectroscopically confirmed WR stars in this association, and both are very strong-lined. They, however, missed the additional WN stars that we have identified. Similarly, in NGC 6822, Moffat and Shara (1983) failed to detect any WR's, while Armandroff and Massey (1985) located 12 WR candidates, of which 7 are extremely probable (all of WN type). Moffat and Shara did not even detect the strong lined WN star in 6822 first discovered by Westerlund et al. (1983) using the grism technique. Based on the overlap between our two surveys, we estimate that the completeness limit of Moffat and Shara's (1985) survey is 150A, which explains the surprisingly low WR surface density and anomalously high WC to WN ratio that they find for M31.

The primary goal of this study is to use the observed WR content to deduce global measures of the massive star population in M31. Two quantities, the present-day star formation rate (SFR) and the initial mass function (IMF), can be used to parameterize this population. As has been discussed earlier, the hypothesis that metallicity controls the WR surface density and the WC to WN ratio fails to explain the results of recent WR surveys. Indeed M31, with a WC to WN ratio of 1:2 and a metallicity somewhat higher than that of the Galaxy, is inconsistent with this hypothesis. The simplest alternative explanation of the WR statistics is that the WC to WN ratio is telling us something about the IMF, given the evidence from binaries that WC stars evolve from more massive progenitors than WN's (Massey 1981). A low value of WC/WN would indicate a steeply sloping IMF, while a ratio near unity would correspond to a flatter IMF. The surface or mass density of WR stars, in this scenario, would be sensitive to the SFR, a higher WR density indicating a more vigorous SFR. M31, with a WC to WN ratio of about 1:2, therefore appears to have a more steeply sloping IMF than the Galaxy or the inner regions of M33. Two of the regions suveyed in M31, NGC 206 and OB 48, have a surface density of WR stars as high as any region in M33. For these two OB associations, the SFR must be comparable to that in the most active regions of M33. The regions in M31 that are actively forming massive stars appear to be as efficient as those in other galaxies, although there may be fewer of them. More OB associations need to be surveyed to answer this question.

REFERENCES

Armandroff, T.E., and Massey, P.: 1985, Astrophys. J., 291, 685.
Massey, P.: 1981, Astrophys. J., 246, 153.
Massey, P., and Conti, P.S.: 1983, Astrophys. J., 273, 576.
Moffat, A.F.J., and Shara, M.M.: 1983, Astrophys. J., 273, 544.
van den Bergh, S.: 1964, Astrophys. J. Suppl., 9, 65.
Westerlund, B.E., Azzopardi, M., Breysacher, J., and Lequeux, J.: 1983, Astron. Astrophys., 123, 159.

IRAS OBSERVATIONS OF WOLF-RAYET STARS

K. A. van der Hucht, SRON Laboratory for Space Research Utrecht
T. A. Jurriens, SRON Laboratory for Space Research Groningen
F. M. Olnon, Netherlands Foundation for Radio Astronomy
P. S. Thé, Astronomical Institute Anton Pannekoek, Amsterdam
P. R. Wesselius, SRON Laboratory for Space Research Groningen
P. M. Williams, United Kingdom Infrared Telecope Unit

ABSTRACT. IRAS PSC, LRS, and CPC observations of Wolf-Rayet stars are used as diagnostics of hot circumstellar dust shells, cool dust in WR ring nebulae, the Ne/He abundance ratio, and the interstellar extinction. In two cases the IR energy distributions are indicative of a WR planetary nucleus status rather than a Population-I WR status.

1. INTRODUCTION

Groundbased IR photometry and spectrophotometry in the seventies had already demonstrated that WR stars can have three important characteristics in the IR: emission spectra of ionic He, C, and N recombination lines, f-f excesses caused by strong stellar winds, and thermal emission radiated by heated circumstellar dust. These data were limited to about 20 μm.

IRAS observations provide photometry at 12, 25, 60, and 100 μm, low resolution spectra (LRS) from 8 to 22 μm, and, with the Chopped Photometric Channel (CPC), 12'x9' maps at 50 and 100 μm.

2. OBSERVATIONS

The observations used are from the IRAS Point Source Catalog (1985), the IRAS Spectral Atlas (1985), and IRAS-CPC observations (Wesselius et al., 1985).

Invaluable for the interpretation of IRAS data are complementary near-IR ground based observations. For this purpose, three of us (KAvdH, PST and PMW) have, over the past five years, observed WR stars at $JHKLMN_1N_2N_3Q_0$ from ESO and UKIRT.

3. THE IR SPECTRUM OF γ^2 VELORUM

The IRAS LRS spectrum of γ^2 Vel shows strong (0.9×10^{-17} W/cm^2) emission of the fine structure line [NeIII]$\lambda 15.5 \mu$m.

Shortward of 13 μm the γ^2 Vel spectrum had been observed and identified by Aitken et al. (1982), who derived an abundance ratio of N(NeII)/N(He)=0.0026. The IRAS LRS spectrum allows an evaluation of the NeIII/He and thus the Ne/He abundance ratio. We find N(Ne)/N(He)=0.009, i.e. 7.5 times the cosmic value. This reflects the evolved status of the Wolf-Rayet stars. Evolutionary calculations by Maeder (1983) yield for WC stars $N(Ne)/N(He)=0.0066$, remarkably close to our result.

If the Ne abundance of γ^2 Vel is representative for all WC stars, then the 38 WC stars within 3 kpc from the Sun provide a ^{22}Ne input of 8.8×10^{-6} $M_\odot/yr \cdot kpc^3$ for the chemical evolution of the solar neighbourhood.

Particulars of this study are given by van der Hucht and Olnon (1985).

4. LOW IONIZATION WR STARS WITH CIRCUMSTELLAR DUST

As the IRAS observations confirm, low ionization WR stars, i.e. WN10-11 and WC8-10 stars, have IR excesses indicative of thermal emission by hot (~900 K) circumstellar dust. For the brightest late WC stars, this has been known since 1974.

Speckle interferometry of WR104 (Allen et al., 1981) has shown that its dust shell radius is of the order of 100 A.U. This means that the grain size is of the order of 0.01 μm and that the dust is formed within the stellar wind sphere. The dust is most probably amorphous carbon. Because no variations have been observed in the past 10 years, the dust must be continually replenished at a distance of 100 A.U. from the star. As an example, the dust shell parameters of WR104 are $T(dust)=900$ K, $M(dust)=10^{-6}M_\odot$, and (with $d = 1.58$ kpc) $L_{IR} = 3.4 \times 10^4 L_\odot \approx 0.14 L_*$. The stellar winds of WR stars are among the most hostile environments in which grains are believed to form.

Particulars of this study are given by Williams et al. (1985).

5. THE 9.7 μm ABSORPTION FEATURE

Superimposed on the smooth energy distributions caused by hot circumstellar dust, the 9.7 μm silicate feature appears in absorption in WR104, WR112, and WR118. These observations confirm groundbased spectroscopy from 8 to 13 μm by Roche and Aitken (1984), who observed the same stars together with three more WR stars. They demonstrated that the silicate absorption features are entirely of interstellar origin. Using the most recent groundbased photometry and the intrinsic WR parameters given by Hidayat et al. (1985), we find for the relation between visual extinction and 9.7 μm absorption strength $A_V = 19.8$ (± 1.7) $\tau_{9.7\mu m}$.

6. TWO NEW WR CENTRAL STARS OF PLANETARY NEBULAE

The IRAS data for WR72 and WR124 reveal energy distributions, which are significantly different from those of all other WR stars. They are indicative of circumstellar dust with temperatures of respectively 85 and 100 K. In that respect they resemble planetary nebulae (Pottasch et al., 1984).

In the case of WR72 (= Sand.3), it was hinted earlier that it is more likely the central star of a planetary nebula than a Pop.I WR star (Barlow and Hummer, 1982). The IRAS data confirm this.

M1-67, the nebula surrounding WR124 (=209 BAC), was for many years considered to be a planetary nebula (PK 50+3 1), till Cohen and Barlow (1975) argued for a Pop.I WR ring nebula status. The IRAS data, however, favour a planetary nebula status again. The fact that the central star WR124 is a single-line spectroscopic binary with a very small mass function (Moffat et al., 1982), suggests that it is a pre-cataclysmic variable, in view of current ideas on the evolution of binary central stars of planetary nebulae.

Particulars of this study are given by van der Hucht et al. (1985).

7. DUST IN WOLF-RAYET RING NEBULAE

IRAS observations of most of the known WR ring nebulae were carried out with the IRAS-CPC instrument (Wesselius et al., 1985), providing 12'x9' maps at 50 and 100 μm with 1!5 spatial resolution. The WR ring nebula RCW58 has 50 μm IR isophots which coincide very well with the dark regions visible in its Hα picture. In a preliminary analysis, we find that the dust temperatures in RCW58, as well as in the other with the CPC observed WR ring nebulae, are of the order of 35 K, in good agreement with size and energy balance considerations.

An elaborate version of this paper has been published by us in: Birth and Evolution of Massive Stars and Stellar Groups, Proc. of a Colloquium in honour of Adriaan Blaauw (W. Boland & H. van Woerden, eds.).

REFERENCES

Aitken, D.K., Roche, P.F. and Allen, D.A.: 1982, Monthly Notices Roy. Astron. Soc. **200**, 69P.
Barlow, M.J. and Hummer, D.G.: 1982, in C.W.H. de Loore & A.J. Willis (eds.), Wolf-Rayet Stars, Observations, Physics, Evolution, Proc. IAU Symp. No. **99** (Dordrecht: Reidel), p. 387.
Cohen, M. and Barlow, M.J.: 1975, Astrophys. Letters **16**, 165.
Hidayat, B., Supelli, K.R., Admiranto, A.G., and van der Hucht, K.A.: 1985, in preparation.
van der Hucht, K.A. and Olnon, F.M.: 1985, Astron. Astrophys. submitted.
van der Hucht, K.A., Conti, P.S., Lundstrom, I., and Stenholm, B.: 1981, Space Science Reviews **28**, 227.
van der Hucht, K.A., Jurriens, T.A., Olnon, F.M., Thé, P.S., Wesselius, P.R., and Williams, P.M.: 1985, Astron. Astrophys. Letters, **145**, L13.
IRAS Point Source Catalog, 1985, Explanatory Supplement, U.S. Government Printing Office.
IRAS Spectral Atlas, 1985, Astron. Astrophys. Suppl., to be submitted.
Maeder, A.: 1983, Astron. Astrophys. **120**, 113.
Moffat, A.F.J., Lamontagne, R., and Seggewiss, W.: 1982, Astron. Astrophys. **114**, 135.
Pottasch, S.R., Baud, B., Beintema, D., Emerson, J., Habing, H.J., Harris, S., Houck, J., Jennings, R., and Marsden, P.: 1984, Astron. Astrophys. **138**, 10.
Roche, P.F. and Aitken, D.K.: 1984, Monthly Notices Roy. Astron. Soc. **208**, 481.
Wesselius, P.R., de Jonge, A.R., Beintema, D.A., Jurriens, T., Kester, D.K., van Weerden, J.E., de Vries, J., Peralt, M.: 1985, Internal Report Ruimteonderzoek Groningen.
Williams, P.M., van der Hucht, K.A., and Thé, P.S.: 1985, in preparation.

HIGH-DISPERSION SPECTROSCOPY OF THE BRIGHTEST EMISSION-LINE STARS OF THE MAGELLANIC CLOUDS

O. Stahl[1], B. Wolf[1], M. de Groot[2], C. Leitherer[1]
1 Landessternwarte, Königstuhl, Germany
2 Armagh Observatory, Northern Ireland

We present an atlas of high dispersion spectra of 24 of the brightest peculiar emission-line stars of the Magellanic Clouds.

Our spectra cover the wavelength range from 3600 to 4900 Å. They have been obtained from 1970 to 1984 with the coudé spectrograph of the ESO 1.52 m telescope at La Silla, Chile. The spectral resolution is 0.4 Å for most of the spectra and 0.2 Å for the very brightest stars. Up to 11 spectra are available for one star. In addition, we have done UBVRIJHK(LM) photometry at several epochs of all stars of our sample.

The largest group in our sample are the stars which we shall call 'classical P Cyg stars'. They are characterized by classical Beals type I P Cyg profiles at H_β and sometimes also at higher Balmer lines. In addition, they show a photospheric spectrum of a B type supergiant. Stars with such properties are R 55, R 62, R 74, R 81, R 85, R 110, R 116 and R 40.

Another group of stars which includes R 84, R 99, S 9 and S 131 are the 'Of-like' objects. They have the Of-features He II 4686 and N III 4634-42 in emission but contrary to normal Of stars they exhibit also strong Balmer emission lines. The photospheric spectrum of an O star is weakly present. We consider the Of-like objects as the hotter counterparts of the P Cyg stars.

The S Dor variables S Dor, R 71 and R 127 are the most extreme variables of the P Cyg- and Of-like stars. Most of the other P Cyg stars and of the Of-like stars are also known to be variable but with smaller amplitude than the S Dor variables.

A group of stars which seems to be fairly homogeneous and well-separated are the B[e] stars with dust shells. R 4, R 50, R 66, R 82, R 126, S 12, S 22 and S 134 belong to this group. In addition to a large IR excess, which is explained by thermal emission from dust with a temperature around 1000 K, they all have Fe II and mostly also [Fe II] emission lines. The photospheric absorption lines are weak or absent in all stars of this group. The stars with dust shells tend to show only small variations, both photometrically and spectroscopically. All this suggests that the supergiant B[e] stars with dust shells form a group of stars that is quite different from the P Cyg and Of-like stars.

A more detailed paper will appear in the Supplement Series of Astronomy and Astrophysics.

MWC 300: A RUNAWAY HYPERGIANT

B. Wolf, O. Stahl
Landessternwarte
Königstuhl
6900 Heidelberg
West Germany

We analysed highly resolved CASPEC spectra with a high S/N-ratio of the peculiar emission-line B-star MWC 300, which is surrounded by a circumstellar dust shell. These high quality spectra enabled us to study the photospheric spectrum of MWC 300. By comparing the absorption spectrum of MWC 300 with the ones of early B supergiants we found a close resemblance with the B1 hypergiant HD 169454. We also found luminosity sensitive fluorescence lines of Fe III 115 and 117 both in MWC 300 and in the most luminous stars of our sample, confirming the hypergiant nature of MWC 300. We estimated an absolute visual magnitude $M_v = -8$ for MWC 300. Assuming a bolometric correction of B.C. = -1.5 mag we derive $M_{bol} = -9.5$. This shows that MWC 300 is very similar to the B e supergiants of the Magellanic Clouds. From its absolute magnitude we derive a distance of $z = 560$ pc from the galactic plane. It is conjectured that MWC 300 is a runaway hypergiant, released via a supernova explosion in a very massive binary star system.

NEW RESULTS ON ETA CARINAE.
EVIDENCE FOR AN ASYMMETRIC, INHOMOGENEOUS WIND.*

R. Viotti, L. Rossi
Istituto Astrofisica Spaziale, CNR, Frascati

A. Altamore, C. Rossi
Istituto Astronomico, Università La Sapienza, Roma

A. Cassatella
IUE Observatory, European Space Agency, Madrid, affiliated
to the Astrophysical Division, Space Science Department

The very peculiar object Eta Car is one of the best laboratory for the study of those physical processes – such as mass loss, superionization, dust condensation, wind interaction with the i.s. medium – that presently are of great astrophysical interest, especially for the study of the most luminous stars. For its light history and high luminosity Eta Car may also be considered as the galactic counterpart of the Hubble-Sandage variables. Eta Car is one of the rare astrophysical objects with evidence of dust condensation from ejected stellar matter (Andriesse et al.78) On the other side the star is also producing a strong, hard X-ray flux (Chlebowski et al. 1984), and the problem is whether there is any physical reason to have these two quite different processes in the same stellar environment. In any case rather extreme physical conditions are required which cannot be verified in a uniformly, spherically symmetric atmospheric enevelope. Andriesse et al. in fact suggested the presence of strong inhomogeneities, such as filaments, possibly related to the presence of a strong magnetic field. This may also explain the X-ray emission. In the following we shall present new optical and UV observations of Eta Car and its small nebula with the aim of clearfying the physical nature of its wind.

Very high spectral resolution ($\lambda/\delta\lambda = 10^5$) line profiles have been obtained in February 1984 at ESO with CAT-CES. The emission lines are characterized by a narrow component and asymmetric broad wings extending to ±500 km/s. The NaI resonance doublet (Fig.1) only displays a very wide emission with broad, absorptions extending to –600 km/s. It is worth noting that during some epochs NaI was more intense than the nearby HeI line.

A large variety of line excitations and profiles is also shown by the UV spectrum (Fig.2). Viotti et al.(1981) have identified the NV doublet

---

* Based on data collected at ESO, La Silla, and with IUE at VILSPA.

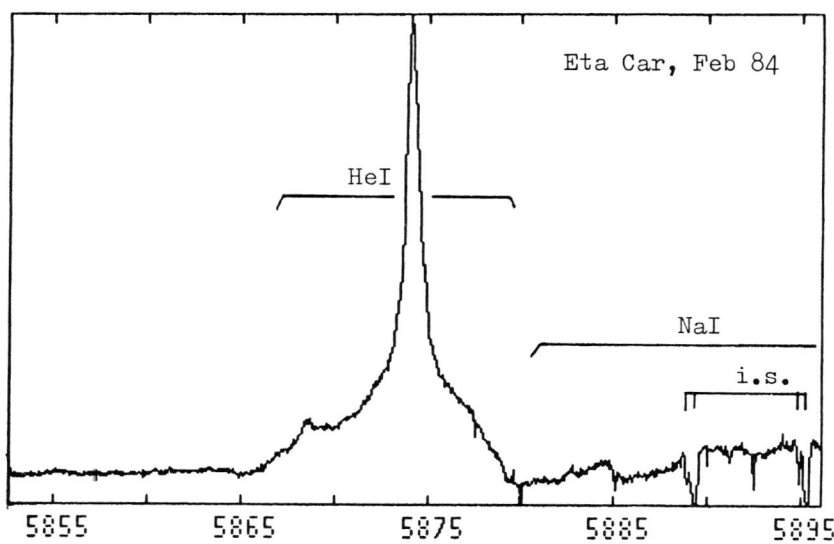

Figure 1. Very high resolution spectrum of Eta Car: line profiles of HeI 5875 A and of the NaI $D_{1,2}$ doublet (ESO: CAT-CES, February 1983).

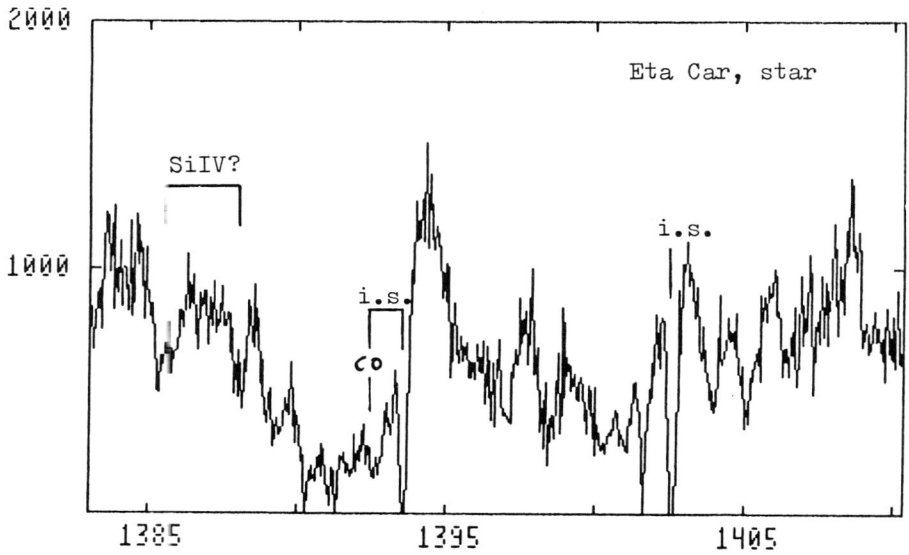

Figure 2. IUE high resolution spectrum of Eta Car near the SiIV resonan doublet, showing the i.s. lines and the broad absorption, with possible high radial velocity components.

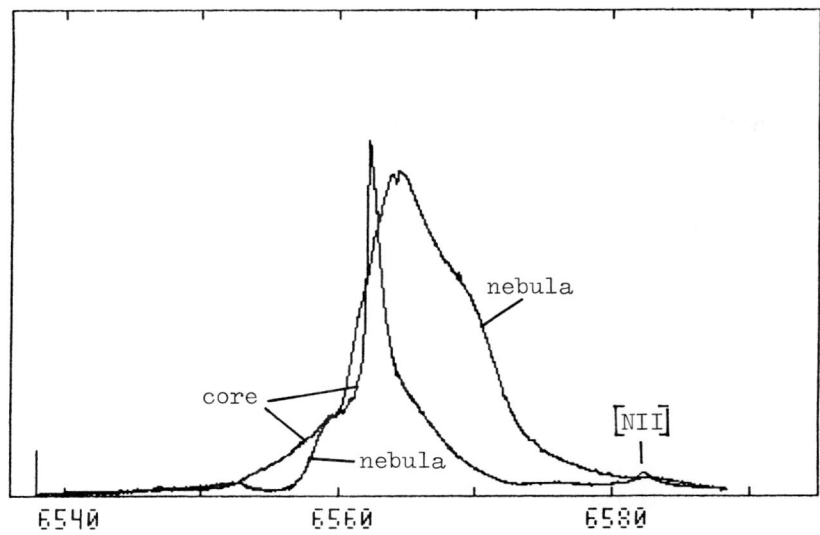

Figure 3. Very high resolution Hα profile in the stellar core and of the nebula 3"-5" South of Eta Car (ESO: CAT-CES, February 1984, April 1985).

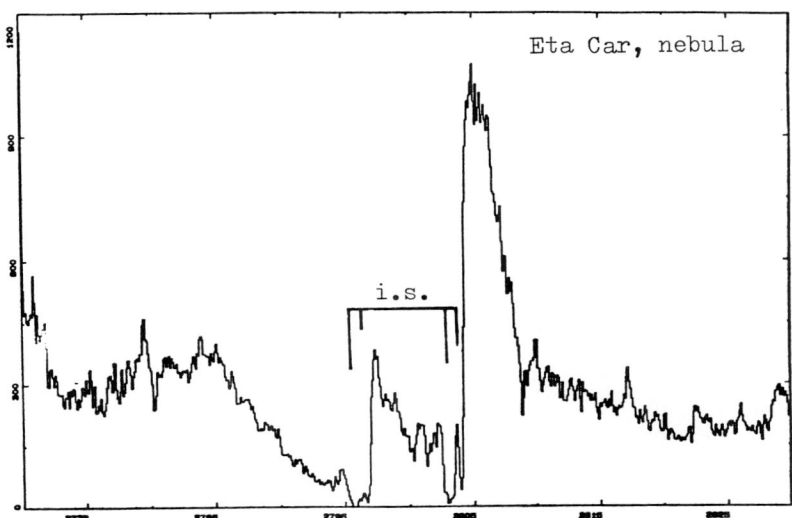

Figure 4. The high resolution UV spectrum of the Eta Car nebula near the MgII doublet (IUE, December 1980).

with a very broad P Cygni profile extending to -600 km/s, the same velocity observed in the low ionization lines of NaI (Fig.1), CaII and MgII (Cassatella et al.1979). This indicates that both high and low ionization lines are present in the whole velocity range of the stellar wind.

Additional information about the process of mass ejection from Eta Car can be derived from the observation of the nebular halo. The spatial structure of the halo is very complex with condensations, shells and diffuse matter all expanding with velocities of up to 800 km/s. High spatial resolution IR imagery of Eta Car by Bensammar et al.(1985) has recently determined a size of about 0".2 for the central core, which is the inner diameter of the dusty nebula where the grains start to condense.
We have observed the high resolution H_α profile 3"-5" South of the star in April 1985 (ESO: CAT-CES). The profile appears broad and asymmetric, shifted by +100 km/s with respect to the stellar velocity (Fig.3). The line width is in agreement with the general expansion of the halo, while the shift suggests this part of the halo to be receding from us. The P Cygni absorption is quite strong with respect to the star, but the main difference is the absence of the central narrow peak seen in the stellar line. Since this peak is normally observed in the higher ionization lines (Fig.1) we conclude that there exists a hot stationary region near the stellar core with a low (projected) expansion velocity.
More informations may be obtained from the IUE high resolution spectroscopy of the nebula. Our observations show that the emission appear broader than in the core spectrum, with less prominent absorptions (compare the MgII profile in Fig.4 with Cassatella et al.1979).

In conclusion these results suggest that in the case of the EtaCar wind we are not dealing with a spherically symmetric structure, with a monotonic density and temperature law. The hot, low radial velocity region may be either matter ejected at right angles with respect to the line of sight or, more probably, a stationary cloud or disk around the stellar core. In any case the expanding envelope of Eta Car is highly asymmetric. High spatial and spectral resolution observations with the Faint Object Camera onboard of the Hubble Space Telescope will be crucial for the understanding of the mass loss, dust formation and X-ray emission process in this unique Hubble-Sandage variable.

REFERENCES

Andriesse C.D., Donn B.D., Viotti R. 1978, Mon. Not. R. A. S. 185, 771.
Bensammar S. et al. 1985, Astronomy Astrophysics 146, L1.
Cassatella A., Giangrande A., Viotti R. 1979, Astron. Astrophys. 71, L9.
Chlebowski T. et al. 1984, Astrophys. J. 281, 665.
Viotti R. et al. 1981, Space Science Reviews 30, 235.

ON THE OCCURRENCE OF WR+O BINARIES.

J.P. De Greve and **P. Hellings**
Astrophysical Institute
Vrije Universiteit Brussel
Pleinlaan 2
B-1050 BRUSSELS, Belgium.

and

E.P.J. van den Heuvel
Astronomical Institute,
Universiteit Amsterdam
Roeterstraat 19
NL-1018 WB AMSTERDAM
the Netherlands.

Using stellar evolution models of two different kinds, one with convective cores determined by the Schwarzschild criterion, one with enlarged convective cores, we discuss the **masses of the WR stars** in a number of double-lined WR binaries. For the main group of systems the mass of the WR-star is between 10 and 20 $M_\odot$.

We further show that the virtual **absence of WR+B** systems can be explained as a result of accretion effects on the inner structure of the companion star (rejuvenation) and of the short timescale of existence of the WR star relative to the dominant timescale of the companion star. Systems with WR components that were tranformed directly by stellar wind may now possibly be observed as WR-SB1 systems with a large mass function.

Examination of the formation mechanism of WR stars in binaries (stellar wind or mass transfer) leads to the conclusion that **mass transfer** is probably **the acting mechanism in** most of the presently **well-known systems.** Especially this must be the case for the WR binaries in the Small Magellanic Cloud.

An extended version of this paper is submitted to Astronomy and Astrophysics. Preprints are available upon request.

Stellar atmospheric instability in the upper part of the
Hertzsprung-Russell diagram

C. de Jager and H. Nieuwenhuijzen
Astronomical Observatory and Laboratory for Space Research,
Utrecht, The Netherlands

The upper limt of stellar luminosity in the Hertzsprung-Russell diagram is a line running approximately from $(\log T_{eff}; \log (L/L_\odot)) = (4.5; 6.3)$ via $(4.0; 5.74)$ to $(3.5; 5.7)$ (Humphreys and Davidson, 1979; Humpreys, 1983). Since Eddington (1921) this limit has been associated with the Eddington condition

$$g_{grav} + g_{rad} = 0, \text{ or } g_{grav}(1-\Gamma) = 0, \qquad (1)$$

where Γ_E is the Eddington factor. But Eddington's criterion appears not to describe the actually observed limit. In addition it does not work for cool stars. One of us (De Jager, 1978, 1980) has therefore introduced another limit

$$g_{grav}(1-\Gamma) + g_{turb} = 0, \qquad (2)$$

with $g_{turb} = \rho^{-1} dP_{turb}/dz$. It can be shown (De Jager, 1984) that condition (2) is equivalent to

$$F_m = (\frac{\Gamma \mathcal{R}}{\mu})^{\frac{1}{2}} g_{grav} T_e^{\frac{1}{2}} (\frac{\bar{T}}{\kappa})(R)(1-\Gamma_E), \qquad (3)$$

if we assume that the mechanical flux F_m dissipates fully within one scale height, an assumption that is justified since in near-unstable stellar photospheres the turbulent velocity v_t is approximately equal to the sound velocity s. Indeed, further elaboration of Equation (3) shows (De Jager, 1984) that this Equation is equivalent with the condition that in the photospheres of stars near the atmospheric instability limit:

$$v_{turb} = s. \qquad (4)$$

In this paper we want to show that Equations (3) and (4) are approximately correct near the stellar instability limit. We will consider this as a justification of the instability condition (2).

1. From literature data on microturbulent velocities in seven very luminous stars we derived a fair correlation between $\log(v_{turb}/s)$ and $\log(L_{star}/L_{limit})$, where L_{lim} is the luminosity limit at the same T_{eff}-value as the star. The regression relation is

$$\log(v_t/s) = 0.54 + 0.85 \log(L/L_{lim}), \qquad (5)$$

(correlation coefficient: 0.6) which shows that the turbulent atmospheric velocity indeed increases strongly with increasing luminosity, and moreover that $v_t = s$ is already reached at $\log(L/L_{lim}) = -0.6$: hence, intrinsically brighter stars and <u>a fortiori</u> stars at the instability limit have supersonic photospheric turbulence, which explains their atmospheric instability.

2. A further check is the following. It is implicitly assumed that an increase of v_t for increasing luminosity causes an increase of the rate of mass-loss. If that is true, stars with the same values of T_{eff} and L but with different v_t values should have different $\dot{M}$-values. From a recent review of mass-loss data (De Jager et al., 1985) it appeared that for all O- to M-type stars $\dot{M}$ is a function $\dot{M}(T,L)$ of T_{eff} and L only; the histogram of deviations in $\log(\dot{M})$ has a sigma of 0.53. We examined if the differences $\Delta \log \dot{M}$ between $\log(\dot{M})_{observed}$ and the function values $\log \dot{M}(T,L)$ are correlated with $\log(v_t/s)$ and found a positive correlation (correlation coeff = 0.7):

$$\Delta \log \dot{M} = -0.04 + 2.35 \log(v_t/s), \qquad (6)$$

which shows that, all other parameters remaining the same, the rate of mass loss appears to increase very strongly with v_t.

3. As a last check we assumed (tentatively, and not fully justifiable) that all dissipated mechanical flux goes into kinetic energy of the stellar wind, hence

$$4\pi R^2 F_m = \tfrac{1}{2} \dot{M} v_\infty^2, \qquad (7)$$

and found that the $\dot{M}$-values thus calculated, with F_m according to Equation (3), agree reasonably well with observed values.

<u>Conclusion</u>: The three evidences listed above under 1, 2, and 3 <u>support</u> the validity of our instability criterion (2), and the consequent relations (3) and (4).

References:

De Jager, C.: 1973, Astrophys. Space Sci. <u>55</u>, 147
De Jager, C.: 1980, The Brightest Stars, Reidel, Dordrecht
De Jager, C.: 1984, Astron. Astrophys. <u>138</u>, 246
De Jager, C., Nieuwenhuijzen, H., Van der Hucht, K.A.: 1985, Astron. Astrophys., submitted
Eddington, A.S.: 1921, Z. Phys. <u>7</u>, 351
Humphreys, R.M.: 1983, Astrophys. J. <u>269</u>, 335
Humphreys, R.M., Davidson, K.: 1979, Astrophys. J. <u>232</u>, 409

INFRARED SPECTROSCOPY OF SOUTHERN P CYGNI STARS

P.J. McGregor[1], A.R. Hyland[1], and D.J. Hillier[2]
[1] Mount Stromlo and Siding Spring Observatories,
The Australian National University, Australia
[2] Anglo-Australian Observatory, Australia

ABSTRACT

Investigations of the 1.0 – 2.5 μm spectra of high luminosity P Cygni stars reveal the presence of several new emission features. All stars measured show emission in the Brackett and Paschen lines of H I, but the emission strengths of He I 1.083, 1.700, and 2.058 μm relative to the H I lines show large variations. Emission lines of [Fe II] are seen at 1.257 and 1.644 μm in several stars and are particularly strong in HR Car. An interesting subset of these stars show the first-overtone CO bands in emission. The presence of this molecular component associated with hot P Cygni stars has not been inferred from previous optical or UV spectroscopy.

INTRODUCTION

Despite their rich optical emission line spectra, the spectra of high luminosity P Cygni stars have until recently received little attention at infrared wavelengths. Circular variable filter spectra of several P Cygni stars have been reported (e.g., Whitelock et al. 1983), but their low resolution makes detailed analysis impossible. Intermediate resolution spectra of P Cygni stars can now be obtained using grating spectrometers and permit detailed study of their infrared emission lines.

OBSERVATIONS AND RESULTS

Infrared spectra of the high luminosity southern P Cygni stars AG Car, CD-27 11944, HR Car, η Car, GG Car, BI Cru, CPD-52 9243, HD 87643, and CPD-57 2874 have been measured in the range 1.0 – 2.5 μm using a cold infrared grating spectrometer. These objects show rich infrared emission line spectra on strong continua due to hot dust emission or free-free emission (AG Car, CD-27 11944, and HR Car). Atomic emission lines of He I, H I, Fe II, [Fe II], and Mg II are present in different objects. The large strengths of high order Brackett lines in many objects indicate clear deviations from optically thin Case B recombination theory due to high optical depth in the H I lines. Low excitation [Fe II] emission lines are present in HR Car, CPD-52 9243, η Car, and possibly AG Car. Densities of 10^4 - 10^5 cm^{-3} are inferred from ratios of these [Fe II] lines in η Car (Allen, Jones, and Hyland 1985) and HR Car.

First-overtone CO emission is present in the spectra of HR Car, GG Car, CPD-52 9243, BI Cru, and CPD-57 2874. The same feature was seen earlier in low resolution spectra of CPD-52 9243 and BI Cru (Whitelock et al. 1983). CO molecules can not survive in the hot photospheres of these stars (T ~ 10^4 K) so this emission must have a circumstellar origin. Either the outer regions of the stellar wind cool sufficiently for CO molecules to associate, or the CO molecules are remnants of an earlier phase of mass loss when the star was a red supergiant (Lamers, de Groot, and Cassatella 1983). Rotational temperatures T_{rot} ⩾ 3000 K are inferred from the band shapes. Vibrational temperatures in the range 2000 K ⩽ T_{vib} ⩽ 5000 K account for the relative strengths of the 2-0, 3-1, and 4-2 vibrational bands.

DISCUSSION

The presence of five emission regions is suggested. He I recombination lines require a He II region (I.P. = 24.6 eV) close to the star. Optically thick H I lines are formed in a part of the stellar wind where hydrogen is ionized (I.P. = 13.6 eV) which probably extends beyond the He II zone. The ionization potential of Fe II (16.2 eV) is such that Fe II can exist within the H II volume, but densities in this part of the stellar wind are probably too high for appreciable forbidden Fe II emission. CO emission must originate in part of the stellar wind where hydrogen is neutral since the dissociation potential of CO is 11.1 eV. Fe II is the predominant ionization state in this excitation range, but two pieces of evidence suggest that the CO emitting region is closer to the star than the [Fe II] emitting region. Firstly, no strong correlation exists between the presence of CO and [Fe II] emission features. Secondly, Scoville, Krotkov, and Wang (1980) conclude that CO emission is a dominant coolant in high temperature molecular gases at densities exceeding 10^7 cm^{-3}. This suggests, although it far from proves, that the CO emission originates from regions at higher densities, and hence closer to the star, than those emitting the observed [Fe II] lines. A cool dust component is seen in several objects. The presence or absence of each component then depends on the precise run of both temperature and density in the stellar wind.

REFERENCES

Allen, D.A., Jones, T.J., and Hyland, A.R. 1985, Ap.J., 291, 280.
Lamers, H.J.G.L.M., de Groot, M., and Cassatella, A. 1983, Astr. Ap., 123, L8.
Scoville, N.Z., Krotkov, R., and Wang, D. 1980, Ap.J., 240, 929.
Whitelock, P.A., Feast, M.W., Roberts, G., Carter, B.S., and
    Catchpole, R.M. 1983 M.N.R.A.S., 205, 1207.

RAPID ULTRAVIOLET SPECTRAL VARIATIONS IN HD 50896 (WN5+?)

A J Willis and I D Howarth
Department of Physics & Astronomy, UCL, London, UK

P S Conti and C D Garmany
JILA, University of Colorado, Boulder, Colorado, USA

HD 50896 has become one of the most promising candidate WR+collapsar systems (Moffat 1982) exhibiting optical light, RV and polarisation variations with a $3^d.763$ period (cf Firmani et al. 1980) yet direct evidence for its postulated n.s. component is lacking. UV spectra of analogous OB X-ray binaries show P-Cygni profile changes as a function of binary phase which are well understood in terms of anisotropic wind ionisation caused by X-ray emission from the accreting n.s. - the Hatchett & McCray (1976) effect. Observations of such UV variability in WR+c candidates with a clear phase dependence would thus provide good evidence for any n.s. companion. To this end we have obtained IUE spectra of HD 50896 in sequential NASA/VILSPA shifts over 6 consecutive days in 1983, together with complementary IUE data secured at various epochs during 1978-82, to search for the Hatchett & McCray effect in this star.

The IUE spectra of HD 50896 do show significant variations in the P-Cygni profiles of NV 1240, CIV 1550, HeII 1640 and, particularly, in NIV 1718 - the line we concentrate on herein. Fig 1 plots the equivalent width (W_a) of the NIV 1718 P-Cygni absorption component measured in each spectrum against binary phase. Although substantial variability is apparent there is no phase dependence analogous to that observed in all OB X-ray binaries studied with IUE. Thus our data do not provide any supportive evidence for the proposed n.s. companion. Rather, the 1983 data show evidence for rapid variability in the P-Cygni absorption (in some cases ~30% in a few hours) occuring at all phases. This strongly suggests that the observed changes are linked to mechanisms intrinsic to the WN star itself rather than binarity effects. Typically the profile changes are confined to a restricted range of -ve velocities (-1000 to -1800 km/s for NIV), considerably less than the wind terminal velocity of about 3000 km/s. This indicates that variations occur in the physical properties of the accelerating part of the wind. Corresponding profile changes in other lines show extensions to different velocities (-2100 km/s for HeII; -2500 km/s for NV) probably reflective of wind stratification in the different ions. Spectra secured during 1978-82 show similar changes in NIV 1718 (cf Fig 1) with evidence for secular effects - some of the 1980 data show markedly stronger values of W_a

Fig 1: Measured values of W_a (see text) for NIV 1718 plotted against binary phase for: (a) IUE data secured in Sept. 1983 ● (b) 1980 data obtained over consecutive days ▫ and (c) IUE data obtained at sporadic epochs during 1978-82 ▲ .

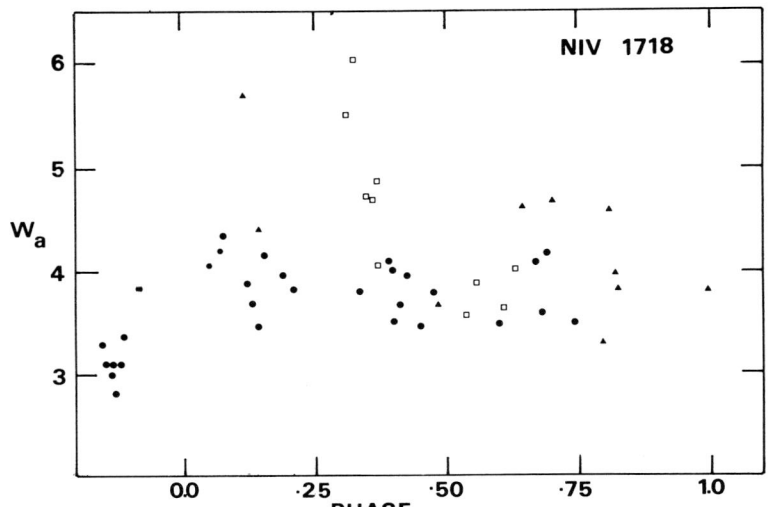

than seen at other times, accompanied by an extension of the P-Cygni absorption edges to higher displaced velocities (making a unique measurement of the terminal velocity for HD 50896 ambiguous).

In our view, current wind models are not capable of explaining the observed P-Cygni profile variations in HD 50896. The time scale of the variability (which can be $\leq$ 3 hours; we have yet to ascertain any periodicity) and the clear indication that we are dealing with changes to the accelerating region of the wind (close to the WN star) leads us to conjecture that they may arise from some kind of sub-wind activity in the WR component itself. Recently Vreux (1984, preprints) has found evidence that the optical variability shown by proposed WR+c candidates may more likely result from single star non-radial pulsations rather than spirally-in n.s. scenarios. Perhaps the UV variability of HD 50896 is also explicable in terms of such pulsational activity.

Hatchett,S.P., McCray,R., 1976, ApJ., 211, 552
Firmani,C., et al., 1980, ApJ., 239, 604
Moffat,A., 1982, in Proc. IAU Symp. No.99 (eds. de Loore & Willis), 263

THE FORMATION OF NITROGEN AND CARBON LINES IN HD 50896(WN5)

D. J. Hillier
Anglo-Australian Observatory
P.O. Box 296, Epping, NSW, Australia 2121

1. INTRODUCTION

Although considerable progress has been made in understanding Wolf-Rayet (WR) stars, the basic mechanisms for producing the nitrogen and carbon emission lines, which characterize a WR's spectrum, are unknown. To determine the mechanisms giving rise to these lines we are currently studying the formation of N and C emission lines in the WN5 star HD 50896. As a basis for our investigation we use a model with –

R_c(core radius)=$2.5 R_\odot$
$\dot{M}$(mass loss rate)=5.0×10^{-5} $M_\odot$/yr
$L=1.0 \times 10^5$ $L_\odot$
T_{wind}=30000K
Wind Velocity(HeII emitting region) < 1800 kms^{-1}

These parameters were previously found to reproduce both the continuous energy distribution of HD 50896, and the HeII emission lines, seen in its spectrum (Hillier 1983). The bulk of the continuous flux is emitted in the region 228Å < λ < 912Å, and is characterized by a very high effective temperature (> 60000 K).

Simplifying assumptions used in the N and C models include –
 (i) Radiation field determined from He and H model.
 (ii) No Auger ionizations, or dielectronic recombinations.
 (iii) Multiplets are treated as a single line in computing the source function but the multiplet structure is taken into account when computing the line profiles. Individual fine structure states within each term are assumed to be populated according to their statistical weights.
 (iv) Sobolev approximation (found to be adequate).

2. CIV LINE FORMATION

In WN stars the CIV spectrum is primarily represented by the CIV(2p–2s) resonance doublet at 1549Å and the CIV(3p–3s) doublet at 5805Å. To interpret the CIV spectrum we used a nine level CIV atom (n≤4). We find the following :

(i) The C ionization structure changes rapidly with radius (Figure 1) with the ionization balance primarily maintained by photoionizations from the 3p level in equilibrium with recombinations to all levels.

(ii) The CIV(3p-3s) doublet is produced by continuum fluorescence via the 3p-2s doublet at 312Å. Fluorescence from the 2p state to the 4d level followed by decay to the 3p level is also important. This latter process becomes more significant as the density increases, and as the C abundance increases. Eaton, Cherepashchuk, and Khaliullin (1984), have also recently recognized the importance of continuum fluorescence in the winds of Wolf-Rayet stars.

For a C/He abundance of 4.0×10^{-5} the theoretical CIV(3p-3s) doublet (Figure 2) is in fair agreement with observation. The line strength is sensitive to C abundance but a change in stellar luminosity from 1.0×10^5 to 5.0×10^4 caused less than a 30% decrease in line strength. This is due to the reduction in luminosity causing the ionization structure to change so that $N_{CIV}/N_{CV} = 1$ at larger electron densities where scattering from the 3p-2s, and in particular, the 4d-2p transitions are more efficient.

(iii) The resonance doublet at 1549Å is primarily produced by collisional pumping from the ground state and consequently is sensitive to the local electron temperature. The strength of the line is dependent on both the C abundance and the star's luminosity. For a C/He abundance of 4.0×10^{-5} the emission component (Figure 3) is a factor of 2 too strong.

3. NITROGEN LINE FORMATION

To interpret the N emission line spectrum in HD 50896 we used a 31 level NIV atom (2snl, n≤4; 2pnl, n≤3) and a 9 level NV atom (n≤4). We find the following:

(i) As for C, the N ionization structure changes rapidly with radius. Again the ionization is maintained by photoionizations from excited levels.

(ii) Little or no emission, like that observed, is produced in the NIV(3p $^1P^\circ$-3s ^1S) transition at 6381Å. Although the 3p level is coupled directly to the ground state by a transition at 247Å, it is effectively drained by a strong transition to the $2p^2$ ^1D level.

(iii) Emission due to NIV(3d ^1D-3p $^1P^\circ$) at 4058Å is produced by continuum fluorescence from the 2s2p $^1P^\circ$ level and indirect fluorescence from the $2p^2$ ^1D level (e.g. 2p3d $^1F^\circ$-$2p^2$ ^1D). For a N/He abundance of 7.5×10^{-4} the predicted strength of the line is a factor of 2 too small, although the line strength is proportional to the N abundance.

(iv) The 3d ^3D-3p $^3P^\circ$ multiplet at 7117Å is produced by fluorescence from the 2s2p $^3P^\circ$ state although indirect fluorescence from the $2p^2$ ^3P level can also be important. The line strength is approximately proportional to the square root of the N abundance. The different behaviour of this line compared to the corresponding singlet transition is related to the larger optical depth in the triplet fluorescing transition. At a temperature of 30000K, the triplet 2s2p state has roughly 60 times the population of the singlet 2s2p level (assuming the same departure coefficients).

(v) 3p $^3P^o$–3s ^3S multiplet at 3481Å is produced by indirect fluorescence from the 2s2p $^3P^o$ level. Of these, decay from the 3d ^3D level is singularly the most important. This line is insensitive to N abundance. For a N/He abundance of 7.5×10^{-4} this line, and NIV(λ7117Å) have theoretical equivalent widths a factor of 2 to 2.5 times smaller than observed.

(vi) The strong UV emission lines due to NV(2p–2s)(λ1240Å), NIV(2s2p $^3P^o$–$2s^2$ ^1S)(λ1486Å) and NIV($2p^2$ ^1D–2s2p $^1P^o$)(λ1719Å) are all formed by collisional pumping. The present model, with N/He = 7.5×10^{-4}, yields line strengths within a factor of two of that observed. The most discrepant line, NIV(1486Å), is a factor of two stronger than that observed. As this line is formed at smaller electron densities than other emission lines (Figure 4) its strength relative to the other N lines can be reduced by changing the electron temperature in its formation region. All lines are sensitive to abundance, but it should be noted that the N abundance is so large that it can influence the atmospheric structure. In particular, at electron densities < 10^{12}, collisional processes between the lowest energy levels of N can be important wind coolants. This cooling process may act to limit the strength of the NIV 1486Å line.

(v) The large strength of the NV(3p–3s) transition at 4609Å is not reproduced in the present models. As there is still a large parameter space to be explored, and numerous model assumptions which may need to be relaxed, we don't consider that it invalidates our model.

4. CONCLUSION

An ongoing investigation into the formation of nitrogen and carbon lines in WN stars has indicated that the optical CIV and NIV lines in the WN5 star HD 50896 are formed by continuum fluorescence. The strong CIV(3p–3s) doublet at 5805Å is pumped by continuum radiation at 312Å through the CIV(3p–2s) transition. This mechanism, together with the requirement that the pumping transition occur longward of the HeII Lyman limit at 228Å, explains the absence of other optical CIV lines in the spectra of HD 50896. Optical NIV lines cannot be fluoresced from the ground state – instead they are fluoresced from the 2s2p singlet and triplet states. Preliminary results suggest C/He=3×10^{-5} (by number) and a N/He abundance of around 10^{-3}.

4. REFERENCES

Eaton J. A., Cherepashchuk, A. M., and Khaliullin, Kh. F. 1984, preprint.
Hillier, D. J. 1983, Thesis, "The Extended Atmosphere of the WN star HD 50896", Australian National University.

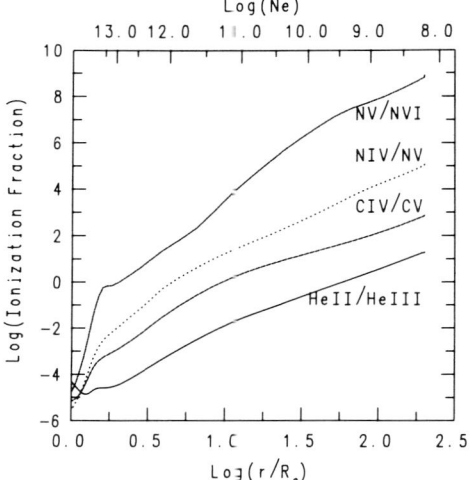

FIG. 1 — The theoretical ionization structure of the stellar wind for a stellar luminosity of 10^5 $L_\odot$.

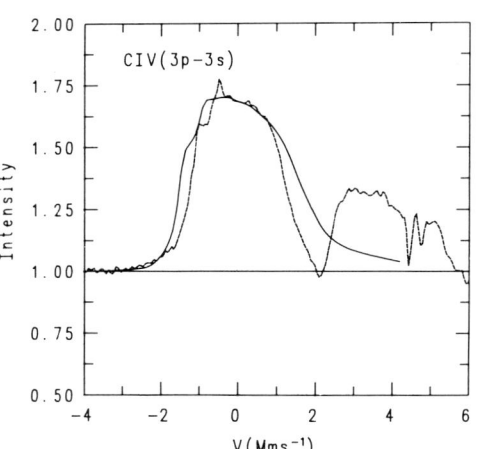

FIG. 2 — The observed CIV(λ5805Å) doublet profile (dashed line) in HD 50896 and a theoretical profile for a C/He abundance of 4.0×10^{-5}.

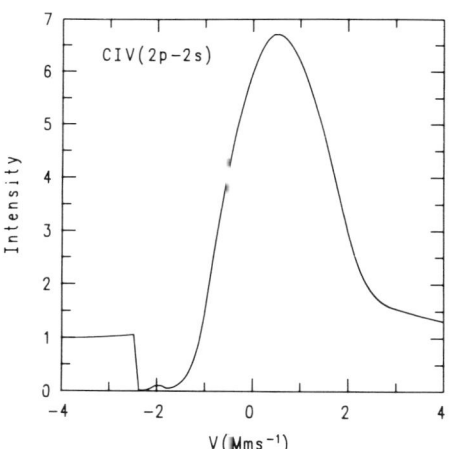

FIG. 3 — The theoretical profile for the CIV resonance doublet at 1549Å. For a C/He abundance of 4.0×10^{-5} the emission component is a factor of two stronger than that observed although the profile is of similar shape.

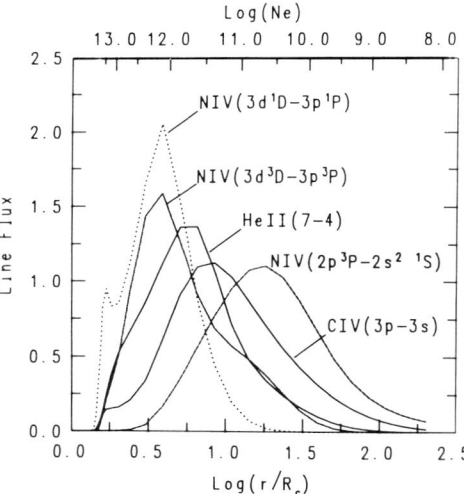

FIG. 4 — Illustration of where the emission lines HeII(7–4), CIV(3p–3s), NIV($2s^2$ ^1S–2s2p $^3P^\circ$), NIV(3d ^1D –3p $^1P^\circ$) and NIV(3d ^3D –3p $^3P^\circ$) originate in the stellar wind. The curves have been normalized to have unit area.

B[e]-STARS OF THE MAGELLANIC CLOUDS

F.-J. Zickgraf, B. Wolf, O. Stahl, C. Leitherer
Landessternwarte Königstuhl
6900 Heidelberg
Germany

We carried out a spectroscopic and photometric study of the 8 presently known B[e]-stars of the Magellanic Clouds (MC), i. e. R4 and R50 of the SMC, R66, R82, R126, Hen S12, Hen S22 and Hen S134 of the LMC. These stars are characterized by the following typical properties: a) strong Balmer emission lines, frequently with P Cygni profiles, b) permitted and forbidden emission lines predominantly of Fe II, [Fe II], [O I] etc., c) strong IR excess due to circumstellar dust. Photospheric absorption lines are usually weak or even not detectable.

In a (J-H)-(H-K)-diagram of MC emission line stars the B[e]-stars form a clearly separated group with H-K > 0.8. The spectral energy distribution from the UV to the IR (Fig. 1) can be interpreted as the sum of radiation from a hot stellar core (visible mainly in the UV), of a contribution of (f-f)-(f-b)-radiation of ionized hydrogen in the visual and near infrared and of thermal radiation of circumstellar dust at about 5 μm with a temperature of about 900 to 1200 K. The discussion of the line spectrum of R126 resulted in a two-component model for the stellar wind which is schematically shown in Fig. 2 (Zickgraf et al. 1985). Stellar rotation is thought to bring the stars close to the break-up-velocity leading to the formation of a dense, cool equatorial disk. R126, Hen S134 and probably Hen S22 present the pole-on case. R50, R82 and presumably R4 are examples for the equator-on case. Particularly R50 and R4 are characterized by sharp shell type absorption lines of Ti II and Cr II, which are typical for classical shell stars.

The remaining two stars Hen S12 and R66 seem to form an own group among the B[e]-stars. The Balmer line P Cygni profiles of Hen S12 show a knee-like transition between absorption and emission component, similar to the profiles of R66. For this particular star Stahl et al. (1983) derived a model with a gravitationally decelerated stellar wind.

Fig. 3 shows the HR-diagram of the B[e]-stars. We conclude that the B[e]-stars are evolved massive post-main-sequence stars. Massive premain-sequence stars are supposed to be only visible as IR-sources. Six objects can be explained by a model with a disk-like structure observed under different inclination angles. Two stars might represent the case of a gravitationally decelerated stellar wind.

A more detailed paper will appear in Astronomy and Astrophysics.

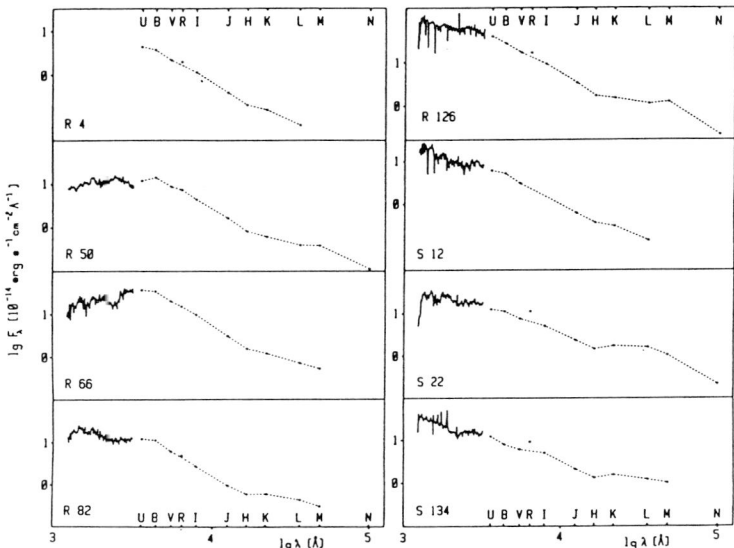

Fig. 1: Spectral energy distributions from the UV to the IR. Note the IR excess with colour temperatures between 900 and 1200 K.

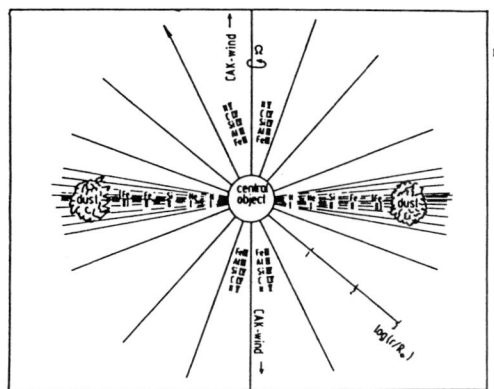

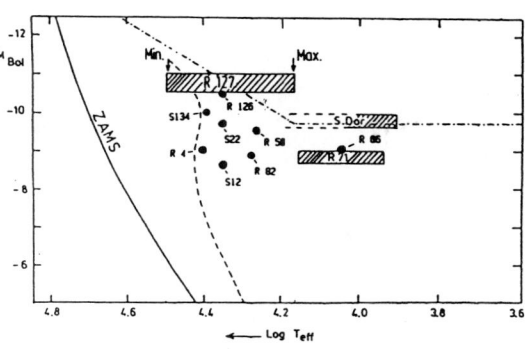

Fig. 2: Proposed schematic model of the two-component stellar wind. Emission-lines and dust are formed in the equatorial disk, broad UV-resonance absorption lines originate in the line-driven wind in the polar region.

Fig. 3: HR-diagram of the B[e]-stars in the MC (dots) together with the three known S Dor variables of the LMC. All B[e]-stars are situated to the right of the main-sequence band.

References:

Stahl, O., Wolf, B., Zickgraf, F.-J., Bastian, U., de Groot, M. J. H., Leitherer, C.: 1983, Astron. Astrophys. 120, 287

Zickgraf, F.-J., Wolf, B., Stahl, O., Leitherer, C., Klare, G.: 1985, Astron. Astrophys. 143, 421

THE MASS LOSS RATE OF SK 80 (O7Iaf) IN THE SMALL MAGELLANIC CLOUD

A J Willis and I D Howarth
Department of Phbsics & Astronomy, UCL, London, UK

K Nandy and D H Morgan
Royal Observatory, Blackford Hill, Edinburgh, Scotland

The star SK 80 in the SMC is classified as O7Iaf by Walborn (1976) who notes that it is the only confirmed Of star in that Galaxy known to date. A knowledge of the mass loss properties of OB stars in the Magellanic Clouds is of interest because of the recent evidence that such stars show reduced mass loss properties than their galactic counterparts (Hutchings 1980) and for Of stars because of the possible link between such stars and Pop I transition WNL stars (Conti 1976).
We have secured HIRES IUE and optical spectra of SK 80 and have attempted to derive the mass loss rate from these data.

The IUE spectrum of SK 80 shows the usual P-Cygni profiles in the resonance lines of NV, SiIV and CIV, and additionally in NIV 1718 and HeII 1640. The resonance lines indicate a terminal velocity of -1550 km/s which is significantly less than the typical values of -2500 for galactic O7 stars (Garmany et al. 1981), consistent with other evidence of systematically lower wind speeds in SMC early-type stars. To derive estimates of the mass loss rate we have used the NV and NIV profiles.

Using the first moment, W_1, of the unsaturated absorption component to the NV 1240 P-Cygni profile, we use the simple expression for estimating mass loss rates developed by Howarth (1984):

$$\dot{M} (M_\odot/y) = 3.1 \times 10^{-13} W_1^{0.373} (R_*/R_\odot)^{1.627} (v_\infty/kms^{-1})^{1.373}$$

The NV 1240 profile yields W_1 = 0.6 which, adopting E_{B-V} = 0.05, T_{eff} = 35000 K and a distance to the SMC of 65 kpc, a comparison Kurucz model atmosphere scaled to V = 12.35 gives R_* = 26 $R_\odot$ and with the above v_∞ we deduce $\dot{M}$ (NV) = 1.1×10^{-6} $M_\odot$/y.
This simple estimate relies on the same scaling between $N(N^{4+})$ and $\dot{M}$ as derived empirically for galactic OB stars and the assumption of a normal SK 80 wind composition. The latter may be particularly suspect and the above rate could be higher by a factor of up to x10 given the SMC metallicity deficiencies quoted in the literature. An additional uncertainty is introduced in the assumption that the wind ionisation structure in SK 80 is similar to galactic counterparts.

This latter uncertainty can partially be circumvented by analysing P-Cygni profiles in excited transitions which almost certainly arise in the dominant parent ion of the species in the wind - eg NIV 1718.

Olsen (1981) has produced an Atlas of such line profiles, specified by a fit parameter T given by the relation:

$$T = \frac{\pi e^2}{m_e c} f \lambda_o \frac{w_i}{w_g} \frac{\lambda_1^3}{2hc} F_{\nu 1}(A_E \dot{M}/4\pi\mu m_H) R_*^{-1} v_\infty^{-2} g_i$$

where λ_1 is the wavelength of the photoexciting line (755Å for NIV 1718) and $F_{\nu 1}$ is the monochromatic stellar flux at λ_1. The observed NIV profile gives $T = 3$ which with the value of $F_{\nu 1}$ taken from Kurucz models gives $\dot{M} = 7 \times 10^{-7} M_\odot y^{-1}$ - at first sight in agreement with the NV result.

However, Garmany et al. (1981) have found systematic differences in mass loss rates derived from excited and resonance lines indicative of a significant oversetimate of $F_{\nu 1}$ in the Kurucz models. Inspection of their data suggests a correction factor of $\Delta\log \dot{M} = 1.3$, implying a mass loss rate of $1.4 \times 10^{-5} M_\odot y^{-1}$ for SK 80 from the NIV line. However the scatter in $\Delta\log \dot{M}$ is large and this result is thus uncertain.

Alternatively (and probably most reliably) we can estimate $\dot{M}$ for SK 80 by comparing its NIV 1718 profile with those of galactic stars of the same type with accurate rates. From Garmany et al. (1981) we choose for comparison HD 167659 (O7I((f))) and HD 190864 (O7III((f))) and use a relation:

$$\dot{M} (SK 80) = \dot{M} (standard) (T R^* v_\infty^2)_{SK80}/(T R^* v_\infty^2)_{std} \Delta A_e$$

where ΔA_E is the relative metallicity, which for simplicity we take as unity. The comparison with the two stars gives $\dot{M} = 2.5 \times 10^{-5}$ and $8.5 \times 10^{-6} M_\odot y^{-1}$ respectively. Of course ΔA_E is probably greater than 1. We are inclined to adopt these latter higher mass loss rates as real for SK 80 given the overall appearance of its wind spectrum and in particular the high wind density implied by P-Cygni absorption in HeII 1640. We thus propose for SK 80, a value $\dot{M} \geq 10^{-5} M_\odot y^{-1}$.

Although the wind velocity of SK 80 is lower than normal, which would be consistent with low SMC metallicity and thus reduced radiative wind acceleration, any such metal deficiency does not appear to have retarded the mass loss rate itself.

Conti,P.S., 1976, Mem.Soc.r.Sci. Liege, 9, 193
Garmany,C.D. et al., 1981, ApJ., 250, 660
Howarth,I.D., 1984, MNRAS., 211, 167
Hutchings,J.B., 1980, ApJ., 237, 285
Olsen,G.L., 1981, ApJ., 245, 1054
Walborn,N.R., 1976, ApJ., 215, 53

How to form a 200$M_\odot$ star

Hans Zinnecker,
Royal Observatory,
Edinburgh,
Scotland

ABSTRACT
Stellar coalescence is suggested as a possible mechanism for doubling the upper stellar mass limit from ~100$M_\odot$ to ~200$M_\odot$ in a moderately dense cluster of a few hundred young massive stars (~$10^5 M_\odot pc^{-3}$). The merger will be between the two components of the dominant central tight binary formed in the core of the cluster by the N-body evolution. This process may occur in some giant extragalactic HII regions.

1. Introduction

Theoretical work on star formation by accretion suggests an upper stellar mass limit of the order of 100 $M_\odot$ (Larson and Starrfield 1971, Appenzeller and Tscharnuter 1974, Kahn 1974, Yorke 1979) in agreement with observations (Humphreys 1982), barring possible exceptions. Formation of an HII region and the reversal of the accretion process due to the radiation pressure on the dust grains (which transfer their momentum to the infalling gas) are thought to stop the growth of stellar masses beyond ~100$M_\odot$. One possible means of exceeding this limit may be a reduction of the dust-to-gas ratio and alteration of the dust properties (Wolfire and Cassinelli 1985). Here I suggest another possibility: a collisional merger between the two most massive stars in the core of a compact cluster of young OB stars. The original motivation for investigating this possibility came from the debate about the object R136a in the core of the central cluster in the 30 Doradus Nebula (see the review by Walborn 1984 and the panel discussion following his paper; see also Walborn 1986 and Moffat et al. 1985). Furthermore, a recent numerical simulation of the collision of two identical massive stars with various non-zero impact parameters has shown stellar coalescence to be quite effective (Benz and Hills 1985).

2. The Physical Idea

Consider a superdense cluster of massive stars which (for simplicity) are assumed to be coeval and of the same mass M (say 100$M_\odot$). A necessary condition for star-star collisions to occur in such a cluster is that the cluster evolution becomes faster than the stellar evolution. In other words: the timescale for cluster core contraction (30-40 initial crossing times), after which a central dominant binary is formed, must be less than the main sequence lifetime of the massive stars (a few times 10^6 yr). The former timescale is based on Aarseth's (1974) numerical simulation of the N-body evolution of isolated star clusters (N=250). The crossing time in a virialized cluster is $t_{cr} \sim (G\rho)^{-1/2}$ where ρ is the mean stellar mass density and G is the gravitational constant. It follows that $\rho \geq 10^5 M_o pc^{-3}$.

As a criterion for collision I adopt the condition that the semimajor axis (a) of the central binary be as small as 10 times the stellar radius (r) or less. In that case the numerical calculations of Benz and Hills (1985) predict that merging between the two binary components is inevitable after several orbits, especially if the binary has a sizable excentricity (as predicted by the N-body simulations). Moreover, the presence of a captured third body generates repeated perturbations of the eccentricity of the binary orbit so that nearly head-on collisions and hence merging will occur. The key point is that the central binary absorbs practically all the energy of the cluster (Aarseth 1974), thus $a = R_o N^{-2}$ for an initial cluster radius R_o and N cluster members of equal mass. Since $r/a \leq 1/10$ (as discussed above) and $r \sim 0.1$ AU for O-stars, $a \leq 1$ AU is required for the merging of the binary. Thus, if $R_o = 0.2$ pc, $N \sim 200$ follows from $a = R_o N^{-2}$; if $R_o = 0.5$ pc, $N \sim 300$; but for larger R_o (and larger N) the above limit $\rho = NM/4R_o^3 \geq 10^5 M_o pc^{-3}$ is violated and the cluster evolution is not fast enough.

3. Discussion

The main, perhaps surprising conclusion is that two massive stars may merge in moderately compact, young clusters. It remains to be shown that the result is valid in more general cases. Use of a realistic stellar mass spectrum at a fixed number N of stars would speed up cluster evolution by a factor ~ 3 (Aarseth 1974), but extension of the mass spectrum to lower mass stars (B-stars), that is increasing N to $\sim 4N$ as well as decreasing the mean stellar mass, would slow down cluster evolution by a factor 3-4. Therefore, as long as the mass spectrum terminates at $\sim$B5 stars, both effects will cancel, and the timescale for cluster evolution as given remains correct. Note that the basic event of stellar coagulation requires the presence of a mass spectrum otherwise the central binary would not become tight enough (Aarseth 1974). In fact, the speckle data on R136a (Weigelt and Baier 1985) do suggest that the central binary a1/a2 is not tight at all (sep. $\sim 0\rlap{.}{''}1$ or 6000 AU) which in turn may indicate that the mass distribution in 30 Dor is quite narrow indeed, i.e. WR and O-stars only (cf. Melnick 1986). The large spatial extent of 30 Dor is consistent with the N-body relaxation of a cluster born in a small volume.

At present the only known remaining candidate for a very massive star ($\sim 2000 M_o$) is (was!) the progenitor of the unique supernova SN 1961v in the galaxy NGC 1058 (Utrobin 1984).

ACKNOWLEDGEMENT

It is a pleasure to thank Dr. D.C. Heggie for very valuable discussion.

REFERENCES

Aarseth, S.G. 1974, Astr. Astrophys. 35, 237
Appenzeller, I. and Tscharnuter, W.M. 1974, Astr. Astrophys. 30, 429.
Benz, W., and Hills, J.G. 1985: priv. communication.
Humphreys, R.M. 1982, in The Most Massive Stars, Proc. ESO-Workshop, eds. S. D'Odorico, D. Baade, and K. Kjär, p.5
Kahn, F. 1974, Astr. Astrophys. 37, 149
Larson, R.B., and Starrfield, S. 1971, Astr. Astrophys. 13, 190
Melnick, J. 1986, this Symp.
Moffat, A.F.J., Seggewiss, W., and Shara, M.M. 1985, Ap.J. 295, 109.
Utrobin, V.P. 1984, Ap. Sp. Sci 98, 115.
Walborn, N.R. 1984, in Structure and Evolution of the Magellanic Clouds, eds. S. van den Bergh and K.S. de Boer, IAU Symposium 108, 243
Walborn, N.R. 1986, this Symp.
Weigelt, G. and Baier G. 1985, Astr. Astrophys (in press)
Wolfire, M.G., and Cassinelli, J.P. 1985, BAAS 16, 960
Yorke, H.W. 1979, Astr. Astrophys. 80, 308

FeII IN THE UV SPECTRUM OF LUMINOUS EMISSION LINE STARS

G. Muratorio[1], M. Friedjung[2], R. Viotti[3]
[1] Observatoire de Marseille, France
[2] Institut d'Astrophysique CNRS Paris, France
[3] Instituto Astrofisica Spaziale CNR Frascati, Italy

Following the excessive mass loss rates we derived in a previous analysis of the FeII emission and absorption lines of some luminous Magellanic Clouds stars, assuming the two components formed in the same region (Muratorio et al., 1984), we again analysed the FeII data using the same method (Muratorio, 1985), but taking into account the presence of high velocity winds detected in some stars (R66, R126) by the study of the high dispersion IUE spectra (Stahl et al., 1983; Zickgraf et al., 1985).

The synthetic UV spectra calculated with such an hypothesis is found to fit very well the observed IUE spectra not only for R66 (Fig. 1), but also for S22 (Fig. 2). For this later star, we derived a wind velocity of 280 km s^{-1} and a density of Fe$^+$ of $6\ 10^4$ ions cm^{-3} at a maximum radius of $1.4\ 10^{13}$ cm. Assuming most hydrogen ionized in the wind, all iron in the Fe$^+$ state, a metastable level excitation temperature of 5700 K derived from the emission line study (Muratorio, 1985), a mass loss rate of $7\ 10^{-6} M_\odot$ y^{-1} is obtained. In such an hypothesis, the optical region emission lines originate from a zone which, most of part, cannot lie on the line of sight of the star, while the above defined wind produces by itself the observed UV dominant absorption line spectra. The emitting region could be a disk, whose radius would lies between a minimum of $4.8\ 10^{13}$ cm and a maximum of $5\ 10^{14}$ cm. These quite high values compared with the stellar radius $R = 3.2\ 10^{12}$ cm, calculated assuming for the star an effective temperature of 18000 K, exclude the formation of the emission lines in a chromosphere. Moreover, an illuminated disk can explain (Friedjung, 1985) the peculiar near-infrared energy distribution of S22 as well as that of various other stars studied (Muratorio, 1985). The presence of dust, generally argued to explain the infrared excess longward of 2 μm is compatible with this model, as the physical conditions in the outer parts of a disk favour the formation of dust grains.

In the case of R66 the figure 1 synthesis is computed assuming the ultraviolet absorption lines formed in a wind of 300 km s^{-1} originating at the photosphere of radius $8.6\ 10^{12}$ cm with a density at that radius of $4\ 10^9$ cm^{-3}; these values lead to a mass loss rate of $6\ 10^{-6}\ M_\odot$ y^{-1}. The optical emission lines may originate in a disk whose radius lies between $1.6\ 10^{13}$ cm and $3.3\ 10^{14}$ cm. The same metastable level excitation temperature (6300 K) is assumed for both media.

In the case of R126, the absence of FeII absorption lines in the IUE spectra suggest that the emission lines can be formed in a disk like region as was pointed out by Zickgraf et al.(1985). The disk radius is estimated to lie between $3.9\ 10^{13}$ and $7.8\ 10^{13}$ cm, while the radius of the star is $5\ 10^{12}$ cm.

FeII emission and absorption lines of the variable P Cygni star AG Car, originate from a variable velocity wind for which at the date of our observations (1981/11/9) we derived the values: $V = 70$ km s^{-1} Metastable level $T_{exc} = 7500$K $4\ 10^{12} < R < 1.2\ 10^{13}$ cm $\dot{M} = 6\ 10^{-6}\ M_{\odot}\ y^{-1}$.

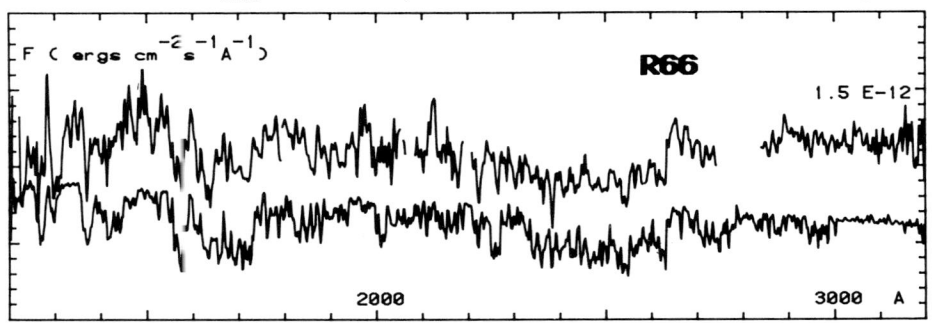

Fig 1 : R66 IUE spectra (SWP 12856,LWR 16619) Dereddened (E(B-V)= .12)
5 E-13 verticaly shifted (upper curve)
Fe II synthetic spectrum (Disk + High velocity wind) (lower curve)

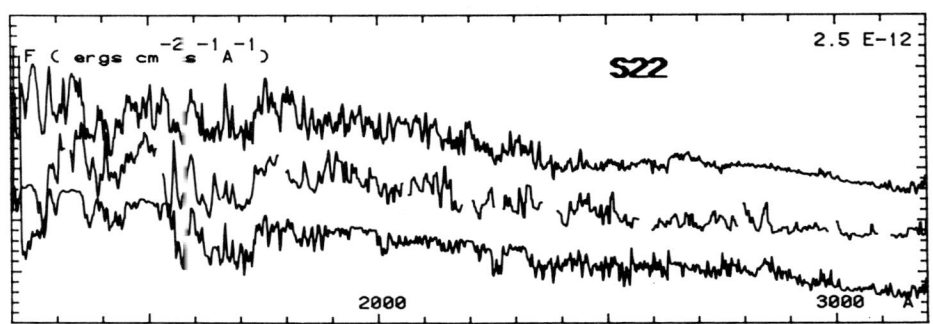

Fig 2 : Fe II synthetic spectrum (Low velocity wind) 1E-12 vert shifted (upper curve)
S22 IUE spectra (SWP 6868,LWR 7881) Dereddened (E(B-V)=.17)
5E-13 vert shifted (intermediate curve)
Fe II synthetic spectrum (Disk + High velocity wind) (lower curve)

REFERENCES

Friedjung, M.: 1985, Astron. Astrophys. (in press)
Muratorio, G., Friedjung, M., Viotti, R.: 1984, Fourth European IUE conference Rome ESA SP 218 p. 309
Muratorio, G.: 1985, Thèse de Doctorat d'Etat, Université Aix-Marseille I
Stahl, O., Wolf, B., Zickgraf, F.J., Bastian, V., De Groot, M.J.M., Leitherer, C.: 1983, Astron. Astrophys. **120**, 287
Zickgraf, F.J., Wolf, B., Stahl, O., Leitherer, C., Klare, G.: 1985, Astron. Astrophys. **143**, 421

POSTER PAPERS - SESSIONS 3 and 4.

Chairman : J. GRAHAM.

Graham:

The two papers on (1) the new results for Eta Carina by Viotti et al. and (2) that by Wolf et al. on Be stars in the MC's both pointed out that one is not always dealing with perfect symmetry in these objects. It bothered me this morning that we seemed to be concentrating on the simplest geometrical case of spherical symmetry in the stellar winds. Is this an observational fact or a necessary thing to adopt in order to make any progress?

Lamers:

For the star P Cyg we have no evidence that the wind is not spherically symmetric. For Be stars there is strong evidence that their mass loss is not spherically symmetric. For instance the mass loss rates derived from UV observations for pole on stars is about 100 times smaller than from the IR observations where we are able to see emission from the disk.

Stalio:

A general comment on the stars we discussed this morning. The structure that has been presented today involves a star, a wind and a cool static shell around the wind - a kind of PN structure except in that case the wind is hot and of high velocity. Now what happens to this static shell when the wind hits it, particularly at a phase where the mass loss rate has increased by say a factor of 100. Is it destroyed or does it act like a barrier around the star? Secondly the kind of structure envisaged is quite contrary to what is observed for a WR star where we have a hot wind with very high velocity. If these stars come from P Cyg stars what happens to the shell?

Graham:

I wanted to highlight the poster paper by Azzopardi and Meyssonier which gives new results on a survey for H-alpha emission line stars in the LMC based on a 4 hour exposure Schmidt plate, which is pretty hard to get. I hope the paper will give details of the plate epoch since it is very important to know when a star starts becoming an emission line object and when it stops being so.

Kontizas M.:

What evidence have you that the H-alpha emission stars are Be stars and how many Be stars have been found compared to the total number of B stars?

Azzopardi:

This survey is at an early stage. At present we have identified all kinds of H emission-line objects which are mainly Be stars, PN, and unresolved HII regions. Nevertheless, if we consider the H emission line objects showing up a continuum we can say that about 80% of discovered field objects are Be stars.

Kontizas E.

Are the H emission objects cluster or field stars and for the cluster NGC 330 do the H emission objects you find coincide with the Be stars found years ago by Feast?

Azzopardi:

It is not known if they are cluster stars, and the objects we are finding are not the same as those found by Feast.

Graham:

I want to turn now to some posters dealing with individual objects. The question I have for Wolf and Stahl concerning MWC300 is: to what extent is the distance above the plane sensitive to the interstellar extinction used; could it be closer and therefore possibly not a runaway star?

Wolf:

We are fairly sure of its distance if the identification with a hypergiant is correct and there are strong indications that it is a hypergiant in that its observed absorption spectrum is almost identical to usual hypergiant spectra.

de Jager:

The results from the paper dealing with V444 Cyg are cearly very important. Do all WR stars have this high temperature?

DISCUSSION

Conti:

I would not be surprised if we had some WN5 stars at one temperature and others at another - it would be scary but it would not surprise me. There are not very many WR eclipsing binary systems suitable for this kind of analysis. We really need to apply better models for these stars to get a handle on meaningful temperatures.

Lamers:

Is it true that one really has to go well above 50000 K for WR temperature to get mass loss rates from radiation pressure to those observed?

Conti:

If you want WR mass loss to be radiatively driven you need these high temperatures. Also if they are He burning then the He burning Main Sequence is generally at these hot values.

Stickland:

I'd throw in the comment that in our recent study of CQ Cep, where we used the binary characteristics and the distance, if we took a temperature of 30000K we got serious over-contact. This could only be relieved by raising the temperature above about 55000K even though it is a WN7 star.

Kudritzki:

A comment concerning the WR stars. The big problem in the past for models of WR atmospheres was that no matter what one did you could never produce an extended atmosphere. Now when you take these high temperatures you come close to the Eddington limit, and you can easily get an extended photosphere with supersonic velocities.

Nugis:

How can you reconcile the adopted high effective temperature with the presence of strong HeI lines in the spectrum of V444 Cyg?

Kudritzki:

Up to now we did not concentrate on line formation calculations. However, assuming grey radiative equilibrium in the wind, the temperature quickly drops to 30000K at $r = 3r*$ from 90000K at r (Tau Ross = 1). In consequence there is a change to explain the HeI lines. However, I agree that it will be crucial for future work to check the temperature structure of the wind against the line spectrum information.

Vanbeveren:

How does the luminosity you obtain and the binary mass of V444 Cyg correspond with the theoretical M-L relations for He burning stars?

Kudritzki:

If we assume 90000K for T(eff) at $r = 2.9R(sun)$ then we are above the Zero Age Helium MS by a factor of two in L. This is not too worrying since we expect the observational He MS to have a certain bandwidth. In addition even if we go to the lower limit of 70000 K given by Cherepaschuk et al. we get a reasonable value of L though the extension of the photosphere is somewhat reduced. However, several quantitative improvements in the calculation of the line force will have to be made before we can become really ambitious and compare the results with evolutionary calculations in detail. Another question is whether the WR stars could be He shell burning?

Chiosi:

The lifetime in He shell burning is extremely short.

Lamers:

I would like to ask a question on the paper by John Hillier. In your thesis you showed that the velocity law in the WR star HD 50896 goes in two steps: first an acceleration to 1800 km/s and then, much further out, a second acceleration region up to 2700 km/s. Do you still require this two-step acceleration in your new models?

Hillier:

The model still requires the wind to reach its terminal velocity at large radii. Constraints on the velocity law come from the optical line profiles of HeII and HeI. For example, the HeII (n - 4) transitions show no evidence for emission at veloxities > 1800 km/s, whereas the terminal velocity derived from the UV lines is about 2700 km/s. From the illustration I have given of the region where the HeII lines are formed it follows that the wind velocity must be lower than 1800 km/s at $r < 20r(core)$. The HeI profiles push the limit to even larger radii in the wind.

DISCUSSION

Graham:

I want to turn now to the papers dealing with searches for WR stars in galaxies. It seems to me that the interpretation of such searches can be very sensitive both to bias and also to completeness, and that we need to put on a quantitative basis estimates of completness.

Moffat:

Surveys of WR stars in galaxies can give meaningful results if treated differentially (e.g. comparing the numbers of stars of a given type down to a limit in absolute magnitude and line strength in one galaxy an another) as opposed to absolute comparisons.

Graham:

The paper by McGregor, Hyland and Hillier on southern P-Cygni stars shows tht we are dealing with a complex situation. Not only do they show FeII and MgII in the environment but strong evidence for molecular CO emission, which of course just to exist has to come from a pretty low temperature regime of a few thousand degrees or so. Again this shows that we are not dealing with a simple situation with these stars.

Lamers:

Where do they think these CO lines are formed? In the star P-Cygni which is not in their sample, we found some evidence for cool dust at a temperature of about 600 K.

Viotti:

I think one important result, derived with two different methods by the Heidelberg and France-Italy groups, is in the evidence for asymmetry in the expanding envelopes of these very luminous line stars in the MC's. This would be an important step to explain some peculiarities such as the presence of circumstellar dust referred to by Lamers. In your list you include some objects which are not proper P Cygni stars, such as the binary GG Car and the symbiotic star BY Car. Actually in the latter star you can find some easier explanation for the CO emission.

Graham:

I would like to ask how the results on emission line stars

Bohannan:

I compared my spectroscopic description of emission-line stars that were in common with the work of Stahl et al. on bright emission-line stars in the LMC. I found that those stars which I found different from the earliest descriptions by Feast, Thackeray and Wesselink, showed variations in the short time interval between my spectra and those of Stahl. Those that showed no variations on the long timescale appeared similar also on the short timescale. If these are to be identified with Hubble-Sandage variables, or whatever we call luminous blue variables with emission lines, it should be noted that some of them have an almost constant spectroscopic appearance over the past 25 years and show none of the obvious photometric or spectroscopic variability of the real H-S variables in M33.

Graham:

Walborn certainly convinced me that it looks like the idea of a supermassive, $> 1000\ M_o$ object in the centre of R136 is not on. I would like to ask Appenzeller and de Jager whether they are happy to consider stars of say 200 M_o.

Appenzeller:

I do not see any insurmountable difficulties for such objects.

de Jager:

I agree with Appenzeller that 200 M_o stars will be stable, showing only (weak) vibration, but their atmospheres may be near-unstable, with the Eddington-gamma close to unity, so that small disturbances, like the vibrations mentioned, will be sufficient to cause strong (episodal) ejection of mass, in extreme cases even comparable to the Eta Carina events. Actually Eta Car may be such a case.

Feast:

I would like to ask Nolan Walborn if the derived stellar content of 30 Doradus, with its large WR population but lack of HS variables rules out the hypothesis that the latter evolved into the former as we have heard discussed several times at this meeting.

DISCUSSION

Walborn:

It is likely that the most luminous, narrow-lined, WN objects found in 30 Doradus and the Carina nebular can be produced from very massive stars without a H-S phase. Eta Carina may be considered an extreme H-S variable; it may represent a short-lived post-WN, pre-supernova stage of a very massive star.

Zinnecker:

What is the age of 30 Doradus?

Walborn:

2-3 million years on the basis of the predicted Main Sequence lifetimes of the most massive stars, which are evolved.

SESSION 5.

MASSIVE STAR EVOLUTION.

Chairman : R. HUMPHREYS.

1. A.MAEDER: Massive Star Evolution: Mass Loss and Mixing.

2. J.SILK: Physical Processes in Massive Star Formation.

3. C.CHIOSI: Effects of Convective Overshooting, Mass Loss (and Chemical Composition) across the HR Diagram.

4. J.P. DE GREVE and C. DE LOORE: Binary Evolution in the Upper H-R Diagram.

MASSIVE STAR EVOLUTION: MASS LOSS AND MIXING

André Maeder
Geneva Observatory
1290 Sauverny / Switzerland

1. INTRODUCTION

At first it may be surprising that mass loss, overshooting and mixing, which are indeed very different physical processes, have similar consequences on stellar evolution. These various processes may increase the Main-Sequence (MS) lifetime, extend the width of the MS, bring CNO-processed materials to stellar surfaces and, in extreme cases, lead to quasi-homogeneous evolution. The physical reason of this similarity is that these processes increase the relative mass fraction of the stellar cores. Thus we understand that, on the basis of their evolutionary consequences, it may not be easy to disentangle the contributions of mass loss, overshooting and mixing processes. The present status of our knowledge on these effects, which appear to have major consequences on the evolution of massive stars, is now examined in detail.

2. EFFECTS OF MASS LOSS ON THE POPULATIONS OF MASSIVE STARS

An impressive number of authors have studied the effects of mass loss on stellar evolution, particularly on main sequence evolution (e.g. Tanaka, 1966a,b; Hartwick, 1967; Simon and Stothers, 1970; Chiosi and Nasi, 1974; de Loore et al., 1977; Dearborn and Eggleton, 1977; Dearborn et al., 1978; Sreenivasan and Wilson, 1978; Chiosi et al., 1978; de Loore et al., 1978; Stothers and Chin, 1978; Dearborn and Blake, 1979; Czerny, 1979; Chiosi et al., 1979; Stothers and Chin, 1979, 1980; Maeder, 1980; Noels et al., 1980; Bressan et al., 1981; Maeder, 1981a,b; Stothers and Chin, 1981; Noels and Gabriel, 1981; Brunish and Truran, 1982a,b; Doom, 1982a,b; Sreenivasan and Wilson, 1982; Maeder, 1983; Doom, 1984; Sreenivasan and Wilson, 1985).

Table 1 summarizes what the present author considers to be the main effects of mass loss on stellar evolution; not included are the changes of surface abundances which are examined in § 3. Firstly, the MS case is considered. It is a noticeable fact that, while a moderate mass loss

TABLE 1: Main effects of mass loss on massive star evolution.

MAIN SEQUENCE

★ M_{core} ↓ , $q_c = \frac{M_{core}}{M}$ ↑ , semi-convection ↓

★ L ↓ , L/M ↑

★ MS lifetime t_H ↑ (by 5-15%)

★ moderate $\dot{M}$: MS widening
   very high $\dot{M}$: MS narrowing (quasi-homogeneous evolution)

HE - BURNING PHASE

★ Large effects in HRD/Very small in log Tc vs. log ρc.
                                    (central conditions)

★ 3 evolutionary sequences according to $\dot{M}$ and $M_{initial}$
   (cf. Table 2)

★ $t_{He} \simeq t_{BSG} + t_{RSG} + t_{WR}$, sharing varies with $\dot{M}$ (cf. Maeder, 1981b)

★ *BLUE SUPERGIANTS (BSG):*

   no $\dot{M}$: $t_{He} \simeq t_{BSG}$

   with $\dot{M}$: $\begin{cases} \text{He - phase moves to red} \\ \text{Blue loops reduced} \end{cases}$ t_{BSG} ↓

★ *RED SUPERGIANTS (RSG):*

   moderate $\dot{M}$ => $\frac{t_{RSG}}{t_{OBA}}$ ↑ $\dot{M}$ ↑ =>

   (for low $\dot{M}$: lack of RSG)

   high $\dot{M}$ => $\frac{t_{RSG}}{t_{OBA}}$ ↓ $\frac{t_{RSG}}{t_{OBA}}$ ∧

★ *WOLF-RAYET STARS (WR):*

   $\dot{M}$ increases t_{WR} / t_{OBA} $\dot{M}$ ↑ =>
   lowers threshold mass for
   forming WR stars (most $\frac{N_{WR}}{N_{OBA}}$ ↑↑
   from $M_{initial} \gtrsim 40 M_\odot$)
                                      (factor 2 on $\dot{M}$ =>
                                       factor 17 on N_{WR}/N_{OBA})

   $\dot{M}$ ↑ => N_{WR}/N_{RSG} ↑↑

   Mass - luminosity relation $\log \frac{L}{L_\odot} = 3.8 + 1.5 \log \frac{M}{M_\odot}$
   for WR stars

produces a MS widening, a high mass loss (as it occurs above 10^2 $M_\odot$) produces a MS narrowing and quasi-homogeneous evolution. For the He-burning phase, the major results are also emphasized. Indeed, according to the mass loss rates and to the initial stellar masses, three different evolutionary sequences may occur in the HR diagram (cf. Maeder, 1981a,b, 1984): always blue; blue-red-blue; blue-red. These three different evolutionary sequences are illustrated in Table 2, and Fig. 1 also shows the corresponding HR diagram. In contrast, we recall (cf. Maeder and Lequeux, 1982) that the effects of mass loss on central conditions are very limited.

The He-burning phase is shared between the blue supergiants (BSG), the red supergiant (RSG) and the WR stages. This sharing is very dependent on mass loss rates and initial mass; quantitative information can be found in Maeder (1981a,b). The time spent as BSG is reduced by mass loss; this is due to the redwards displacement of the horn-shaped band corresponding to the He-burning phase in the HR diagram. The RSG lifetime firstly increases with moderate mass loss and then decreases with higher rates. As to the relative frequency of WR stars, it strongly increases with mass loss, because the peeling-off by stellar winds reveals the underlying bare core earlier, and also because the threshold mass for the formation of WR stars is lowered.

According to Conti et al. (1983b), the observed ratio (WR number) /

TABLE 2: Three different evolutionary sequences for massive stars.

For $M \geq 60$ $M_\odot$ *Always blue*

O star - Of - BSG and Hubble-Sandage variables -
WN - WC - (WO) - SN

For 25 $M_\odot \leq M \leq 60$ $M_\odot$ *Blue-red-blue*

O star - BSG - yellow and RSG - BSG - WN - (WC) - SN - *high $\dot{M}$*
                                         ↘
                                          SN   - *moderate $\dot{M}$*

For $M \leq 25$ $M_\odot$ *Blue-red*

O star - (BSG) - RSG - yellow supergiant and Cepheid -
RSG - SN

(OB stars with $M \geq 40\ M_\odot$) is 0.23, while evolutionary models usually predict a ratio t_{He}/t_H of the helium- to H-burning phases of 0.10. This difficulty seems to be considerably reduced by models (Maeder, in prep.) using the new, higher $^{12}C(\alpha,\gamma)^{16}O$ rate (cf. Kettner et al., 1982). The physical reason of the longer He-lifetime is that in addition to the conversion of ^{4}He to ^{12}C, we also have an almost complete fusion of ^{12}C to ^{16}O, making thus more energy available than previously considered. In addition, as the He-burning reactions are more energetic, the mass fraction of the convective core is larger and more fuel is available, which also increases the He-burning lifetime. (One must notice that these effects are smaller for stars of lower masses, because less ^{12}C is turned into ^{16}O). Another effect, which also sizeably increases the WR lifetimes, is the decline of the stellar luminosity resulting from the strong decrease of the stellar mass in the WR stage. Even with the old $^{12}C(\alpha,\gamma)^{16}O$ cross-section we obtained (Maeder, 1983), due to that effect, a ratio $t_{WR}/t_H = 0.3$ for a star with an initial mass of 120 $M_\odot$.

We also emphasize in Table 1 that the expected number ratio of WR stars and RSG very strongly increases with mass loss. As the bulks of these two kinds of massive stars do not originate from the same mass interval, the ratio t_{RSG}/t_{WR} changes extremely steeply (cf. Fig. 6 in Maeder, 1981b) in function of the M_{bol} at MS turnoff or in function of cluster ages. Thus, at a given age, a cluster contains in general either WR stars or RSG, and therefore both rarely cohabit in the same cluster. In order to observe a possible change in the Galaxy of the ratio N_{RSG}/N_{WR} (very sensitive to $\dot{M}$-rate) for these two kinds of massive and easily identifiable stars, it is very necessary to consider in the sample all stars brighter than a given luminosity ($M_{bol} = -6, -7$ or -8). However, by restricting the sample only to clusters and associations containing WR stars (cf. Humphreys et al., 1985) one a priori excludes RSG-rich groups.

The last topic mentioned in Table 1 is the predicted mass-luminosity relation for WR stars. This relation results from the fact that WR stars, whatever their initial masses, are nearly homologous stars consisting of a large He, C, O core. It is to be emphasized that a stability analysis indicates that WR stars generally are vibrationally unstable due to the Eddington ε-mechanism.

3. MASS LOSS AND SURFACE CHEMICAL ENRICHMENTS

The surface abundances in massive stars may change during evolution as a result of mass loss by stellar winds and due to dredge-up by deep convective envelopes, as they occur in red supergiants. We also note that the intermediate fully convective zones (which usually exist in massive stars with no excessive mass loss) produce an averaging of composition over some fraction of the star, also influencing surface abundances when the concerned layers are revealed by mass loss.

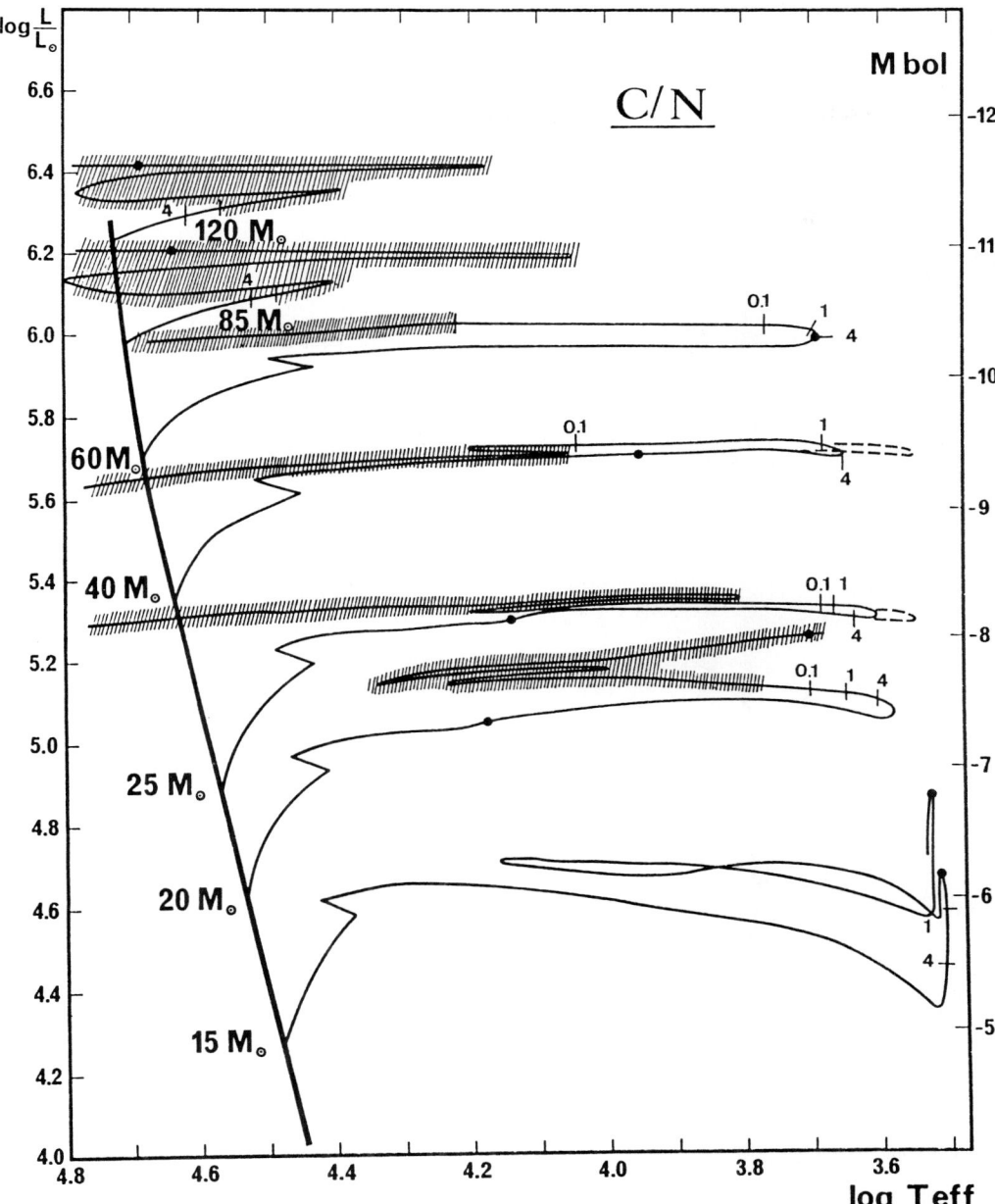

Figure 1. Values of the surface ratios $X(^{12}C) / X(^{14}N)$ along the evolutionary tracks of massive stars. The hatched parts indicate where the equilibrium value of 0.025 occurs.

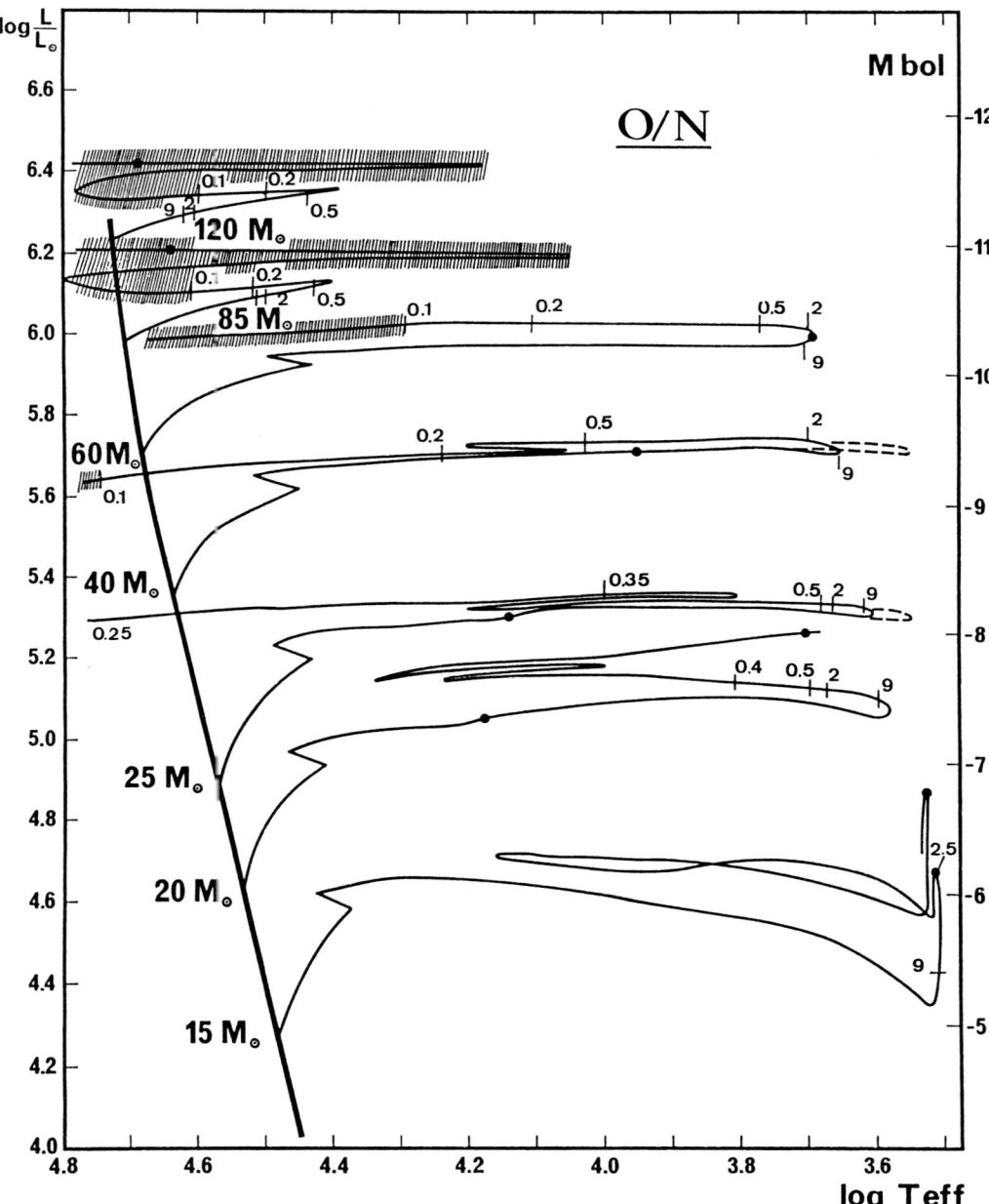

Figure 2. Values of the surface ratios $X(^{16}O) / X(^{14}N)$ along the evolutionary tracks of massive stars. The hatched parts indicate where the equilibrium value of 0.1 or smaller occurs.

Calculations taking the above effects into account have been made by Noels et al. (1980), Noels and Gabriel (1981), Maeder (1983), Greggio (1984) and Maeder (1985). The effects of more sophisticated, but uncertain hydrodynamical processes have generally not been included, as it is firstly necessary to have a standard basis for comparison with observed CNO abundances (cf. also § 4). Spectroscopic observations of CNO elements constitute a powerful test of stellar evolution, particularly with regard to the possibility of deviations from the standard case. Detailed evolution of the surface abundances of ^{1}He, ^{3}He, ^{4}He, ^{12}C, ^{13}C, ^{14}N, ^{15}N, ^{16}O, ^{17}O, ^{18}O, ^{20}Ne, ^{22}Ne, ^{24}Mg, ^{25}Mg and ^{26}Mg in stars of initial masses 120, 85, 60, 40, 25, 20 and 15 $M_\odot$ have been calculated by the author (1983, 1985). For 60 $M_\odot$ and lower masses, the surface abundances of CNO processed elements change step-like rather than continuously as in higher masses. The "plateaux" are due to the convective zones, which average the chemical composition over some parts of the stars. When mass loss unveils these chemically homogeneous regions, the composition keeps rather constant for some time.

Of all chemical changes at stellar surfaces, those of C/N and O/N ratios are among the largest and observationally the most accessible ones. As an example, the C/N ratio may change from 4 (cosmic value in mass fraction) to $2.5 \cdot 10^{-2}$ (equilibrium CNO value) and to 10^{14} in WC stars which, as is known, exhibit products of partial He-burning.

Fig. 1 shows the changes of the ^{12}C/^{14}N ratio during the evolution in the HR diagram. On the tracks, the value of $X(^{12}C) / X(^{14}N) = 4$ is indicated at the latest point where it occurs; the values 1 and 0.1 are also given (all data in mass fraction). The hatched area indicates where the equilibrium values of about 0.025 are expected. We notice that for stars brighter than $M_{bol} = -10.5$, equilibrium C/N ratio can be found everywhere in the HRD. For initial masses between 60 and 25 $M_\odot$, the stars keep their standard abundances until the red supergiant stage is reached and departures only occur in later phases. Below 20 $M_\odot$, equilibrium C/N no longer occurs.

Fig. 2 shows the distributions of the ^{16}O/^{14}N ratios in the HRD (data also given in mass fraction). Equilibrium O/N ratios only occur for initial masses larger than 60 $M_\odot$, in particular in the LBV stars and in further evolutionary phases. For lower masses, deviations from the standard O/N ratio appear in stages later than the supergiant stage; the smaller the initial masses, the more limited these deviations are, as normally expected.

Turning now to the observations, we notice that the LBV variables and the WR stars are the best examples for changes of surface abundances. In the case of the LBV star η Carinae, Davidson et al. (1982, 1984) determined the abundances in some of the outer condensations surrounding the central nebula and found ratios C/N < 0.05 and O/N < 0.15 - 0.5 (in

Zahn (1983) has made a detailed discussion of the instabilities generated by stellar rotation, which is a long standing problem since the time of Eddington. Other recent works have also been made by Sreenivasan and Wilson (1982, 1985). Let us consider the main steps of Zahn's developments:

- The meridional circulation resulting from thermal imbalance generates a small differential rotation.
- This differential rotation is insufficient for generating shear instabilities between adjacent layers. However, the small differences in angular velocity Ω create some horizontal turbulence of meteorological nature. This turbulence is two-dimensional and in itself it leads to no vertical mixing.
- The 2-D turbulence cascades, as turbulence always does, towards small scales until inertial terms become larger than Coriolis terms. Then, this small scale 2-D turbulence becomes 3-D.
- The small scale tail of the 2-D turbulence produces vertical exchanges and a diffusion mixing characterized by a diffusion coefficient $D = Re^* \nu$, with

$$Re^* = \frac{K}{\nu} \frac{\Omega^2 r}{g} (\nabla_{ad} - \nabla_{rad}) \qquad (1)$$

Re^* is the modified Reynolds number, ν the viscosity and K the radiative diffusivity. In order to obtain the value of Re^* at each level r in the star, we need to know $\Omega(r)$. Following a suggestion by Schatzman, we may suppose that the star is just at the verge of the axisymmetric baroclinic instability (Knobloch and Spruit, 1983), which is the first instability met by a differentially rotating star. In this case, Re^* becomes

$$Re^* = \frac{8r}{(\frac{\partial \ln \Omega}{\partial \ln r})^2 H_p} \quad \text{with} \quad \frac{\partial \ln \Omega}{\partial \ln r} = (\frac{8\nu N^2}{K \Omega^2})^{\frac{1}{2}} \qquad (2)$$

where N is the Brunt-Väisälä frequency. Account must also be given to stabilization by μ-gradients.

It is generally believed that mild diffusion processes may be significantly acting only in low mass stars, where the MS lifetimes are long enough. However, the radiative viscosity is so large in massive stars that the diffusion timescales may become shorter than the MS lifetime (Maeder, 1982).

Models with different angular velocities have been computed for a 40 $M_\odot$ star according to the above scheme, including also mass loss and overshooting. The results (Maeder, 1986) essentially show two different types of evolution:

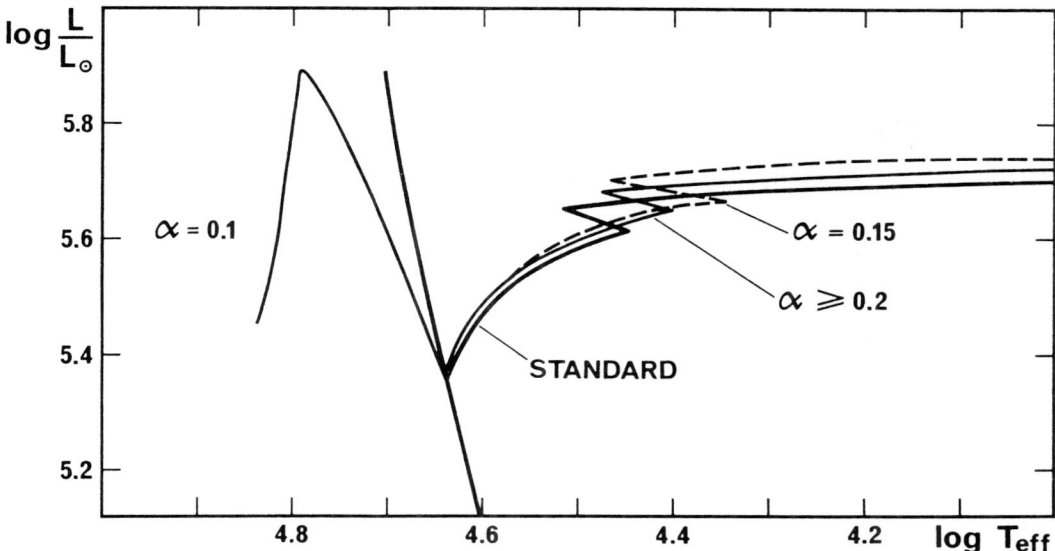

Figure 3. Evolutionary tracks of an initial 40 $M_\odot$ star for different values of $\alpha \equiv \partial \ln \Omega / \partial \ln r$.

A) <u>Classical evolution</u>: In case of low and moderate rotation, the stabilizing effect of the μ-gradient severely limits the extension of turbulent diffusion and all models nearly follow the same tracks close to the classical ones (with overshooting). The evolution is essentially inhomogeneous; the MS lifetime and the surface CNO abundances undergo only very limited changes.

B) <u>Mixed models</u>: For fast rotation (or low $\partial \ln \Omega / \partial \ln r$ according to expression 2), the stabilizing effect of the μ-gradient is unable to prevent mixing. This evolution resembles the homogeneous one giving bluewards tracks in the HR diagram (cf. Fig. 3). Processed CNO elements rapidly appear at the stellar surface. The luminosity firstly increases strongly. In view of its composition, the star becomes a WR star and then undergoes a decline in luminosity due to the rapid decrease of its mass. Mainly because more nuclear fuel is available, the MS lifetime is larger by about 55%.

The switching from one behaviour to the other occurs quite rapidly in terms of $\partial \ln \Omega / \partial \ln r$; the critical value lies between 0.10 and 0.15. Observationally, the mixed models seem to be interesting for explaining those of OBN stars, which often occur as blue stragglers in clusters (cf. Schild, 1985). Future works, both theoretical (since expr. 1 may contain uncertain numerical factors, cf. Zahn, 1983) and also observational, will allow us to determine the critical limiting rotational velocity, which we provisionally estimate to be of the order of a few hundredths of km/sec.

REFERENCES

Bertelli, G., Bressan, A., Chiosi, C., 1985: Astron. Astrophys., in press
Bressan, A.G., Bertelli, G., Chiosi, C., 1981: Astron. Astrophys. 102, 25
Brunish, W.M., Truran, J.W., 1982: Astrophys. J. 256, 247
Brunish, W.M., Truran, J.W., 1982: Astrophys. J. Suppl. 49, 447
Chiosi, C., Nasi, E., 1974: Astron. Astrophys. 34, 355
Chiosi, C., Nasi, E., Bertelli, G., 1979: Astron. Astrophys. 74, 62
Chiosi, C., Nasi, E., Sreenivasan, S.R., 1978: Astron. Astrophys. 63, 103
Conti, P.S., Garmany, C.D., de Loore, C., Vanbeveren, D., 1983b: Astrophys. J. 274, 302
Conti, P.S., Leep, E.M., Perry, D.N., 1983a: Astrophys. J. 268, 228
Czerny, M., 1979: Acta Astronomica 29, 1
Dearborn, D.S.P., Blake, J.B., 1979: Astrophys. J. 231, 193
Dearborn, D.S.P., Blake, J.B., Hainebach, K.L., Schramm, D.N., 1978: Astrophys. J. 223, 552
Dearborn, D.S.P., Eggleton, P.P., 1977: Astrophys. J. 213, 448
Doom, C., 1982a: Astron. Astrophys. 116, 303
Doom, C., 1982b: Astron. Astrophys. 116, 308
Doom, C., 1984: Astron. Astrophys. 138, 101
Doom, C., 1985: Astron. Astrophys. 142, 143
Eggleton, P.P., 1983: Mon. Not. R. Astron. Soc. 204, 449
Davidson, K., Dufour, R.J., Walborn, N.R., Gull, T.R., 1984: in "Observational Tests of the stellar evolution theory", IAU Symp. 105, Ed. A. Maeder and A. Renzini, Reidel Publ. Co., p. 261
Davidson, K., Walborn, N.R., Gull, T.R., 1982: Astrophys. J. 254, L47
Greggio, L., 1984: in "Observational Tests of the stellar evolution theory", IAU Symp. 105, Ed. A. Maeder and A. Renzini, Reidel Publ. Co., p. 329
Hartwick, F.D.A., 1967: Astrophys. J. 150, 953
Humphreys, R.M., Nichols, M., Massey, P., 1985: Astron. J. 90, 101
Iben, I., Renzini, A., 1983: Ann. Rev. Astron. Astrophys. 21, 271
Kettner, K.U., Becker, H.W., Buchmann, L., Görres, J., Kräwinkel, H., Rolfs, C., Schmalbrock, P., Trautvetter, H.P., Vlieks, A., 1982: Z. Physik A308, 73
Knobloch, E., Spruit, H.C., 1983: Astron. Astrophys. 125, 59
Langer, N., Sugimoto, D., 1985: Astron. Astrophys., in press
de Loore, C., de Grève, J.P., Lamers, H., 1977: Astron. Astrophys. 61, 251
de Loore, C., de Grève, J.P., Vanbeveren, D., 1978: Astron. Astrophys. 67, 373

Luck, R.E., Lambert, D.L., 1981: Astrophys. J. 245, 1018
Maeder, A., 1974: Astron. Astrophys. 32, 177
Maeder, A., 1976: Astron. Astrophys. 47, 389
Maeder, A., 1980: Astron. Astrophys. 92, 101
Maeder, A., 1981a: Astron. Astrophys. 99, 97
Maeder, A., 1981b: Astron. Astrophys. 102, 401
Maeder, A., 1982: Astron. Astrophys. 105, 149
Maeder, A., 1983: Astron. Astrophys. 120, 113
Maeder, A., 1984: Adv. Space Res. 4, 55
Maeder, A., 1985: in ESO Workshop on CNO isotopes in astrophysics, Ed. J. Danziger, in press
Maeder, A., 1986: Astron. Astrophys., in prep.
Maeder, A., Lequeux, J., 1982: Astron. Astrophys. 114, 409
Maeder, A., Mermilliod, J.-C., 1981: Astron. Astrophys. 93, 136
Matraka, B., Wassermann, C., Weigert, A., 1982: Astron. Astrophys. 107, 283
Mermilliod, J.C., Maeder, A., 1986: Astron. Astrophys., in press
Noels, A., Conti, P.S., Gabriel, M., Vreux, J.M., 1980: Astron. Astrophys. 92, 242
Noels, A., Gabriel, M., 1981: Astron. Astrophys. 101, 215
Nugis, T., 1982: in "Wolf-Rayet stars: observation, physics and evolution", IAU Symp. 99, Ed. C. de Loore and A.J. Willis, Reidel Publ. Co., p. 127 and p. 131
Roxburgh, I., 1978: Astron. Astrophys. 65, 281
Schild, H.R., 1985: Astron. Astrophys. 146, 113
Simon, N.R., Stothers, R., 1970: Astron. Astrophys. 6, 183
Smith, L.J., Willis, A.J., 1982: Mon. Not. Roy. Astron. Soc. 201, 45
Sreenivasan, S.R., Wilson, W.J.F., 1978: Astrophys. Space Sci. 30, 57
Sreenivasan, S.R., Wilson, W.J.F., 1982: Astrophys. J. 254, 287
Sreenivasan, S.R., Wilson, W.J.F., 1985: Astrophys. J., in press
Stothers, R., Chin, C.W., 1978: Astrophys. J. 226, 231
Stothers, R., Chin, C.W., 1979: Astrophys. J. 233, 267
Stothers, R., Chin, C.W., 1980: Astrophys. J. 240, 885
Stothers, R., Chin, C.W., 1981: Astrophys. J. 247, 1063
Stothers, R., Chin, C.W., 1983: Astrophys. J. 264, 583
Stothers, R., Chin, C.W., 1985: Astrophys. J., in press
Tanaka, Y., 1966a: Publ. Astr. Soc. Japan, 18, 47
Tanaka, Y., 1966b: Progr. Theoret. Phys. Kyoto, 36, 844
Willis, A.J., 1982: in "Wolf-Rayet stars: observations, physics and evolution", IAU Symp. 99, Ed. C. de Loore and A.J. Willis, Reidel Publ. Co., p. 87
Xiong Da Run, 1985: Astron. Astrophys., in press
Zahn, J.P., 1983: in "Astrophysical processes in Upper MS stars", 13th Saas-Fee Course, Ed. B. Hauck and A. Maeder, p. 253

Discussion : MAEDER.

SREENIVASAN :

We have recently shown that massive stars (WR stars and supergiants) are overstable for non-radial pulsations (retrograde modes) due to Kelvin-Helmholtz instability, driven by differential rotation. We have also shown that significant rotation produces "blue stragglers" due to rotational mixing. An extended core and the peeling off of the outer layers due to mass loss reveals "abundance anomalies" much earlier in the evolution of a star.

I therefore believe that a confrontation of observations with the prediction of these new models should be more profitable and is along the right direction (as you have also observed).

KONTIZAS M. :

At which ages of star clusters does one expect to find blue stragglers?

MAEDER :

Usually, blue stragglers occur in old clusters with an age of the order of a few billion years. A good example is NGC 7789. However, as I mentioned, some N-enriched O-type stars in very young clusters are located to the blue side of the considered cluster turnoff.

PHYSICAL PROCESSES IN MASSIVE STAR FORMATION

Joseph Silk
Department of Astronomy
University of California
Berkeley, Ca 94720, USA

SUMMARY: The gravitational fragmentation theory of star formation is reviewed. Theoretical arguments are presented which suggest that the lower stellar mass cut-off to the IMF in giant HII regions may be as high as 10 $M_\odot$. Mechanisms for bimodal star formation are described in the context of a coagulation model for formation of the giant molecular clouds, and application is made to starbursting galaxies.

I. INTRODUCTION

Star formation is poorly understood, despite the fact that modern observational techniques can probe the very cores of nearby molecular clouds in which stars currently are forming. The reason is simply that a large number of physical parameters are involved in determining the fate of a collapsing cloud. Apart from specifying the cloud mass and initial geometry, these include gas density, temperature, turbulent velocity field, magnetic field, rotation, ionization, dust grain properties, molecular and heavy element abundances. One must also consider feedback from the presences of newly formed stars, and global effects, such as the effect of the environment or the molecular cloud. Different (and occasionally even the same) authors have argued at various times that different subsets of these parameters play a dominant role. Not even fully three-dimensional numerical computations, now or in the foreseeable future, seem capable of unravelling these many complexities. Hence it is perhaps presumptuous to attempt to even speak of star formation theory. Rather, there are certain observational facts, some secure, others less so, from which one can try to reconstruct a theoretical framework within which one can seek to test various speculations involving collapse and fragmentation. That is the philosophy underlying this review, in which I shall examine some aspects of massive star formation.

To commence, I discuss the fragmentation of collapsing clouds, and argue for the existence of a critical mass scale (§II). I show that this leads in a natural way to a reasonable form for the initial mass function, and that the lower stellar mass cut-off may increase in giant HII regions (§III). The mechanism of secondary star formation is described in §IV, where I argue that this naturally leads to bimodal star formation. The observational evidence for bimodality is summarized, and then in §V, I describe the role of giant molecular cloud formation in initiating starbursts. A final section draws implications for future research, both theoretical and observational.

II. FRAGMENTATION

A popular view of star formation is that it occurs via the hierarchical fragmentation of a collapsing gas cloud. Hoyle (1953) originally proposed that during the isothermal contraction phase characteristic of a diffuse cloud, the Jeans mass M_J would decrease and lead to fragmentation on decreasing scales. This mass scale is the mass contained within a sphere of diameter equal to the critical wavelength for gravitational instability in an infinite uniform medium, namely the Jeans length $\lambda_J \equiv \pi V_s (G\rho)^{-1/2}$. A more sophisticated analysis in a uniform, isotropically collapsing cloud (Hunter 1962) confirmed this result in linear perturbation theory: all wavelengths exceeding λ_J are unstable to small perturbations in densities

However these simple analyses in linear theory have been seriously questioned (Layzer 1963; Tohline 1980). Fluctuations acquire angular momentum by tidal torques and may undergo disruptive interactions. The level of initial fluctuations δ is crucial to the outcome of the instability, which has a secular growth rate. In spherical collapse, the mean density must increase by a factor of order δ^{-2} (Mestel 1965) and in aspherical collapse by a factor δ^{-1} (Silk 1982), before fragmentation can occur. Growth only commences on a given scale once this scale exceeds the instantaneous value of the Jeans length.

Numerical simulations have not hitherto had adequate resolution to verify whether fragmentation indeed occurs in the dynamic collapse phase. Collapses subject to large amplitude initial perturbations are found to fragment (Rozyczka 1983). Most hydrodynamical simulations study the effects of initial pressure and rotation. Typically, collapse occurs to a quasi-equilibrium sheet or ring which subsequently fragments (Bodenheimer et al. 1980). In essentially all numerical studies, the fragments contain many Jeans masses, but this is undoubtedly an artifact of the low resolution obtainable because of grid limitations (Tohline 1982).

It seems plausible that some combination of thermal pressure, rotation and magnetic support will greatly slow the clouds from free-fall collapse. Observational studies strongly suggest that clouds are not in free-fall collapse, otherwise, given the observed efficiencies of star formation, far too high a rate of star formation would ensue. In view of the unacceptably low resolution of hydrodynamical three-dimensional studies, it is therefore useful to pursue an analytical treatment. The stability of thin, self-gravitating sheets of gas has been extensively studied, including pressure support, rotation and magnetic fields. A general conclusion from linear stability analyses is the following, shown most simply for an infinite isothermal sheet, but characteristic of more general conditions (Larson 1985). The dispersion relation for the growth of exponential perturbations $exp(\omega t - i\underline{k}.\underline{x})$ is

$$\omega^2 = V_s^2 k^2 - 2\pi G \mu k, \tag{1}$$

where the first term on the right hand side represents the pressure force and the second the self-gravitational force for a sheet of surface density μ. Equation (1) explicitly demonstrates that $\omega(k)$ has a maximum value at k_c, corresponding to the most rapidly growing mode: $k_c \approx (2H)^{-1}$, where H is the scale-height of the sheet. The corresponding mass-scale may be written

$$M_c = 2.4\ T^2 \left[\mu(M_\odot pc^{-2})\right]^{-1}\ M_\odot \tag{2}$$

To apply equation (2) to the interstellar medium, one may note that $\mu \approx 100 - 200 M_\odot pc^{-2}$ over a wide range of clump or cloud masses. The empirical relation $n \propto \ell^{-1}$ is satisfied over scales from ~ 0.1 to $\sim 30 pc$ (Larson 1982). Gas temperatures span the range $\sim 5K$ to $\sim 100K$, so that $0.3 M_\odot \lesssim M_c \lesssim 200 M_\odot$.

Since M_c is derived from linear perturbation theory, it is important to examine the non-linear criterion for a fragment to collapse. This is best understood for an isothermal gas sphere embedded in a medium of pressure P (Ebert et al. 1960). If the surface pressure on the spherical cloud exceeds a critical value, it will collapse. Hence for newly formed fragments to be unstable and collapse, one requires

$$P/k > P_{crit}/k = 10 T^4 M^{-2}\ cm^{-3} K, \tag{3}$$

where the cloud mass M is in units of $M_\odot$. For the previously cited temperature and mass ranges (if $M = M_c$), the critical pressures inferred from (3) are common in molecular clouds.

Once fragments do collapse, they are likely to form stars with high efficiency. This is because magnetic field lines thread gas fragments to the surrounding cloud. Magnetic torquing enforces approximate corotation as the fragments collapse (Mestel 1965; Mouschovias and Paleologou 1979), and this outward transfer of angular momentum enhances the efficiency of star formation: essentially all of the mass in a fragment should be able to contract to high density. This argument presupposes that magnetic fields are well coupled to the molecular gas. Ionization studies, as well as indications that ionizing X-ray photons are produced by protostellar flares deep within molecular clouds, suggest that the magnetic field diffusion time-scale (for ions and charged grains which carry the field and slip relative to the neutrals) is at least 10^7 yr (Langer 1984). The inefficiency in exhausting the gas supply of a molecular cloud lies rather in the existence of stable, non-collapsing fragments.

The preceding results for critical mass and pressure are exceedingly sensitive to temperature. This motivates the ensuing discussion: heat input is a crucial ingredient in determining the protostellar mass range.

III. THE INITIAL MASS FUNCTION

Suppose that for some set of initial conditions determining surface density and temperatures, fragments form protostars of specified mass M. Now not all of the cloud will fragment simultaneously. Either in a reasonably chaotic collapse or because of magnetic support, occasional dense cores will run away in density, while there will be a continuing reservoir of gas that is capable of ongoing collapse and fragmentation. Hence once the first protostars form, their subsequent energy output can affect the energy balance in the cloud. In general, one might plausibly expect that more massive stars form as the cloud heats up, but decreasingly smaller numbers of stars are formed because the heat input per star will rise with increasing mass.

It is possible to estimate in a schematic way the shape of the resulting initial stellar mass function. Suppose that the luminosity of a protostar scales as $L \propto M^{1+\beta}$, and that the dust grain emissivity varies as $T_d^{4+\alpha}$ when integrated over the spectral energy distribution. Here T_d is the grain temperature, and α depends on the specific grain properties: typically $1 \lesssim \alpha \lesssim 2$. Then suppressing all coefficients, the fragment mass spectrum satisfies

$$N\frac{dL}{dM} - L\frac{dN}{dM} = \frac{d}{dM}(T_d^{4+\alpha}) = \frac{d}{dM}M^{2+\alpha/2},$$

whence

$$\frac{dN}{dM} = M^{-2-\alpha/2-\beta}. \tag{4}$$

Now one expects that $\beta \approx 0$ for massive protostars, and one therefore infers an IMF with slope about -2.5. A more detailed treatment with allowance for gas-grain coupling was given by Silk (1977). This derivation does assume that all masses form with similar efficiency: hence one can convert from gas fragment to stellar masses. It is encouraging that (4) approximates the observed IMF, but the comparison does not merit being taken too seriously.

A more quantitative comparison with stellar data may be made by focusing on the lower mass cut-off of the IMF. When a cloud begins to form its first stars, the proto-stellar masses are likely to be inhibited below a particular value M_* that coincides with the critical mass corresponding to fragments that can collapse at pressure P_{crit}. Lower mass stars would require larger pressure at given T and μ or ρ to have contracted from newly formed fragments, whereas all fragments more massive than M_* would have collapsed. Suppose the fragments which are just going non-linear have the same temperature as the surrounding gas. Then since $M_* \propto T^2 P_{crit}^{-1/2}$, we infer that the critical gas density to form stars more massive that $M_{10} \equiv M_*/10 M_\odot$ is

$$\rho_{crit} = 6000\, T_{100}^3 M_{10}^{-2}\ M_\odot\ pc^{-3}, \tag{5}$$

where $T_{100} = T/100K$.

It is of interest to compare this theoretical prediction with the stellar density observed in the cores of the most compact clusters of massive stars. The comparison is valid provided that the stellar density observed now reflects that at birth: a necessary condition for this is that the relaxation time for core collapse be at least comparable to the hydrogen-burning lifetime of massive stars. This condition is satisfied for R136a in 30 Doradus, which contains at least 8 massive stars within a region 0.25 pc across (Weigert 1985). The core collapse time-scale is $\sim 7 \times 10^6\ M_{10}^{-1/2}\ yr$, and the stellar mass density is $\sim 3 \times 10^4\ M_{10}^{-1/2}\ M_\odot\ pc^{-3}$, where the IMF has been assumed to have slope $dN/dM \propto M^{-2.5}$ and to cut off below mass $M_{10} \sim 1$ (Moffat et al. 1985). Comparison with the prediction (5) suggests that $M_* = 13(T/200K)^2\ M_\odot$ in R136a. Application to the even more compact stellar core HD 97950 in NGC 3603 is not so clear-cut, since the core collapse time-scale is only $10^6\ yr$ in this object.

The previous discussion is not self-consistent, since an arbitrary value has been adopted for the temperature, which admittedly is likely to exceed $100\ K$

for dust grains in R136a. To improve this, let us assume that dust and gas temperatures are well coupled. Then for grains with emissivity inversely proportional to wavelength, the dust temperature $T_d^5 \propto \rho_*$, where ρ_* is the star density and it has been assumed that O stars dominate the contribution to the local radiation field. The approximate normalization of this equation is

$$T_d = 3(1 + \rho_*)^{1/5} \; degrees \; K, \tag{6}$$

where ρ_* is measured in units of $M_\odot \; pc^{-3}$. To apply this, we use equation (2) which expresses the critical stellar mass-scale in terms of only one parameter, namely T_d, if we make the observationally motivated assumpion that all potentially star-forming fragments have a universal surface density, taken to be 150 $M_\odot \; pc^{-2}$. Combining (6) and (2), and identifying M_{crit} with the lower mass cut-off M_*, then yields

$$M_* = 0.1(1 + \rho_*)^{2/5} \; M_\odot. \tag{7}$$

Note that ρ_* is a weak function ($\propto M_*^{-1/2}$) of M_*. This is the desired relation for M_*: it yields $M_* \approx 7 M_\odot$ for R136a. This should correspond to a peak in the IMF.

IV. SECONDARY STAR FORMATION AND BIMODALITY

Hitherto, only radiative heating by newly formed massive stars has been shown to provide sufficient feedback to ensure continuing massive star formation. However dynamical effects may play an even greater role. Ionization fronts and stellar winds can compress ambient condensations as well as induce the agglomeration of larger condensations by sweeping up molecular gas fragments (Silk 1985). For compression of gas clumps by shock propagation to result in gravitational instability and collapse of an initially stable configuration, it is necessary that the compression be quasi-three-dimensional. A one-dimensional compression is stable, since the pressure gradient is proportional to R^{-1} but the gravitational restoring force is constant. Only three-dimensional compression (with the gravitational force proportional to R^{-2}) is guaranteed to destabilize ambient clumps. Numerical simulations of destabilization of clumps embedded in an HII region have been performed by Klein et al. (1983). In order to have such three-dimensional effects play an important role, it is necessary for a number of massive stars to simultaneously be present in a molecular cloud. This requirement poses a minimum mass requirement on the cloud. If a mass fraction f of a cloud forms O stars within a cloud lifetime t_{cl}, then for a number N of O stars to be present, one needs

$$M_{cloud} > 5 \times 10^4 (N/10)(f/0.1)(t_{cl}/3 \times 10^7) \; yr \; M_\odot. \tag{8}$$

Massive molecular clouds are therefore likely to undergo much more pronounced bursts of star formation than low mass clouds: star formation becomes self-reinforcing once a number of O stars are present.

This means that any remaining low mass globules should be triggered into forming stars. Even the lower mass stars are known to drive winds during their pre-main-sequence evolution. A dramatic consequence of the dynamical interactions will be that the molecular cloud is disrupted. HII regions, wind interactions, and supernovae can destroy a molecular cloud within a few tens of millions of years (Whitworth 1979; Norman and Silk 1980). All of this is greatly compounded by the argument given in the previous section, namely that continuing fragmentation, once a few massive stars have already formed, produces exclusively more massive stars because of the increase in M_*. In a low mass molecular cloud, however, where $M_* \sim 0.1 \; M_\odot$, the typical star will be of low mass, the heating effects will be negligible apart fom dynamical wind interactions, and low mass star formation should continue until the gas supply is exhausted. The occasional massive star may form but only if the cloud accretes sufficient mass can enough O stars form to stimulate a star formation burst.

This suggests the following model for galactic star formation. The interarm regions contain small molecular clouds (SMC), which form stars with an IMF extending down to $\sim 0.1 \; M_\odot$. Individual SMC make very few O stars, hence there is negligible feedback. Coagulation of the SMC as they orbit around the galaxy is enhanced by the spiral density wave, and giant molecular clouds develop predominantly in the spiral arms (see following section). A GMC forms many O stars, feedback is important, and the low mass cut-off increases to a few $M_\odot$ for newly forming fragments. At the same time, some lower mass stars form by implosion of existing globules. This distinction between the character of star formation in SMC and GMC favors a bimodal origin for low mass and massive stars.

Observational evidence for bimodality is well known. Herbig (1962) originally cited T associations as sites of exclusively low mass star formation, and argued that this must precede a phase of massive star formation. Observations of the Pleiades open cluster show that while the nuclear age inferred from the main sequence turn-off is $\sim 3 \times 10^7 \; yr$, low mass star formation must have been continuing to occur for $\sim 10^8$ yr (Stauffer 1984). That young as well as old mass stars are present is inferred from the presence of both rapid and slow rotators (Stauffer et al. 1984). A study of the Pleiades IMF using a complete sample from proper motion studies shows that there is no detectable turn-over in the IMF: stars appear to have formed down to $\sim 0.1\ M_\odot$. A similar conclusion is reached from isochrone fitting of NGC 2264, where there is evidence that low mass star formation occurred over a longer period than the recent phase of massive star formation, and that the rate of massive star formation appears to be increasing with time (Iben and Talbot 1966; Adams et al. 1983; but see Stahler 1985 for a different interpretation).

Other arguments suggest that there are regions which can form predominantly massive stars. A study of the diffuse galactic thermal radio emission suggests that this is true for the spiral arms, as opposed to the interarm regions (Güsten and Mezger 1983). Studies of starburst regions strongly indicate that these regions are forming only stars above a few solar masses (Rieke et al. 1980). This follows from the observed ratio of far infrared luminosity, which measures the mass in OB stars, to the available stellar mass, together with independent constraints on the minimum duration of the starburst (for example, from the presence of spectral features associated with giants or supergiants).

Bimodal star formation is taken here to mean that one can in some regions form almost exclusively low mass stars, and in other regions almost exclusively massive stars. The same gas clouds could be involved, and make the transition from one mode to another. Indeed the general mix in open clusters of low mass and massive stars suggests that stars of all masses form in the same site. Bimodality argues that they have not formed coevally, however.

V. STARBURSTS

The coagulation theory for the origin of the GMC provides a natural mechanism for understanding starbursts, not merely of the mild variety encountered in our galaxy but also in much more active galaxies. An early difficulty encountered in the coagulation theory was that the time-scale for GMC formation was too slow, taking in excess of 2×10^8 yr (Scoville and Hersh 1979, Kwan 1979). It was pointed out, however, that the inelastic response of cloud-cloud collisions to the spiral density wave gives a considerable boost to the coagulation rate and also favors GMC formation in the spiral arms. (Cowie 1980; Norman and Silk 1980; Kwan and Valdes 1983). Detailed computations show that up to 30 percent of all molecular gas clouds can coagulate into GMC within a fifth of a galactic rotation period (Tomisaka 1984).

The cloud mass spectrum that results from coagulation of many SMC should evolve self-similarly according to simple coagulation theory (Nakano 1976; Silk and Takahashi 1979). The only scale is the characteristic mass M_{char}, determined by the product of the mean number of collisions and the mean initial cloud mass, and the resulting self-similar spectrum has the form

$$\frac{dN}{dM} \propto M^{-\nu} \, exp(-M/M_{char}). \qquad (9)$$

Unfortunately, the theoretical investigations, involving either analytic studies or numerical simulations (Pumphrey and Scalo 1983), cannot constrain the power-law index ν of the low mass spectral tail to better than

$$1 \lesssim \nu \lesssim 3.$$

It is noteworthy that $\nu = 2$ is a critical value that distinguishes two regimes. If $\nu > 2$, one has most of the mass being stored in low mass clouds, whereas if $\nu < 2$, most of the mass is stored in the most massive clouds.

This has the following implications. If $\nu < 2$, one might expect that since the massive clouds are most rapidly eroded by OB star formation according to our earlier discussion, one would have a series of bursts of massive star formation and use up essentially the entire gas reservoir within the GMC build-time, roughly the time spent of time between spiral arms, or $\sim 10^8 \, yr$. On the other hand, if $\nu > 2$, the mass would mostly be stored in the long-lived SMC, and there would be a relatively uniform star formation rate.

The mass spectrum of molecular clouds in our galaxy has been determined observationally, and corresponds to $\nu > 1.5$ (Sanders et al. 1985; Dame 1983). Most of the molecular gas is therefore stored in the most massive clouds. Hence our galaxy should be subject to bursts of star formation. Another prediction of a coagulation origin for the GMC is that they should be in equipartition: this is precisely what is observed for clouds of mass in excess of $\sim 10^5 \, M_\odot$ (Stark 1983). The empirically determined collision time between GMC ($0.6/kpc, \sigma_r \sim 3km \, s^{-1}$) is $\sim 2 \times 10^8 \, yr$, but this is considerably decreased within the spiral arms. GMC disruption occurs when enough OB stars have formed.

The observed efficiency of OB star formation is estimated to be ~ 0.3 percent in the λ Orionis OB associations and the surrounding molecular shell (Duerr et al. 1982). It will take a GMC of $3 \times 10^5 \, M_\odot$ at least one free-fall time or about 10^7 yr to even commence forming OB stars. For winds to develop, the stars must evolve past hydrogen burning, and this takes an additional 3×10^6 yr for a massive star. With enough O stars forming, disruption of a GMC should occur by propagation of the ionization front within $\sim (1-2) \times 10^6$ yr. This would be the case for a giant HII region. For a more typical galactic cluster, one might have to wait until B stars have evolved, and produced supernovae before disrupting a GMC. This could take up to $\sim 4 \times 10^7$ yr. In fact, Bash (1979) has used a ballistic model of cloud trajectories to analyze the galactic CO distribution, and he infers a phenomenological lifetime of $\sim 4 \times 10^7$ yr.

These two time-scales for GMC formation and disruption are the principal ingredients for an estimate of the star formation rate in a starbursting galaxy. If M_H is the reservoir of gas and ξ denotes the efficiency at which it is converted into stars of all masses, then the star formation rate can be expressed as

$$\dot{M}_* = M_H \xi \left[max(t_{disrupt}, t_{form}) \right]^{-1}$$
$$= 4 \left(\frac{\xi}{0.2} \right) [M_H/4 \times 10^9 \ M_\odot] \ M_\odot \ yr^{-1}, \qquad (10)$$

The efficiency ξ has been set equal to 20 percent in order to give the known galactic star formation rate, and t_{form} has been set equal to 2×10^8 yr. Note that far infrared observations indicate an efficiency of 20 - 40 percent for effective cloud lifetimes. Observations of cold cloud cores directly show that an efficiency of $\xi \gtrsim 0.1$ is attainable in forming solar mass stars (Wilking and Lada 1983). Equation (10) is consistent with the hypothesis that our galaxy is a ministarburster, about to effectively tap its gas reservoir over the GMC formation time-scale.

Now one can immediately rescale equation (10) to infer the star formation rate in more active galaxies. The star formation rate per unit mass of gas is sensitive to the efficiency factor. However there is no obvious reason why it could be significantly larger than the value inferred for our galaxy. The remaining freedom is in the ratio of cloud formation to disruption times. A tidal interaction with a close companion or a merger could greatly reduce t_{form}. Toomre (1981) has found that even small ($\sim 1 - 2$ percent) tidal perturbations of a cold disk galaxy amplify by a factor of ~ 20 over 4 rotation periods. His calculations did not include any allowance for inelastic gas cloud interactions, and one could readily imagine that the response of molecular clouds would be far stronger. However saturation of the star formation rate is reached if $t_{form} < t_{disrupt}$. Hence, allowing for uncertainty in ξ, one would expect the star formation rate per unit mass of gas to increase by up to a factor of 10 in closely interacting galaxies.

Far infrared observations of starbursting galaxies seem to be in accord with this prediction. The ratio of far infrared to blue luminosity L_{IR}/L_B measures the ratio of star formation rate to the total stellar mass. For our galaxy this ratio is approximately unity. Galaxies with high values of L_{IR}/L_B are generally found to be interacting galaxies or have close companions (de Jong 1986). One of the most extreme cases known is Arp 220, with $L_{IR}/L_B \sim 100$. Our model predicts that the maximum enhancement in star formation per unit stellar mass expected is $(5-10) \times$ (gasfraction/0.03) $\sim 150-300$. Provided that the GMC properties are similar to those in our own interstellar medium, a higher ratio is not possible, unless the galaxy is genuinely young and lacks a gravitationally dominant old star population, that is to say, unless $M_* \ll M_H$.

VI. CONCLUSIONS

The low mass cut-off in the IMF is sensitive to heat input from massive stars because the critical mass above which molecular cloud fragments collapse and form stars depends sensitively on the temperature. A simple model suggests that in the stellar cores of giant HII regions, the low mass cut-off may be increased from ~ 0.1 to $\sim 10\ M_\odot$. An extension of the argument suggests that positive feedback from continuing massive star formation can account for the approximately power-law slope of the IMF.

Feedback effects are especially important in GMC, where a number of massive stars simultaneously coexist. Dynamical interactions of ionization fronts and winds will further stimulate star formation: The outcome is a star formation burst, destroying the GMC after about $4 \times 10^7\ yr$. In SMC, massive star formation is relatively infrequent, feedback effects will be unimportant, and one may expect low mass star formation to continue until continuing coagulation between SMC produces a GMC after $(1-2) \times 10^8\ yr$. The mass spectrum of molecular clouds in our galaxy shows that starbursts can occur over $\sim 2 \times 10^8\ yr$. The star formation per unit mass rate in an active galaxy undergoing vigorous bursts of star formation will be enhanced by up to a factor of 100 due to the increased gas content relative to our galaxy, and to the enhancement in rate of SMC coagulation triggered by tidal interaction with a close companion galaxy.

A search for the low mass IMF cut-off in R136a and 30 Doradus and in other giant HII regions would help confirm some of the ideas presented here.

REFERENCES

Adams, M.T., Strom, K.M., and Strom, S.E. 1983; *Ap. J., Suppl.*, **53**, 893.
Bash, F.N. 1979, *Ap. J.*, **233**, 524.
Bodenheimer, P., Tohline, J. and Black, D. 1980, *Ap. J.* **242**, 209.
Cowie, L. L. 1980, *Ap. J.*, **236**, 868.
Dame, T. 1983, Ph. D. Thesis, Columbia University
Duerr, R., Imhoff, C.L. and Lada, C. J. 1982, *Ap. J.*, 261, 135.
Ebert, R., von Hoerner, S. and Temesvary, S. 1960, *Kondensationen Diffuser Materie* (Berlin: Springer Verlag). p. 311.
Güsten, R. and Mezger, P. 1983, *Vistas in Astronomy*, 26, 159.
Herbig, G. 1962, *Ap. J.*, **135**, 736.
Hoyle, F. 1953, *Ap. J.*, **118**, 513.
Hunter, C. 1962, *Ap. J.*, **136**, 594.
Iben, I. and Talbot, R. 1966, *Ap. J.*, 144, 968.
de Jong, T. 1986, in *Spectral Evolution of Galaxies*, ed. C. Chiosi.
Klein, R. I., Sandford, M. T. and Whitaker, R. W. 1983, *Ap. J. Letters*, **271**, L69.
Kwan, J. 1979, *Ap. J.*, **229**, 567.
Kwan, J. and Valdes, F. 1983, *Ap. J.*, **271**, 604.
Langer, W. D. 1984, in *Protostars and Planets II*, (University of Arizona Press, Tucson, (in press).
Larson, R. B. 1981, *M.N.R.A.S.*, **194**, 809.
Larson, R. B. 1985, *M.N.R.A.S.*, **214**, 379.
Layzer, D. 1963, *Ap. J.*, **137**, 351.
Mestel, L. 1965, *Q.J.R.A.S.*, **6**, 161.
Mouschovias, T. C. and Paleologou, E. V. 1979, *Ap. J.*, **230**, 204.
Moffat, A.F.J., Seggewiss, W., and Shara, M. M. 1985, *Ap. J.*, (in press).
Nakano, T. 1976, *P.A.S.J.*, **6**, 161.
Oort, J. H. 1954, *B.A.N.*, **12**, 177.
Pumphrey, W. A. and Scalo, J. M. 1983, *Ap. J.*, **269**, 531.
Norman, C. and Silk, J. 1980; in *Interstellar Molecules*, ed. B.H. Andrew (D. Reidel: Dordrecht), p. 137.
Norman, C. and Silk, J. 1980, *Ap. J.*, **238**, 158.
Rengarajan, T. N. 1984, *Ap. J.*, **287**, 671.
Rieke, G. H., Lebofsky, M. J., Thompson, R. I., Low, F. J., and Tokunaga, A. T. 1980, *Ap. J.*, **238**, 24.
Sanders, D. B., Scoville, N. Z. and Solomon, P. M. 1985, *Ap. J.*, (in press).
Scoville, N. Z. and Hersh, K. 1979, *Ap. J.*, **229**, 578.
Silk, J. 1977, *Ap J.*, **214**, 718.
Silk, J. 1982, *Ap. J.*, **256**, 514.
Silk, J. and Norman, C. 1983, *Ap. J. Letters*, **272**, L 49.

Silk, J. 1985, in *Birth and Infancy of Stars*, ed. R. Lucas, A. Omont, R. Stoner (North-Holland: Amsterdam) (in press).
Silk, J. and Takahashi, T. 1979, *Ap. J.*, **229**, 242.
Stahler, S. 1985, *Ap. J.*, **293**, 207.
Stark, A. A. 1983, in *Kinematics, Dynamics and Structure of the Milky Way* ed. W. L. H. Shuter (D. Reidel: Dordrecht), p. 127.
Tohline, J. 1982, *Fund. Cos. Phys.*, 8, 1.
Tomisaka, K. 1984, *P.A.S.J.*, **36**, 457.
Toomre, A. 1981, in *The Structure and Evolution of Normal Galaxies*, eds. S. M. Fall and D. Lynden-Bell (Cambridge University Press: Cambridge), p. 111.
Van Leeuwen, F. 1980, in *Star Clusters*, ed J. E. Hesser, (D. Reidel: Dordrecht), p. 157.
Weigert, A. 1985, private communication.
Whitworth, A. 1979, *M.N.R.A.S.*, **180**, 59.
Wilking, B. A. and Lada, C. J. 1983, *Ap. J.*, **274**, 698.

Discussion : SILK.

WALBORN :

I was interested by the application to 30 Doradus suggesting a lack of lower mass stars, because one argument that had been advanced against a cluster with many very high mass stars was that the normal IMF would then predict an improbable number of lower mass stars, whereas I never saw any reason to expect that a normal IMF should apply to a region like 30 Dor.

SREENIVASAN :

An important qualitative aspect that must be considered is shown in the interstellar medium. Fortunately, the Jeans' criterion is still valid, so your discussion may hold. But, Jeans' criterion is only a necessary condition when shear is present. A sufficient condition is that field lines have to be sheared much more efficiently perpendicular to the axis of a protostar than parallel to it. This makes for a slightly higher density in the Jeans' criterion.
One can thus have star formation only in those regions where the shear is favourable.

ZINNECKER :

I don't understand your claim that hierarchical fragmentation is not important. After all, observations show that there are filaments or sheets which contain dense knots of gas which in turn contain young stellar objects.

SILK :

Numerical collapse simulations suggest that multiple fragmentation during dynamical collapse is rather rare. The sheets and filaments seem to form first and then fragment and form protostars.

McGREGOR :

Gatley, Jones and Hyland have found at least four massive protostars near 30 Dor where there is no large molecular cloud and the dust temperature is much higher than in galactic star formation regions.

SILK :

It would be interesting to see quantitatively if this result provides evidence for heating by the nearby O stars.

APPENZELLER :

I fully agree that the suppression of the formation of low mass stars after high mass stars have formed may lead to different IMF's in different associations. However, I am not quite convinced that this must lead to a bimodal distribution. From data on young associations of low mass stars, we know e.g. that in the Lup dark cloud we observe only very low mass stars, in Orion also medium and some high mass stars, and in other associations a mix between the Lup and Orion IMF. Could there not simply exist a smooth transition from massive-star-poor to massive-star-rich associations?

SILK :

In the solar neighbourhood, molecular clouds appear to quisciently form low mass stars for 10^7 yrs or longer. I agree that the data you mention suggests these clouds subsequently form OB stars. This may well be a stochastic effect: a very large cloud should be able to form more OB stars than a small cloud, and the strong heating associated with these massive stars is what I have argued will induce a starburst of primarily more massive stars. Such giant clouds are found in the inner galaxy and presumably they are also responsible for such giant HII regions as 30 Doradus.

EFFECTS OF CONVECTIVE OVERSHOOTING, MASS LOSS (AND CHEMICAL COMPOSITION) ACROSS THE HR DIAGRAM

C. CHIOSI
Institute of Astronomy, Padova, Italy

ABSTRACT. The far reaching consequences of convective overshooting during the core H and He-burning phases of stars in the mass range 1.3 M$_\odot$ to 100 M$_\odot$ are discussed. In addition to this, the effects of mass loss in luminous stars of all spectral types, and in the red giant and asymptotic giant branch stars are briefly outlined. Furthermore, the effects of the novel $^{12}C(\alpha\,\gamma)^{16}O$ reaction rate are also illustrated. The main purpose of this review resides however in lending convincing support to the idea that convective cores of real stars are greater than commonly supposed in classic models. To this aim, several observational embarrassments that could not be explained by classic models are reanalyzed in the light of the new ones. Since a much better agreement between theory and observations is now possible, we are inclined to conclude that convective overshooting may be of paramount importance in stellar structure theories and that convective cores in real stars ought to be larger by approximately one pressure scale height than predicted by classic models.

INTRODUCTION

The amount of observational data available today on star clusters and associations of different ages and chemical compositions makes it possible to study stellar evolution from an empirical point of view. Even if the general properties of stellar evolution are known from long time, still it has become evident over the recent years that many points of contradiction between classic theory and current observation exist demanding a deep revision of stellar evolution. In this paper we will limit ourselves to discuss those observational facts that led us to revise the current models of massive and intermediate mass stars. Populous young clusters of the Magellanic Clouds (LMC and SMC) are particularly suited to this purpose as they are sufficiently populated throughout the various evolutionary stages, thus allowing us to compare them safely with theoretical predictions, even for the very short lived evolutionary phases. On the contrary, this is not easily feasible with galactic open clusters and associations as they contain much fewer stars.Therefore, the above comparison can be made only by collecting data for many individual clusters and by deriving composite HR diagrams. Since many review papers exist describing the properties of massive and intermediate mass stars (Chiosi 1982a,b; Maeder

1984; Chiosi and Maeder 1985; Iben 1974; Iben and Renzini 1983,1984), we will not go into any detail relative to classic evolutionary models. On the contrary, we will focus on four main lines of work, which have (or potentially may) deeply changed (change) the classic scenario. In fact, over the past years, our appreciation of convective overshooting, mass loss by stellar wind, nuclear reaction rates ($^{12}C(\alpha\gamma)^{16}O$ in particular), and true stellar opacity (more precisely the one due to ionization of CNO and heavier elements in the middle temperature regions, see below) has been changed either by upgraded observational information, laboratory experiments and/or theoretical considerations. It goes without saying that not all of the above physical processes are known with the same degree of confidence. In fact, while the occurrence of mass loss is indicated by observations in spite of the uncertainty still affecting the mass loss rates, only very indirect arguments can be put forward to lend support to the existence of convective overshooting. Furthermore, if the novel rate for the $^{12}C(\alpha\gamma)^{16}O$ reaction is supported by laboratory measurements, modification to current opacities has not yet received widespread acceptance nor clearcut confirmation or disproval by theoretical calculations due to the complexity of such a task. The evolutionary scenarios developed insofar, in which one or more of the above physical processes have been incorporated, are reflective of the underlying uncertainty. In fact, while the effects of mass loss and new $^{12}C(\alpha\gamma)^{16}O$ have been thoroughly investigated, those of convective overshooting all over the life of a star now begin to be assessed, those of varied opacity are still in an exploratory stage. In the following, we will mainly concentrate on convective overshooting, firstly describing the far reaching effects of this deeply seated phenomenon, and secondly searching as many as possible observational facts which, discrepant with classic models, may now be interpreted by the new ones. The great advantage shown by the new models over the old ones will be taken as strongly indicating the existence of convective overshooting in real stars. Furthermore, since many important details of this phenomenon are still poorly known, this way of proceeding will implicitly tell us more about the physical processes in which convective overshooting roots. Since the effects of opacities different from the standard ones have been occasionally investigated in the domain of massive and intermediate mass stars, they will be shortly summarized without emphasizing the possible implications. The plan for the remainder of this paper is as follows. In section I, we concisely review several observational facts that cannot be easily interpreted by classic models. In section II, we discuss the fundaments of the four physical ingredients of model construction we have been referring to insofar. In section III we present the main results for models with convective overshooting, new $^{12}C(\alpha\gamma)^{16}O$ reaction rate and mass loss by stellar wind (when appropriate), all over the mass range in which convective overshooting may be effective. Section IV applies those results to massive stars and evidenciates those points of uncertainty that still exist with the new models. Section V deals with several important consequences of the new models in the range of intermediate mass stars and in particular it discusses the capability of greatly improving upon the observational embarrassments presented in section I. Section VI reviews some preliminary results obtained with modified opacities. Finally, some concluding remarks are drawn in section VII.

1. SHADES OF THE OBSERVATIONAL SCENARIO

From the large body of literature dealing with observations of star clusters in our own galaxy and nearby galaxies (LMC and SMC), we have selected the following points, which in our opinion cannot be reasonably explained by common theories of stellar evolution.

1.1 Young Luminous (massive) Stars

The catalogue of all known supergiants, O type stars and less luminous B type stars in our galaxy with MK spectral types and luminosity classes compiled by Humphreys and McElroy (1984) provides the most extended source of data for galactic luminous stars. Since WR stars are commonly understood as the descendents of O type stars via the mechanism of mass loss by stellar wind (see Chiosi 1982a for a recent review of the subject), before comparing stellar counts with theoretical predictions, the Humphrey and McElroy catalogue has to be complemented by the list of known galactic WR stars (van der Hucht et al 1981). Upon transformation of Mv's and spectral types into Mb's and Teff's by means of a given Sp:Teff:BC scale, several important features of the resulting composite HR diagram are soon evident. As they have been amply discussed by Humphreys (1982), Chiosi (1982b) and Chiosi and Maeder (1985), they will not be illustrated here. I shall rather begin with a few comments of general interest and then concentrate on a result of stellar counts, which has driven most of the theoretical work done in this context. First of all, many of the features shown by that composite HR diagram, in particular the location of O type stars and related apparent decline of the maximum luminosity reached by these stars with decreasing Teff, depend od the adopted Sp:Teff:BC scale. Humphreys widely used the so-called "hot scale of Teff", which, if on one hand has received widespread consensus, on the other hand amplifies the above features. Other scales of Teff are known to exist, which associating a cooler Teff to a given spectral type make the above trend less pronounced. Secondly, the question may arise whether the Teff resulting from evolutionary computations is comparable with Teff given by observation. Strong stellar winds may in fact produce a pseudo photosphere in the flow itself, thus indicating a Teff which may be significantly cooler than the one derived from hydrostatic atmospheres (de Loore et al 1982; Bertelli et al 1984). Thirdly, the catalogues of supergiant and O-type stars suffer from a certain degree of incompleteness which is difficult to assess. This may be particularly severe for the earliest O type stars for which the bolometric corrections are the greatest. Somewhat related to this, there is another intricacy which makes the comparison with theory even more complicated. Rough stellar counts per spectral type and given luminosity interval indicate that the star frequency distribution seems not to mimic the distribution of relative lifetimes one would expect from models. It appears as there is a deficiency of stars near the zero age main sequence. Is this indicating that stellar models are in error or that the majority of O type stars are already evolved ? (see also Vanbeveren 1986). The deficiency of O type stars seems to begin at Mb < -8 or equivalently for masses greater than about 30 M☉, see for instance the HR diagram of Fig 1 in Humphreys and McElroy (1984). We will touch upon this point later on. Despite of those uncertainties, stellar counts by Meylan and Maeder (1982) and Bertelli et al (1984) have indicated an excess of A to G stars and the

iii) <u>Lack of bright AGB Stars</u>. The remarkable absence in NGC 1866 and other similar clusters of LMC of very bright AGB stars, which on the contrary are expected to occur in a cluster with an age of about 86×10^6 yr and a turnoff mass of 5 M☉ (Becker and Mathews 1983). It has been suggested that a rate of mass loss (either in stationary or superwind mode) much greater than customarily assumed for stars in this phase (see Iben and Renzini 1983, 1984 for all details) may result in an early termination of the AGB phase, thus accounting for the observed lack of very luminous AGB stars. As it will be discussed later on, this is not a viable explanation, or at least inadequacy of the mass loss rates is not the sole cause of disagreement.

iv) <u>Quasi Old Clusters</u>. The peculiar morphology of galactic as well as LMC clusters with age in the range 1 to 2×10^9 yr is another puzzling problem. In fact, Barbaro and Pigatto (1984) found that, while clusters older than than 2 to 3×10^9 yr generally agree with theoretical predictions for their red giant star luminosity function, theory fails in interpreting the red giant distribution in clusters of slightly lower age. In fact, their behaviour is typical of even younger clusters, in that a well developed red giant branch is not observed. On the theoretical side, this can be explained supposing that those red giants, which on the basis of the cluster turnoff mass (< 2 M☉) are expected to be in shell H-burning phase, to develop a highly degenerate He core and therefore to eventually undergo core He flash, actually evolve as more massive stars. Since the minimum core mass for non degenerate He ignition is 0.33 M☉, to which an initial mass of 2.2 to 2.3 M☉ is customarily assigned, everything occurs as if stars of initial mass as low as 1.3 M☉ or thereabouts were able to build He cores more massive (or as massive as) the above limit without passing through a phase of degeneracy in the core. The explanation of this dilemma, suggested by Barbaro and Pigatto (1984) and confirmed by Bertelli and Bressan (1985), was attributed to convective overshooting during the core H-burning phase.

v) <u>Age Discrepancy</u>. It has been pointed out (Hodge 1983) that ages of LMC clusters derived from the terminal AGB luminosity (Mould and Aaronson 1982 are in disagreement with ages derived from the main sequence turnoff or termination luminosity (Hodge 1983) and/or red giant clump luminosity (Flower 1984). While the ages derived from the last methods are in satisfactory agreement, they are too low when compared to Mould's and Aaronson' estimates for most of the clusters in common. Furthermore, the discrepancy is the highest for young clusters and it gets negligible for the oldest ones. To get rid of the difficulty both Mould (1983) and Hodge (1983) suggested that more mass has to be lost during the ascent of the AGB and/or the planetary nebula ejection phase. However, the same arguments against more substancial mass loss advocated for the AGB termination problem, hold even in this case and other causes of disagreement are likely to exist.

vi) <u>AGB Star Luminosity Function</u>. Another facet of the lack of very luminous AGB stars resides in the disagreement between theoretical and observational luminosity function for field stars of LMC studied by Reid and Mould (1984). In brief, the observed luminosity function not only shows very few stars brighter than $M_b = -6$, but also decreases with increasing luminosity steeper than predicted by classic models of AGB stars under any plausible assumption for the star formation rate and initial mass function. To overcome the difficulty Reid and Mould (1984) suggested that more mass has to be lost by AGB stars. As discussed by Renzini (1984a), Bertelli et al (1985) and Chiosi et al (1985), this is not a viable solution.

vii) Number Frequency - Period distribution of Cepheids. In a recent paper, to which the reader should refer for a better understanding of what follows, Serrano (1983) has analyzed the number frequency-period distribution of Cepheids in the light of classical models. The distribution is specified by three parameters: the short period cutoff Po, the period P1 of the maximum of the distribution and the rate at which Cepheids with P>P1 decay in number with respect to the period. These two periods, weighted on the initial mass function correspond to the minimum masses whose He-burning loops enter the instability strip at the red side and spend the whole He-burning in the loop within the instability strip. Equivalently, the period Po and P1 depend on the location and inclination in the HR diagram of the blue loop band with respect to the Cepheid instability strip. The blue loop band is known to depend critically on the chemical composition. Limiting the discussion to the sole solar vicinity to minimize effects of chemical composition, the observed difference LogP1-LogPo is in the range 0.3 to 0.6 while theory predicts 0.06 (Becker et al 1977). Furthermore, there is an excess of long period Cepheids which could be explained only by invoking a two component birth rate, more efficient for massive stars (Becker et al 1977). It is easy to show that Becker's (1981) models yield too a small LogP1-LogPo difference for any combination of chemical abundances. To overcome the problem we may either make wider the instability strip (Pel and Lub 1978) or to increase the slope of the blue loop band. This latter alternative will turn up to be the most reasonable solution. In such a case, not only the period difference is matched, but also the excess of long period Cepheids is accounted for.

2. PHYSICAL FUNDAMENTS OF EVOLUTIONARY MODELS

In this section, we briefly summarize the main ideas relative to a few points of stellar structure which are at the base of the most recent computations, even if they have not yet received general consensus.

i) Convective Overshooting

In classic stellar models, the boundary of the convective core is defined by the condition $\nabla_R = \nabla_A$, which is equivalent to say that the core may extend up to the layer where the buoyancy acceleration of convective elements vanishes. The velocity however does not get zero at this layer, implying that convective elements may penetrate (overshoot) into the formally stable radiative zones, thus increasing the mass of the convective core. Due to well known uncertainties in the physics of convection and mixing processes, contrasting conclusions have been reached by different authors, which go from considering convective overshooting negligibly small to claiming that it is of paramount importance. Among others, we recall Maeder (1975, 1976), Cogan (1975), Roxburgh (1978), Cloutman and Whitaker (1980), Maeder and Mermilliod (1981), Matraka et al (1982), Doom (1982a,b, 1985), Bertelli et al (1985a). In particular, Bressan et al (1981) have shown the importance of convective overshooting in massive stars, while Bertelli et al (1985a) have investigated its far reaching effects in the domain of intermediate mass stars. These authors describe convective overshooting by means of the mixing length theory of convection and propose a formalism containing the mixing length scale of motions (l =λ Hp) as an adjustable parameter (λ) to be eventually fixed by comparing model results with observations. In this formalism, when λ = 0 the classic condition is recovered.

ii) Mass Loss by Stellar Wind
Luminous early type stars and late type giants and supergiants are known to lose mass from the surface at rates that may significantly affect their evolution. The subject has been reviewed so many times that a detailed presentation of current mass loss rates, rate parametrizations in terms of basic stellar quantities, and physical processes powering the wind of different types of star is superfluous. Quite an exhaustive review of those topics and appropriate referencing to original sources can be found in Chiosi and Maeder (1985), to whom we refer for more information. It suffices to recall here the mass loss rate parametrizations that have been used in model computations we are going to describe. a) Massive stars have been evolved taking into account mass loss according to the following prescription for the rates: Chiosi and Olson (1984) for O-B type stars, Barlow and Cohen (1977) for A to M supergiants, the rates however scaled to the results of Jura and Morris (1981), and Barlow et al (1981) for the so called WR stage. This is assumed to begin when the following conditions are satisfied: surface abundance of hydrogen less than 0.1 (in mass) and Teff of the model greater than 20000 °K. b) A few exploratory sequences for intermediate mass stars have been computed using either Waldron's (1984) parametrization, however rescaled as described in Bertelli et al (1985a) to obey the constraint imposed by globular clusters (Fusi Pecci and Renzini 1976), or Reimers' (1975) formula with his parameter η comprised in the range 0.3 to 1. Using Bertelli's et al (1985a) notation, Waldron's mass loss rates multiplied by the parameter α = 0.2 to 0.3 are equivalent to Reimers' rates for η = 1.

iii) Nuclear Reaction and Neutrino Emission Rates
The major novelty in this context is the recent determination of the cross section for the $^{12}C(\alpha \gamma)^{16}O$ reaction by Kettner et al (1982) together with the theoretical reanalysis by Langanke and Koonin (1982). The new rate runs 3 to 5 times faster than the classic rate of Fowler et al (1975). In the usual notation this rate is parametrized by θ_α^2. Here the new value for θ_α^2 is taken to be three times greater than the canonical value of 0.078. All other reaction rates pertinent to the evolutionary phases under consideration are as in Fowler et al (1975). Whereever appropriate, the neutrino energy losses have been included. These are from Beaudet et al (1967), using their interpolation formula which is accurate within 20%.

iv) Radiative Opacities
Opacity of the stellar material was derived by interpolation (in temperature, density and chemical composition) of the opacity tables of Cox and Stewart (1970) for most of the models computed. However, there are several independent arguments (Simon 1982; Iben and Renzini 1984; Bertelli et al 1984) suggesting that current opacity calculations may actually underestimate the true opacity in the middle temperature region (5×10^5 to 5×10^6 °k) where the main sources of opacity are the bound-bound and bound-free

transitions involving occupied levels in highly ionized species of elements from carbon to iron. Starting from the modification proposed by Bertelli et al (1984) and a suggestion advanced by Renzini (1984b), Bertelli and Bressan (1985) adopted the following relation

$$\kappa = \kappa_{XY} + A\,\Delta Z \left[1 + \chi\, f(\rho,T)\, \exp(-\alpha\, \mathrm{Log}(T/T_o)^4)\right] \quad (1)$$

where κ_{XY} is the opacity of a metal free mixture of H and He. A ΔZ gives the contribution of heavy elements to the classical opacity. This is evaluated subtracting at each value of ρ and T the opacity of a metal free mixture with given X and Y from the opacity of a mixture having the same X and Y but metal abundance Z. Furthermore, $f(\rho,T)$ is a suitable function defining in the ρ-T plane a band along which the opacity enhancement given by the exponential term is allowed to occur. T_o is the central value of the temperature interval (10^6 °K). The request of matching the opacity increase proposed by Simon (1982) and Bertelli et al (1984) allows us to determine α, χ and $f(\rho,T)$. Since all this is highly speculative, only a few models have been computed. They will be discussed separately.

v) <u>The Models</u>
Evolutionary sequences in the mass range 1 M☉ to 100 M☉ have been computed from the main sequence to the red giant tip for masses in the range 1 M☉ to 1.6 M☉, to the beginning of the thermally pulsing AGB phase for masses in the range 1.6 M☉ to 6 M☉ (Z = 0.02) or to 5 M☉ (Z = 0.001), and beyond the C-ignition stage for larger masses. Two sets of chemical abundances have been considered, namely X = 0.700 and Z = 0.02 (Bertelli et al 1985a and 1985b) and X = 0.700 and Z = 0.001 (Angerer et al 1985). Although many sequences have been evolved with different choices of the overshooting parameter, we will present here only the set of models for $\lambda = 1$ as according to Bertelli et al (1984,1985a) it gives the most interesting results.

3. EFFECTS OF CONVECTIVE OVERSHOOTING AND NEW θ_α^2 ON STELLAR STRUCTURE

To fully appreciate the differences with respect to standard theories of stars given by convective overshooting it is worth introducing five critical values of the initial mass which in turn define six mass ranges in which stellar evolution proceeds differently.

1) $M_i > M_{mas}$ (massive stars): above this limit stars proceed through a series of nuclear burnings in non degenerate conditions towards the construction of an iron core, subsequent photodissociation instability with core collapse and supernova explosion (see Woosley et al 1984 for all details). The current value of M_{mas} is about 12 M☉.

2) $M_{up} < M_i < M_{mas}$ (quasi massive stars): stars in this mass range ignite carbon under mildly degenerate conditions, suffer a mild carbon flash but burn carbon non violently. Their subsequent evolution is rather complicate, but eventually terminated by core collapse leading to a supernova explosion (Nomoto 1984). These stars and those of the previous range fail to develop a highly degenerate CO core, and do not experience the thermally pulsing phase of AGB (TP-AGB). The current value of M_{up} is about 9 M☉ for Pop I and lower than this (= 7 M☉) for Pop II chemical composition (Becker and Iben 1979).

3) $M_{HeF} < M_i < M_{up}$ (intermediate mass stars): stars in this mass range

ignite helium nondegenerately but following He-exhaustion develop a highly degenerate CO core. They experience a long thermally pulsing AGB phase, terminated either by envelope ejection and formation of a white dwarf (for $M_{HeF} < M_i < M_W$) or carbon ignition and deflagration in a highly degenerate core which has grown to the Chandraseckar mass of 1.4 M☉. This event is usually referred to as a supernova of type I1/2 (Iben and Renzini 1983). The critical mass M_W above which this may occur is determined by the efficiency of mass loss by stationary wind or so-called "super wind". Current estimates set M_W around 5 M☉ (Iben and Renzini 1983, 1984).

4) $M_i < M_{HeF}$ (low mass stars): stars in this range of mass develop a highly degenerate He core shortly after central hydrogen exhaustion. They develop the red giant branch along which they suffer significant mass loss by stellar wind. The prolonged red giant branch phase is terminated by the violent ignition of He-burning in the core (He-flash), when the core mass has grown to 0.4 to 0.5 M☉ depending on the chemical composition. Their subsequent evolution is quite similar to that of intermediate mass stars. The classic value of M_{HeF} is 2.2 to 2.3 M☉ for Pop I and about 1.8 M☉ for Pop II chemical composition. Within this mass range it is worth defining another mass, M_{LC}, above which stars possess convective cores on the main sequence. This mass represents the minimum value for stars being affected by convective overshooting during their core H-burning phase. A provisional estimate sets M_{LC} at about 1.2 M☉ for Pop I composition (Bertelli et al 1985b).

i) Core H-burning Phase

Since the major effects of overshooting on core H-burning models are already known (Maeder 1976; Maeder and Mermilliod 1981; Bressan et al 1981; Bertelli et al 1985a), the discussion will be kept very short. Models with overshooting possess more massive convective cores, run at higher luminosity and live longer than classical models. They also extend the main sequence band over a wider range of Teff. The above effects depend on λ and on the star mass. The relative increase in the mass of the convective core is $\Delta(Mr/M) = 0.16$ when λ varies from 0 to 1, while at given λ the increase in core mass by overshooting is greater at lower masses. The increase in the lifetime mimics the dependence of the increase in the core mass on the star mass. The same is true for the increase in the stellar luminosity. The variation in the range of Teff's covered by core H-burning models, namely the extension of the main sequence band, increases with the stellar mass. Massive stars ($M_i > 40$ M☉) would spread all across the HR diagram, it were not for the contrasting effect of mass loss.

ii) Core He-burning Phase

The overluminosity caused by overshooting during the core H-burning phase still remains during the shell H and core He-burning phases. The mass of the H-exhausted core, M_{He}, and the mass of the CO rich He-burning convective core are increased by $\Delta(Mr/M) = 0.13$ when λ increases from 0 to 1. As a consequence of the higher luminosity, the lifetime of the He-burning phase (t_{He}) gets shorter in spite of the increase in the core caused by overshooting. This combined with the longer H-burning lifetime t_H makes the ratio t_{He}/t_H rather low (from 0.12 to 0.06 when the star mass varies from 1.6 M☉ to 9 M☉. In Table 2 we summarize the lifetimes for the set of models with $X = 0.700$ and $Z = 0.02$.

Table 2

(Lifetimes of models with overshooting and X = 0.700 and Z = 0.02)

M/M☉	t	t(H)	t(He)	t(Heb)	t(Her)	t(He)/t(H)
1.2		4.26 (9)				
1.4		3.24 (9)				
1.5		2.73 (9)				
1.6	2.59 (9)	2.30 (9)	2.76 (8)	No loop	2.76 (8)	0.12
1.7	2.18 (9)	1.95 (9)	2.30 (8)	No loop	2.30 (8)	0.12
3.0	4.60 (8)	4.27 (8)	3.26 (7)	No loop	3.26 (7)	0.08
5.0	1.25 (8)	1.18 (8)	7.06 (6)	3.59 (6)	3.20 (6)	0.06
6.0	8.34 (7)	7.88 (7)	4.62 (6)	2.37 (6)	2.07 (6)	0.06
7.0	8.04 (7)	7.70 (7)	3.35 (6)	1.98 (6)	1.37 (6)	0.05
9.0	3.55 (7)	3.34 (7)	2.10 (6)	1.20 (6)	9.00 (5)	0.06

ages are in years; overshooting is for $\lambda = 1$; new θ_α^2

The location in the HR diagram of core He-burning models can be schematically summarized as follows. In massive stars, where mass loss may occur even during the main sequence phase, it is almost entirely dominated by this phenomenon and it may take place either partly in the red and partly in the blue, or entirely in the red, or entirely in the blue. A detailed description of this phase can be found in Chiosi and Maeder (1985). In the range of quasi massive and intermediate mass stars, it is well known that extended loops may develop. Their extension however depends on chemical composition and details of the model structure (see the discussion of Renzini 1984c on this subject), and furthermore on overshooting, θ_α^2 and mass loss during the red giant phase. In brief, convective overshooting strongly decreases the loop extension, mass loss along the Hayashi line makes the loops even redder, while an increase in θ_α^2 acts in the opposite sense. Finally, loop extension depends on the star mass, in general they begin and get bluer at increasing mass. Fig. 1 shows the HR diagram of intermediate mass stars evolved at constant mass whose lifetimes are given in Table 2.

iii) <u>The Critical Masses M_{mas}, M_{up} and M_{HeF}</u>
In virtue of the larger helium core and carbon-oxygen core left over at the end of core H and core He-burning respectively, the relation between the initial mass and M_{He} and M_{CO}, which defines the above critical masses, is different with the new models. The new critical masses are summarized in Table 3 for the two sets of chemical composition. The most important result is that both M_{up} and M_{HeF} are significantly lower in models with overshooting. This means that no AGB phase is now expected above the new M_{up}, while no prolonged RGB phase occur for stars as low as 1.6 M☉. The impact of this finding on observational front is straightforward and of paramount importance. Lower values of M_{up} and M_{HeF} have been also suggested by Renzini et al (1985), Castellani et al (1985) and Barbaro and Pigatto (1984) respectively. Remarkably, in models with overshooting, M_{HeF} has been found not to depend on Z, while its possible dependence on X has not yet been tested. Finally, we recall that a different $M(M_{He})$

Table 3

(Critical Masses M_{mas}, M_{up} and M_{HeF} : in M⊙)

X	Z	M_{mas}	M_{up}	M_{HeF}
0.700	0.020	9	6	1.6
0.700	0.001	8	5	1.6

relation now holds all over the mass range in which convective overshooting is in operation. This is particularly relevant in the domain of massive stars in conjunction with the problem of chemical enrichment per stellar generation (see Chiosi and Matteucci 1984 for a complete discussion of the topic). Furthermore, the drastic lowering of M_{up} towards M_W makes the occurrence of type I1/2 supernovae very unlikely.

iv) <u>Effects of θ_α^2 on the chemical structure of the CO core</u>
With the old value of θ_α^2 almost equal abundances of C^{12} and O^{16} are expected in the carbon-oxygen core, the abundance of C^{12} being moreover a decreasing function of the star's mass (Arnett 1972). With the new θ_α^2 very little carbon is left in the core, the final abundance of it being lower than 10%. This result is preliminary and has to be confirmed by more extensive calculations. Furthermore, if confirmed by other measurements of the $^{12}C(\alpha\gamma)^{16}O$ cross section, it may have strong implications for the subsequent evolution of intermediate and massive stars. We will touch upon this subject later on.

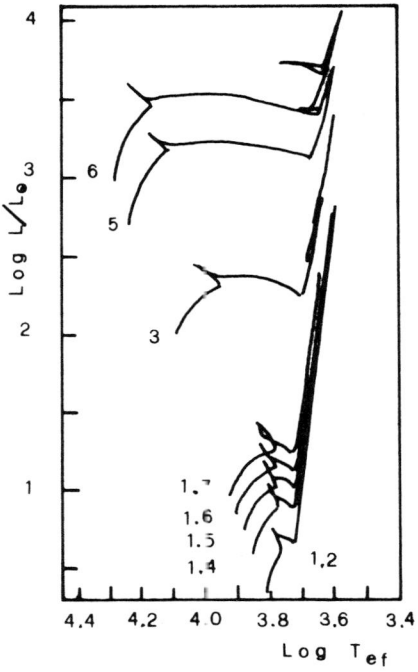

Fig. 1 The HR diagram of intermediate mass stars and low mass stars with convective overshooting ($\lambda = 1$) and new θ_α^2. The transition mass M_{HeF} is at 1.6 M⊙, while M_{up} is at 6 M⊙. The chemical composition is X = 0.700 and Z = 0.020. The models are evolved at constant mass

4. OBSERVATIONAL IMPLICATIONS FOR MASSIVE STARS

Under the combined effects of overshooting and mass loss, the band of core H-burners may now extend up to the spectral type B1 in the luminosity range correspondent to stars of initial mass from 20 M☉ to 60 M☉. At higher masses the core H-burning band shrinks towards the zero age main sequence as it occurred for models with classical convective cores but losing mass at substantial rate. The advantage here is that the same result is obtained without enormous losses of mass, owing to the larger convective cores, which favour the appearance of CNO processed material at the surface. The lowering of the opacity (mostly electron scattering) will produce a smaller radius. The results for the mass range 20 M☉ to 60 M☉ are particularly interesting. In fact the model stars may now spend in the spectral range B0 - B1 about 20 % of their total core H-burning lifetime, which approximately amounts to three times the core He-burning lifetime. As a consequence of this, we expect in this spectral range, which approximately corresponds to the Kelvin-Helmotz gap of classical models as many stars as in all other later spectral classes. As shown by Bertelli et al (1984) this greatly alleviates the difficulty of too many stars falling outside the formal main sequence band of standard models. The core He-burning phase takes place partly in the red and partly in the blue supergiant regions for initial masses in the range 20 M☉ to 40 M☉ or thereabouts. Approximately 30 % of the He-burning lifetime is spent by a typical 20 M☉ star as a red supergiant. Stars more massive than 40 M☉, even though they may reach the red supergiant region, spend there very short time and soon run back into the blue side of the HR diagram (WR progenitors ? as suggested by Humphreys et al 1985). The behaviour of stars in the mass range 10 M☉ to 20 M☉ in not entirely clarified. Likely they will spend the whole core He-burning lifetime as red supergiants. 10 M☉ to 12 M☉ stars likely constitute the upper limit for loops to develop with Pop I composition. Table 4 shows the stellar counts by Bertelli et al (1984) for models with overshooting in the luminosity range - 7 > Mb > - 9, using the same material we presented in section I. Now 80 %

Table 4

Sp	O-B1	B2-B9	A	F	G	K	M	WR
N	432	49	10	4	1	1	29	20
N/Nt	0.79	0.09	0.02	0.01	~0	~0	0.05	0.04

of stars fall in the formal main sequence band. However, since t_{He}/t_H = 0.06, there seems to be a residual deficiency of stars amounting to about 15 %. Whether the arguments of Garmany et al (1982) and Vanbeveren (1986), in that a fraction of potential O type stars is not yet observable, or other causes of further extension of the main sequence band (up to the spectral type A0) are to be found, is not very clear. Bertelli et al (1984) analyzed the second alternative, advancing the idea that only a suitable combination of atmospheric effects caused by mass loss on the radius and a gentle increase in the middle temperature opacity could be able to reproduce the observed stellar frequencies. Their conclusion rested on and was somewhat vitiated by the assumption that stars down to about 20 M⊙ should have been able to generate WR stars. More recent analyses of WR star progenitors based on cluster membership (Schild and Maeder 1984 and Humphreys et al 1985) by moving upwards the progenitor mass have alleviated the discrepancy. Even if the main properties of the evolution of massive stars in occurrence of mass loss are reasonably well understood, still they are hampered by the uncertainties on mass loss rates and their parametrization. Perhaps the most interesting advancement in the theory of massive stars is related to the new cross section for $^{12}C(\alpha,\gamma)^{16}O$ reaction. In fact as a consequence of the very little carbon left in the core following He-exhaustion, massive stars may be able to skip the C-burning phase. As found by Wilson et al (1985) this has profound consequences in subsequent evolution. In fact, their 25 M⊙ star with the new rate develop a much larger iron core. The reason of it resides in the nature of carbon and neon burnings and how they affect the entropy structure of the core. In brief, with the old reaction rate C-burning ignites as a well developed, exoergic convective burning stage. While C-burning goes on, neutrino losses cool the core and cause loss of entropy. Later on a C-burning shell ignites at the border of a semidegenerate core close in mass to the Chandraseckar limit. Due to the entropy barrier set up at the border of this core, the C-burning shell cannot migrate outward. So that when oxygen and silicon burning are completed and the core eventually collapses, it does so with an iron core of about 1.33 M⊙. With the new rate, C-burning and also Ne-burning never ignite in the centre as exoergic convective burning stages. The little traces of C and Ne burn out radiatively on a very short time scale. Because there is no sufficient cooling stage, the core does not degenerate and becomes insensitive to the Chandraseckar mass. When the star reaches the collapse, it has built an iron core of about 2.2 M⊙. A black hole is likely to form in this event. It is clear that the particular value of the initial mass above which this may occur, depends on the amount of carbon present in the core which in turn depends on the star's mass.

5. OBSERVATIONAL IMPLICATIONS FOR INTERMEDIATE MASS STARS

In this section we briefly touch upon the use of models with overshooting in a few of the topics presented in section I, namely the problem of the HR diagram of intermediate age and quasi old clusters, the problem of clusters dating and age discrepancy, and finally the problem of AGB star luminosity function. Although the adopted chemical composition may not be fully suited, we shall consider only LMC clusters as they are the best laboratory where such a comparison can be successful.

i) <u>Intermediate Age Clusters (NGC 1866)</u>
NGC 1866 is particularly suited to this purpose, because it is well populated and because it has been studied recently by Becker and Mathews (1983) on the basis of classic models. Those authors assigned NGC 1866 a turnoff mass of about 5 M⊙, a chemical composition Y = 0.273 and Z = 0.016, and an age of $86 \pm 5 \times 10^6$ yr. However they failed in reproducing the relative stellar frequencies (main sequence versus red giant stars, blue versus red giants) and the lack of bright AGB stars. Table 5 contains the star counts derived from the photometric study by Robertson (1974) plus the list of super luminous stars by Flower (1981). Since the main sequen-

Table 5
(Stellar Counts in NGC 1866)

Note	MS	BG	RG	AGB	SLS
(1)	206 + 14	$46 ^{+10}_{-7}$	$47 ^{+29}_{-7}$	$14 ^{+13}_{-5}$	6
(2)	116	60	137	40	
(3)	21.1 (7)	0.79 (7)	0.79 (7)	<< 5 (6)	

(1) Star counts from Becker and Mathews plus the super luminous stars (SLS) of Flower; (2) Star numbers predicted by Becker and Mathews; (3) Theoretical lifetimes for a 4 M☉ star with convective overshooting ($\lambda = 1$) and chemical composition X = 0.700 and Z = 0.020

ce stars may be affected by incompleteness in that the majority of them fall below the survey limit (18 mag.), we prefer to use the giant star population in comparing theory with observations. Omitting all the details of the discussion which can be found in Bertelli et al (1985a) we propose the following scenario: turnoff mass of about 4 M☉, age of about 210 x 10^6 yr, an equal percentage of blue and red giants (cfr. the lifetimes of Table 5). With the above choice, the number of main sequence stars in the observational sample would amount only to 10 % of the total pertinent to a turnoff mass of 4 M☉. The rest falls below the survey limit. The lack (paucity in general) of very bright AGB stars can be understood by the much lower M_{up} indicated by models with overshooting. The few superluminous stars are compatible (in number) with being core C-burners (cfr. Flower et al 1980). Mass loss by stellar wind might help to remove them away from the AGB area. If all this is correct, it would imply that M_{up} for chemical composition pertinent to NGC 1866 is down to about 4 M☉. Models of ours with X = 0.700 and Z = 0.001 yield M_{up} = 5 M☉. Other chemical compositions have not yet been explored. Direct measurements of the chemical abundance in NGC 1866 stars would be highly useful.

ii) Quasi Old Clusters (NGC 2190 and NGC 2162)

The HR diagrams of clusters like NGC 2190 and 2162, recently obtained by Schommer et al (1984) with CCD photometry, are reflective of the lower M_{HeF} given by models with overshooting, since these two clusters turn up to be similar to those studied by Barbaro and Pigatto (1984). In fact, in the study of Chiosi and Pigatto (1985) an age of about 1 x 10^9 yr has been assigned and more important it has been shown that they possess a single peak luminosity function for red giants. The correspondent mass is about 1.8 M☉. This well to models with overshooting which predict M_{HeF} as low as 1.6 M☉ and core He-burning to occur along the Hayashi line, to which a clump of red giants corresponds on observational side. With classical models of the same age and chemical composition, there would be a prolonged red giant branch followed by core He-burning in the horizontal branch (red in this case). A double peak luminosity function for red stars is therefore expected, contrary to what observed.

iii) Clusters Dating and Age Discrepancy

As already anticipated in section I, three methods exist to date clusters, namely the luminosity of the main sequence turnoff (and/or termination), the luminosity of the He-burning red giants, the luminosity of the bright AGB stars. Each of those rests upon suitable luminosity versus age relationships derived from theoretical models. Unfortunately, among other factors, the theoretical relations depend on the chemical composition, which has to be specified a priori. Good abundance determinations are available only for very few clusters. Despite of it, age compilations have been derived (Hodge 1983; Flower 1984; Mould and Aaronson 1982). The results are very discouraging as ages derived by different methods in general disagree by large factors. It goes without saying that each method competes with intrinsic observational as well as theoretical difficulties that often invalidate the whole results.
a) The turnoff method is hampered by the lack of very good photometric data down to magnitudes faint enough to delineate the unevolved portion of the main sequence. This is particularly severe for LMC clusters older than a few 10^8 yr or thereabouts.
b) The red giant luminosity method requires that the luminosity of core He-burners be a monotonic function of the age. If this is true (the luminosity decreases with increasing age) for stars more massive than M_{HeF}, it does no longer hold below this limit, firstly because the relation flattens out, secondly because the luminosity of core He-burners generated by stars less massive than M_{HeF} is higher than that of stars in the range $M_{HeF} < M < M_{HeF} + 0.5\ M_\odot$ (Renzini and Buzzoni 1986). The relation is somewhat bivariate at M_{HeF}. Therefore, before using the red giant method, care has to be payed to ascertain the type of cluster we are dealing with.
In Fig 2 we present the above calibrating relationship derived from models with overshooting and two values of metallicity.

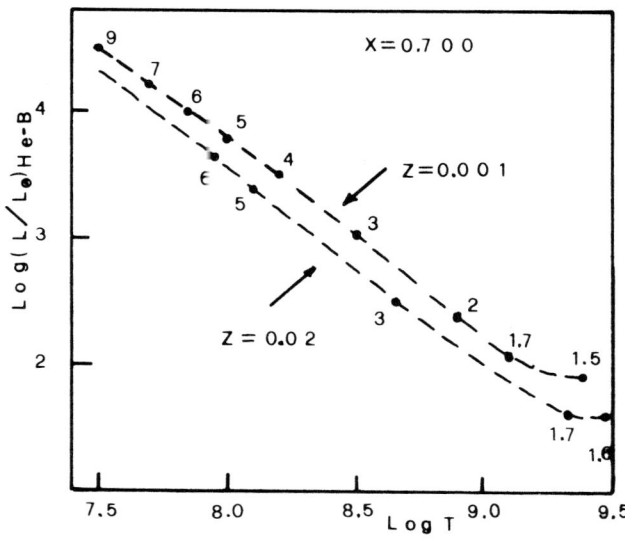

Fig. 2 Luminosity of core He-burning models versus age relationships. The evolutionary models allow for convective overshooting as described in the text. Along each curve the stellar masses are indicated. The chemical compositions are indicated in the panel

c) The AGB luminosity method can be applied only to clusters with turnoff mass below M_{up}. However the greatest uncertainty with this method resides in the low number of stars in AGB which makes the identification of the maximum AGB luminosity quite uncertain.

Chiosi et al (1985) revisited the whole problem of clusters dating in the light of models with overshooting. The results can be summarized as follows:

1) Due to the higher luminosity and longer lifetime of core H-burning of the new models, we expect that for any given turnoff (or termination) magnitude of clusters whose turnoff mass has a convective core, the correspondent age is greater than classic estimates. Similar increase in the age is also expected from the red giant method for all clusters which have a turnoff mass greater than M_{HeF}. The relations below may be used to convert ages derived from classical models into the new ones,

$$\text{Log } t = 0.78 \text{ Log } t_{old} + 2.337 \quad \text{(turnoff method)}$$
$$\text{Log } t = 1.05 \text{ Log } t_{old} + 0.092 \quad \text{(red giant method)}$$

where ages are given in years.

2) Following the procedure outlined by Iben and Renzini (1983, 1984) however adapted to the new models, a novel relation between maximum AGB luminosity and age is derived. The rate of mass loss during the red giant and AGB phases is from Reimers (1975) with $\eta = 1$ or equivalently from Waldron (1984) with $\alpha = 0.2$ (cfr. section II). Ages derived from the novel relation are only modestly changed by overshooting. This surprising result can be understood as due to the fact that AGB evolution is mainly driven by the CO core mass and it depends little on the past history. What is actually changed is the correspondence between initial mass and total lifetime.

The most important result is that ages based on the main sequence turnoff get closer to those from AGB luminosity. Even with all reservations caused by the uncertainties discussed insofar, the so-called Age Discrepancy if not completely ruled out is greatly alleviated when overshooting is taken into account.

iv) <u>The AGB Luminosity Function For Field Stars In LMC</u>

Here we discuss the AGB star luminosity function obtained by Reid and Mould (1984) for a selected area of LMC. The luminosity function (number of stars per magnitude bin) is shown in Fig 3 and compared to the prediction based on the standard theory of AGB stars. All details relative to the procedure and assumptions made concerning the star formation rate and initial mass function are given in Reid and Mould (1984) and therefore omitted here. It suffices to recall that the particular case shown in Fig 3 is derived assuming a constant star formation rate and the Salpeter mass function. Chiosi et al (1985) have repeated the whole analysis using models with convective overshooting in the previous evolutionary phases. The mass loss rates were the same as reported in the previous paragraph. All other assumptions are as in Reid and Mould (1984). Nevertheless, two major changes are important. First, the contribution of early AGB stars neglected by Reid and Mould (1984) was taken into account. Second, it has been found that also normal red giants (core He-burners) with initial mass in the range 4.5 M☉ to 7 M☉, being overluminous because of overshooting, may contaminate the lower magnitude bins. No such contamination was expected with classical models because the stars in question were either too faint or too short lived to appreciably alter the statistics. After

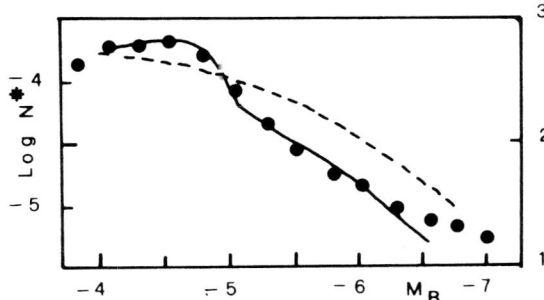

Fig. 3 The luminosity function for AGB stars. N and N* are the observed and predicted number of stars per magnitude bin. They differ by a normalization factor. The dashed line visualizes the standard case by Reid and Mould, while the continuous line shows the result for models with overshooting (see text)

normalization of the theoretical stellar numbers per magnitude bin at the second 0.25 mag. bin, we find the relation shown in Fig 3. The agreement with observation is remarkably good. The net improvement is mostly due to the novel features of models with overshooting, and only partly to the slight increase in the mass loss rate we have adopted. The same result would not have been possible by varying the mass loss rate alone.

6. MODELS WITH VARIED OPACITY

In addition to the calculations presented by Bertelli et al (1984) with opacity given by eq. (1), a few exploratory sequences have been computed by Bertelli and Bressan (1986) in the domain of intermediate mass stars. The results are as follows:
i) In absence of convective overshooting the models perform very extended loops in the HR diagram, as already shown by Robertson (1972), who first proposed such a modification of standard radiative opacities.
ii) In models with overshooting the blue loops are strongly suppressed for all masses greater than 5 M☉ when ΔZ in eq. (1) is 0.02. This almost independently of λ. On the contrary, when $\Delta Z = 0.01$, loops though reduced are still there. They tend however to get narrower at increasing mass.
iii) Those numerical experiments bring an interesting feature to the morphology of blue loop band of intermediate mass stars, namely a C-shaped behaviour whose blue apex is much more sensitive to metallicity than with standard opacity. This may have important observational implications. In particular, the Cepheid instability strip could be crossed two times by the blue loop band on quite long time scales or even be overlapped by this under favourable circumstances (efficient overshooting and/or high metal content). It is easy to foresee how the number frequency versus period relation (cfr. section I) would change in this case. Amazingly enough, the bimodal distribution of SMC Cepheids (Becker et al 1977) could be accounted for in a simple way. This is however highly speculative and more work has to be done to cast light on the need (if any) of opacities different from those currently in use.

7. CONCLUDING REMARKS

Before concluding this paper, we like to touch upon two more points that may be affected by models with overshooting.

i) The Number Frequency - Period Distribution Of Cepheids

In section I, we reported on the difficulty encountered by classic models for core He-burning stars in reproducing the observed number frequency - period distribution of Cepheids, and the need of a steeper band of core He-burning stars in the HR diagram. This problem has been addressed by Bertelli et al (1985a), who showed how models with overshooting possess the required feature and can almost match the observed period difference $LogP_1 - LogP_0$.

ii) Integrated Colour Versus Age Relation For Star Clusters

With the aid of the new models, Bertelli et al (1986) have calculated the integrated colours - $(B-V)_o$ and $(U-B)_o$ - versus age relationship for clusters having turnoff mass down to 1.6 $M_\odot$. All phases, from the zero age main sequence to either the termination of the AGB (for stars with initial mass in the range 1.6 $M_\odot$ to 6 $M_\odot$) or beyond the core He-exhaustion stage (for stars more massive than 6 $M_\odot$), have been included. Starting from isochrones, stars are distributed along a given isochrone by means of a "Montecarlo" technique weighted on the initial mass function. A huge number of stars per model cluster has been considered in order to simulate and approach the analytical case. The resulting colour versus age relationships seem to better fit the colours of a few clusters of LMC for which the age was independently derived from the turnoff magnitude (Chiosi et al 1985). Particularly interesting is the abrupt change in the slope of the $(B-V)_o$ versus age relation occurring at about 1×10^9 yr, which could be related to the discontinuity seen in the $(B-V)_o$ versus cluster type classification parameter of Searle et al (1980). See also the thorough discussion on this topic by Renzini and Buzzoni (1986). More work is under way to cast light on this problem.

In this paper we reported on studies of the effects of convective overshooting on stars of all masses and evolutionary phases in which this phenomenon may be effective. The aim was not only to discuss convective overshooting from the viewpoint of general interest toward this particular physical process, but also to find astrophysical tests of its occurrence in real stars and hopefully to indirectly assess the actual extension of convective cores. Looking at the results we have presented insofar, convective overshooting turns up to be a very promising tool for removing or at least alleviating some of the discrepancies that were known to exist between current theories of stellar structure and crucial observational facts. Whether or not the arguments presented in this paper have been convincing and the goal achieved is difficult to say. Certainly this line of work deserves more careful studies.

ACKNOWLEDGEMENTS

I like to express my deepest gratitude to Drs. G. Bertelli, A. Bressan, C. Forieri, E. Nasi and L. Pigatto for the continuous assistance, encouragement, hard work and unvaluable help shown to me over the many years we have been working together. This work has been supported by the National Group of Astronomy (GNA) and The Italian Space Research Program (PSN) of the National Council of Research of Italy (CNR) under contracts n. 8302422-02 and n. 83-018

REFERENCES

Angerer, K., Bertelli, G., Bressan, A., Chiosi, C., 1985, in preparation
Appenzeller, J., 1980, in Star Formation, 10th Advanced Course of the Swiss Society of Astronomy and Astrophysics, Saa-Fee, 1980, p. 3
Arnett, W. D., 1972, Astrophys. J., 176, 681
Barbaro, G., Pigatto L., 1984, Astron. Astrophys., 136, 355
Barlow, M. J., Cohen, M., 1977, Astrophys. J., 213, 737
Barlow, M. J., Smith, L. J., Willis, A. J., 1981, M. N. R. A. S., 196, 101
Beaudet, G., Petrosian, V., Salpeter, E. E., 1967, Astrophys. J. 150, 979
Becker, S. A., 1981, Astrophys. J., Suppl., 45, 475
Becker, S. A., Iben, I. Jr., 1979, Astrophys. J., 238, 831
Becker, S. A., Iben, I. Jr., Tuggle, R. S., Astrophys. J., 218, 633
Becker, S. A., Mathews, G. J., 1983, Astrophys. J., 270, 155
Bertelli, G., Bressan, A., 1986, this conference
Bertelli, G., Bressan, A., 1985, in preparation
Bertelli, G., Bressan, A., Chiosi, C., 1984, Astron. Astrophys., 130, 279
Bertelli, G., Bressan, A., Chiosi, C., 1985a, Astron. Astrophys., in press
Bertelli, G., Bressan, A., Chiosi, C., Angerer, K., 1985b, in preparation
Bertelli, G., Bressan, A., Chiosi, C., Nasi, E., Pigatto, L., 1986, this conference
Bressan, A., Bertelli, G., Chiosi, C., 1981, Astron. Astrophys., 102, 25
Castellani, V., Chieffi, A., Pulone, L., Tornambè, A., 1985, Astrophys. J., in press
Chiosi, C., 1982a, in Wolf Rayet Stars: Observation, Physics, Evolution, eds. C. de Loore and A. Willis, Reidel P. C., p. 323
Chiosi, C., 1982b, in The Most Massive Stars, ESO Workshop, eds. S. D'Odorico, D. Baade, K. Kajar, p. 27
Chiosi, C., Bertelli, G., Bressan, A., Nasi, E., 1985, Astron. Astrophys. submitted
Chiosi, C., Maeder, A., 1985, Ann. Rev. Astron. Astrophys. (1986)
Chiosi, C., Matteucci, F., 1984, in Stellar Nucleosynthesis, eds. C. Chiosi and A. Renzini, Reidel P. C., p. 359
Chiosi, C., Nasi, E., Sreenivasan, S. R., 1978, Astron. Astrophys., 63,103
Chiosi, C., Olson, G. L., 1984, unpublished
Chiosi, C., Pigatto, L., 1985, Astrophys. J., submitted
Cloutman, L. D., Whitaker, R., 1980, Astrophys. J., 237, 900
Cogan, B. C., 1975, Astrophys. J., 201, 637
Cox, A. N., Stewart, J. N., 1970, Astrophys. J. Suppl., 19, 243
de Jager, C., Nieuwenhuijzen, H., van der Hucht, K., 1985, preprint
de Loore, C., Hellings, P., Lamers, H., 1982, in Wolf Rayet Stars: Observation, Physics, Evolution, eds. C. de Loore and A. Willis, Reidel P. C., p. 53
Doom, C., 1982a,b, Astron. Astrophys., 116, 303, 308
Doom, C., 1985, Astron. Astrophys., 142, 143
Flower, P. J., 1981, Astrophys. J., 249, L11

Flower, P. J., 1984, Astrophys. J., 278, 582
Flower, P. J., Geisler, D., Hodge, P., Olszewski, E. W., 1980, Astrophys. J., 235, 769
Fowler, W. A., Caughlan, G. R., Zimmermann, B. A., 1975, Ann. rev. Astron. Astrophys., 13, 69
Fusi Pecci, F., Renzini, A., 1976, Astron. Astrophys., 46, 447
Garmany, C. D., Conti, P. S., Chiosi, C., 1982, Astrophys. J., 263, 777
Hodge, P. W., 1983, Astrophys. J., 264, 470
Humphreys, R. M., 1978, Astrophys. J. Suppl. 38, 309
Humphreys, R. M., 1982, in The Most Massive Stars, ESO workshop, eds. S. d'Odorico, D. Baade, K. Kajar, p. 5
Humphreys, R. M., McElroy, D. B., 1984, Astrophys. J., 284, 565
Humphreys, R. M., Nichols, M., Massey, P., 1985, Astron. J., 90, 1, 101
Iben, I. Jr., 1974, Ann. Rev. Astron. Astrophys., 12, 215
Iben, I. Jr., Renzini, A., 1983, Ann. Rev. Astron. Astrophys., 21, 271
Iben, I. Jr., Renzini, A., 1984, Physics Reports, 105, n. 6, 329
Jura, M., Morris, M., 1981, Astrophys. J., 251, 181
Kettner, K. U., Becker, H. W., Buchman, L., Gorres, J., Kravinkel, H., Rolfs, C., Schmalbrok, P., Trauttvetter, H. P., Vlieks, A., 1982, Z. Phys., 308, 73
Langanke, K., Koonin, S. E., 1982, Nuclear Physics, A410, 334
Lindoff, U., 1969, in Mass Loss from Stars, ed. M. Hack, Reidel P. C., p. 106
Maeder, A., 1975, Astron. Astrophys., 40, 303
Maeder, A., 1976, Astron. Astrophys., 47, 384
Maeder, A., 1984, in Observational Tests of Stellar Evolution Theory, eds. A. Maeder and A. Renzini, Reidel P. C., p. 299
Maeder, A., Mermilliod, J. C., 1981, Astron. Astrophys., 93, 136
Matraka, B., Wassermann, C., Weigert, A., 1982, Astron. Astrophys., 107, 283
Mermilliod, J. C., 1981, Astron. Astrophys., 97, 235
Meylan, G., Maeder, A., 1982, Astron. Astrophys., 108, 148
Mould, J., 1983, in Structure and Evolution of the Magellanic Clouds, eds. S. van den Bergh and K. S. de Boer, Reidel P. C., p. 195
Mould, J., Aaronson, M., 1982, Astrophys. J., 263, 629
Nomoto, K., in Stellar Nucleosynthesis, eds. C. Chiosi and A. Renzini, Reidel P. C., p. 239
Pel, J. W., Lub, J., 1978, in The HR Diagram, eds. A. G. D. Philip and D. S. Hayes, Reidel P. C., p. 229
Reid, N., Mould, J., 1984, Astrophys. J, 284, 98
Reimers, D., 1975, Mem. Soc. Roy. Sci. Liège, 8, 369
Renzini, A., 1984a, in Stellar Nucleosynthesis, eds. C. Chiosi and A. Renzini, Reidel P. C., p. 99
Renzini, A., 1984b, private communication
Renzini, A., 1984c, in Observational Tests of Stellar Evolution Theory, eds. A. Maeder and A. Renzini, Reidel P. C., p. 21
Renzini, A., Bernazzani, M., Buonanno, R., Corsi, C. E., 1985, Astrophys. J. Letters, in press
Renzini, A., Buzzoni, A., 1986, in Spectral Evolution of Galaxies, eds. C. Chiosi and A. Renzini, Reidel P. C.
Robertson, J. W., 1972, Astrophys. J., 173, 631
Robertson, J. W., 1974, Astron. Astrophys. Suppl. 15, 261

Roth, M., 1984, preprint
Roxburgh, I., 1978, Astron. Astrophys., 65, 281
Schommer, R. A., Olszewski, E. W., Aaronson, M., 1984, Astrophys. J.,285,L5
Searle, L., Wilkinson, A., Bagnuolo, W. C., 1980, Astrophys. J., 239, 803
Schild, H., Maeder, A., 1983, Astron. Astrophys., 127, 238
Simon, N. R., 1982, Astrophys. J., 260, L87
Vanbeveren, D., 1986, this conference
van der Hucht, K., Conti, P. S., Lundstrom, I., Stenholm, B., 1981, Space Sci. Rev. 28, 227
Waldron, W. L., 1984, preprint
Wilson, J. R., Mayle, T., Woosley, S., Weaver, T., 1985, preprint
Woosley, S., Axelrod, T. S., Weaver, T., 1984, in Stellar Nucleosynthesis, eds. C. Chiosi and A. Renzini, Reidel P. C., p. 263

with low Z-value. This implies that in a low Z region, mass transfer is the dominant formation mechanism, the stellar wind mass loss rate probably being too small to modify the atmospheric abundances. This corresponds with the results of Maeder (1980), who found that the ratio N(WR)/N(M), M refering to M-type supergiants, was very sensitive to metallicity. The correlation between M adn Z was also discussed by Prevot et al. (1980) and Abbott (1982).

TABLE 4

Binary frequency of WR-stars in external galaxies.
(Lequeux, 1983)

GALAXY	Z	N(WR BIN) / N(WR TOT)	N(WC) / N(TOT)
M 31	0.04		1
Galaxy (Z=0.03)	0.03	0.15	0.52
Galaxy (Z=0.02)	0.02	0.24	0.44
M33	0.02	–	< 0.50
Galaxy (Z=0.014)	0.014	0.25	0.41
LMC	0.009	0.29	0.20
SMC	0.003	0.63	0.12

3. MODELS FOR MASSIVE BINARIES.

The next question concerning the establishment of generic relations between the different groups is 'what models should we use to describe the evolution?'. The outlook of the system and its evolution depend on the adopted model for the constituent stars, and here a wide variety of possibilities is available (cfr. the review of Maeder, this volume).

a) The very classical massive star, with a convective core determined by equalization of the two dominant temperature gradients responsible for the transport of energy, with constant stellar mass during non-interacting phases. These models were used in close binary evolution up to 1977.

b) Mass decreasing stars, mimicking stellar wind mass loss with a formalism essentially in agreement with observed mass loss rates. Such models were applied from 1978 on and were considerably improved on the initial mass ratios necessary to obtain the present short period WR-systems (Vanbeveren et al., 1979).

c) Stars with different inner structure (diffusion, overshooting, extended mixing).

The influence on close binary evolution may best be shown by the difference in initial-to-final-mass relations, connected with these models (initial = ZAMS, final = after RLOF). Let's put it the other way round. Consider a WR star observed as member of a binary system, with an estimated mass of 10 M_o. **What is its initial mass?** The different possibilities are depicted in **Figure 1**.

For simplicity we have assumed that the system has evolved through a case B of RLOF, hence mass transfer occurring after the moment the most massive star has left the main sequence (Kippenhahn and Weigert, 1966). We considered two cases:

1 - the star has just entered the WR-phase after a phase of mass transfer : $M_{WR} = M_f$ (dotted lines).
2 - the star has lost about 1/3 of its mass during part of its WR-phase : $M_{WR} = 10$ M_o, $M_f = 13$ M_o (full lines).

The results with the different models a, b, c, described at the beginning of this section are labeled a, b, c in the figure.

Figure 1.

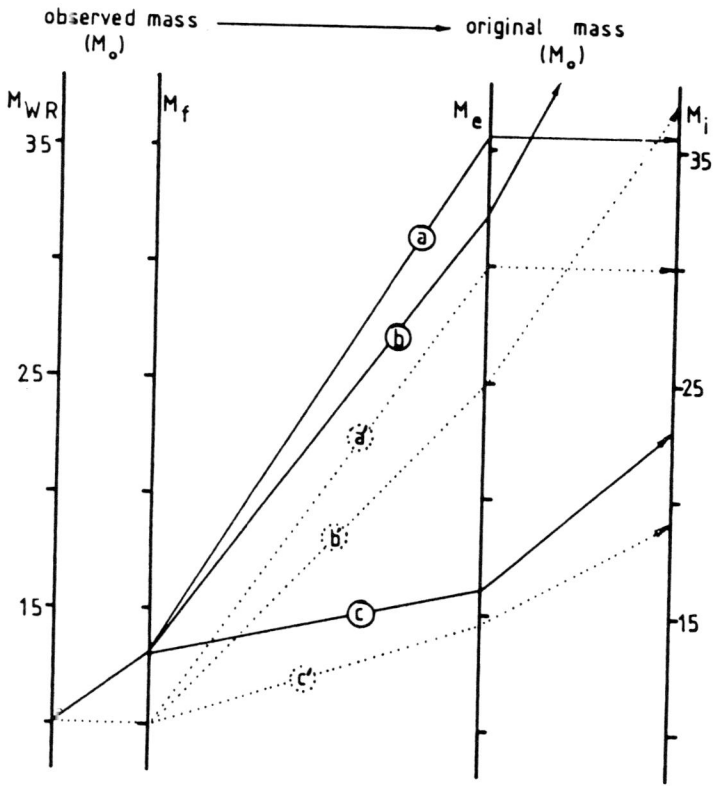

Summarizing we find that a presently observed WR star of $10M_o$ may have originated from an initial mass of 19 to 48 M_o, depending on the stellar model used in the computation. We emphasize that with model b simple stellar wind mass loss will not transform the massive component into a WR-star as proposed by Conti (1976).

In the following we will use the models b and c to explore the evolution of massive systems.

4. CLOSE BINARY EVOLUTION.

The earlier importance of computations on massive close binary evolution is undoubtedly found in the confirmation of the hypothesis of van den Heuvel and Heise (1972) on the origin of massive X-ray binaries. Schematically the scenario occurs as follows (de Loore, De Greve and De Cuyper 1975).

Process	:	RLOF		SN	
SYSTEM	:	O + O	WR binary		X-ray binay

The intial masses considered in the computations carried out in the mid-seventies, range from 15 to 30 M_o. As a consequence the derived masses of the WR companions were estimated in the range 5 to 10 M_o, values that were not contradicted by observations at that time!

Examining this scenario in somewhat more detail we can ask 'Is that scenario true for all O-type systems?', i.e. will each O-type system be transformed into a WR-system and later on into an X-ray binary?

Let us consider a massive binary system which is evolving from the ZAMS up to an advanced stage. As system characteristics we adopt the mass of the initially more massive star M_1, the mass ratio $q = M_2/M_1$, and the period variation P/P_i. The mass range considered for M_1 is 15 to 100 M_o. To simplify the picture we adopt $q_i=0.7$ for the whole mass range. We will discuss the different bifurcations in the evolution resulting from the use of two different models and assumptions on the mass transfer.

The following stages are considered :
A : ZAMS
B : End of core hydrogen burning of the most massive star
C : End of RLOF and/or onset of WR stage
D : End of WR stage (=end core helium burning)
D': End SN-explosion ($M_1 = 2 M_o$, $P = P \times 1.5$)
E : End of core hydrogen burning of the companion star.

From computations with models **b** and **c** respectively, one may infer that :
1- the evolution leads to an overall period increase with almost one order of magnitude
2- with extended cores the mass of the companion star remains nearly unchanged (as the influence of mass transfer, if present, is small)

or is slowly decreasing to a value of 65% of the initial value as a result of stellar wind.

3- unless very severe angular momentum loss is considered, it is not possible to obtain advanced systems with small periods (P < 10 d) starting from periods $P_i > 10$ d.

It seems therefore unlikely that O-type systems with large masses and periods evolve into X-ray binaries, at least what is concerned the samples of well-known binaries.

Now we may try to answer the question: what are the general characteristics of massive close binary systems at the various stages of their theoretical evolution? Figure 2 shows the results for model **b**, with a classical convective core and stellar wind, and model **c**, with an extended core and stellar wind, with the following constraints:

1- the stellar wind mass loss (for O and WR stars) is assumed to be spherical by symmetric. The period increase is computed from

$$\frac{P}{P_o} = \left(\frac{M_{10} + M_{20}}{M_1 + M_2}\right)^2$$

(Hadjidemetriou, 1967)

with subscript o referring to starting values.

2- when mass transfer is involved, case B is assumed (after core hydrogen burning of the most massive star). In the non-conservative case the period variation is calculated using the formalism of Vanbeveren et al. (1979), adopting a value of alpha = 2 (corresponding to moderate angular momentum loss from the system, typically 50%).

3- as mass loss rate during the WR stage we adopted $3\ 10^{-5}\ M_\odot\ yr^{-1}$. The mass decrease of the Wolf-Rayet is truncated at $4\ M_\odot$. The lifetime of the WR star is taken equal to the core helium burning timescale.

4- the remnant mass after a SN-explosion is $2\ M_\odot$. A period increase of 1.5 times the pre-SN period is assumed after resynchronisation of the orbit (Sutantyo,1974; De Greve, de Loore and van Dessel, 1978).

A very remarkable result appears in the diagrams of the models with extended cores. First we note that the interval $15 - 35\ M_\odot$ contains some 5 times more stars than the interval $35 - 100\ M_\odot$. Assuming that the fraction of binaries is constant over the mass range $15 - 100\ M_\odot$ this implies the same difference in number of binary systems. Moreover we know that in the case of a spherical symmetric SN explosion the system remains bound if the removed mass is less than half the total mass of the system. Neglecting impact and influences of asymmetric explosions we find that for $q_i = 0.7$ or smaller the removed mass during the SN explosion is roughly half the mass of the system (within 30%) for $M_i > 40\ M_\odot$.

Figure 2.

Characteristics of massive close binary evolution with $15 < M_i / M_\odot < 100\ M_\odot$ and $c_i = 0.7$.
Left : classical models with Schwarzschild convective cores.
Right : models with extended cores (Roxburgh criterion).

Dotted area : main sequence evolution of the most massive star, losing stellar wind.

Vertically hatched area: mass transfer (conservative case B), the dashed curves in the mass ratio diagram (left) show the final positions in case 50% (upper curve) or all (lower curve) the transferred mass is lost from the system.

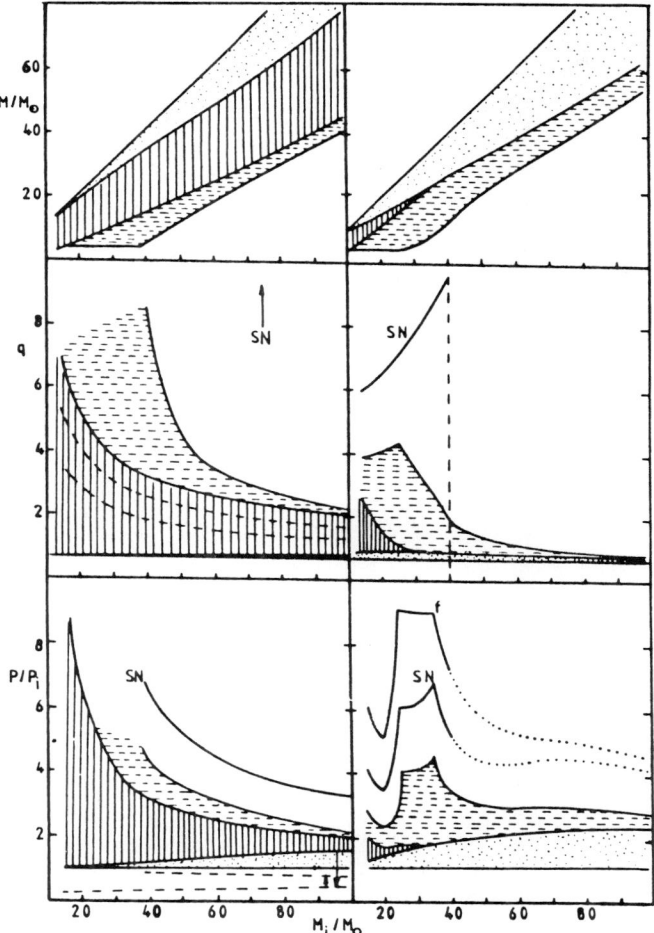

The dashed curves in the period diagram represent the end of mass transfer when all the expelled mass leaves the system (lower curve) and the end of the ensuing WR phase (upper curve).

Horizontally dashed area : Wolf-Rayet phase of the originally more massive star ($\dot{M}=3 \; 10^{-5} \; M_\odot \; yr^{-1}$, with timescale equal to the core helium burning phase). A lower boundary cut-off of $4 \; M_\odot$ is adopted.

The curve labeled SN is the period increase after the SN explosion and synchronisation of the orbit (xb x 1.5); the dotted parts denote the low probability; f is the final period increase at the end of the main sequence of the companion star.

The vertical dashed line (q-diagram, right) also denotes a probable boundary for bound systems.

Combining both effects we obtain a large disruption probability for $M > 40\ M_o$. **Hence in that mass range the appearence of massive X-ray binaries is very unlikely.** This conclusion should be added to the one obtained from period and secundary mass considerations.

The classical models predict large mass ratios during the WR-stage (overall $q > 2$, and for $M_i < 40\ M_o$: $q > 3$) unless severe mass loss from the system is assumed. A mass ratio smaller than 1 is not encountered. For models with extended cores the mass ratio remains small for $M_i > 40\ M_o$ ($q\ 1$, with $q < 1$ for $M_i > 70\ M_o$). For smaller masses the mass ratio is 4.5 at maximum (given the initial value $q_i = 0.7$).

The classical models predict X-ray binaries with small periods only in the case of extensive angular momentum loss from the system. For extended core models with initial mass $M_i > 40\ M_o$ (in case the system should remain bound after the SN explosion) a massive X-ray binary results with a period 4 to 6 times the initial period.

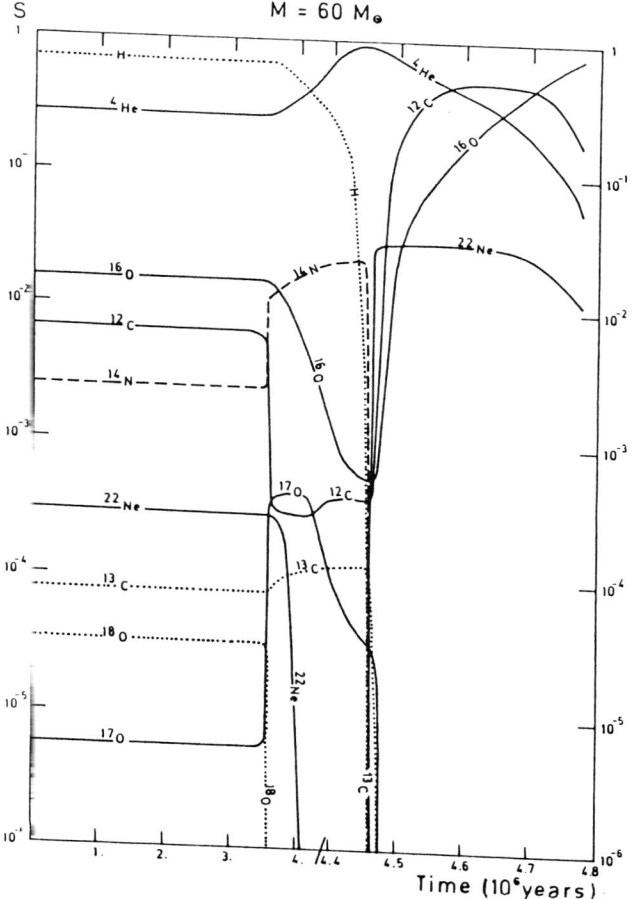

BINARY EVOLUTION ON THE UPPER HRD   349

Summarizing we believe that all O-type systems evolve into WR+O (or very rarely B) systems, the more massive ones ($M_{1i} > 40 M_o$) through direct transformation by stellar wind mass loss (O Of WR), the less massive ones ($M < 40 M_o$) through mass transfer. Only systems of the latter group will further evolve into massive X-ray binraies.

In view of the short periods of the X-ray binaries (1.5 to 9 days) they are likely the result of a previous mass transfer process. As such the chemical composition of the surface of **both** components offers an opportunity to study the nucleosynthesis in the stellar interior as well and to test models of close binary evolution.

Exceptions to this general scheme may result from very massive systems with small periods (implying case A of mass transfer; Doom, 1984) or systems with rather extreme mass ratios.

5. Recommended future work in the field.

From the observational side 2 large regions of interest need extensive further investigation
- an extended survey of unevolved massive systems with reliable dimensions and with correlation to the single stars in the same region would deliver the necessary data to predict statistical properties of advanced stages of those systems.
- detailed abundance analysis of members of the different groups of massive binaries (Unevolved, WR, X-ray, ...) will lead to improved models of mass transfer, surface mixing and mass loss from the system.

Related to theoretical evolution it would be interesting
1) To carry out computations of case A of mass transfer in massive systems, taking into account the changes of the two components. The results should be checked against to improved observations in WR- and X-ray binaries. (cfr. Nakamura and Nakamura, 1984, Sybesma, 1985, Hellings, 1985). Such computations might explain well observed systems such as V444 Cyg.
2) To investigate in detail the chemical evolution

Central and surface abundances for evolving stars have been calculated by Maeder (1983,a,b) and compared with observations, and more recently evolutionary computations for massive single stars with extended mixing, stellar wind and detailed nucleosynthesis (including the new - reaction) have been presented by de Loore et al. (1985) and Prantzos et al.(1985). Abundance ratios N/He,C/He, O/He have been derived which agree with earlier computations, except for O, where the use of improved reaction rates gives an overabundance of O during the final phases. Nitrogen is burned rapidly during the final phases of He-burning, so that no N is left at the surface during the WC stage.

As an example the surface abundances of C, N and O for a $60 M_o$ star are shown in Figure 3. The total amount of the most abundant elements, returned to the interstellar medium, $1H$, ^4He, ^{12}C, ^{14}N, ^{16}O, ^{22}Ne is shown in **Table 5**.

Here again the importance of massive binaries for the enrichment of the interstellar medium has to be stressed. A massive binary with a not too short period will evolve without mass exchange, and each of the two components will restore a large amount of processed material to the ISM, due to stellar wind mass loss. On the other hand, if mass transfer occurs, part of the matter, expelled by the mass losing component, will be accreted by its companion, and remain stored in its outer layers, until mass loss by stellar wind in this component returns also part of this material to the ISM.

Table 5.

$100\ M_o$

t	Phase	Mass	Prot	He4	C12	N14	O16	Ne22
2.06	O	91.8	5.7	2.2	0.049	0.026	0.12	0.002
4.04	Of	76	11.8	11.6	0.055	0.34	0.17	0.002
4.29	WN	69	12	19	0.059	0.50	0.175	0.003
4.39	WC	66	12	19.5	0.930	0.50	2	0.06

$50\ M_o$

t	Phase	Mass	Prot	He4	C12	N14	O16	Ne22
4.62	O	42	5.7	2.2	0.055	0.02	0.12	0.002
5.57	Of	37.4	8.6	4.5	0.058	0.11	0.16	0.003
5.69	WN	33.9	9.1	7.4	0.145	0.17	0.17	0.003
6.08	WC	22.3	9.1	12.4	4.16	0.17	3.73	0.117

References.
Abbott, D., 1982, Astrophys. J., **263**, 723. Avni, A., Bahcall, J.N., 1975, Astrophys.J. **197**, 675. Azzopardi, M., Breysacher, J., 1979, Astron.Astrophys. **75**, 120
Blaauw, A., 1961, Bull.Astron.Inst.Neth. 15, 265. Conti, P.S., 1976, Mem. Soc.Roy.Liege, 6e serie, tome IX., 193.
Conti, P.S., 1978, Astron. Astrophys. **63**, 225. Conti, P.S., Niemela, V.S., Walborn, N.R., 1979, Astrophys. J. **228**, 206.
Conti, P.S., Garmany, C.D., de Loore, C., Vanbeveren, J., 1983, Astrophys. J. **274**, 302.
De Greve, J.P., de Loore, C., Van Dessel, E.L., 1978, Astrophys.Space Sci. **53**, 105.
de Jager, C., 1980, The brightest stars (Dordrecht, Reidel).
de Loore, C., De Greve, J.P. de Cuyper, J.P., 1975, Astrophys.Space Sci, **36**, 219.
de Loore, C., Altamore, A., Baratta, et al. G.B., 1979, Astron. Astrophys. **78**, 287.
de Loore, C., Prantzos, N., Arnould, M., Doom, C., 1985, 5th. Morion, Astrophys. Meeting, Audouze, J. Tran Thanh Van, (eds.).
Doom, C., 1984, Astron. Astrophys. **138**, 101.
Doom, C. De Greve, J.P., Astrophys. Space Sci. **80**, 369.
Doom, C., De Greve, J.P., de Loore, C., 1986, Astrophys. J. (in press)

Garmany, C.D., Conti, P.S., Massey, P. 1980, Astrophys. J. 242.
Gies, 1985, Ph.D.
Giuricin, Mardirossian, Mezztti, 1983, Astrophys.J.Suppl., **52**, 35.
Hadjidemetriou, J., 1967, Adv.Astron.Astrophys. **5**, 131.
Hellings, P., 1985, Ph.D. Thesis, V.U.B. Brussels.
Hidayat, B., Admiranto G., van der Hucht, K.A., 1984, Astrophys. Space Sci **99**, 175.
Kippenhahn, R., Weigert, A., 1967, Z.Astrophys. **65**, 251.
Lequeux, J., 1983, in M.C. Lortet and A. Pitault (eds.) 'Wolf-Rayet Stars : Progenitors of Supernovae?, p. 19.
Maeder, A., 1980, Astron.Astrophys. **92**, 101.
Maeder, A., 1983, Astron. Astrophys. **120**, 113.
Maeder, A., 1983, Astron. Astrophys. **120**, 135.
Massey, P., 1982, in C. de Loore and A. Willis (eds.), 'Wolf-Rayet Stars, Observations, Physics and Evolution', IAU Symp. 99 (Dordrecht, Reidel) p. 251.
Nakamura, M., Nakaruma, Y., 1982, Astrophys. Space Sci. **104**, 163.
Niemela, V.S., 1979, in P.S. Conti and C. de Loore (eds.) Mass Loss and Evolution of O-type Stars', IAU Symp. 83, (Dordrecht, Reidel), p. 291.
Prantzos, N., de Loore, C., Doom, C., Arnould, M., 1985, 5th Moriond Astrophys. Meeting, J. Audouze, J.Tran Thanh Van eds.
Prevot, L., Laurent, C., Paul, J., Vidal-Madjar, A., Audouze, J., Ferlet, R., Lequeux, J., Maucherat-Joubert, M., Prevot-Burnichon, M.L., Rocca Volmerange, B., High ionized species in the spectra of SMC stars, A.A.A. 90, L13, 1980.
Primini F., Rappaport, S., Joss, P.C., 1977, Astrophys. J. **217**, 543.
Rappaport, S.A., Joss, P.C., Stothers, R., 1980, Astrophys. J. **235**, 570.
Ruffini, R., 1982, in Z. Kopal and J. Rahe (eds. 'Binary and Multiple Stars as Tracersof Stellar Evolution (Dordrecht: Reidel), p. 373.
Sutantyo, W., 1974, Astron. Astrophys 35, 251.
Sybesma, C.: 1985, Astron. Astrophys. **142**, 171.
Taylor, J.H., Hulse, R.A., 1974, IAU Circ. No 2704.
Vanbeveren, D., De Greve, J.P., van Dessel, E.L., de Loore, C., 1979, Astron.Astrophys. **73**, 19.
van den Heuvel, E.P.J., 1968, Bull.Astron.Inst.Neth. **19**, 326.
van den Heuvel, E.P.J., 1983, in W.H.G. Lewin and E.P.J. van den Heuvel (eds.) 'Accretion Driven X-ray Sources' (Cambridge:University Press), p. 303.
van den Heuvel, E.P.J., Heise, J., 1972, Nature Phys. Sci. **239**, 67.
Van Paradijs, J., Zuiderwijk, E.J., Takens, R.J., Hammerschlag-Hensberge, G., van den heuvel, E.P.J., de Loore, C., 1977, Astron. Astrophys. Suppl. **30**, 195.

Discussion   :   C. DE LOORE

DE JAGER

You show that OB binaries with initial masses exceeding 35 to 40 M_o can evolve, via the WR stage into O stars with a compact core, I assume without a disruption phase in between. But this does not mean that such stars could not eventually, evolve into a supernova, is it not?

SREENIVASAN:

I believe that you have shown in your presentation, using the logic employed in arriving at your conclusions, that for stars more massive then 40-50 M_o at Zero age do become WR stars, binarity is not necessary and for those that are less massive RLOF is required to make WR stars. This does not conflict with the conclusions from single star evolution that the minimum mass for stars to become WR stars is around 40 M_o.

Could we then agree that binarity is not a necessary prerequisite for WR star formation?

DE LOORE:

You are absolutely right. Wolf-Rayet stars can be produced by various channels, either by stellar wind mass loss, where the mass loss rate, hence the peeling off of the outer layers leading to helium-enhanced atmospheres, is dependent on the stellar mass, or in binaries, where similar changes in the atmosphere occur, for the most massive ones again by stellar wind, for less massive ones, and depending on their orbital periods, according to a different mechanism, by Roche lobe overflow. In this case the mass transfer rate is independent on the stellar mass, but rather on the mass ratio. So for the production of single Wolf-Rayet stars one needs stars of very high mass, for binaries, the stellar mass can be lower. But in any case, single Wolf-Rayet stars, as well as Wolf-Rayet binaries can be produced by mass loss in stars. Hence it is exact that binarity is not a necessary condition for the production of Wolf-Rayet binaries.

SESSION 6.

LARGE STELLAR COMPLEXES IN GALAXIES.

Chairman : C. DE LOORE.

1. M.ROSA and S.D'Odorico : The Exciting Stars of Giant Extragalactic HII Regions.

2. P.HODGE: Systems of Stellar Associations in Galaxies.

THE EXCITING STARS OF GIANT EXTRAGALACTIC H II REGIONS

Michael Rosa (1,*) and Sandro D'Odorico (2)
(1) The Space Telescope European Coordinating Facility,
European Southern Observatory, D-8046 Garching, F.R. Germany
(2) European Southern Observatory, D-8046 Garching, F.R.
Germany

ABSTRACT. Recent investigations of the stellar content of giant H II regions and star formation bursts in nearby galaxies are summarized. The preliminary results of a spectroscopic survey of 78 H II regions in 15 galaxies are presented. A minimum frequency of 30 percent is found for the occurrence of luminous Wolf-Rayet stars in such bursts of star formation. Observations of the positions of historical type II supernovae revealed the presence of WC stars, indicating very high ZAMS masses for the SN progenitors. The application of population synthesis models to the analysis of the integrated stellar spectra in terms of IMF parameters is discussed. An example of such an analysis for a giant H II region in Cen A is presented.

I. INTRODUCTION

Giant H II regions host significant portions of the population of the most massive stars in galaxies. The study of their exciting objects therefore provides direct insights into the mechanisms of star formation on large scales, the evolution of stellar associations, the evolution of the most massive stars and the chemical evolution of galaxies.

The stellar component in regions of star formation activity in spiral arms, irregular galaxies and blue compact galaxies (excluding nuclear activity) has been subject of a variety of investigations recently completed or under way. In going from nearby galaxies in the Local Group to remote objects as far as 50 Mpc, the observational material changes from the detailed investigation of single stars or groups of stars into spectra or photometry integrated spatially over areas measuring tens of parsecs up to several kiloparsecs in diameter. Walborn (this book), Moffat et al. (1985) and Melnick (1985) have obtained an amount of detailed photometry and spectral classification for some 100

(*) Affiliated to the Astrophysics Division, Space Science Department, European Space Agency stars in the 30 Doradus cluster, that allows one to construct realistic HR diagrams and to address the questions of stellar evolution, age and nature of the burst and the IMF parameters. The investigation of a large sample of giant H II regions in Local Group members and galaxies out to the M 101 distance is the subject of the remainder of this paper. On an even larger scale, both in distance and size or mass of the star forming complexes, the group Melnick, Terlevich, Campbell and Smith (see contribution in this book and references therein) is working on giant star formation bursts or "violent star formation" activity, mainly addressing objects where the mass and luminosity of the SF bursts is comparable with the remainder of the underlying galaxy. Recent studies on large samples of irregular, blue compact and dwarf galaxies have been presented by Gallagher and Hunter (1984) and Kunth and Joubert (1985). Though these objects show the presence of SF bursts of the 30 Dor type, the integrated light is dominated by a population in the age interval from a few 10^7 yrs to 10^8 yrs, thus giving clues on more global properties of long term SF activity on galaxy wide scales, dealt with in the contribution of J. Hoessel in this book.

From the above mentioned work it becomes clear that the terms star formation activity, and in particular star formation burst, are applied to a wide variety of phenomena in the parameter space of time, linear scale, mass involved and scales in relation to the underlying galaxy. All these objects contain considerable amounts of extreme population I stars with masses above 20 $M_\odot$, but for a varying degree of significance in relation to the whole stellar population. In the following we will present an overview of our investigations concerned with star formation activity on scales comparable to the 30 Dor cluster size and a few 10^6 years in age.

II. WOLF-RAYET STARS IN GIANT H II REGIONS

In the period 1980 to 1984 we have collected about 200 deep spectra of star formation bursts in nearby galaxies in order to detect the presence of WR stars through their characteristic broad emission features centered at 4650 Å and 5812 Å and to establish quantitative limits to the strengths of these emission bands. Data were obtained with Cassegrain spectrographs and IDS detectors at the 1.5 m and 3.6 m telescopes in the wavelength range 3600 to 7200 Å at a resolution of 5 to 10 Å and with the Lick 3 m telescope in the range 4000 to 6200 Å at a resolution of 3 Å. Data obtained till 1982 have been published in D'Odorico, Rosa and Wampler (1983). To complement the optical data, 14 regions have been observed with IUE and additional spectra have been collected from the IUE data bank. Most of the spectra have been presented in an atlas by Rosa, Joubert and Benvenuti (1984).

The sample of giant H II regions that we have surveyed is affected by a number of selection biases. Galaxies have been chosen which have conspicuous H II regions (mainly Sc and Irr) or to which our intention has been drawn otherwise (e.g. Centaurus A). Within a galaxy we general-

Table 1. Wolf-Rayet stars in giant H II regions

Galaxy	Type	Surveyed Regions	Regions with WR Stars	Reference
LMC	Irr	1	1	Walborn (this book)
NGC 55	Sc	8	2	
NGC 300	Sd	15	2	DRW 1983
M 33	Scd	2	2	DR 1981, CM 1981
NGC 625	IM	2	-	
NGC 4038	Sc	10	1	
NGC 4216	Sb	5	2	
NGC 5068	SBc	5	2	
NGC 5128	Epec	6	2	
M 83	Sc	8	3	
NGC 5253	pec	4	-	
NGC 5396	pec	1	-	
NGC 5408	pec	2	-	
M 101	Scd	7	5	DRW 1983
IC 4662	Irr	2	2	
Sum		78	24	(= 30 percent)

CM 1981 = Conti and Massey 1981
DR 1981 = D'Odorico and Rosa 1981
DRW 1983 = D'Odorico, Rosa, Wampler 1983

ly surveyed the H II regions in order of decreasing luminosity, though morphology and location of the H II complex within the host galaxy played a role as well. Finally, the signal to noise ratios in the stellar continua differ by large factors due to the gain obtained in repeated observations of more interesting objects. Table 1 summarizes the results of our survey. Details of the objects not covered by D'Odorico et al. (1983) will be given elsewhere.

As a detection we took those cases in which the WR emission, in particular the blue feature between 4580 Å and 4730 Å, is elevated above the continuum at least by 3 times the noise in the continuum and is broader than the nebular emission lines. Non-detections generally have equivalent widths of less than 10 Å. In summary, Table 1 contains 78 H II regions surveyed in 15 galaxies, out of which 24 regions (or 30 percent) showed detectable emission from WR stars. Among the population of the largest H II regions this probability of detection is actually a lower limit. The emission, when present, is not spread uniformly about the entire cluster of O stars but is concentrated in subgroups (D'Odorico and Rosa 1981a). A vivid example is R 136 in 30 Dor which emits more strongly in the WR features than all the remaining WR stars in 30 Dor combined (D'Odorico and Rosa 1981b). Figures 1 and 2 illustrate this effect on a vigorous star forming region in NGC 55 (cf.

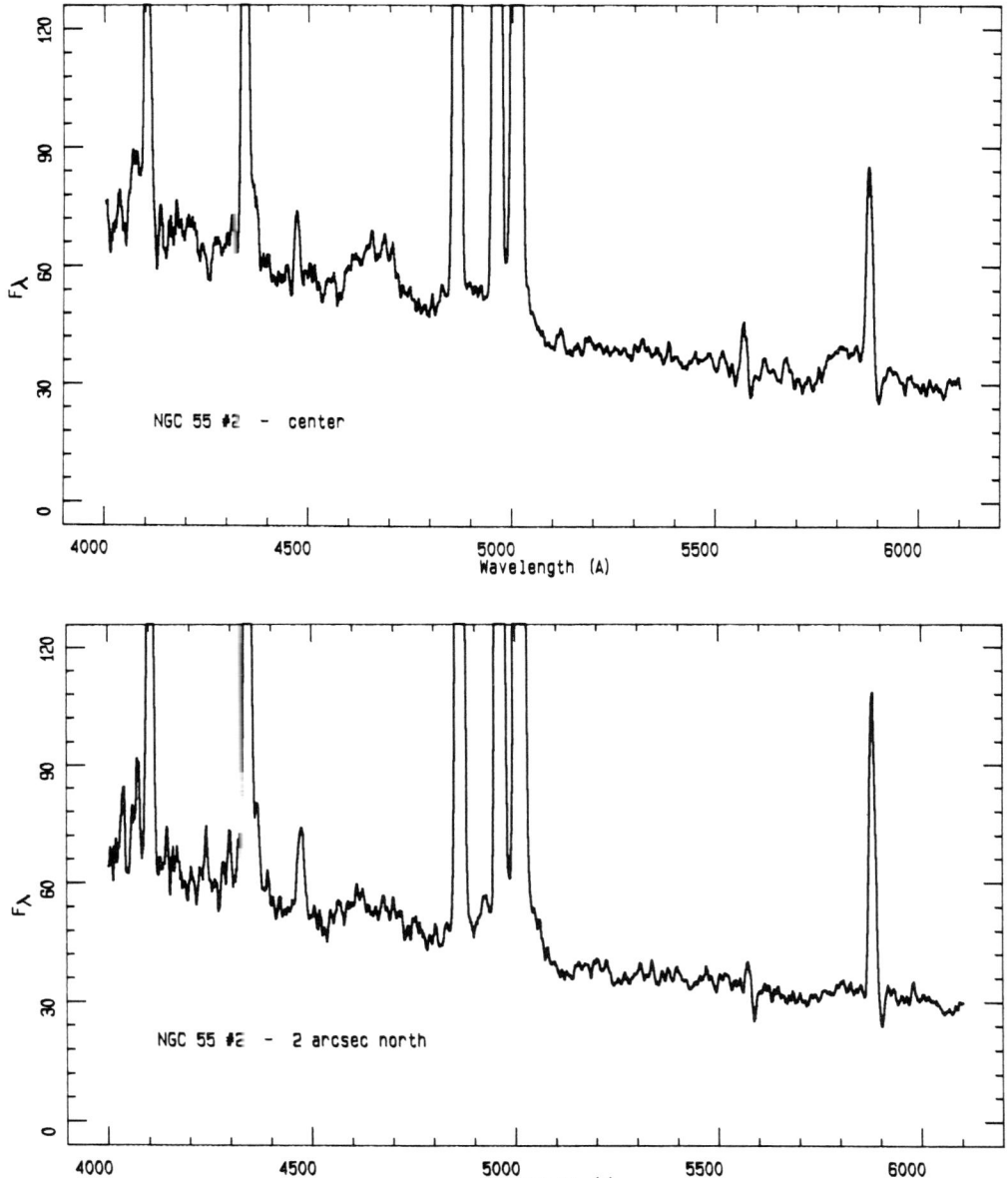

Figure 1 and 2. Spectra of a star burst in NGC 55, 6 arcsec in diameter. The 4 x 4 arcsec aperture was shifted by 2 arcsec between spectrum 1 and 2. Note the similarity of the continuum and nebular lines (shape and absolute scale) and the drastic change in the WR emission features between 4500 Å and at 5812 Å.

Graham 1979). The ionizing stellar cluster has a diameter of about 6 arcsec. The two spectra have been obtained through an aperture of 4 by 4 arcsec, No. 1 centered on the cluster, No. 2 offset by 2 arcsec or half the aperture width. Both spectra are surprisingly similar in shape and absolute scale, but for the reduced equivalent widths and shape of the WR emission bands. Spectrophotometry through apertures of constant size on similar objects will yield either non-detections or detections with rather large equivalent widths in nearby galaxies and a large detection probability but low equivalent widths in more remote galaxies. Qualitatively we can conclude that the presence of luminous WR stars in giant bursts of star formation is a common phenomenon.

The variety of shapes in the WR emission features observed can be seen in the spectra published by D'Odorico et al. (1983). All spectra are composite and the luminosity in the WR emission bands corresponds to rather large number of WR stars equivalent to the most luminous WR stars in the 30 Dor cluster, rendering a classification of WR spectral types difficult. Our previous finding that the luminous types WN 6, 7, 8 and (more rarely) WC 5 are the main contributors is in general confirmed and supported by the emission line spectra in the IUE UV range.

For the reasons discussed above, any quantitative assessment of our data sample in statistical terms is difficult and conclusions cannot be drawn without a detailed analysis of each individual object. Questions to be addressed to are for example the possible variation of WR to RSG number ratios with metal abundance, or the evolutionary sequence of WR spectral types.

Nevertheless, some interesting trends can be seen at this stage of the analysis already. Figure 3 shows a plot of the equivalent widths (EQW) of the WR emission (corrected for nebular contamination) versus EQW of Hβ for all "detections". The EQW of Hβ is a measure of the ratio of ionizing luminosity versus the luminosity in the visual wavelength range, depending on the IMF and the age of the cluster. Models of star formation bursts discussed below predict a decline of the EQW(Hβ) from above 300 Å at zero age to below 50 Å at about $6 \cdot 10^6$ yrs, a result of the rapid evolution of the most massive stars away from the main sequence. Consequently, one would expect an increase in the number of WR stars going hand in hand with the decrease in EQW(Hβ). Our data actually show the presence of WR stars at all ages with a factor 3 enhancement around $4 \cdot 10^6$ yrs, corresponding to WR stars stemming from 50 to 100 $M_\odot$ progenitors. In Figure 4 we plot the EQW(WR) versus the line ratio (4959+5007)/(3727), which is primarily a measure of the ionization in the H II region and hence the effective temperature of the ionizing radiation. This ratio depends however as well on the chemical composition and the gas density (Stasinska, 1980).

Our data indicate that WR stars are relatively more common in H II regions of high ionization. A straight forward interpretation of this effect would imply that the WR stars appear at an evolutionary phase of the cluster at which the ionizing radiation becomes harder. Synthetic

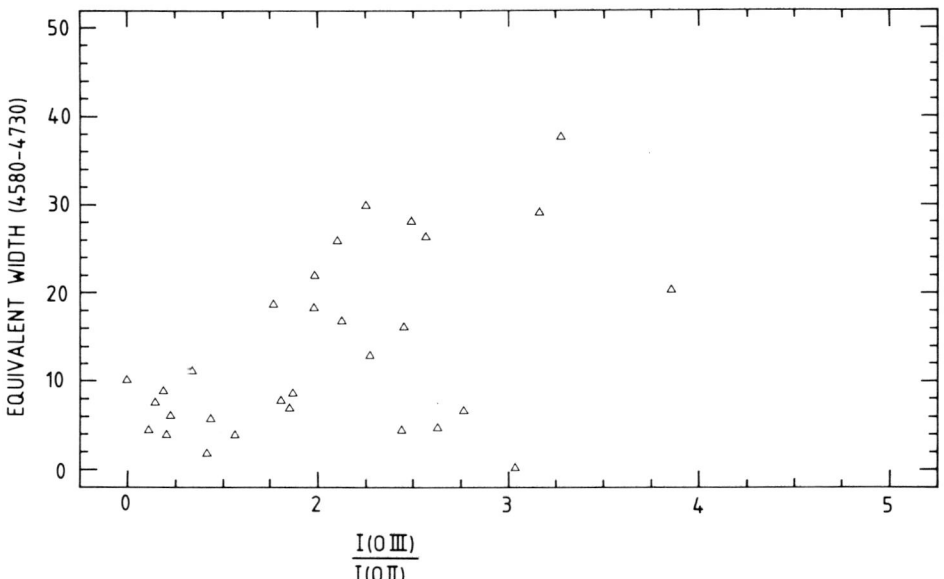

Figure 3. The equivalent widths of the WR emission versus EQW of nebular Hβ emission, which is sensitive to the age of the burst.

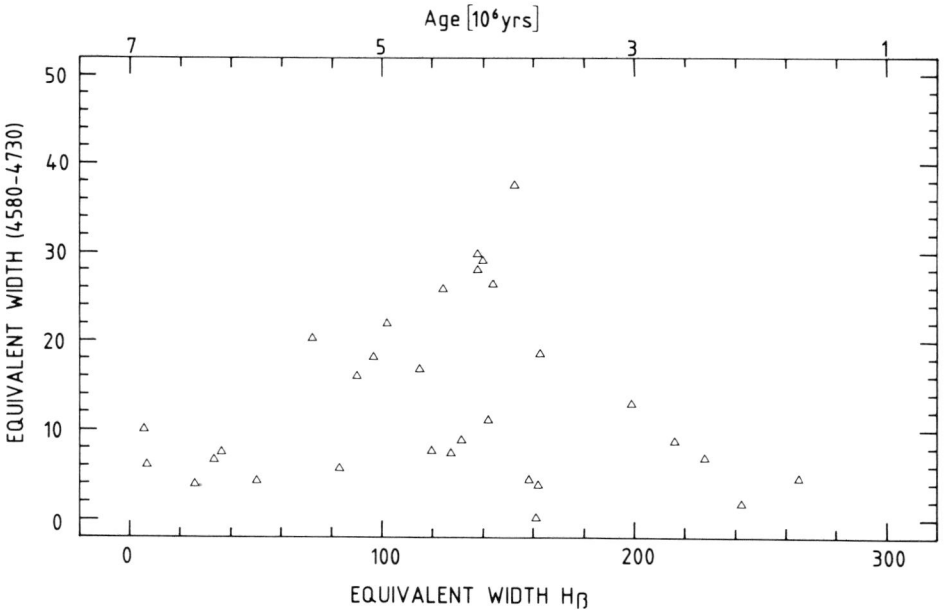

Figure 4. The EQW of WR emission versus the ratio of I(OIII) over I(OII) sensitive to the effective temperature of the ionizing radiation.

clusters (see below) using the (currently believed) low effective temperatures of about 30 000 K for WR stars will not show this effect. There are some indications from Non-LTE, extended atmosphere calculations (cf. contribution by Kudritzki in this book) for high and very high effective temperatures in the Lyman continuum of WR stars, but our data need a more thorough analysis of each individual spectrum in order to support the explanation by high effective temperatures of the WR stars. Such an analysis has been carried out on some objects and an example is described in section V.

III. SUPERNOVAE IN GIANT H II REGIONS

If the very massive stars in giant H II/OB complexes end their evolution in supernova explosions of the type II then a correlation between giant H II regions and the positions of historical type II SNe events has to be expected. Indeed, Richter and Rosa (1984) have shown that all historical SNe that could be classified as type II in M 83 and M 101 were located within the boundaries of giant H II regions. Surveys of SN remnants in nearby galaxies (e.g. D'Odorico et al. 1980) have revealed quite a number of objects located in giant H II regions, in spite of selection effects against finding non-thermal radio sources or faint emission objects on top of the high surface brightness H II region background.

Pennington et al. (1982) have estimated the progenitor masses of 2 historical SNe in M 83, based on an age and IMF analysis of the spatially integrated UBV colours of the underlying stellar population. As discussed below, IMF parameters of populations younger than 10^7 years determined from UBV data are subject to large errors stemming from the inability to completely separate the effects of reddening, age, shape and mass limits of the IMF and contributions of the underlying disk population. Accordingly, the lower mass limits of Pennington et al. (1982) are rather low (18 and 11 $M_\odot$).

In our search for WR stars in giant H II regions we took spectra of 8 regions in M 83, including those hosting the (certain or very probable type II) SNe 1923a, 1950b and 1957d (cf. Richter and Rosa 1984). We found WR emission features solely at the positions of those supernovae and no WR stars in the remaining 5 H II regions. This is a quite unexpected result in view of WR stars being found at random (with a 30 percent probability) in giant H II regions. Most interestingly, the spectra are among the very few in our large sample that unambiguously display the signatures of WC 5 stars. Figure 5 shows the combined spectrum of all 3 SN positions in M 83. The characteristic lines of WC 5 stars are indicated. We interpret our findings with a late (4 to 7 · 10^6 yrs) evolutionary stage of the OB associations (WC 5 stars present in large numbers). If the progenitors of the SNe were coeval with those of the WC 5 stars, then they could have had ZAMS masses well above 30 $M_\odot$.

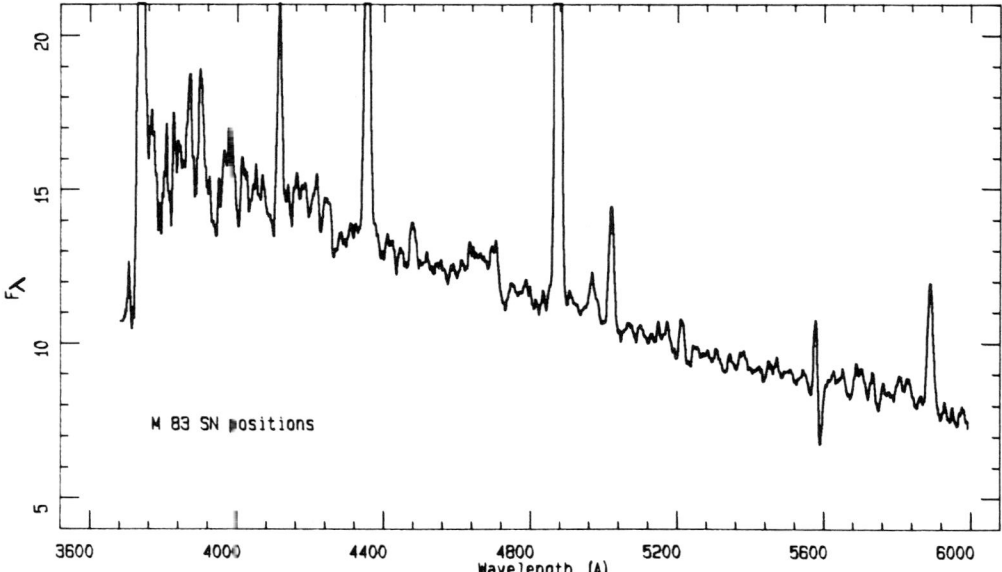

Figure 5. The combined spectrum of the positions of 3 historical SNe in M 83. The presence of luminous WC (5-6) stars is indicated by the broad emission lines of CIII/CIV at 4650 Å shortward of 5696 Å and at 5812 Å, WN 6 by NV at 4620 Å and HeII at 4686 Å.

IV. SYNTHETIC SPECTRA OF LARGE OB ASSOCIATIONS

The foregoing discussion of the observational material has been hampered mainly by the fact that the spectra are integrated over large assemblages of stars populating quite different parts of the HR diagram. A powerful method to disentangle the information from the integrated spectra is the synthesis of energy distributions using either observed spectra or model atmospheres together with evolutionary tracks and prescriptions of the IMF. An example of the straight forward modelling of an IUE spectrum with observed spectra has been presented by Benvenuti (1983). It does however not include a selfconsistent evolutionary scenario of an initial mass spectrum.

Melnick et al. (1985) have computed population synthesis models for the ionizing clusters of giant extragalactic H II regions. They are based on the model atmospheres of Kurucz (1979), evolutionary tracks with different metallicities and mass loss rates over a wide range of masses and power law initial mass functions. They assume single star burst without age spread, i.e. coevality of all stars in the HR diagram. Models are calculated for the zero age and the $3 \cdot 10^6$ yrs isochrone only, reasoning that the most massive stars do not contribute to the ionization after that stage any more. However, our observations show that luminous WR stars, hence evolved massive stars, are observed over a wide range in EQW(Hβ). In the models of Melnick et al. (1985) very low values of EQW(Hβ) are only reached under extreme assumptions on the IMF. We therefore extended their calculations to ages of up to $7 \cdot 10^6$ yrs, thus covering the entire lifetime of stars more massive than 30 $M_\odot$ and the core hydrogen exhaustion of stars more massive than 25 $M_\odot$, corresponding to 08 main sequence stars. The largest uncertainties in these population synthesis models arise from the combination of evolutionary tracks for the stellar core with model atmospheres in the late evolutionary stages, in particular WR stars for which self constistent model atmospheres do not exist at all and for which the luminosity in the Lyman continuum is essentially unknown.

The main results from the populations synthesis models concerning the analysis of integrated spectra of single bursts of star formation can be summarized as follows. In the age range from 0 to $7 \cdot 10^6$ years the main effect of the evolution of the massive stars is the decrease in the Lyman continuum luminosity, leading to a significant drop in the EQW of nebular Balmer lines, while the slope of the continuum between 1200 Å and 10 000 Å remains almost unchanged if evolution to the red supergiant phase is prohibited by stron mass loss. However, models at constant age with variations in the IMF parameters (slope and upper mass limit) and the evolutionary tracks produce energy distributions quite similar to those of clusters att different ages. Broad photometry is generally inadequate to separate out the effects of variations in age, upper mass limits and slope of the IMF. The combination of extreme values of these parameters produces and effect less than 0.2 mag in U-B or 0.8 mag in (1400 - 5500 Å) colours. Such an accuracy cannot be reached in view of the large uncertainties in interstellar extinction. The effective temperature of the ionizing cluster radiation, very sensitive to the composition of the upper part of the HR diagram, could in principle be used in diagnostic diagrams together with a long base line colour or the EQW(Hβ) to separate age effects from IMF parameters. However, the required accuracy of about 1000 K cannot be obtained from the analysis of the nebular spectrum (cf. Stasinska 1980, Mathis 1982). All these limitations are valid only for ages up to about 10^7 years. As soon as the intermediate mass stars evolve and low mass stars reach the main sequence the near UV and visual energy distribution is reflecting more significantly the composition in the HR diagram. At this stage, however, the O stars and giant H II regions are gone and the contrast of the star formation burst over the underlying populations in host galaxies starts to be lost.

A way out of this dilemma is the complete use of the large amount of information contained in the observable energy distribution, i.e. the spectra between 1100 Å and 10 000 Å. Comparison with model energy distributions over that wavelength range using the equivalent width of nebular lines, the strength and shapes of stellar absorption lines and the interstellar absorption features can in fact lead to a simultaneous solution for all the free parameters concerned, i.e. reddening, age and IMF parameters, as will be shown below. When applied to a large body of observations the method should permit to test additional parameters as for example duration and multiplicity of the burst of star formation and the validity of the theoretical evolutionary tracks.

V. THE STELLAR COMPONENT IN H II REGION NO. 13 OF CEN A

As an example of the combination of observations and modelling of the OB clusters in giant H II regions we outline here the main features of the analysis of the largest H II comples in Cen A (= NGC 5128), i.e. #13 in the notation of Moellenhoff (1981). The H II region is located just north of the nucleus of Cen A very close to the dust lane. Several properties make this object a prime candidate for the analysis of the integrated spectrum using synthetic energy distributions of star bursts. At the comparatively large distance the aperture sizes employed in the spectrophotometry cover the entire cluster of stars. The very low equivalent widths of Hβ results in a negligible contamination of the UV part by nebular continuum emission. The background at the position of Cen A #13 is composed of an old E galaxy population and a blue population with an age of about 1 to $4 \cdot 10^7$ yrs (Dufour et al. 1979 and Pennington this symposium). And finally, our high signal to noise spectra allow for the spectral classification of the most luminous stars present. Figure 6 (upper curve) shows the combined dereddened spectrum of the object in logarithmic flux scale in the wavelength range 1100 Å to 3200 Å from IUE low dispersion observations and 3600 tr· 7200 Å from spectra obtained at the ESO 3.6 m telescope. The two parts have been adjusted using surface

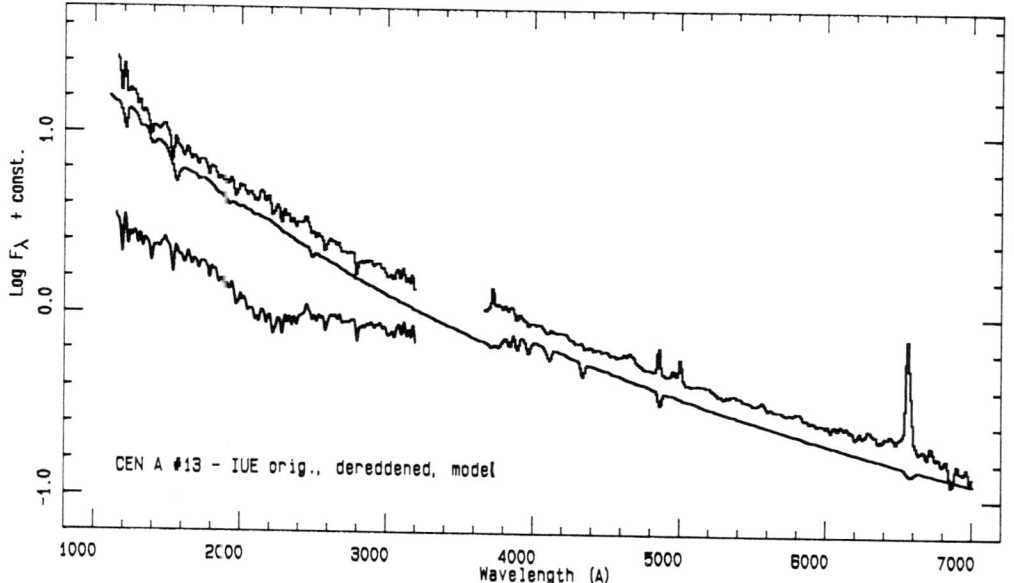

Figure 6. The cbserved (IUE part only) and the dereddened spectrum of the H II region Cen A #13 in logarithmic flux scale. Superimposed, shifted down 0.1 in log (Fλ) for clarity, is a population synthesis model (see text).

photometry on CCD images to correct for the different sizes in apertures used and to scale the background spectra observed on either side of Cen A #13.

The prominent 2200 Å feature seen in the original UV spectrum (lower tracing) was used in combination with the constraints on the range of possible shapes of the IUE UV continua as deduced from the spectral synthesis to determine the amount of reddening in the spectrum of Cen A #13. Balmer decrements from the nebular emission are not useful because of contamination by stellar absorption lines and because the reddening of the nebula may not be applicable at all to the embedded stellar clusters. We found $E(B-V) = 0.15$ mag from a galactic extinction law and an additional $E(B-V) = 0.08$ from an LMC type law, in agreement with an average galactic foreground extinction for the galactic coordinates of Cen A and the idea that the northern ring of H II regions in Cen A may be located in front of the dust lane.

The smooth tracing offset by 0.1 in log $F\lambda$ from the dereddened spectrum is a synthetic spectrum that corresponds to a star burst with a solar neighbourhood IMF (Miller and Scalo 1979) with an age of 5 to 6 10^6 yrs. This corresponds to a current upper mass limit for fully evolved stars of about 40 tr. 60 $M_\odot$. The lowest mass limit depends very much on the assumptions made about the duration of the burst. If stars of all masses were formed within a few 10^5 yrs, then only stars with $M/M_\odot > 2$ can have reached the main sequence, otherwise the energy distribution would be consistent with an IMF extending down to 0.1 $M_\odot$. Using a distance of 6 Mpc to Cen A, the cluster contains 10^6 stars in the 2 to 50 $M_\odot$ mass range, 500 of which had ZAMS masses larger than 30 $M_\odot$. The latter group of stars contains about 300 WN, WC stars as estimated from the luminosity in the WR emission features. In the case of a small age spread in the burst, most of the flux in the red has to be provided by about 100 red supergiants with progenitors in the 30 to 60 $M_\odot$ mass range. Additional information on the composition of the upper HR diagram is provided by the analysis of the optical and UV absorption and emission line spectrum. In conclusion, the dominance of O9 to B2 supergiants in the blue spectrum, the presence of WN 7 and WC 5 stars and the numerical agreement with the numbers observed and required by the IMF solution for RSGs again indicate a Hydrogen core burning age of the cluster of 5 to 6 10^6 yrs with a very small age spread.

The example presented above demonstrates how much information can be obtained simultaneously on the star forming processes, stellar evolution and the IMF parameters from the analysis of the integrated spectra of extreme population I star bursts. The requirements are high signal to noise data at sufficient spectral resolution over a wavelength range as large as possible and, most important, the selection of regions of star forming activity well isolated in space and time (about $2 \cdot 10^7$ yrs) from other star formation activity in the underlying galaxies.

REFERENCES

Conti, P.S., Massey, P.: 1981, Ap. J. **249**, 471.
D'Odorico, S., Dopita, M.A., Benvenuti, P.: 1980, Astron. Astrophys. Suppl. Series **40**, 67.
D'Odorico, S., Rosa, M.: 1981a, Ap. J. **248**, 1015.
D'Odorico, S., Rosa, M.: 1981b, in "ESO Workshop: The Most Massive Stars", S. D'Odorico, D. Baade and K. Kjär (eds.), ESO Garching, p. 191.
D'Odorico, S., Rosa, M., Wampler, J.E.: 1983, Astron. Astrophys. Suppl. Series **53**, 97.
Dufour, R.J., Van den Bergh, S., Harvel, C.A., Martins, D.H., Schiffer, F.H., Talbot, R.J., Talent, D.L., Wells, D.C.: 1979, Astron. J. **84**, 284.
Gallagher, J.S., Hunter, D.: 1984, Am. Rev. Astron. Astrophys. **22**, 37.
Kudritzki, R.P.: this book.
Kunth, D., Joubert, M.: 1985, Astron. Astrophys. **142**, 411.
Kusucz, R.L.: 1979, Ap. J. Suppl. Series **40**, 1.
Mathis, J.S.: 1982, Ap. J. **261**, 195.
Melnick, J.: 1985, to be published in "Workshop on Star Forming Dwarf Galaxies and Related Topics", IAP Paris, July 1985.
Melnick, J., Terlevich, R., Eggeton, P.P.: 1985, MNRAS in press.
Miller, G.E., Scalo, J.M.: 1979, Ap. J. Suppl. Series **41**, 513
Moellenhoff, C.: 1981, Astron. Astrophys. **99**, 341.
Moffat, A.F.J., Seggewiss, W., Shara, M.M.: 1985, Ap. J. in press.
Pennington, R.L., Talbot, R.J., Dufour, R.J.: 1982, Astron. J. **87**, 1538.
Richter, O.-G., Rosa, M.: 1984, Astron. Astrophys. **140**, L1.
Rosa, M., Joubert, M., Benvenuti, P.: 1984, Astron. Astrophys. Suppl. Series **57**, 361.
Stasinska, G.: 1980, Astron. Astrophys. **84**, 320.
Walborn, N.R.: this book.

Discussion   :   ROSA

PENNINGTON :

There is no question that population deconvolutions for HII regions can best be done using spectral synthesis. I would like to point out the usefulness of broad band colors for large scale mapping. The ages derived by Rosa for Mollenhoff's region 13 of $< 6 \times 10^6$ yrs is consistent with the results of digital surface photometry, $< 6.5 \times 10^6$ yrs. I would also like to point out that the necessary red continuum can very easily be due to the old stellar population of the elliptical component, both behind and in front of the dust lane, without requiring red supergiants.

The positions of two of the three SNe sites studied in M83 lie on the edges of HII regions, not in the cores. The inference that SN1957d was a type II is drawn only from its location on the edge of an HII region. The presence of a coincident radio point source with this position suggests that this may not have been a type II.

MASSEY :

I have a comment and a question : first off, why are the WC stars so absent? None of these seem to have strong WC features compared to HeII 4686A. Secondly, why use models for your synthesis? There are tons of IUE data on stars of all types, and Hunter and Jacoby have a very fine spectrophotometric atlas in this year's Ap.J.Suppl. (Jacoby, Hunter, Christian, 1985). You still would need models for fluxes, perhaps, but lines like OV in the UV must give you a very powerful handle.

SANDAGE :

In your spectral syntheses you need red stars added to the OB component to fit the observed I (..). Do you think these are red supergiants or main sequence G, K, M stars. The statement at this meeting by Humphreys and by Massey is that WR stars and RSG are anticorrelated, presumably because the WR star progenitors are more massive (younger) than the RSG progenitors. If so can we expect any red supergiants in those HII regions that show WR features?

SYSTEMS OF STELLAR ASSOCIATIONS IN GALAXIES

Paul Hodge
University of Washington
Seattle, Washington 98195
USA

This paper begins with an attempt to examine the problem of identifying stellar associations in galaxies in a consistent way, so that meaningful physical comparisons can be made for the population of stellar associations of different galaxies. A compilation of the existing data on associations in other galaxies is given and their properties compared. Questions relating to star formation in stellar associations are discussed, and then the issue of the initial mass function of core clusters, especially those located in giant HII regions, is briefly examined.

1. IDENTIFYING STELLAR ASSOCIATIONS IN OTHER GALAXIES

Stellar OB associations are not particularly easy to identify in the solar neighborhood without extensive kinematic and spectroscopic data. In other galaxies, where we do not yet have the ability to measure individual candidate stars' proper motions and radial velocities, the problems are severe. Even for the Magellanic Clouds, where we can at least get precise spectral types and colors, it is not yet possible, especially in crowded regions, to be sure that apparent stellar associations are truly physical entities in every case. This question is a key consideration before we can use populations of associations in attacks on such problems as the distance scale or galaxy evolution.

1.1 The Case of the SMC

A good example of some of these concerns is provided by the OB associations of the SMC. In spite of the extensive amount of work that has been done and published about the associations of the LMC, almost nothing has been said about those of the SMC, except for some early pioneering work in the SMC wing by Westerlund (1961, 1964). Considerable spectroscopy of SMC luminous stars has been done, allowing the mapping of OB stars and their use in exploring the structure of the SMC (Azzopardi and Vigneau 1977, Dubois 1980, Florsch, Marcout and Fleck 1981). However, the associations as a class of objects have been largely ignored. Azzopardi and Vigneau (1977) divided the surface distribution of O-B2 stars into five relatively large high-density regions, but did not report on the fainter stars in them, which might have helped to distinguish any smaller subdivisions that would have physical meaning. If we interpret their five groupings as the complete sample of OB associations in the SMC, then we find that they are both surprisingly few in number and unusually large in size, compared to the sample in the LMC or in our Galaxy.

To try to understand this situation better, I carried out some experiments in identification of stellar associations for the SMC. Using the published OB surveys as a base, I attempted to see what kind of consistency I could achieve by doing independent searches on a large variety of different kinds of observational material, including large and small scale plates, plates in ultraviolet, blue and visual colors, and with both deep and shallow limiting magnitudes. The result is a list of 70 OB association candidates (Hodge 1985a). But the important point for this review is the fact that many of the possible associations were identified in only one or two of the surveys, especially in the dense core of the Cloud. One might suggest that this means that stellar associations are difficult to find and that all kinds of observational material are needed to rout them out. However, I believe that the experiment has a more discouraging message. I think that the problem of distinguishing real associations in a matrix of irregularly-spaced bright stars is not trivial and that it must be carried out with proper controls and statistical techniques to be sure that we are dealing with physical groupings. For example, one can use pattern-recognition analysis (Feitzinger and Braunsfurth 1984) or other statistical tests for clumping (Hodge 1985b). Therefore, I consider my catalog of objects to be suspect in the SMC core, where at least 16 of the candidates could be asterisms. Good photometry and fainter spectroscopy could help to settle the question of their nature. The point is that even for a galaxy as near as the SMC it is not a simple matter to distinguish stellar associations reliably.

1.2 Existing Samples of Stellar Associations in Resolved Galaxies

Table 1 summarizes the properties of stellar associations in resolved galaxies for which data have been published. Clearly there is a large spread in the properties of the samples. Considering the case of the SMC, however, it is not obvious whether the differences are physical or are the result of different identification criteria and techniques. The rest of this section attempts to answer that question.

Table 1. Stellar Associations in Resolved Galaxies

	Total No.	No./L_B (x$10^8 L_o$)	Mean Diam. (pc)	Abs. Mag of Brightest, M_B	Reference
LMC	122	4	78	-11.8^a	Lucke & Hodge 1970
SMC	70	8	77	---	Hodge 1985a
M31	319	0.5	480^b	-12.8^c	van den Bergh 1964; Nikolov and Ivanov 1986
M33	197	2.4	321	-12.4^a	Humphreys & Sandage 1980; Kunchev & Ivanov 1984
NGC 6822	16	10	163	---	Hodge 1977
IC 1613	20	21	164	---	Hodge 1978
NGC 2403	88	1.0	348	-13.8^d	Hodge 1985c
NGC 4303	235	0.3	290	-13.9^a	This paper
NGC 7331	142	0.1	440	---	This paper

[a] From Wray and de Vaucouleurs (1980)
[b] From van den Bergh's data only
[c] From an unpublished paper by Schwartz
[d] From an unpublished paper by Wegner; Wray and de Vaucouleurs (1980) give M = -12.5.

Consider the contrast between M31 and the LMC. The associations in the former are very large and comparatively few (in terms of the normalized number per unit galaxy luminosity), compared to those in the latter. Is this because M31 is a Sb galaxy and the LMC a Irr? Or is this merely due to different selection criteria? Fig. 1 shows that it is not just the mean sizes that differ, but that the entire size distribution is markedly different.

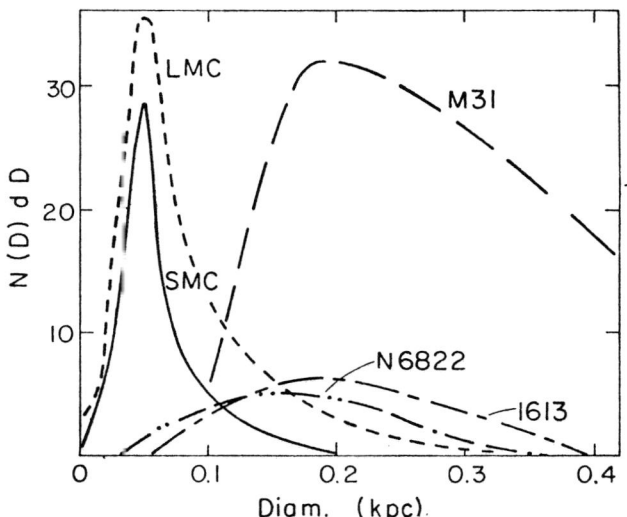

Figure 1. A comparison of the size distributions of OB associations in several local group galaxies.

In an attempt to distinguish between these possibilities, I have examined a series of plates of M31 that have nearly the same linear scale, color, and limiting absolute magnitude as the plates from which the LMC sample was distinguished (Lucke and Hodge 1970), without referring back to the original paper on M31 (van den Bergh 1964). Fig. 2 shows a sample of such a comparison, illustrating the result that the difference is probably not real. At least this particular test provided evidence that the differences in the published characteristics of the samples arise from the use of different observational material and selection criteria.

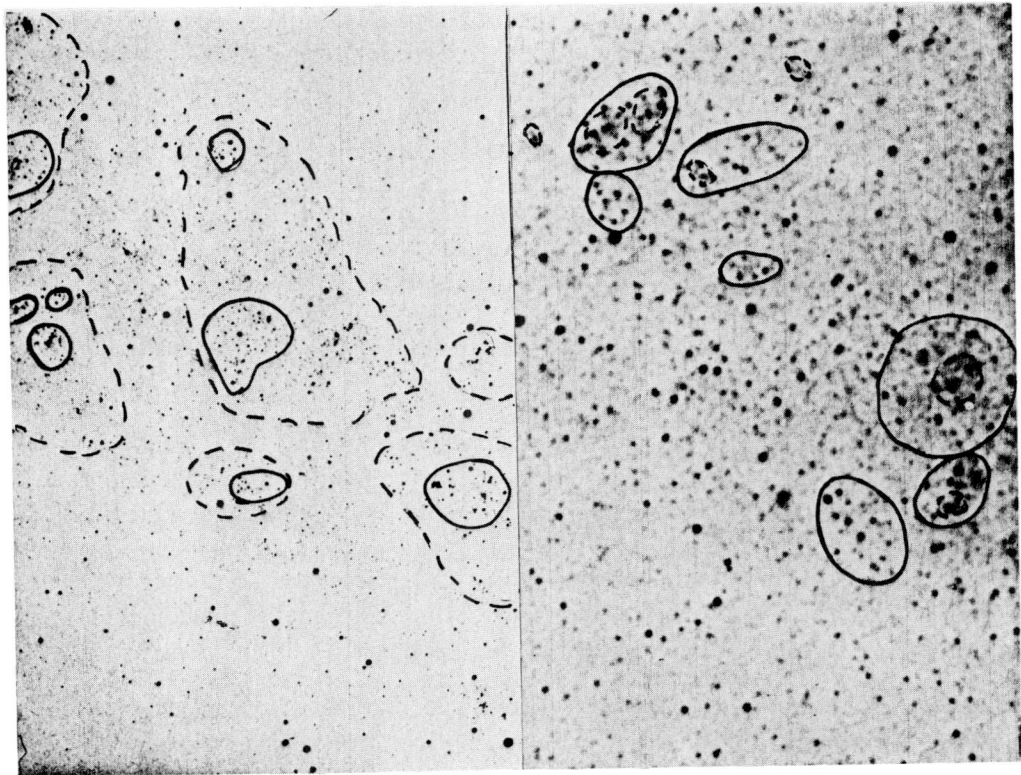

Figure 2. A comparison of photographs of portions of M31 (left) and the LMC (right) from plates that are similar in intrinsic scale and limiting magnitude. Solid lines define the boundaries of associations identified in this experiment, while dashed lines outline those identified in the original publications.

The data in Table 1 show that the normalized numbers of associations in galaxies might be a function of the distances to the galaxies. In each distance realm galaxies have a nearly constant relative population of associations, except for M31 and NGC 7331, and the smaller numbers for them can be interpreted as the result of their earlier Hubble type and consequent smaller content of extreme Population I objects. Although the sample is inadequately small to generalize reliably, the more distant galaxies have decidedly fewer associations per unit galaxy luminosity. I believe that this fact can be understood as a result of their greater distance, which makes the recognition of small associations more difficult, and which leads, as Figure 2 illustrates, to preferentially interpreting complexes of associations as single objects.

1.3 Unresolved Associations in More Distant Galaxies

Identifying OB associations in more distant galaxies, in which the individual stars are not even resolved, is a difficult task. With only the integrated color and patchiness of the images as criteria, the danger of having severe selection effects in the sample is acute.

To test for these problems, I have conducted two simple experiments. In the first I used two small-scale plates of M31 (in B and V), chosen to match in scale and degree of resolution the best available plates of galaxies at the distance of the Virgo cluster. Individual stars, for the most part, are not resolved. Without reference to the existing maps of associations in M31, I then attempted to identify the population of blue patches that might correspond to OB associations. A total of 42 objects was selected and their diameters were measured, with the mean calculated to be 300pc. When I compared my map with that of van den Bergh (1964), I found that every one of the 42 objects corresponded to an association chosen by him. However, using much larger scale plates and including ultraviolet plates, he located many more objects (188) and found a larger mean diameter (480pc). I consider that this experiment provides a demonstration of the fact that reasonable reliability in identifying such objects can be achieved even for unresolved galaxies. The efficiency of detection, however, decreases with distance and the sizes may be measured differently.

In the second experiment, I chose a KPNO 4-m plate of the galaxy NGC 7331, very similar in appearance to M31, of the same Hubble type, similar though somewhat brighter in absolute magnitude, and with a nearly identical inclination angle. Its distance is approximately 20 Mpc. Attempting to duplicate the procedures I used for the M31 experiment, I produced a catalog of candidate OB associations in NGC 7331. Table 2 shows the result. The total number of associations, 142, when normalized by the galaxy's luminosity, compares well with the number found in M31. They are somewhat larger than those in M31, however; this may be due to the greater intrinsic scale of NGC 7331. The experiment appears to confirm the fact that Sb galaxies have intrinsically larger OB associations than later Hubble types and that a reasonable fraction of the population of a galaxy's associations can be distinguished out to 20 Mpc.

Table 2. M31 versus NGC 7331

Associations	NGC 7331	M31 (small scale plates)
No. detected	142	42
No./L_B (x$10^8 L_\odot$)	0.09	0.07
Size range (pc)	100-950	100-500
Mean diameter (pc)	440	300

Beyond these distances, however, it is expected that problems can arise. For example, while the stellar associations in nearby galaxies are measured to have absolute magnitudes in B that reach a maximum at about -12 (see Tables 1 and 3 for several values), Shakhbazyan (1968) describes more distant galaxies that have stellar complexes with absolute magnitudes of -16.

Wray and de Vaucouleurs (1980), who attempted to identify the brightest blue patches in galaxies for their use as distance indicators, apparently found that they could compensate for any selection effects. On the other hand, various other studies of associations (e.g., Humphreys 1977 and Efremov 1982) show us that difficulties remain in finding consistent criteria that will allow unbiased comparisons of associations in different galaxies.

Table 3. Photoelectric Measures of Integrated Light from the Brightest Stellar Associations in M101

Association*	V	B-V	U-B
A1	13.70	0.79	-0.80
A2	15.92	0.36	-0.83
A3	14.65	0.62	-0.83
A4	14.85	0.52	-1.01
A5A	15.87	0.17	-0.88
A5B	15.83	0.22	-0.89

* All measures were made with a 20 arcsec diaphragm with the KPNO 0.9m telescope.

In concluding this section of the paper, I will summarize the various possible conclusions, which can be made with different levels of confidence. First, it can be said with certainty that the derived characteristics of a sample of extragalactic OB associations can depend on various selection effects, including the distance of the galaxy, the resolution of the images, the depth of exposure, and the identification criteria adopted. With considerable confidence we can say that their properties also depend somewhat on the total luminosity of the host galaxy, especially the total number of associations, which averages approximately 10 per 10^8 solar luminosities for late type galaxies. We conclude tentatively that the characteristics also depend on the Hubble type of the galaxies, with Sb galaxies having fewer (for a given total luminosity) OB associations than Sc and Irr galaxies.

2. STAR FORMATION IN STELLAR ASSOCIATIONS

The surveys listed in Table 1 provide a useful pool of data for looking at certain questions regarding star formation in OB associations. This is not the right place for an exhaustive review of this important topic, but it might be appropriate to point out a few facts that emerge and that seem to relate to the basic questions of how star formation is triggered and how it proceeds, once begun.

First, we can compare the positions of the recognized associations with the locations of HII regions, which represent the loci of (visible only) massive star formation processes. Table 4 summarizes the statistics for several galaxies. On the average, only about 50% of stellar associations contain HII regions, but the dispersion in this figure is surprisingly large. The LMC's have 79%, while those in the SMC and IC 1613 have only a little over 30%. These two latter galaxies are in a relatively quiescent phase, as far as star formation is concerned, but their numbers of stellar associations are not especially small. The fact that they have small numbers of HII regions imbedded in them may reflect the possibilty that the associations are relatively old compared to those of the LMC and NGC 6822. This fact may also explain the observation that the SMC associations, although the same size as those of the LMC, possess notably fewer supergiants (Hodge 1985a).

Table 4. HII Regions and Core Star Clusters in Associations

	LMC	SMC	NGC 6822	IC 1613
No. Associations	122	70	16	20
No. containing bright HII regions	96	22	10	7
Percent with bright HII regions	79	31	62	35
Percent bright HII regions not in associations	79	79	62	12
No. young star clusters in associations	14	21	0	4
Percent with core star clusters	11	30	0	20
Percent of young star clusters not in associations	91	70:	100	0

[a] With $M < -4.5$ for their brightest stars
[b] Several clusters included with unknown ages

It is also of interest to note that, on the average, about 60% of the bright HII regions do not occur in stellar associations. IC 1613, with 12%, is an exception to this rule, but the general trend seems to indicate clearly that massive star formation in irregular galaxies can occur on a sufficiently small temporal or spatial scale that it need not involve a large complex, as represented by a stellar association.

Core clusters are fairly rare in stellar associations. The exact figures are difficult to obtain because of the difficulty of knowing whether coincidences are physical or accidental (a real problem for clusters, especially in the MC's, where clusters are extremely abundant). I have attempted to limit consideration to only very young clusters, those with brightest stars brighter than -4 in absolute blue magnitude, in order to avoid chance coincidences with the large numbers of older clusters. On the average, only 15% of the associations contain possible core clusters, without any correction made for chance superpositions. This implies that there will be no stable star concentration left at the position of the association after it disperses, in most cases.

For three of the galaxies, the overwhelming majority of young clusters are found outside of stellar associations. Does this mean that they formed as small-scale, dense objects without a large-scale star formation region around them? Conceivably each cluster once belonged to a giant gas cloud in which star formation went on at the 100pc scale level, and subsequent events have caused the surrounding, less dense group to disperse. Ages of the clusters in question, however, are on the order of 10-20 million years. To disperse in that time, a surrounding association would have to have had a high velocity dispersion. One might instead suggest that dense clusters can form in these galaxies independently of large-scale associations. The large number of isolated HII regions and lone O and B stars seem to substantiate this hypothesis.

2.1 The Mass Function of Young Clusters in HII Regions

In this brief section of the paper, I report on a somewhat related result of our study of the core clusters in giant HII regions. Measurement of the H-alpha luminosities of complete samples of HII region populations in several galaxies has now been complete and we find that the form of the luminosity function seems to be nearly universal. The universality of this shape implies a common cluster formation function (the exciting stars must be in the form of a cluster, because the models indicate that the brighter HII regions require hundreds of O and B stars). This somewhat surprising result seems to indicate that, whatever causes star formation to occur, the mass function of condensing regions is (or becomes) always the same. Can this really be true for all star-forming complexes, whether they are triggered by the passage of a density wave, started by the compression of gas from a supernova, or somehow otherwise involved in a GMC? The further detailed study of OB associations, star clusters, and populations of HII regions in different parts of galaxies and in galaxies of different types should eventually help answer this important question.

3. REFERENCES

Azzopardi, M. and Vigneau, J. 1977, A. and Ap. 56, 151

Dubois, P. 1980, D.Sc. Thesis, U. Louis Pasteur, Strasbourg.

Efremov, Y. N. 1982, Astron. Zh. 8, 663.

Feitzinger, J.V. and Braunsfurth, E. 1984, in "Structure and Evolution of the Magellanic Clouds" (van den Bergh and de Boer, eds.), Reidel, Dordrecht, p. 93

Florsch, A., Marcout, J. and Fleck, E. 1981, A. and Ap. 96, 158.

Hodge, P.W. 1985a, P.A.S.P., in press.

Hodge, P.W. 1985b, P.A.S.P., in press.

Hodge, P.W. 1985c, in preparation.

Humphreys, R. 1977, in I.A.U. Symp. No. 84.

Humphreys, R. and Sandage, A.R. 1980, Ap.J. Suppl. 44, 319.

Kennicutt, R.C. and Hodge, P.W. 1980, Ap.J. 241, 573.

Kunchev, P.Z. and Ivanov, G.R. 1984, Ap.Sp. Science, 106, 371.

Lucke, P. and Hodge, P.W. 1970, A.J. 75, 171.

Nikolov, N.S. and Ivanov, G.R. 1986, this Symposium.

Shakhbazyan, R.K. 1968, Astrofizika, 4, 273.

van den Bergh, S. 1964, Ap.J. Suppl. 9, 65.

Westerlund, B. 1961, Uppsala Astr. Obs. Ann., 5, No. 2.

Westerlund, B. 1964, Mon. Not. Royal Astron. Soc. 127, 429.

Wray, J. and de Vaucouleurs, G. 1980, A.J. 85, 1.

Discussion   :   HODGE.

KAUFMAN :

1) Is the turnover in the M101 luminosity function just the effect of incompleteness?
2) Do your luminosity functions for the 3 galaxies shown include only giant HII regions?
3) How do you explain the high percentages of bright HII regions not in associations - are these just one or two O stars?

HODGE :

1) Yes.
2) They extend down to 3 orders of magnitude below the brigthest in the case of M101.
3) Yes, or a compact cluster.

ZINNECKER :

Have you made a comparison between your (universal) luminosity function with van den Bergh's cluster luminosity function (van den Bergh and Lafontaine 1985, PASP $\underline{96}$, 880) i.e. the frequency distribution of star clusters as a function of their integrated light?

HODGE :

No.

POSTER PAPERS 3.

Chairman : P.S. CONTI

1. P.HELLINGS: The Average X-Ray Lifetime of Massive X-Ray Binaries.

2. R.SREENIVASAN and W.J.F. WILSON:
   Consequence of Rotational Mixing in late Type Massive Stars.

3. Y.N.EFREMOV, G.R.IVANOV and N.S. NIKOLOV:
   New Stellar Associations in M31.

4. P.Z.KUNCHEV and N.S.NIKOLOV:
   The Associations OB 110 and OB 112 in M33 Galaxy.

5. S.A.SILICH: On HI Shells and the Mass Spectra of OB Associations.

6. R.PENNINGTON: Star Formation in NGC 5128.

7. G.BODIFEE: Oscillating Star Formation.

8. C.H.B.SYBESMA: The Evolution of Massive Close Binaries with Mass Loss and Overshooting. An Application to V 729 Cyg (= BD +40 4220)

9. M.C.LORTET, M.HEYDARI-MALAYERI and G.TESTOR:
   CCD Observations of Young Stellar Associations and Multiple Systems in the Magellanic Clouds.

10. D.H. MORGAN and A.R. GOOD:
    Two Faint WC Stars near 30 Doradus.

11. E.KONTIZAS, A.DAPERGOLAS and M.KONTIZAS:
    Bright Stars in the SMC Clusters.

12. M.KONTIZAS, M.CHRYSOVERGIS, E.KONTIZAS and D.HADJIDIMITRIOU:
    Tidal Radii and Masses of the Clusters in the LMC.

13. E.KONTIZAS, E.THEODOSSIOU and M.KONTIZAS:
    The Distribution of Bright Stars in the SMC Clusters.

14. M.KAFATOS and R.McCRAY: Propagating Star Formation Induced by Superbubbles.

15. T.GEHREN, D.HUSFELD, R.P.KUDRITZKI, P.S.CONTI and D.G.HUMMER: Non-LTE Analysis of Massive Stars in the Magellanic Clouds.

16. P.DUBOIS: Metallicity Effect on Absolute Magnitude Determination.

17. E.KONTIZAS, E.XIRADAKI and M.KONTIZAS:
    Bright Stars in Five LMC Star Clusters.

18. G.R.IVANOV and P.Z.KUNCHEV: On the Arm Pattern of M33.

19. A.BRESSAN and G.BERTELLI: Convective Overshooting and the Upper Mass limit for Stars undergoing Core helium Flash.

20. H.SCHILD: The Final Fate of Massive Stars.

THE AVERAGE X-RAY LIFETIME OF MASSIVE X-RAY BINARIES

Paul Hellings, Astrophysical Institute, Free University of
Brussels, Pleinlaan 2, B-1050 Brussel, Belgium

Massive X-ray binaries may be powered by two mechanisms generating X-radiation: accretion of stellar wind material of the O-star, or Roche lobe overflow (RLOF). The evolution of RLOF powered X-ray binaries has been studied by Savonije (1978). For massive binaries the duration of the RLOF powered stage is less than 100 years for binaries evolving through case B of mass transfer, and 5000 to 10 000 years for case A. On this basis Savonije concluded that the majority of the X-ray binaries should evolve through case A of mass transfer, in order to explain the observed number of active sources in massive binaries. In this study we compute the transition from compact+O systems into X-ray binaries, with emphasis on the wind powered stage.

Evolutionary series from ZAMS to the supernova explosion have been determined by Hellings (1984) for a sample of WR+O binaries. We have calculated the effects of the supernova explosion on the pre-SN models for the binaries WR 31, 79, 97, 139 and 151, adopting the formalism for asymmetric explosions by De Cuyper (1984). The mass of the SN-remnant is taken $1.5 M_\odot$. Just after the supernova, $M(OB) = 20 \sim 27 M_\odot$, with in their core $X_c = .33 \sim .50$. The resulting compact+O binary is assumed to evolve in corotation, after synchronisation and circularisation. For systems evolving through case A, the wind powered stage, defined as the time during which the Röntgen luminosity is stronger than 10^{35} erg per second, takes about 800 000 years, before the core H burning O-star fills its Roche lobe, suffocating the X-radiation. The wind powered stage starts after the ratio semi-major axis/ radius of the OB-star, has dropped below 2.6. For the case B systems, the active stage starts during the rapid expansion during the shell H burning stage, lasting about 5000 to 8000 years. This is quantitatively a factor 100 less than for case A. Taking WR151, WR31 and WR79 representative for the classes of WR binaries separated by the conditions (P<3 days), (3<P<7 days) and (P>7 days), the probability for case A, including the possibility of asymmetric explosions, of these three classes is respectively : 0.621, 0.058 and 0.006. Taking into account the number of binaries in each class, the combined probability for case A is about 0.126, for case B 0.845 and for disruption 0.029. This corresponds to an average lifetime of about 10^5 years. The number of non transient massive X-ray sources within a distance of 3 kpc is 7 (Bradt,

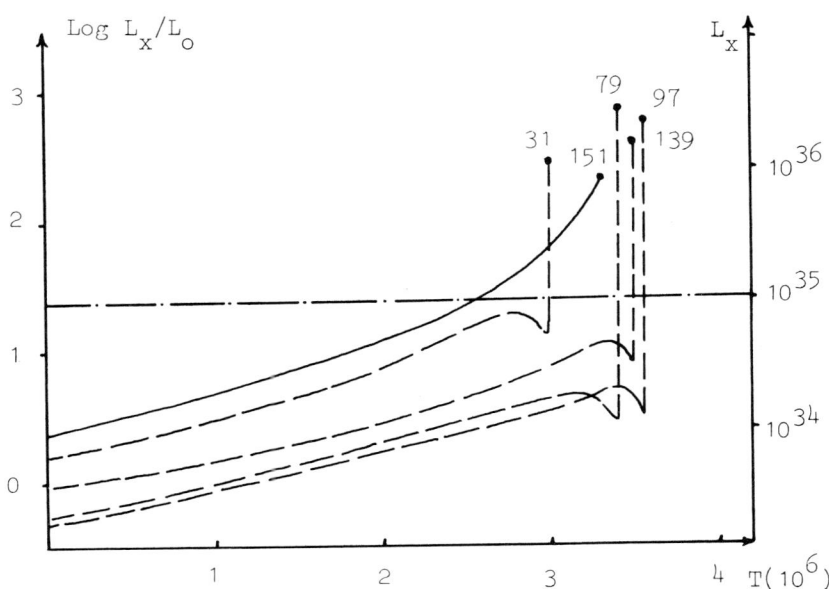

Figure 1 : The wind powered X-ray luminosity as a function of time for WR 31, 79, 97 and 139, evolving through case B (dashed lines) and for WR 151, evolving through case A (full line). The dots at the end of the tracks mark the start of RLOF.

Mc Clintock, 1983). The number of WR stars in the same volume is 50 (Hidayat et al. 1982). With a typical WR lifetime of $5 \ 10^5$ years, this corresponds to a typical X-ray lifetime of 70 000 to 140 000 years, depending on the fraction of binaries in the WR sample ranging from 0.5 to 1. The observed value is of the same order as the rough estimate obtained with the model calculations.

We conclude that the wind powered stage is an important part of the total X-ray lifetime. However, the X-ray luminosity attained during this stage never reaches 10^{37} erg/s (fig.1). During the short RLOF powered stage the sources reach a brightness of several times 10^{38} erg/s (Savonije, 1978). We also conclude that the majority of the observed X-ray binaries with optical OB component evolve indeed through case A, but that this is a selection effect of the lifetime. Against a probability for case A evolution of order 0.1 stands a lifetime 100 times larger than for case B. A complete version of this poster is submitted to A&A.

References
Bradt H., Mc Clintock J.(1983) Ann.Rev.Astron.Astrophys. 21, 13
De Cuyper J.P. (1984), Ph.D. Thesis, University of Brussels
Hellings P. (1984) Astrophys.Space Sc. 104, 83
Hidayat B., Supelli K., Van der Hucht K.A. (1982) in "Wolf Rayet stars : observations, physics, evolution" C. de Loore, A. Willis (eds) IAU symp. 99, p27, D. Reidel, Dordrecht
Savonije G.J. (1978) Astron.Astrophys. 62, 317

CONSEQUENCES OF ROTATIONAL MIXING IN LATE TYPE MASSIVE STARS

S.R. Sreenivasan and W.J.F. Wilson
Department of Physics, The University of Calgary
Calgary, Alberta, Canada T2N 1N4

We have examined further the consequences of mixing induced by rotation in massive stars (Pop I composition: $X = 0.7$; $Y = 0.27$) in the mass range $40 - 100$ $M_\odot$. The basic physics and a detailed examination for a 60 $M_\odot$ model has recently been published (Sreenivasan and Wilson 1985b), as has a detailed examination of supergiants without such mixing (Sreenivasan and Wilson 1985a). We shall therefore not go into the background of our computation here, but merely summarize our present findings.

The first result is that rotational mixing due to an extended core in the main sequence phase has the consequence that in all cases but one ($M_0 = 80$ $M_\odot$, where M_0 = Zero Age Main Sequence mass) the models exhaust their core helium on the red side of the HR diagram.

We present a table below to show the changes in Blue to Red Supergiant ratios. This is to be compared with our discussion of B/R ratios in our Supergiants paper (Sreenivasan and Wilson 1985a) and the Table 5 there. Also, the new $M_{BOL} - \log T_{eff}$ diagram of Humphreys and McElroy (1984) reveals a less steep slope for the envelope of observed stars, showing the absence of red supergiants above it.

TABLE I

M_0(ZAMS)	40	60	80	100
M(X=0)	30.0	43.7	55.4	76.9
M(Y=0)	27.0	33.7	52.6	66
V_s(ZAMS)	140	265	210	98
f	0.157	0.281	0.214	0.099
$\dot{M}$(ZAMS)	1.2	3.0	4.3	6.3
$\dot{M}$(RGB)	8	30	–	50
τ_B/τ_R	38.1	0.41	∞	0.75
$X_s(Y_c=0)$	0.57	0.38	0.22	0.31
$q_{He}(Y_c=0)$	0.69	0.95	0.96	0.77

UNITS: $M(M_\odot)$; $V(\text{km s}^{-1})$; $\dot{M}(10^{-6} M_\odot \text{ yr}^{-1})$

We have found from our model calculations that higher spins on the ZAMS result in greater mass-loss and a tendency for models to turn towards the blue as they age. There is also to be expected a distribution of spin speeds for a ZAMS model of given M_0. Thus a comparison of theoretical B/R ratios with those observed should incorporate the rotational history of models.

We therefore conclude that our spinning models predict higher effective temperatures for WR stars (Sreenivasan and Wilson 1982), a much broader main sequence due to mixing resulting from an extended core and the possibility of predicting the observed Blue/Red supergiant ratios when the rotational history of the stars is taken into account.

The lower slope of the envelope of observed stars demonstrated recently by Humphreys and McElroy (1984) is consistent with the consequence of rotational mixing in massive stars.

Fuller details of this investigation will be published elsewhere. Our work is supported by an NSERC grant (SRS) and The University of Calgary.

REFERENCES

Humphreys, R.M., and McElroy, D.B. 1984, Ap.J. 284, 565-577.
Sreenivasan, S.R., and Wilson, W.J.F. 1982, Ap.J. 254, 287-296.
_____ 1985a, Ap.J. 290, 653-659.
_____ 1985b, Ap.J. 292, 000.

NEW STELLAR ASSOCIATIONS IN M31

YU. N. EFREMOV, STERNBERG INSTITUTE, MOSCOW
G. R. IVANOV AND N. S. NIKOLOV, DEPARTMENT OF
ASTRONOMY, UNIVERSITY OF SOFIA, SOFIA, BULGARIA

About twenty years ago van den Bergh (1964) recognized 188 OB associations in the Andromeda Nebula. He used plates taken by the 52-inches Tautenburgh Schmidt telescope in GDR. Later on Richter (1971) added 7 new associations in the south-western periphery of M31. Now we have the opportunity to continue the search of stellar associations in M31 with the 2m Ritchey-Chrétien (RC) telescope of the Bulgarian National Astronomical Observatory. The limiting magnitude of this telescope is approximately the same as that of the Tautenburg 2m Schmidt telescope but it possesses somewhat smaller field ($1^\circ \times 1^\circ$) with plate-scale 12.8 mm^{-1}. That is why a new search of stellar associations in M31 by means of the 2m RC telescope is very efficient. Indeed, some previous inspections of the RC plates (Efremov, 1982) indicated some new associations in M31.

Within the frames of an investigation project of ours of nearby galaxies we had at our disposal mainly B plates of M31 using 103aO Kodak + GG 385 and 30 - 90 minutes exposure time. As Ivanov (1985) showed, in the spiral arms of M31 one background star occurs on the average at 2 square arc min. The mean dimension of an M31 association is 2.4 arc min and therefore the background stars do not influence significantly the search of associations.

We covered the Andromeda galaxy with four plates. Figure 1 shows part of an M31 field. The associations of van den Bergh are shown surrounded by continuous line with their numbers. The new associations are marked by a dashed line. We extended van den Bergh's numeration from No. 189 up to No. 312, i. e. we distinguished 123 new objects. A large amount of subgroups does not enter in this number. This might be noted in many of the associations outlined by van den Bergh, as he himself remarked.

A not very detailed inspection of the Andromeda Nebula associations shows that a lot of the newly recorded associations are smaller than about 1 - 2 arc min. Another

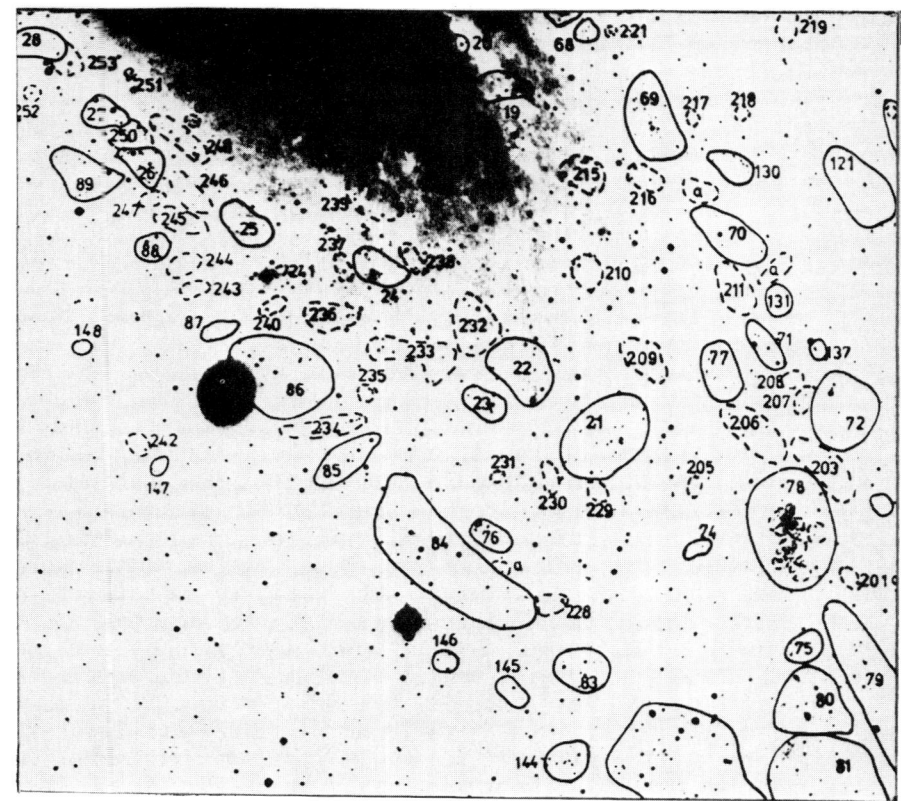

Figure 1. The stellar associations indicated by van den Bergh (1964) are outlined with line and ours - with dash.

part of them contains B stars near the limiting magnitude of $21^m.5$. Both facts once more underline the statement that the 2m RC telescope is suitable for investigation of nearby galaxies.

REFERENCES :
Bergh, S., van den, 1964, Astroph. J. Suppl., 9, 65.
Efremov, Yu. N., 1982, Pisma v Astron. Zh., 8, 585.
Ivanov, G. R., 1985, Astrophysics and Space Sci., (in press).
Richter, G. A., 1971, Astron. Nachr., 292, 275.

THE ASSOCIATIONS OB 110 AND OB 112 IN M 33 GALAXY

P.Z.Kunchev, N.S.Nikolov
Astronomy Department, University of Sofia

The associations OB 110 and OB 112 in M 33 galaxy (Humphreys and Sandage, 1980) form a powerful group of young stars. This rich star cloud is situated far (~6 kpc) from M 33 nucleus. The abundance of hot stars and its huge dimensions (900 x 1500 pc) make it similar to the association OB 78 (400 x 800 pc) in the Andromeda galaxy. We use distance modulus $(m-M)_V = 25.35$ given by Sandage and Carlson (1983) and $E_{B-V} = 0.03$.
The observations of OB 110 and OB 112 were obtained in the period 1983 - 1984 by 2 m Ritchey-Chretien telescope in the Rhodopes Mountains (Bulgaria). Because of its large field (1° x 1°), plate-scale ($12".8$ mm^{-1}) and slow focal ratio (f/8) this telescope is suitable for observations of exceedingly crowded stellar fields in which the background density of the unresolved stellar images is high.
The U, B, V photometry of more than 150 stars allowed the building up of diagrams colour - magnitude (Fig. 1 a,b) and colour - colour (Fig.2 a,b). The lines of the intrinsic colours of the main sequence stars and the supergiants (Ia) were shown on Fig.2 as well. The error of our photometry does not exceed $0^m.1$ in V and B and $0^m.15$ in U. The absence of strong H II regions facilitates the star photometry.
The diagrams gave the possibility to separate very confidently the high luminosity stars and the red supergiants from the foreground stars of our Galaxy. Probably all stars with B-V < 0.35 and U-B < -0.2 (indicated by closed circles on Fig.1 and Fig.2) are members of the upper part of the main sequence which appears in 17.5 - 18 magnitude. The stars situated higher and on the right from the main sequence (crosses on Fig.1 and Fig.2) are on the whole stars of the foreground. Some of the objects marked by crosses and situated on the right and down from the main sequence have non-stellar nature. Their characteristic is B-V > 0.4 and at the same time they have large negative

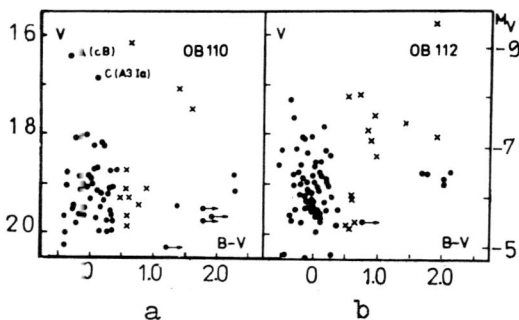

Fig.1. Colour - magnitude diagrams of the associations OB 110 (a) and OB 112 (b) in M 33.

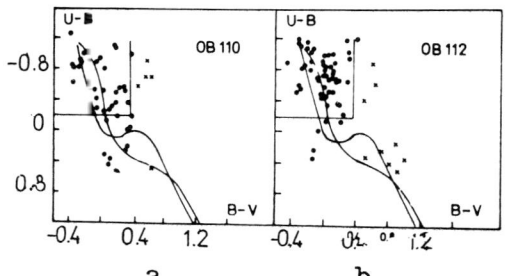

Fig.2. Two-colour diagrams of the associations OB 110 (a) and OB 112 (b) in M 33.

U-B. Four objects of this kind are seen on Fig.2a and three - on Fig.2b. Six stars belonging to OB 110 association and five from OB 112 are probably red supergiants. They appear when $V \approx 19^m$ ($M_V \approx -6.3$) and $B-V \approx 2.0$ which coincides with the results obtained by Humphreys and Sandage(1980) concerning the bright blue and red stars in M33. The stars marked by A and C in OB 110 are undoubtedly M 33 members having in mind their spectrums determinated by Humphreys (see Humphreys and Sandage, 1980). This gives $M_V = -8.9$ for A and $M_V = -8.5$ for C.

The stellar magnitudes of the brightest blue stars enabled to estimate the ages of the associations. Using the method of Hodge(1983) we obtained 4.1×10^6 years for OB 110 and 6.5×10^6 years for OB 112.

A more detailed study of this star cloud and the adjacent neighborhood as well as a chart of the measured stars would be published later.

References

Hodge, P.W. 1983, Ap.J., 264, 470.
Humphreys, R.M., and Sandage, A. 1980, Ap. J. Suppl., 44, 319.
Sandage, A., and Carlson,G. 1983, Ap. J. Lett., 267, L25.

THE DYNAMICAL EVOLUTION OF EXPANDING HI SHELLS AND INITIAL MASS FUNCTION OF OB ASSOCIATIONS

S.A.Silich
Main Astronomical Observatory Academy of Sciences of the
Ukrainian SSR
252127 Kiev-127, Goloseevo
USSR

ABSTRACT. The evolution of expanding supershells produced by SN explosion in OB associations is considered. The expansion velocities of the shells and OB associations IMF spectral index are obtained. The results agree well with observational data.

The distribution of the Galactic HI column density in small velocity ranges shows the existence of many curved HI filaments. In some cases filaments change their sizes with velocity as expading shells. Some shells are huge. Their radii vary between 0.1kpc - 1kpc approximately and their masses reach $10^6 M_\odot$ – $10^7 M_\odot$. But their expansion velocities vary within narrow velocity range $10 \text{km s}^{-1} \leq U \leq 24 \text{km s}^{-1}$ (Heiles, 1979).Hence, it follows that expansion velocities of these shells are almost costant (at least for the late evolutionary stage).

We consider the evolution of OB association within HI supercloud (Elmegreen and Elmegreen, 1983) accompanied by supernovae explosions and expanding HI shell origin. It is shown that the shell expansion velocity will be constant as the total supershell energy will increase with a rate $\mathcal{E}_0(t) \sim t^2$. Then by assuming that supernovae explosions are the main energy source in OB associations one can find OB association stars initial mass function and shell expansion velocity U_0.

We have studied the propagation of a strong radiative shock wave in infinitely thin layer approximation. It have been assumed that all swept-up interstellar gas collapses in a thin shell radius R and the gas pressure is uniform within cavity. The motion of the shell is described by equation

$$\ddot{R} + \frac{3(\gamma + 1)}{2} R^{-1}\dot{R}^2 = \frac{9(\gamma -1)}{4\pi \rho R^4} E(t) , \qquad (1)$$

where $E(t)$ is the total shell energy, γ is adiabatic index, ρ is the gas density. From equation (1) it follows that the shell expansion velocity is constant as the supernovae add energy to cavity at a rate

$$\mathcal{E}_{SN} = 4\pi \gamma (\gamma - 1)^{-1} \rho U^5 t^2 . \qquad (2)$$

Taking into account the analytical expression by Bisincehi et al.(1983) for the massive stars main sequence lifetime $t=Am^{-\alpha}$, where $A=5.3 \cdot 10^7$ yr $\alpha \approx 0.6$, it is easy found that in OB association with IMF $n(m)=dN/dm=Cm^{\delta}$ supernovae release energy at a rate

$$\mathcal{E}_{SN} = CE_0 \alpha^{-1} A^{\frac{\delta+1}{\alpha}} t^{-\frac{\delta+\alpha+1}{\alpha}}, \qquad (3)$$

where E_0 is the supernova explosion energy. Equating the right-hand parts (2) and (3) we get the IMF spectral index $\delta = -1-3\alpha \approx -2.8$ and the value of the shell expansion velocity:

$$U_0 = \left[\frac{3(\gamma - 1)NE_0 m_1 \frac{3\alpha}{\alpha}}{4\pi \gamma \rho A^3} \right]^{0.2}, \qquad (4)$$

where N is the total number of massive stars in association, m_1 is minimum mass of stars exploding as supernovae. From (4) it follows that U_0 depends weakly on parameters of OB association and superclouds. Substituting into (4) $\gamma = 5/3$, $E_0 = 10^{51}$ erg, $N = 100$, $m_1 = 9$ and $\rho = 10^{-23}$ g cm^{-3} we have $U_0 = 16$ km s^{-1}, for the same values N, E_0, m_1 and $\rho = 10^{-24}$ g cm^{-3} we have $U_0 = 25$ km s^{-1}.

The expansion velocity of the shell will be constant up to the time $t_c = Am_1^{-\alpha} \approx 1.5 \cdot 10^7$ yr, when the last sufficiently massive star becomes supernova. When $t = t_c$ the shell radius reaches $R_c = U_0 t_c = $ 250pc - 400pc. After such a time the energy pumping stops and shell decelerates up to the random velocity of the interstellar clouds $U_c \approx$ 10 km s^{-1}. The maximum radius of the shell is higher by a factor $2 \div 3$ than R_c and can reach 0.5 kpc - 1.0 kpc.

Our calculations show that for undestanding dynamics of expanding HI shells it is necessary to take into account that supernovae release energy continuously up to the time t_c. The present modification of the theory by Bruhweiler et al. (1980) and Tomisaka et al. (1981) provides a natural explanation of the fact that expansion velocities of the Heiles shells vary within the narrow velocity range and is also confirmed by a good agreement of the obtained values of shell expansion velocities and IMF spectral index with the observational data.

We believe that the self-consistent theory including entire history of star formation regions can be analyzed now. Superclouds formation in unstable galactic disk, origin of the molecular clouds, massive stars and OB associations within superclouds, the interaction of stellar winds and supernovae with the surrounding interstellar medium, formation of HI shells, theirs development and destruction seem to be the main elements of such theory.

REFERENCES
Tomisaka,K.,Habe,A.,Ikeuchi,S. : 1981, Astrophys. Spase Sci.,78,273.
Heiles, C. : 1979, Astrophys.J., 229, 533.
Elmegreen, B.G., Elmegreen, D.M. : 1983, MNRAS, 203, 31.
Bisincehi, G.F., Fermani, G., Sarmiento, A.F.; 1983, Astron. Astrophys., 119, 167.
Bruhweiler, F.G., Gull, T.R., Kafatos, M., Sofia, S. : 1980, Astrophys. J., 238, L27.

STAR FORMATION IN NGC 5128

Robert L. Pennington
University of Minnesota

Star formation in the dust lane of NGC 5128 (Cen A) has been studied using digitized CTIO 4m plates. The digital images were calibrated to the standard U, B and V passbands using photoelectric photometry (van den Bergh 1976). Ages for the dominant stellar component were derived by de-reddening each pixel along an R = 3.3 reddening law to its intercept with a theoretical cluster evolution track in the U-B, B-V plane (Davis 1979). The HII regions of Hodge and Kennicutt (1983) were used as tracers for the most recent star formation.

The dust lane of NGC 5128 is probably a tall, thin ring that is presently subject to infall from the IGM (Pennington 1984), rather than a disk-like structure, as had previously been assumed. The lane is orbiting from the SE to the NW at ~250 km/s. Recent star formation is not distributed evenly along the dust lane, with the SE part of the lane noticeably lacking in HII regions while the northern edge of the lane downstream from the infall point is the region most intensely undergoing star formation. This suggests that star formation is initiated near the infall region on the northern edge of the lane and dies out as the lane orbits around. The travel time from the infall region to the first of the major HII regions is 4×10^6 years and the orbit time is 5×10^7 years, assuming the galaxy is at 3 Mpc.

The northern edge of the lane was extracted from the rasters to examine the possibility of an age gradient downstream from the infall region as the newly formed HII regions age. A gradient was found with the average age increasing from log (age in years) = 7.18 to 7.33±0.07. This may be due to either an aging stellar population or to anomalous reddening as the NW end of the lane is approached. The absence of a similar gradient at the SE end of the lane and the magnitude of the gradient, ~1/4 orbit, suggests that this is an age gradient.

Although star formation is initially most intense on the northern edge of the lane, young stars become evenly distributed across the height of the lane is less than one orbit. Examination of the dust lane to the SE, where it orbits back around to the near side of the

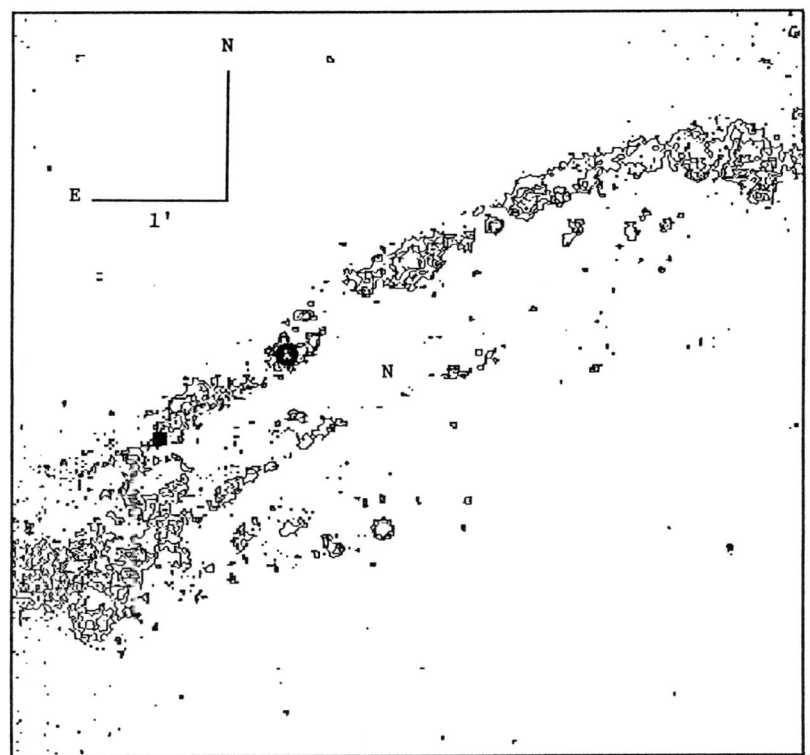

Figure – This is a contour map of NGC 5128 showing regions bluer than U-B = -0.20. The infall region is marked by a square east of the nucleus (N) and the first major HII region is marked by an open circle. The SE end of the lane is blue, but has few HII regions, unlike the northern rim, which is sharply defined both by its color and numerous HII regions.

galaxy, shows no age or U intensity gradient between the northern and southern edges. This is probably due to star formation spreading across the height of the lane rather than a dispersal of the young stars and HII regions. A dispersal would require that these young clusters move only to the south at velocities ≥ 20 km/sec. While this is possible if the lane is turbulent, it is not considered likely because many of the HII regions and young clusters that are present away from the northern edge appear to lie in chains that parallel the edges of the dust lane. This suggests a star formation mechanism that acts along the flow lines of the lane.

References
Davis, J.: Ph.D. Thesis, Rice University, 1979.
Hodge, P.W. and Kennicutt, R.C.: 1983, A.J. 88, 296.
Pennington, R.L.: 1984, Ph.D. Thesis, Rice University, 1984.
van den Bergh, S., 1976, Ap.J. 208, 673.

OSCILLATING STAR FORMATION

G. BODIFEE
Astrophysical Institute
Vrije Universiteit Brussel
Brussels, Belgium.

ABSTRACT.

As a consequence of positive feedback effects in interstellar chemical and star forming processes, a star formation region may undergo nonlinear oscillations.

A model has been built of a star formation region in which mass transformation processes take place, regulated by a throughflow of matter and the interactions between the components of the system. In this approach, a star formation region is regarded as a galactic dissipative structure (Nicolis and Prigogine,1977), sustaining itself as a more or less stable, ordered, non-equilibrium system, independent of the of the environment, except for supply of material and the removal of waste. The system of the model includes three components:
Cool atomic gas. This gas does not lead to collapse; no star formation is possible.
Cool and dusty molecular clouds. Due to efficient molecular cooling, this gas can collapse. However, external pressure by expanding HII regions applied for a collapse to set in.
Young stars with their HII regions. The stars are able to trigger further star formation in molecular clouds (Elmegreen & Lada, 1977).
It is assumed that the total mass of the system is constant:
The system is attached to two "reservoirs":
- cool atomic gas is available to replenish the mass that is removed from the system.
- matter that is permanently buried in stellar remnants as low-mass main sequence stars, is removed from the system ("waste" reservoir).

Mass transformations between components are described by a system of parameterized equations.

$$\frac{dX}{dt} = F(X)$$

where X is a vector of the component concentrations, and F a vector of nonlinear polyonomial and autonomous equations, that describe the transformation processes.

	X_c	X_a	M_1/M_o	M_2/M_o	P	X_c	X_a	M_1/M_o	M_2/M_o	P
initial	.7	.7	40	20	2	.7	.7	80	40	1.5
start RLOF	.4	.7	37	19.5	2.3	.65	.7	71	39.7	1.5
end RLOF	.4	.66	32	24.3	1.8	.2	.63	30	61	3.2
start WR(*)	0	.14	24*	22.9	2.6	.2	.63	30	61*	3.2
end WR(*)	0	0	9*	22.6	5.7	.15	.61	29	49*	4.2

Table 1. parameters in the evolution of very massive close binaries undergoing case A of mass transfer.

The Evolutionary status of V729 Cyg

Recent observations of the spectroscopic binary V729 Cyg have cast doubts on its classification as an Of +Of binary. Vreux, 1985, has studied the H profiles and proposed a model consisting of a WR component (the lower mass, most evolved component) accreting mass from the more massive Of companion.

The system has been spectroscopically studied by Bohannon and Conti, 1976, and Massey and Conti, 1977, and found to consist of a 47 M_o primary and a more evolved 15 M_o secondary both classed as Of stars.

In order to test the models from a theoretical evolutionary standpoint the system has been compared to evolutionary tracks for single stars. (taken from Pylyser et.al., 1985). If the secondary is an Of star it must still have hydrogen at the surface. Hence the H exhausted core at the end of core H burning must be smaller than 15 M_o. This means that the initial mass of the current secondary must have been no larger than 25 M_o (comparison to the tracks). The total current mass however is 62 M_o so that the current secondary, which is the more evolved, must have an initial mass of at least 31 M_o. This is of course a contradiction and therefore we conclude that the 15 M_o component is indeed a He star, and that V729 Cyg is a post mass transfer binary with a WR component accreting in the wind of the more massive Of star.

References

Bohannon, B., Conti P.S.,1976: Astrophys. J. **204**,797
Doom, C., 1984: Astron. Astrophys. **138**,101
Doom, C., 1985: Astron. Astrophys. **142**,143
Doom, C., De Greve, J.P., 1983: Astron. Astrophys. **120**,97
Massey, P., Conti, P.S., 1977: Astrophys. J. **218**,431
Pylyser, E., Doom, C., de Loore, C., 1985: Astron. Astrophys. (in press)
Sybesma, C.H.B., 1985a: Astron. Astrophys. **142**,171
Sybesma, C.H.B., 1985b: in: "Birth and Evolution of Massive Stars and Stellar Groups" H. van Woerden and W. Boland eds. Reidel, Dordrecht, p. 183.
Sybesma, C.H.B., 1985c: Astron. Astrophys. (submitted)
Vreux, J.M., 1985: Astron. Astrophys. **143**,209

CCD OBSERVATIONS OF YOUNG STELLAR ASSOCIATIONS AND MULTIPLE SYSTEMS IN THE MAGELLANIC CLOUDS (1)

M. C. Lortet[1], M. Heydari-Malayeri[2], G. Testor[1]
Observatoire de Paris-Meudon, France
(1) D.A.P.H.E.
(2) D.E.M.I.R.M.

1. Introduction

The nebular and stellar observations were carried out by G.T. at the Cassegrain focus of the Danish 1.50m telescope at the ESO La Silla Observatory in August and September 1984, using a RCA CCD (512 x 320 pixels) with a pixel size of 0.47". The frames were reduced using the VAX/MIDAS software at ESO Garching and Meudon Observatory (INVENTORY automatic program).

2. The brightest stars : single stars or tight clusters ?

One of the most remarkable results of CCD imagery is its ability to point to multiple stars. A number of double or mutiple stars were found.

Table 1 is a list of stars for which our observations definitely indicate a multiple system. We also include the integrated magnitudes and colours of two of the clusters first imaged with the electron camera in N 157C (Lortet and Testor, 1984). These are to be compared with V = 11.14 to 11.90, B-V ~ -0.03 to -0.08 for young (poor) clusters in N 51 and N 59 (LH 51, 55, 88, Vuillemin, 1985) in the LMC and V = 11.2 for R 136a (diaphragm 1"), indeed comprised of at least 8 stars (Walborn, 1986). Similarly in SMC (which distance modulus is ~ 0.5m larger than for LMC), several clusters have integrated magnitudes between 12.0 and 13.3 (Gordon and Kron, 1983).

4. Conclusions

The implications of recognizing multiple systems among the brightest stars of an association or a galaxy bears on several important problems :
- nature of the so-called transition stars Of/WN and WN/WC : are they rather <u>always two</u> separate stars ?
- existence of supermassive stars. Though it is now clear, after the resolution by speckle of R 136a into 8 components, that the concept of supermassive star is lost, yet it is up to now not at all known how frequent are tight clusters (or cluster remnants) of hot stars : appropriate observations should first detect and then discover their detailed content, up to the less massive stars.

---

(1) Based on the observations obtained at ESO La Silla, Chile.

Table 1 : Multiple Systems among Bright Stars

Identification	V	B-V	Spectral Type	Estimated number of stars	Ref., Note
LMC					
N11B : star α	12.00	- 0.25	$O4 - 5^a$	6 in 9" x 6"	b
N11B : BI 42	13.5^c		OB	5 in 10" x 8"	b
N11B : Sk-66°33	11.9^c		$O3 - 4^a$	> 2 in 3" x 2.6"	b
N11C : Sk-66°41	11.4 pg		$O5^a$	4 in 6" x 5"	1
N158C : Sk-69°249	11.13	- 0.26	O9 or $B0.5I^e$	3 in 4" x 3"	2
N159A : DD 13	13.84	- 0.01	early O	2 in 3" x 2.6"	d, 3
	13.19	$+ 0.53^d$			
N157C : Cluster β	10.71	- 0.26		14 in 12" x 10"	f, 4
N157C : Cluster δ	11.07	+ 0.02		15 in 12" x 12"	f
SMC					
N66 : star 2	12.59	- 0.32	$O8.5^a$	> 2 in 3" x 2.6"	5

The B-V measurements except otherwise specified, are the present measurements (preliminary)
a) V. Niemela, 1985, private communication
b) Heydari-Malayeri and Testor, 1983, Astron. Astrophys. 118, 116
c) Brunet et al., 1975, Astron. Astrophys. Suppl. 21, 109
d) Dufour and Duval, 1975, P.A.S.P. 87, 769, field of LMC X-1
e) Nandy et al., 1984, M.N.R.A.S. 210, 131
f) Lortet and Testor, 1984
1 The star Sk-66° 41 = HD 268743 is among the "candles" selected as the brightest stars in LMC (Humphreys, 1983). It is a multiple system (at least 3 stars) embedded in a nebular condensation.
2 4" apart is the WR star Brey 91, with V = 12.68, B-V = - 0.40.
3 Contamination by nebulosity may explain that Dufour found this star to be so red.
4 This cluster contains the WN7 star Brey 65, with V = 13.12, B-V = - 0.18.
5 The second brightest star in NGC 346, as designed by Walborn, 1978, Ap. J. Lett. 224, L134. Outside NGC 346 though in the same nebula N66 are the two bright stars HD 5980 (binary, OB ? + WN3, V = 11.77, B-V = - 0.20) and Sk 80, a candle in SMC (O7Iaf, V = 12.36, B-V = - 0.21).

- blue stars as distance indicators. The frequent occurrence of multiple systems among the brightest blue stars may partly explain why problems arise as to their ability as extragalactic distance indicators (Humphreys, 1983).

References

Gordon, K. C., Kron, G. E., 1983, *P.A.S.P.* **95**, 461
Humphreys, R., 1983, *Ap. J.* **269**, 335
Lortet, M.C., Testor, G., 1984, *Astron. Astrophys.* **139**, 330
Lucke, 1974, Ph. D. Thesis
Vuillemin, A., 1985, Thèse d'Etat, Université de Marseille and private communication
Walborn, N., 1986, I.A.U. Sympos. **116**. Luminous Stars and Associations in Galaxies. Edit. C. de Loore, A. Willis, P.G. Laskarides, Reidel, this volume.

TWO FAINT WC STARS NEAR 30 DORADUS

D.H. Morgan and A.R. Good
Royal Observatory,
Edinburgh, EH9 3HJ

"Wolf-Rayet stars in the LMC : how faint do they go?" That was the title of a recent paper by Massey and Conti (1983) who found that the faintest WN star in Breysacher's (1981) catalogue of LMC WR stars has m_v = 17.1. The faintest WC star in that catalogue is Br 74 which has m_v = 15.6. We have now detected two faint WC stars at $m_v \sim 18.6$, three magnitudes fainter than Br 74.

These stars are 3 and 5 arcmin from the centre of the 30 Doradus nebula and were detected by blinking a pair of photographic plates taken with the UK 1.2m Schmidt Telescope (UKST). The plates were hypersensitized Kodak IIIa-J emulsion taken through narrow ($\Delta\lambda \sim 70\text{Å}$) and broad ($\Delta\lambda \sim 1600\text{Å}$) band filters centred on HeII $\lambda 4686$, and covered an area $\sim 2.3^\circ \times 2.3^\circ$ centred near 30 Doradus. We found that with this filter combination we could easily detect WC stars but not WN stars.

The brighter of these two faint WC stars was later observed with the IPCS on the AAT in the wavelength range 3900-4900Å and was confirmed to be an early WC star. The fainter candidate was not observed. Estimates of the apparent continuum magnitudes near 5100Å made from the IPCS spectrum and other UKST plate material are $m_v \sim 18.3$ and 18.9. These are ~ 3 mag fainter than Br 74.

There are several explanations of the surprising faintness of these stars. One is simply extinction by dust; but an extinction of 3 magnitudes or more is extremely large, even for the 30 Doradus region. A second is that the stars are the bright nuclei of planetary nebulae. However, we could detect no emission lines in the IPCS spectrum other than or stronger than those seen in the sky nearby. Moreover, we found both these objects close to 30 Doradus where there are many WR stars, but we found no similar objects on a second plate pair centred at $5^h 37^m$, $-66^\circ 35'$. We hope to be able to establish the nature of these stars from further observations. Details of these two objects, including finding charts will be given by Morgan and Good (1985) along with details of a further 4 new WR stars ($m_v \sim 13.3 - 15.6$) not in Breysacher's (1981) catalogue, detected by us during searches of UKST objective prism plates.

ACKNOWLEDGEMENTS

We are grateful to the staff of the UK Schmidt Telescope Unit for the provision of the plate material, and to the staff of the Anglo-Australian Observatory, especially Dr J.R. Walsh who made the observations, for the provision of the IPCS spectrum through the PATT/AAO service observing scheme.

REFERENCES

Breysacher, J., 1981. Astron. Astrophys. Suppl., 43, 203.
Massey, P. and Conti, P.S., 1983. Astrophys. J., 264, 126.
Morgan, D.H. and Good, A.R., 1985. Mon.Not.R.astr.Soc., In press.

BRIGHT STARS IN THE SMC CLUSTERS

E. Kontizas*, A. Dapergolas* and M. Kontizas**

*Observatory of Athens, **University of Athens.

The bright stars for 15 SMC clusters were classified in order to derive the distribution of various spectral types. The studied clusters represent all evolutionary ages (disk, intermediate and halo) and are located at various places of the parent galaxy. The spectal classification of the stars was carried out using film copies of the 1.2 m Schmidt telescope objective prism plates. Low dispersion (2440 Å at H_γ) and medium dispersion (830 Å at H_γ) unwidewed UJ and RI spectra were examined by means of a binocular microscope. Short exposure plates were used as well for the most bright stars and particularly for the stars at the central areas where crowding is more severe. More details about the used material and the criteria used for the classification are described by Kontizas et al (1985). For each cluster a circular area was examined inside its tidal radius. (Kontizas, 1984).The stars in the innermost part of the populous clusters were not classified because of the overlapped images. Stars of fields in the vicinity of each cluster were also classified to find the contribution of field stars in the cluster area. The magnitude range of the studied stars is $14.5 < V < 17.50$.

Fig. 1, shows the number of stars per spectral type for (a) the seven halo clusters, (b) the three intermediate and (c) the five disk clusters respectively. The stars classified as B represent stars O, B and A spectral types since the material used does not permit us to distinguish them. The dashed line represents the distribution of the various spectral types of the adjoining fields normalised to the cluster area.

From these diagrams it can be seen that for the halo and intermediate clusters the bright stars are mainly M stars whereas for the disk clusters they are almost equally distribution into B and M stars.

Another interesting result that comes out from these dia-

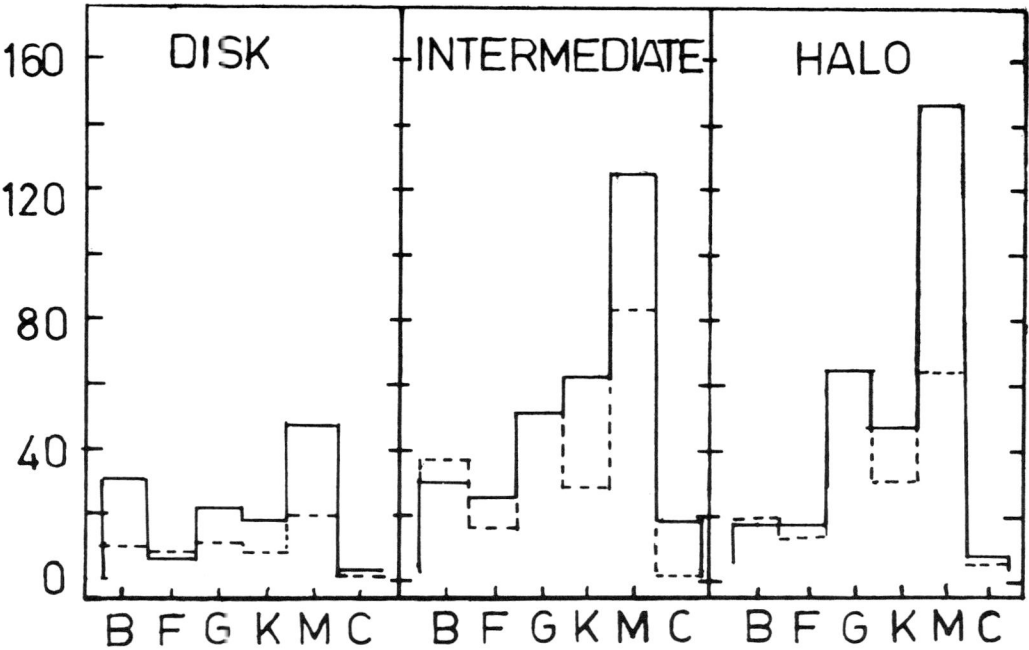

Figure 1. Number of bright stars per spectral type (solid line) for (a) disk (b) intermediate and (c) halo SMC clusters and their adjoining fields (dashed line).

grams is that the carbon stars are mainly found in the intermediate age clusters, as it is expected.

Acknowledgements

We would like to thank the 1.2m U.K. Schmidt Unit for loan of the material.

References

Kontizas, E., Dapergolas, A., and Kontizas, M. : 1985 (in preparation)
Kontizas, M. : 1984, Astron and Astrophys., 131, 58.

TIDAL RADII AND MASSES OF THE CLUSTERS IN THE LMC

M. Kontizas*, M. Chrysovergis*, E. Kontizas** and
D. Hadjidimitriou**.
* University of Athens, ** Observatory of Athens.

ABSTRACT

Masses and tidal radii of star clusters in the LMC have been derived by means of star counts from U.K. Schmidt plates. Two groups of clusters according to their distance from the rotation centre of the LMC were measured. The tidal radii of the central clusters vary from 58 to 85 pc and those of the most distant clusters from 33 to 86 pc whereas masses were found to vary from 10^5 to $4\times10^5 M_\odot$ and from 10^4 to $2\times10^5 M_\odot$ respectively.

OBSERVED DYNAMICAL PARAMETERS

Star clusters are an important key of understanding the stellar evolution and the dynamical properties of the stars in a galaxy. LMC is our nearest galaxy and gives the opportunity to study many clusters in detail. Plates (covering all the LMC area) taken with the 1.2m U.K. Schmidt Telescope were measured on IIIaJ, IV-N and IIaD emulsion. Star counts were carried out on the screen of a magnifying system. The homogeneous observational material allowed star counting far beyond the tidal radii of the clusters so that the background is reached safely in all cases. Two groups of cluster were selected according to their distance from the rotation centre of this galaxy. One group of 26 clusters is located in a ring within the radii of 0.5 and 3 degrees and a second group of 41 clusters (occupying the outer part of the LMC) at distances from 5° to 7°. Therefore this second group represents the most remote LMC clusters in all directions.

The tidal radii were found by the method described by King (1962) and the various sources of errors give an uncertainty of 15% in the derived values. The tidal radii of the central group of clusters were found to be from 58 to 85 pc whereas for the outermost clusters the derived values are 33 to 86 pc. From these values it can seen that the tidal radii of the studied clusters are in the same range of the galactic globulars (Peterson and King, 1975) and the SMC clusters (Kontizas, 1984). The central clusters classified by Freeman et al (1983) as kinematically disk clusters are all found to have tidal radii systematically larger that the SMC disk clusters whereas the remote clusters show a range of radii similar to those of the halo SMC clusters (Kontizas 1984).

The masses of the central clusters were found by the method described by Chun (1978) and the derived values are from 10^5 to $4 \times 10^5 M_\odot$ whereas the masses of the outermost clusters calculated using the King formula (1962), were found to be from 10^4 to $2 \times 10^5 M_\odot$. These values show that the LMC clusters are about 10 times less massive that the galactic globulars and 10 times more massive that the SMC halo clusters.

ACKNOWLEDGEMENTS

The authors would like to express their sincere thanks to the U.K. Schmidt Telescope Unit for loan of the Observational material.

REFERENCES

King, I. R. : 1962, Astron. J. 67, 471.
Chun, M. S. : 1978, Astron. J. 83, 1062.
Peterson, C. J., King, I.R. : 1975, Astron. J., 80, 427.
Kontizas, M. : 1984, Astr. and Astrophys. 131, 58.
Freeman, C.K., Illingworth G. and Oemler A Jr. 1983. Ap. J. 272, 488.

THE DISTRIBUTION OF BRIGHT STARS IN THE SMC CLUSTERS

E. Kontizas*, E. Theodosiou** and M. Kontizas**
* Observatory of Athens, ** University of Athens

Star counts can be used to investigate radial distribution of stars of different mass. Relaxation through stellar encounters is a mechanism that does make a distrinction between stellar masses, so systems that have undergone such relaxation should show differences in distribution between stars of high and low mass. That does not happen for systems that have undergone an initial violent relaxation since this type of relaxation treats all masses equally.

Twenty five disk, three intermediate and eighteen halo clusters of the SMC were measured for studying the segregation effect by means of star counts. Short and long exposure red plates (IIIaF+RG630) were used, taken with the 1.2m U.K. Schmidt telescope. On the short exposure plates (15min) the counted stars reach a limiting magnitude of $V\sim19.30$, whereas the deep plates (90 min) reach a magnitude of $V\sim21.00$ mag.

Then to compare the distributions of the stars of different masses, the counts from the short exposure plates have been shifted vertically until they overlapped with the counts from the deep plates in the outer region of the cluster. This has been done because the tidal cut off, that determines the tidal radii of these clusters, treats all stars equally and so that in the outer regions of the clusters the distribution of stars should be the same (Da Costa, 1982).

Most of the clusters studied here do not support the mass segregation effect and their diagrams resemble the diagram of the cluster L57 illustrated in Fig. 1a. Only two disk and two halo clusters show evidence of mass segregation and one of those L9 is illustrated in Fig. 1b.

As it was pointed out by King (1975) the bright stars are expected to be studied in the core but are two few to give good statistic in the envelope, while the faint stars are well observed in the envelope but cannot be resolved in the core. Only in globular clusters of low central concentration we can observe faint stars in the centre and this is

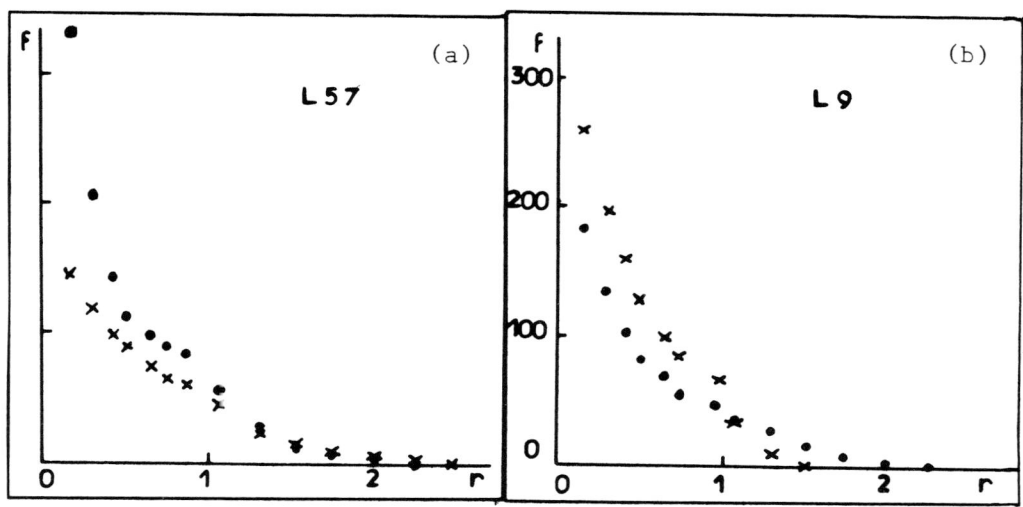

Fig. 1: Radial distribution of stars, $f = \frac{N}{A} - b$ for the clusters: (a) L57 and (b) L9. Circles represent counts from plate (R) and crosses from plate (SR).

rather the case for the SMC clusters. The concentration parameters of these clusters (Kontizas, 1985; Kontizas et al 1985) do not show any relation with mass segregation, at least up to the central area of the cluster with $r<0.10\ r_t$.

From theoretical models it is known that the segregation effect is most striking for the highly concentrated clusters but this is not the case for the SMC clusters, where $\log r_t/r_c$ has values 1.0 to 1.5. From the above it can be assumed that the observed values of the SMC cluster's central concentrations are rather low to show this effect and/or most of them have undergone violent initial relaxation.

Acknowledgements

We would like to thank the U.K. Schmidt Telescope Unit for loan of the observational material.

References

Da Costa, S.G. : 1982, Astron. J. 87, 990.
King, R. I. : 1975, IAU Symp. No 69, p. 99.
Kontizas, E. : 1985, Mem. Nat. Obs. Athens Ser. I, No 21, p.11
Kontizas, M., Theodosiou, E. and Kontizas, E. : Astron. and Astrophys. Suppl. Ser. (submitted).

PROPAGATING STAR FORMATION INDUCED BY SUPERBUBBLES

Minas Kafatos
Department of Physics, George Mason University, Fairfax, VA
Richard McCray
Joint Institute of Laboratory Astrophysics, University of
Colorado, Boulder, CO

A supernova explosion in the galactic disk creates a hot, low density cavity which persists much longer than typical lifetimes of supernova remnants. McCray and Snow (1979), Bruhweiler et al.(1980), Tomisaka, Habe and Ikeuchi (1980) and Kafatos et al. (1980) pointed out that repeated supernovae from a stellar association will produce an expanding shell of gas ($R \gtrsim 100$ pc) and offered extensive evidence for such shells -primarily in the form of H I shells.

A young association imparts mechanical power to the interstellar medium (ISM) via ionizing photons from O stars, stellar winds-primarily-from O stars and supernova explosions from O and early B-type stars. The input from B stars dominates after a few million years. An association typically contains ~ 30 stars earlier than B3 (Humphreys 1978; Garmany, Conti and Chiosi 1982). We find that typically 20% of the total energy available from the association is delivered during the first 10^7 years of its lifetime.

A supershell will expand with its radius given by the relation

$$R_S = 100 \text{ pc } (N_* E_{51}/n_0)^{1/5} t_7^{3/5} \qquad (1)$$

(cf. Weaver et al. 1977) where N_* is the number of stars formed by the association earlier than B3, E_{51} is the supernova energy in units of 10^{51} egrs, n_0 is the ambient number density of the ISM and t_7 is the age in 10^7 years.

When a supernova explodes inside a supershell its ejecta expand freely for $\sim 10^4$ years. Then an adiabatic blast wave is established (for $r \gtrsim 30$ pc) until the supernova shell encounters the supershell at which time it rapidly deaccelerates and merges with the supershell. The supershell itself expands adiabatically until radiative cooling becomes important (typically after $\sim 10^7$ years). Moreover, if the radius of the supershell is greater than the density scale height z_0, the supershell will burst through the H I medium and equation (1) is no longer valid.

Multiple supernovae from OB associations can accelerate the formation of gravitationally bound clouds. This happens because of the packing of the interstellar gas into dense supershells. Such clouds could be the sites of new star formation. Following the theory of Ostriker and Cowie (1981) we find that progressively smaller fragments become gravitationally unstable. Typically such fragments have a mass of 3×10^4 $M_\odot$ for timescales of a few tens of millions years. The minimum mass of the fragments depends strongly on the value of the magnetosonic speed in the shell a_S (to the 29/8 power of a_S). This is likely reduced in the supershell as result of the elevated density ($a_S \lesssim 1$ km s^{-1}). The development of the supershell accelerates the formation of gravitationally unstable fragments.

There is abundant evidence for giant shells in the Milky Way galaxy (cf. Heiles 1979). Such structures are also found in M31 and the Magellanic Clouds. In the irregular galaxies-such as the Magellanic Clouds-the metallicity is lower and radiative cooling would set in much later. Equation (1) would then be valid for values much larger than a few hundred parsecs. Moreover, the scaleheight in the irregulars is probably much greater than the corresponding scaleheight in a large spiral. We expect much larger supershells to be found in irregular galaxies than typically found in the Milky Way. Such a structure is loop IV which surrounds Constellation III in the LMC (Davies, Elliott and Meaburn 1976). It contains about 700 O stars as well as a number of supernova shells around its rim. There is probably extensive star formation associated with this supershell. Discussion of the detailed theory and some relevant observations is carried out elsewhere (McCray and Kafatos 1985, submitted).

REFERENCES

Bruhweiler,F.C., Gull, T.R., Kafatos, M., and Sofia, S.: 1980, Ap. J. (Letters) 238, p. L27.
Davies, R.D., Elliott, K.H., and Meaburn, J.: 1976, Mem. R. Astr. Soc. 81 p. 819.
Garmany, C.D., Conti, P.S., and Chiosi, C.: 1982, Ap. J. 263, p. 777.
Heiles, C.: 1979, Ap. J. 229, p. 533.
Humphreys, R.M.: 1978, Ap. J. Suppl. 38, p. 309.
Kafatos, M., Sofia, S., Bruhweiler, F., and Gull, T.: 1980, Ap. J. 242, p. 294.
McCray, R., and Snow, T.P.: 1979, Ann. Rev. Astr. Ap. 17, p. 213.
Ostriker, J.P., and Cowie, L.L.: 1981, Ap. J. (Letters) 243, p. L127.
Tomisaka, K., Habe, H., and Ikeuchi, S.: 1980, Progr. Theor. Phys. 64, p. 1587.
Weaver, R., Castor, J., McCray, R., Shapiro, P., and Moore, R.: 1977, Ap. J. 218, p. 377; (errata, 220, p. 742).

Non-LTE Analysis of Massive Stars in the Magellanic Clouds*

T. Gehren[1], D. Husfeld[1], R.P. Kudritzki[1], P.S. Conti[2] and D.G. Hummer[1,3]

[1] Institut für Astronomie und Astrophysik der Universität München
[2] Joint Institute for Laboratory Astrophysics, University of Colorado, Boulder
[3] Staff Member, Quantum Physics Division, National Bureau of Standards; Permanent address, JILA

The massive stars of the Magellanic Clouds are of considerable current interest with regard to questions of initial mass function, star formation mechanisms, stellar evolution with mass loss and the chemical evolution of galaxies. The effective temperatures, surface gravities and helium abundances of 6 main sequence O-type stars, obtained by fitting non-LTE model atmospheres to high quality spectra, are presented here; these are the first results from a long-term program to determine accurately the parameters and chemical abundances of massive stars in the Magellanic Clouds. The program stars were selected to be main sequence objects, according to the classification of Conti et al. (1985, in prep.), with He II λ 4686 Å in absorption, and to have minimal reddening and nebular emission. Spectra were obtained in 1984 December with the Cassegrain echelle spectrograph (CASPEC) and a CCD detector at the ESO 3.6 m telescope. A preliminary analysis of these spectra has been carried out by fitting the equivalent widths of He I λ 4471 Å and the profiles of Hγ and the Pickering lines (for details of the technique, see Kudritzki, 1980). The resulting values of effective temperature and gravity are given in Table I, along with the identification, spectral type and m_V of each star.

Table I

	Star	Sp. Type	m_V	T_{eff}	log g	M_V	log $L/L_\odot$
LMC:	Sk172-66°	O3/O4	13.11	50000±3000K	4.20±0.2	−6.20±0.1	6.18±0.1
	Sk 69-70°	O3/O4	13.90	46500K	4.10	−4.82±0.1	5.54±0.1
	LH 81-43°	O7	13.6:	45000K	4.05		
SMC:	AV 388	O4	14.12	47000K	4.00	−5.21±0.1	5.71±0.1
	AV 243	O5/O6	13.87	45000K	4.00	−5.43±0.1	5.74±0.1
	AV 239	O8.5	13.77	35000K	3.75	−5.53±0.1	5.45±0.1

The absolute magnitudes M_V are obtained using distance moduli of 18.6 for the LMC and 18.9 for the SMC and A_V = 3.1 E_{B-V}. The stellar radii

* Based on observations collected at the European Southern Observatory, La Silla, Chile

In Figure 1 we compare values of T_{eff}, log g and L with the evolutionary tracks of Pylyser, Doom and de Loore (1985), which have been computed for LMC and SMC abundances (Z = 0.0083 and 0.003 respectively) allowing for mass loss and overshooting. The upper panel contains the L, T_{eff} diagram, while the bottom panels gives the log g, T_{eff} form. The degree of agreement between the mass of a particular star inferred from the two diagrams gives a measure of the consistency between the spectroscopic determination of T_{eff} and g, the photometric data, and the evolutionary theory. Within the error bars, we see that we have substantial agreement, which was not the case using the earlier tracks of Brunish and Truran (1982), Maeder (1980) and Hellings and Vanbeveren (1981).

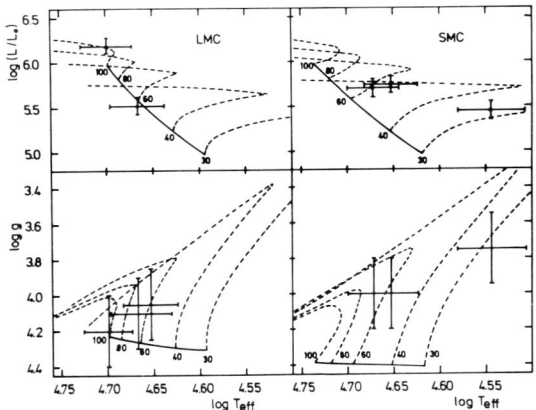

Fig. 1 (see text)

The preliminary results given here may be modified slightly when the observations are re-analyzed using models containing more than one free level of He and He$^+$ in NLTE, and accounting for wind blanketing, which according to Abbott and Hummer (1985) can change the inferred temperatures by several thousands degrees for the stars considered here. As the two effects cause changes of opposite sign, and since the decrease in T_{eff} from wind-blanketing will become smaller with Z, we expect the temperature change to be small, and of unknown sign.

A more thorough analysis of these stars and of other, more highly evolved, O-stars in the MC will be published in due course.

D.G. Hummer acknowledges a "Senior U.S. Scientist Award" from the von-Humboldt Stiftung. We are grateful to Dr. P. Hellings for providing the numerical values of the evolutionary tracks.

References
Abbott, D.C., and Hummer, D.G., 1985, Ap.J. (in press)
Brunish, W.M., and Truran, J.W., 1982, Ap.J. Suppl. **49**, 447
Hellings, P., and Vanbeveren, D., 1981, in *Effects of Mass Loss on Stellar Evolution*, ed. Chiosi and Stalio, Reidel, Dordrecht
Kudritzki, R.P., 1980, Astron. Astrophys. **85**, 174
Maeder, A., 1980, Astron. Astrophys. **92**, 101
Matthews, T.A., and Sandage, A.R., 1963, Ap.J. **138**, 30
Pylyser, E., Doom, C., de Loore, C., 1985, Astron. Astrophys.(in press)

METALLICITY EFFECT ON ABSOLUTE MAGNITUDE DETERMINATION

P. Dubois
Observatoire de Strasbourg

In our Galaxy, it is possible to determine the absolute magnitude of a star by mean of the intensity of the hydrogen lines, if we know the spectral type of the star or its temperature (see for instance Petrie, 1965; Hutchings, 1966; Balona and Crampton, 1974; Crawford, 1973, 1978; ...).

$$M_V = f(W_H, Sp.T.)$$

Such a relation (full lines) is represented in the lower part of the fig. 1 for galactic cluster stars (triangles). The absolute magnitude of a star is obtained from the distance modulus of the cluster corrected by the mean absorption of the cluster. The distance moduli are only determined by fitting the main sequence with the ZAMS. A diagram can be obtained for the Small Magellanic Cloud also (upper part of the fig. 1 - dotted lines) taking into account the apparent visual magnitude of the stars (circles) corrected for the absorption of the stars and a mean distance modulus of 19.0 magnitude for the SMC. The photometric data are from Azzopardi and Vigneau (1975) and the spectroscopic data from Dubois et al. (1977) and Dubois (1982).

We observe that the two sets of relations are not the same. The temperature effect is much higher in the SMC stars than in the galactic ones.

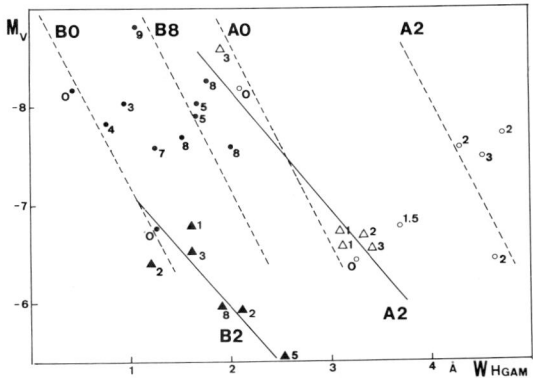

Figure 1. Comparison of the SMC relation (dotted lines) with the galactic one (full lines). Numbers are subclasses of spectral types.

This means that the hydrogen lines are generally stronger in the SMC than in the galactic ones. We may suspect errors in the equivalent width, but many verifications have been made which exclude such an explanation. A similar effect can be seen in the data of Hutchings (1966). Moreover, the photometric narrow-band index H-beta gives similar results with the data of Osmer (1973).

If we use only a restricted range in spectral type to make the fit of the two calibrations, in order to obtain the distance modulus of the SMC, we obtain different distance moduli as a function of the mean spectral observed in the SMC and in the reference stars. The results are presented in table 1 for the preceding calibrations.

Table 1. Distance modulus of the SMC as a function of the mean spectral type used.

Gal.	H-gamma			H-beta		
	B2	A0	A2	B2	A0	A2
SMC						
B0	19.3			20.0		
B3	18.5			19.5		
B5				19.0		
B8	17.6			18.0	19.2	
A0		19.0			18.4	
A2		16.8				16.8

This effect could explain that some spectroscopic distance moduli of the SMC were too small compared with those determined by other methods (Buscombe and Kennedy, 1962; Westerlund et al., 1963; Hutchings, 1966).

Previous mentions of abnormal strong hydrogen lines in the Magellanic Cloud stars have been given by Fehrenbach and Duflot (1972), van den Bergh (1976) or Humphreys (1983). Does this come from the small metal content of the SMC ? It is one of the first explanation -see for instance Tully and Wolff (1984)- but no atmospheric model has yet explained such a difference in the stellar spectra.

REFERENCES

Azzopardi, M. and Vigneau, J. : 1975, **Astron. Astrophys. Suppl. Ser. 22,** 285
Balona, I. and Crampton, D. : 1974, **Mon. Not. Roy. Astron. Soc. 166,** 203
Buscombe, W. and Kennedy, P.M. : 1962, **J. Roy. Astron. Soc. Canada 56,** 113
van den Bergh, B. : 1976, **IAU Coll. no.37,** p. 13
Crawford, D.L. : 1973, **IAU Symp. no. 54,** p. 93
Crawford, D.L. : 1978, **Astron. J. 83,** 48
Dubois, P., Jaschek, M. and Jaschek, C. : 1977, **Astron. Astrophys. 60,** 205
Dubois, P. : 1982, **Astron. Astrophys. Suppl. Ser. 48,** 375
Fehrenbach, Ch. and Duflot, M. : 1972, **Astron. Astrophys. 21,** 231
Humphreys, R.M. : 1983, **Astrophys. J. 265,** 176
Hutchings, J.B. : 1966, **Mon. Not. Roy. Astron. Soc. 132,** 433
Petrie, R.M. : 1965, **Publ. Dom. Astrophys. Obs. 12,** 317
Tully, R.B. and Wolff, S.C. : 1984, **Astrophys. J. 281,** 67
Westerlund, B.E., Danziger, I.J. and Graham, J. : 1963, **Observatory 83,** 74

BRIGHT STARS IN FIVE LMC STAR CLUSTERS

E. Kontizas*, E. Xiradaki**, and M. Kontizas**
* National Observatory, ** University of Athens

The bright stars of five LMC clusters were classified for deriving the distribution of various spectral types. The studied clusters are very young (NGC 2098) young (NGC 1818, NGC 2157) intermediate (NGC 1831) and old (NGC 1806) (Van den Bergh 1981). The spectral classification of the stars was carried out using film copies of the 1.2 m Schmidt telescope objective prism plates. Medium dispersion (830 Å at Hγ) unwidened YJ and widened UJ and low dispersion (2440 Å at Hγ) UJ were examined by means of a binocular microscope. Details of the criteria used for the classification are described by Kontizas et al (1985).

For each cluster a circular area was examined inside its cluster tidal radius. The magnitude range of the studied stars is $11.5 < m_r < 17.5$. The stars in the innermost part of the clusters were not classified because of the overlapped images. The figure 1 shows the number of stars per spectral type a) NGC 2098, b) NGC 1318, c) NGC 2137 d) NGC 1831, e) NGC 1806. The dashed line represents the distribution of the various spectral types of the adjoining fields normalized to the cluster area. The stars classified as B represent stars O and B since the method used does not permit us to distinguish them with confidence.

From the diagrams it can be seen that all clusters in our sample show a relatively large number of M type stars even for the very young cluster. On the contrary the fields have not early type stars, within the observed magnitude range in all five regions. The bright late type stars found here confirm the existence of evolved stellar component even for the very young and young clusters, that may mean that the studied LMC cluster stars are either older than those of the young galactic clusters or much more massive.

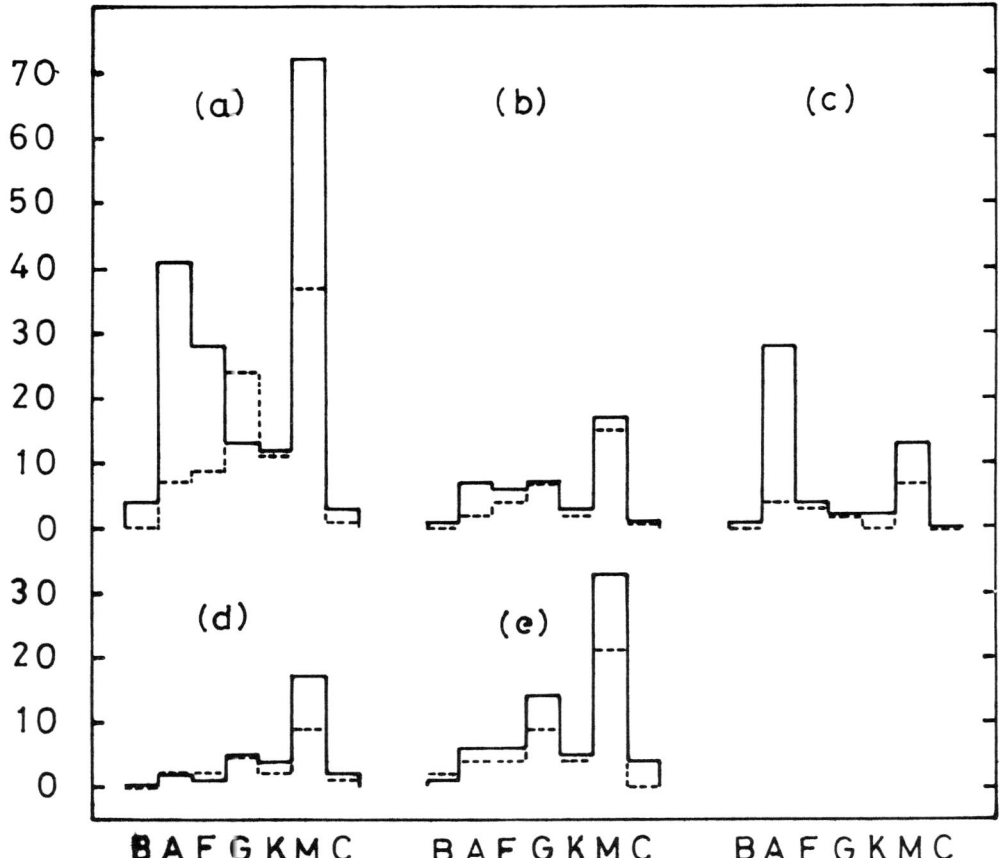

Fig. 1: Number of bright stars per spectral type (solid line) for a) NGC 2098, b) NGC 1818, c) NGC 2157, d) NGC1831 and e) NGC 1806 and their adjioning fields (dashed line).

AKNOWLEDGEMENTS

We would like to thank the 1.2 m UK Schmidt Unit for loan of the material.

REFERENCES

Kontizas, E., Dapergolas. A., Kontizas, M. 1985 (in preparation).
Van den Bergh, S. :1981, Astron. Astrophys. Suppl. 46, 78.

ON THE ARM PATTERN OF M33

GEORGI RAIKOV IVANOV, PETER Z. KUNCHEV
UNIVERSITY OF SOFIA, DEPARTMENT OF ASTRONOMY, SOFIA, BULGARIA

By comparison of U and B plates of M33 we extended the numerations of Humphreys and Sandage (1980) adding 54 new OB associations. The associations identified by HS are outlined with continuous line and ours are represented with dash on Figure 1. Many of the identified by us associations are near the nucleus of M33. Due to U plate used by us the background in the central region is fainter than that on the B plate and the associations are distinguished very well on U plate (Kunchev and Ivanov,1984). Our associations of the outer region of M33 consist mainly of relatively fainter young B stars.

The position angle of the major axis PA = $22° \mp 5°$ and the inclination of the plane of the galaxy i = 57° are derived using 197 associations. The logarithmic spirals with a pitch angle i = arc tg 0.69 = 34°.6 well fit the distribution of the associations. The spiral arms in the southern and in the northern part are symmetrically disposed one another except arm N4.

We explained the spiral arm structure without any assumption for warp plane. It is possible for the plane of M33 to be slightly warped in the central region (Maucherat et all., 1984).

REFERENCES :

Humphreys, R. M., Sandage, A. 1980, Ap. J. 44, 319.
Kunchev, P. Z., Ivanov, G. R. 1984, Astrophysics and Space Sci., 106, 371.
Maucherat, A. J., Dubout-Crillon, R., Monnet, G., Figon, P. 1984, Astron. Astrophys. 133, 341.

C. W. H. De Loore et al. (eds.), Luminous Stars and Associations in Galaxies, 419–420.
© 1986 by the IAU.

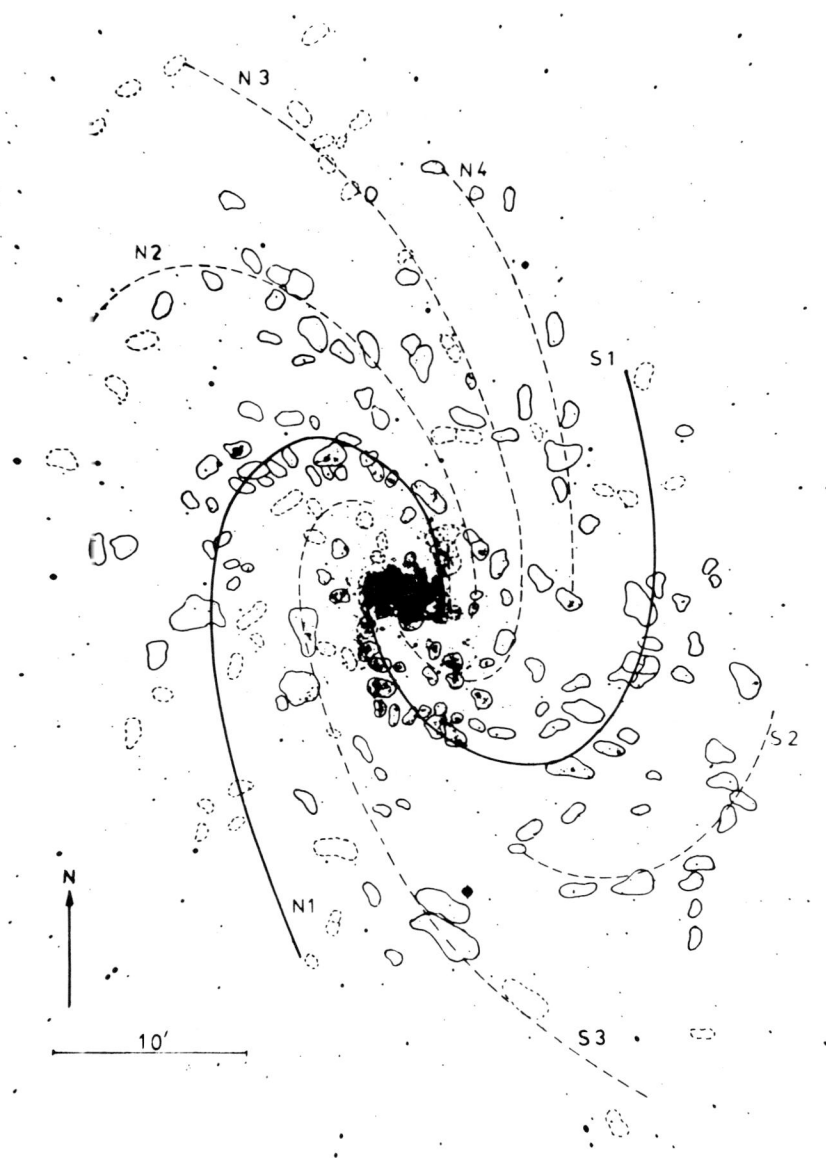

Figure 1. The systems of logarithmic spirals superimposed on B plate. The stellar associations indicated by Humphreys and Sandage (1980) are outlined with line and ours - with dash.

CONVECTIVE OVERSHOOTING: THE UPPER MASS LIMIT FOR STARS UNDERGOING CORE HE-FLASH

G. Bertelli [1,2] and A. Bressan [3]

1) National Council of Research of Italy, C.N.R. - G.N.A.
2) Institute of Astronomy, Padova, Italy
3) International School for Advanced Studies, Trieste, Italy

Theoretical Rationale

With classic models of stars, in which overshooting from convective cores is not taken into account, the transition mass, Mtr, separating stars which undergo core He-flash from stars which ignite He-burning nonviolently in a non degenerate core, is set around 2.2 M☉ (Iben 1967) for Pop I chemical composition (X = 0.700, Z = 0.020). This value is known however to depend on the chemical abundances, being lower at increasing Y and decreasing Z (Wagner 1974; Sweigart and Gross 1978). However, in recent years many independent arguments have indicated that convective overshooting from the central cores may play an important role in stellar evolution (Bertelli et al 1985). In order to assess the dependence of Mtr on overshooting, we have computed evolutionary sequences of 1.4 M☉, 1.5 M☉ and 1.6 M☉ with chemical composition X = 0.700 and Z = 0.020 adopting the description of convective overshooting formulated by Bressan et al (1981) for their parameter $\lambda = l/Hp = 1$. As shown in the T_c vs ρ_c diagram of Fig 1, while the tracks of 1.4 M☉ and 1.5 M☉ stars deeply penetrate into the region of high degeneracy and likely undergo core He-flah, the 1.6 M☉ star succeeds in igniting helium in non degenerate conditions. This means that the classic value of 2.2 M☉ for Mtr is now lowered to the mass range 1.6 to 1.5 M☉. The HR diagram of the 1.4 M☉, 1.5 M☉ and 1.6 M☉ stars in presence of convective overshooting is shown in Fig 2. The region of stationary core He-burning for the 1.6 M☉ and the extension of the red giant branches for the 1.4 M☉ and 1.5 M☉ stars are also indicated.

Observational Counterpart

The luminosity function of evolved stars of clusters with age in the range 1 to 2 x 10^9 yr may constitute a powerful test on the actual value for Mtr. In fact, the luminosity function of red stars is expected to suddenly change passing from clusters having a turn-off mass lower than Mtr to clusters with a turn-off mass greater than this. In the recent study by Barbaro and Pigatto (1984) of a large number of old galactic open clusters with about solar metallicity, it has been suggested that the above transition should occur at approximately 1.5 M☉ instead of the classic 2.2 M☉.

Those authors indicated in convective overshooting during the core H-burning phase the way out of the above disagreement between standard theory and observation. This is strongly supported by the present evolutionary computations. In addition to this, since the dependence of Mtr on overshooting during the core H-burning phase, is not biased by other uncertainties affecting subsequent evolution (neutrino cooling and semiconvection and/or overshooting during the core He-burning phase), this finding may be used to infer the mass size of convective cores, hence the efficiency of convective overshooting, in real stars.

A more complete description of the above results will be published elsewhere (Astron. Astrophys.).

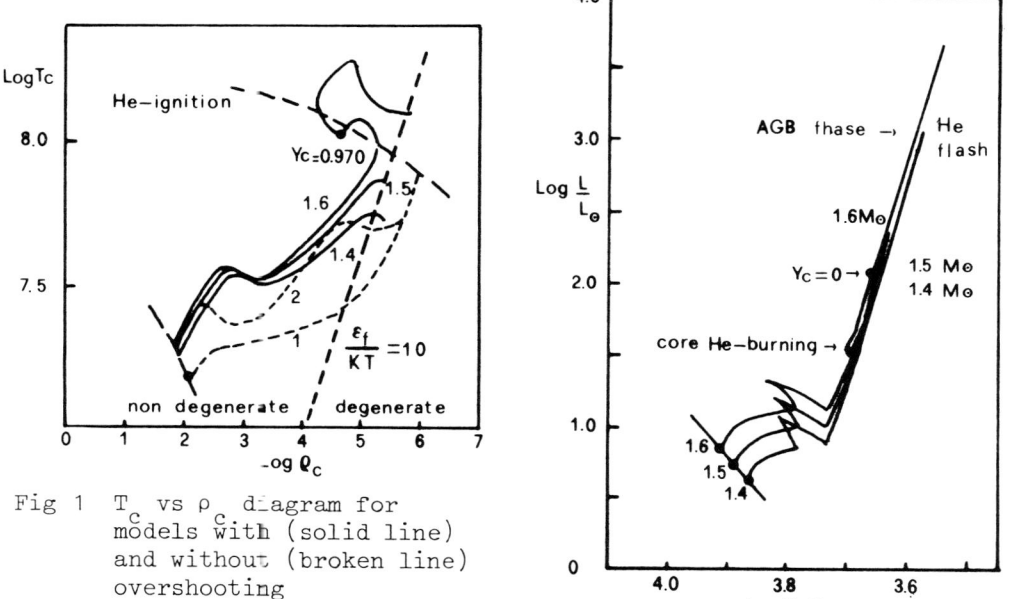

Fig 1 T_c vs ρ_c diagram for models with (solid line) and without (broken line) overshooting

Fig 2 The HR diagram for models with overshooting

References

Barbaro, G., Pigatto, L., 1984, Astron. Astrophys., 136, 355
Bertelli, G., Bressan, A., Chiosi, C., 1985, Astron. Astrophys., in press
Bressan, A., Bertelli, G., Chiosi, C., 1981, Astron. Astrophys., 102, 25
Iben, I. Jr., 1967, Ann. Rev. Astron. Astrophys., 5, 571
Sweigert, A. V., Gross, P. G., 1978, Astrophys. J., 36, 405
Wagner, R. L., 1974, Astrophys. J., 191, 173

THE FINAL FATE OF MASSIVE STARS

H. Schild
Geneva Observatory
CH-1290 S A U V E R N Y
Switzerland

PROCEDURE

The cluster membership of Wolf-Rayet stars, supernova remnants and pulsars is used to study the late stages of stellar evolution of massive stars.

In a hypothetical stellar cluster with strictly coeval members, we find only those stars in these evolutionary stages for which the lifetime ℓ roughly equals the cluster age t. The evolved cluster members must therefore originate from a small range of initial mass M_i for which $\ell(M_i) \approx t$. We have calculated this relation between the initial mass of cluster members near the SN explosion and the cluster age (Schild and Maeder, 1984).

Information on the cluster membership of individual WR stars, SNR's and pulsars can be found in the literature (e.g. Lundström and Stenholm, 1984; Allakhverdiyev et al.,1983; Schild and Maeder, 1985). The age of the parent associations can be obtained e.g. by isochrone-fitting. Clearly, an age spread present in the parent clusters leads to increasing error bars in the initial mass determinations.

RESULTS

The results on the initial mass of WR stars, SNR's and pulsars are summarized in Fig. 1. It should be kept in mind that the emerging picture of the evolution of massive stars is tentative and may be incomplete.

We fistly note that WR stars can form from $M_i \gtrsim 20\ M_\odot$, whereas WC stars originate from $M_i \gtrsim 35\ M_\odot$. We also find that WNL and WCE stars are initially more massive than WNE and WCL stars (Schild and Maeder, 1984). At the present time we know of WR stars with initial mass as high as $\sim 100\ M_\odot$ (Humphreys et al., 1985).

A remarkable feature in Fig. 1 is the presence of both SNR's and neutron stars for initial mass up to $\sim 50\ M_\odot$. There seems to be a tendency for morphological C-type remnants (cf. Weiler, 1983) to originate

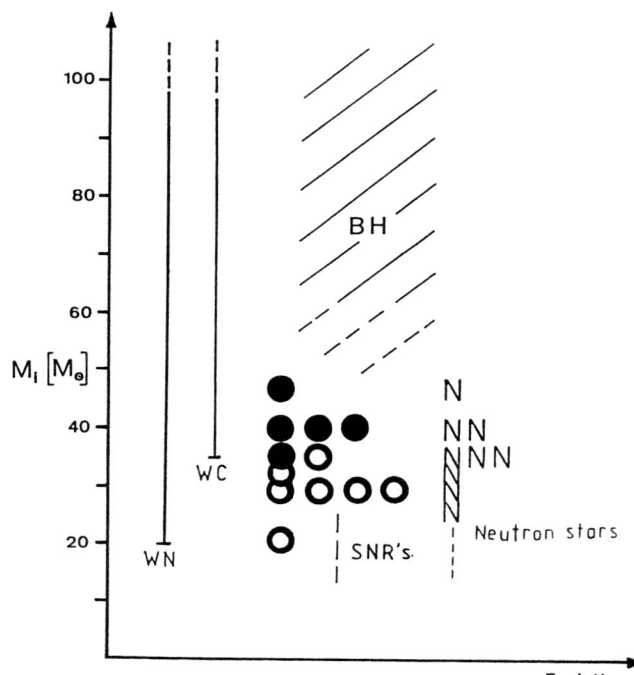

Figure 1:

The initial mass of WR stars, neutron stars and the progenitors of supernova remnants. The filled circles represent C-type remnants, the rings shell-type remnants. The N's stand for neutron stars (pulsars and compact X-ray sources in SNR's). BH indicates the range where possibly black holes are the final outcome of stellar evolution.

from initially more massive progenitors than shell-type remnants (Schild, 1985). It is tempting to speculate that there is an evolutionary connection between WR stars and C-type supernova remnants.

The final phase of stellar evolution for stars with $M_i \gtrsim 50\ M_\odot$ is difficult to establish. From Fig. 1 we see that they do not seem to produce SNR's and neutron stars. Possibly the internal evolution leads them to the formation of a black hole. This is in agreement with the findings of van den Heuvel and Habets (1984) who concluded from the study of two X-ray binaries, that the initial mass limit for black hole formation has to lie between 40 and 80 $M_\odot$.

REFERENCES

Allakhverdiyev, A.O., Amnuel, P.R., Guseinov, O.H., Kasumov, F.K.: 1983, Astrophys. Spa. Sci. 97, 261
Humphreys, R.M., Nichols, M., Massey, P.: 1985, Astron. J. 90, 101
Lundström, I., Stenholm, B.: 1984, Astron. Astrophys. Suppl. 58, 163
Schild, H.: 1985, in preparation
Schild, H., Maeder, A.: 1984, Astron. Astrophys. 136, 237
Schild, H., Maeder, A.: 1985, Astron. Astrophys. 143, L7
van den Heuvel, E.P.J., Habets, G.M.H.J.: 1984, Nature, 309, 598
Weiler, K.W.: 1983, IAU Symp. 101, 299

POSTER PAPERS   -   SESSIONS 5 and 6.

Chairman: P. CONTI.

P. Conti.

   We have a large number of papers dealing with stars and associations in nearby galaxies. In particular there are papers dealing with the stellar content of giant HII regions, for instance 30 Doradus and NGC 3603 and we have already seen other results on this. In the case of 30 Dor we have seen that it is made up of normal stars although at the extreme upper mass range of say up to 200 solar masses.

J. Graham.

   I was impressed by the paper on CCD observations of young stellar associations in the MC's by Lortet et al. which states" it is clear that a number of multiple systems, at any rate those with a separation of 0.5 arcsec or less, are still unrecognised" and this made me feel that is just like R136a - we have to learn a lot more about stellar multiplicity in the MC's. We just cannot assume that every object we see is single.

M.C. Lortet.

   I would add that among the multiple systems discovered is one of the candles for distance determination in the LMC. A significant fraction of these candles might be found to be composite, though we did not yet observe them.

A. Sandage

   The question of whether low mass stars ($M < 2 M_o$) are formed along with the O stars in associations and HII regions is, of course, central for the problem of how much time it will take for a given region of a galaxy to convert all its gas to stars and therefore whether galaxy evolution has taken place yet in a major way along the Hubble sequence from Sc to Sa types.
   But the question cannot be answered by looking for faint MS stars accompanying O star formation because, in any given cluster of a given age, there is only 7 magnitudes of main sequence. Fainter than 7 magnitudes below the brightest O star, the MS terminates due to not enough time for the lower mass stars to contract, first along the Hyashi track and then along the nearly horizontal radiative track to the MS. Hence, if low mass objects are indeed present they would not be MS

objects but rather T-Tauri stars, cocoons or other pre-stellar objects, not seen at optical wavelengths. Merle Walker's 1950's study of NGC 2264 is the prototype. Hence, mere brute force methods of faint searches for red stars in giant HII regions will not answer this crucial point of whether bimodal (Eggen's term) star formation exists, or whether there is a low-mass cut-off, say at 3 M_o as Silk expects. If it is correct that no such low mass stars accompany each O star formed (in numbers given by the "general" luminosity function) then M33 can last as a star-producing galaxy for at least one more Hubble time. But if, in M33 a number G, K and M main sequence dwarfs are produced for each O star, M33 will tie up its remaining gas in such stars (given its present SFR) in only 10^9 more years. If this seems impossible, the conclusion via this argument would be that stars with mass > 2 M_o accompany each O star formed -i.e. there is a low mass cut-off to the formation LF for massive stars.

Conti:

I am worried a little about the background given by the galaxy itself.

Appenzeller:

I have two points. Firstly nobody mentioned T Tauri stars. Two of my colleagues have proposals in for ESO to try to detect these objects in 30 Dor and it should be possible to do this. Secondly, to what extent is it possible to learn about the low mass content of 30 Dor from its integrated spectrum.

Rosa:

From the integrated spectrum it is almost impossible to disentangle the contribution from low mass stars and the red supergiants. Spectra taken in the 8000A region can be used to look for the signatures of RSG's. Also one can try to use 7 colour narrow band photometry excluding the nebula emission lines simulating Strömgren colours going into the near infrared.
This can hopefully be used to search for the low mass stars and whether these are on the MS and possibly have been formed earlier, or whether those are near-IR objects. I hope this can be done.

Moffat.

On the basis of what I saw in the Chu et al. paper the luminosity function is rising rapidly down the 17th magnitude and then it drops. This is a clear case of the incompleteness problem and I think that as it stands now, there is just no indication that the numbers drop

DISCUSSION

off for faint stars. There is no indication of a seven magnitude difference at all, we are only going down to about 3 magnitudes below the turnoff.

Conti.

One of the posters sits a little bit by itself in character, and this is the one by Zinnecker on how to make a superstar. The idea here is that in the central core of a dense cluster you will begin to have star-star interactions and you can eventualy have coalescence and thus build up that way. Also it seems that you can make runaways at the same time as you make the very massive stars since you need three-body interactions.

Appenzeller:

I should point out that he is only dealing with stars up to 200 solar masses and not up to thousands of solar masses that have been postulated.

Zinnecker:

I think that 200 $M_\odot$ is "super" enough! I would like to add that in relation to the dense clusters like 30 Dor and NGC 3603 one has to be aware that N-body evolutionary effects may be potentially very important.

Conti:

A number of papers have dealt whit the evolution of single stars and binaries. One of these papers said something about V729 Cyg.

Sybesma

You are referring to my paper on the influence of overshooting on the evolution of binaries. The problem with V729 Cyg is the nature of the smaller star, whether it is an Of star or a He star accreting from the more massive component showing up as a WR star with a strong hydrogen component in the spectrum. If we take evolutionary tracks including overshooting and you consider the masses which are determined by Conti and Bohannan, and Massey and Conti, you have to conclude that it has to be a He star, that is a leftover He core which was an overshooting core. So if the model proposed by J.M. Vreux in A and A this year is correct, the smaller star is not an Of star.

Conti:

I was very amused at this because I have also seen Vreux's paper where from his better H-alpha data for this system, he finds in fact what is now suggested theoretically, that the lower mass star, which is very luminous but shows H, is in fact a helium burning star, with an accretion of hydrogen such that it masquarades as an Of star. The irony of this is that when Bruce and I wrote that paper the title was "BD+40 4220 , a star on its way to become a Wolf-Rayet", the irony is that it seems it's already been there and it's now on its way back! It's nice that there is now theoretical confirmation of this.

Sybesma:

I would also stress that the theory confirms that the He star is accreting in the wind and not as the result of Roche lobe overflow. The system is too large for Roche lobe overflow given the considered mass range.

Conti.

In fact, if this is correct, then this is the first evidence that I know that we really have wind accretion.

Vanbeveren:

Some eight years ago a couple of papers appeared in the literature critically discussing the physical meaning of the Roche equipotentials for massive binaries and especially the influence of radiative forces which are extremely important in massive stars. The papers on massive binary evolution presented in this poster session are all using the classical Roche lobes. I wonder what is the meaning of the results, in particular the results of accretion. Do they have to be considered as "interesting" theoretical results" without any real application?

Conti.

I have heard this type of worry expressed at several of our recent conferences as Danny has done today, and on each occasion nobody really responds! Moving on, I think there may be some comments on the poster dealing with "Observed dynamical properties of star clusters in the LMC".

DISCUSSION

Kontizas.

The very important thing that has come out of the observed dynamical study of the populous young star clusters in the LMC is that they are found to be very extended. In other words the young disk clusters are found very much larger than the disk clusters in the SMC and both are larger than corresponding clusters in our Galaxy.

Graham.

I expect this difference is also telling us about the large burst of star formation at intermediate ages (say 10^8 years) in both Magellanic Clouds.

Conti.

I would draw attention to the important paper on metallicity effects on absolute magnitude determination by Dubois and this calls our attention to the fact that using H-gamma calibration, which is one of the methods used to determine the distance of the Magellanic-Clouds, the line strength will depend a little bit on the metallicity and thus the calibration that is derived from stars in our Galaxy may not be strictly applicable, particularly for the SMC. The stars we are concerned with are B and A supergiants. This paper thus raises the issue of whether the spectral classification in fact has a dependence on metal abundance.

Kudritzki.

I think the problem is not only the one you mention but also normally in these supergiant stars the hydrogen lines may be affected by the stellar wind, so what you normally calculate is photospheric absorption lines which will now be partially filled in by emission from the wind. If for example one takes the case of the SMC the lower metal abundance will give a less strong stellar wind and thus one would really be seeing mainly just the photospheric lines.

Feast:

The problem encountered by Dubois appears to be related to the abnormally strong H lines found in some early type stars in the Clouds (but not in all early type stars) first by Fehrenbach and later by Humphreys. The effect in these stars is so strong that they were first thought to be foreground objects. Do any of the theorists present have any ideas on how to explain this interesting effect?

Kudritzki.

I think the explanation I gave can work for these stars. The galactic objects will have the more filled in H-lines because they would have stronger winds. In the SMC the winds are smaller and so would be the filling in. In my thesis on AI supergiant non-LTE model atmospheres 12 years ago I found the strange effect that the Balmer jump and the H lines became stronger if one slightly increases the helium abundance. This is due to the increase of mean atomic weight, which causes an increase of the density in the photosphere. In consequence one has more free electrons and more hydrogen atoms around unit optical depth, although the H abundance is reduced. Now according to what we heard on stellar evolution, in the B and A supergiants phase an enrichment of helium in the photosphere is reasonable and could explain the effect. I would like to see the acquisition (which is now feasable) of high quality CASPEC spectra of these objects and a detailed atmosphere analysis. This should help to clear up the problem.

Conti:

There are several posters dealing with star formation which I am sure will inspire comments.

Graham:

I would like to comment on Pennington's paper about NGC 5128. You will remember that it is the closest of the giant radio galaxies that we can see in detail, which is not available for other examples of this type. The paper shows very interestingly that star formation is occuring across the whole of the dark lanes and I would emphasise that in places like radio galaxies one will find star formation occuring in rather bizarre circumstances, which will not be found in normal galaxies. I would also mention in connection with star formation in 5128 that we also find these loose chains of blue supergiants stars which are about 50 kpc out from the center of the galaxy round to the direction of the radio lobes. It is clear that these stars must have been made out there and somehow they are connected with some sort of compression of material by whatever it is that is powering the radio lobes.

Zinnecker:

Concerning star formation in Cen A; this galaxy is often thought to be a merger. Does your work add support for the merger hypothesis?

Pennington:

Cen A seems to be the result of a merger between a large gas

DISCUSSION

cloud and an elliptical galaxy. My observations suggest that the cloud is still augmenting the dust lane. The importance of Cen A is that it seems to present a simple example of star formation in that HII regions first form downstream from the infall region and later form across the entire height of the lane. Star formation, or at least the HII regions, dies in one rotation. This is a much more simple picture than that present in spiral galaxies.

Graham:

The evidence for a merger rather favours the merger of a whole galaxy with NGC 5128 rather than just a gas cloud, because its very hard (and I don't know how you would do it) to produce a beautiful system of ripples or shells that we have round 5128 without having a stellar component for the infall.

Pennington:

But those ripples are not the same colour as the O-stellar component, they are slightly redder than the O stars. I dont know how we would produce the ripples, but if that were the case why are there ripples in about 123 other elliptical galaxies many of which show no evidence for mergers in terms of dust or star formation. You see them in apparently very normal elliptical galaxies not just in anomalous cases like Cen A.

Graham:

They are normal elliptical galaxies - merged!

Kaufman:

I would like to comment on the oscillating star formation. You showed that under certain conditions, self-sustained oscillations develop in the star formation rate and you plotted this as a function of a dimensionless time variable. For comparing your results with starbursts in dwarf galaxies one would like to convert to time in years to know what the physical variables are involved in converting from dimensionless to dimensional time, so that one can do this for different environments.

Bodifee:

The dimensionless time can be converted to a time variable expressed in years by multiplying by a factor that is essentially the mean lifetime of stars triggering further star formation. If it is assumed that only massive stars are able to induce star formation then the factor should be approximately 10^7 y.

Conti:

We also have time for further discussion of the review papers.

Walborn:

I think it is important to pursue the discussion of the relationship between WN and WC stars and the population of the most luminous HII regions because we seem to have a disagreement between the evolutionary scenarios and the observational data. From the observations established first by Roberts many years ago, the WN stars appear more concentrated in clusters than WC stars and it is observed that the stellar content of the luminous HII regions are completely dominated by WN stars. My understanding of the models is that the most luminous stars should get to the WC phase, whereas the data suggest that the most massive WN stars do not get to become WC stars. So I think there may be a fundamental problem between the observational morphology and the expectations from models.

Vanbeveren:

A question to the observers. Is it more diffuclt to recognise a WC star as association member? Is there any selection effect problem?

Walborn:

No, in fact it should be easier since the emission lines are stronger.

Massey:

I think I should point out that the number of WR stars has approximately doubled since the earlier results referred to. Stenholm and Lundström have published papers giving lists of association members and I cannot remember them having any difference between the WN and WC stars in this respect. Certainly in M33 the OB associations show no difference between the number of WN and WC's found.

McGregor:

WC9 stars are usually heavily reddened so we may not be picking them up in the Magellanic Cloud.

DISCUSSION

Conti:

OK, the suggestion is that we are missing all the WC9 stars in the LMC. I suppose that is possible but there are effectively no WC8, WC7 or WC6's. I do not think those have been missed on the basis of your argument. I suppose it is just possible that the WC9's have been missed.

Niemela:

In our galaxy we have WC stars which have blown away their HII regions. Neutral H radio maps show a huge hole surrounding the WC stars, so we do not see them within the HII region.

Conti:

My personal belief is that it seems to me that the most massive stars may look like WN stars during a good fraction of the H-burning lifetime. Some of the stars Nolan showed the other day clearly have H present yet we identify them as WR stars. What I do not know, and I don't think there is any quantitative work on this, is whether those stars would show up in the integrated light or not. This goes back to the question of what is a standard WR star - there is a whole range of emission line strengths in both WN and WC stars. The very luminous stars, for the most part, have rather weaker lines than a star like HD 50896 or V444 Cyg which are clearly more highly evolved.

de Loore:

If you consider stars of all masses, then we have claimed that you have first formation of lower mass stars in an association and afterwards, say twenty million years later, bursts of more massive star formation occurs. It all depends on what time you are looking at the association. In my opinion you can have all ratios of WC/WN and WR/O-star depending on the moment of observation.

Sreenivasan:

Is it definitely established thah WN-WC is an evolutionary sequence?

Conti:

Oh no, I thought I'd stressed that in my talk. I think you can say the following: the WC because of their composition are more highly evolved than WN stars.

Sreenivasan:

That's fair enough, but does every WC have to go through a WN stage?

Conti:

Nobody knows the answer to that.

Morgan:

Are the evolutionary models capable of explaining WC stars with an absolute visual magnitude of -1, such as the couple of stars I have found in 30 Dor?

Conti:

Apparently David has found one WC star in the LMC with an apparent magnitude of about 18 but you don't have the reddening yet. It seems there are three possible explanations out of this: (i) it is anomalously heavily reddened which is not unprecedented for a WC star in our Galaxy (ii) it is the central star of a Planetary nebula and the nebula has not yet been detected, and the third (iii) is that we had all better go back to our drawing boards!

Maeder:

I would like to comment on a possible simple way to explain the lack of WR stars in some HII regions. Firstly, remember that in most cases the central conditions of stars "ignore" what happens at the surface and the central evolution inexorably leads the star to a supernova explosion. The observational status of the star may be WNL, WNE, WC or WO according to the mass loss rates involved. Remember that there is some considerable scatter in mass loss rate for all the kinds of stages considered. For relatively low mass loss rate (say 10^{-5} M_o/y in the WR stage) the star may reach the SN explosion when its surface conditions make it still a WN star, while for larger mass loss (say 10^{-4} M_o/y in the WR phase) the peeling-off of the outer layers will have largely preceded the WC stage.

Walborn:

I think that may well be the answer of the whole problem. There may be some mechanism whereby these most massive stars become SN after the WN stage, and this is not included currently in models. I think

DISCUSSION

what we observe in the Carina nebula is possibly direct evidence for that. It seems to be possible that Eta Car has already been a narrow-lined WN star.

Renzini:

I do not think Andre's explanation is going to give the answer for the discrepancy between the theory and the data. The point is that if you have a lower mass above which stars can become WR stars you will also have, according to all the parametrisation of mass loss used to date, another mass limit above which WN stars evolve into WC stars. That is so in all evolutionary scenarios so far proposed - the most massive stars become WN and then WC, at intermediate mass range you have only WN stars and the SN takes place during the WN phase. This is in disagreement with the observational evidence in other galaxies, according to which the youngest associations do not have WN or at least in the proportion that they are found in the solar neighbourhood, where it is 1:1.

SESSION 7.

LUMINOUS STELLAR CONTENT OF GALAXIES-INTEGRATED
PROPERTIES-I.

Chairman : R.KUDRITZKI.

1. J.G.HOESSEL: Luminosity Functions in Dwarf Irregulars.

2. J.M.SCALO: Initial Mass Function of Massive Stars in Galaxies: Empirical Evidence.

3. C.WHEELER: Luminous Content of Galaxies: Inferences from Supernovae.

LUMINOSITY FUNCTIONS IN DWARF IRREGULARS

John G. Hoessel
Space Telescope Science Institute

Observing luminosity functions and color-magnitude diagrams is a first step towards understanding the luminous stellar populations in galaxies. Dwarf irregular galaxies are of special interest for a variety of reasons. They are common at distances where they can be resolved, and generally have low background surface brightnesses which makes photometry of the resolved stars relatively easy. Dwarf irregulars exhibit a fairly wide spread in metal abundance allowing the study of massive star formation and evolution under a variety of conditions. Also, the dynamics of such systems is in general simple, without spiral density waves or other effects which might complicate understanding of the star formation process.

The closest and best studied dwarf galaxies are of course the Magellanic Clouds, where the work has progressed to detailed investigation of clusters, associations and individual stellar objects. After the Clouds the next best known objects are the local group irregulars NGC 6822 and IC 1613. Data on these galaxies is less comprehensive, although several studies of individual clusters and luminous stars have been done. This work is described elsewhere in this volume. The present review will concentrate on observations of less well known systems. The galaxies discussed below in no way constitute a proper statistical sample. Data are taken from the literature where they exist and include unpublished results available to the author. There are certainly strong selections applied, including an emphasis on galaxies which have very low surface brightness.

AVAILABLE DATA

Photometry exists for the brightest stars in at least 13 galaxies nearby enough to be resolved. Before summarizing the published data, new CCD photometry in the form of color-magnitude (CM) diagrams in the Thuan-Gunn (1976) photometric system will be described. These are preliminary reductions of data which will be discussed in detail elsewhere. In Figure 1 the G, G-R CM diagram is shown for 3475 measured stars from a mosaic of 9 CCD frames covering the field of NGC 6822. These data will be described in detail by Hoessel and Anderson

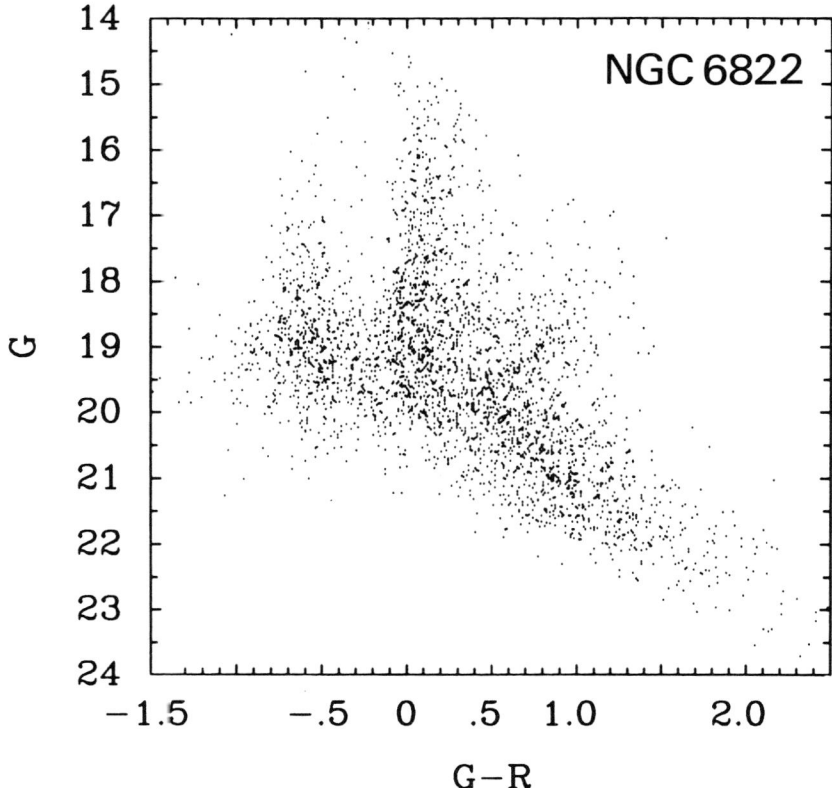

Figure 1--The color-magnitude diagram for NGC 6822. Nine CCD fields were observed or the Thuan-Gunn photometric system. Data are shown for 3475 stellar objects reduced with the aid of RICHFLD.

(1985) along with comparisons star by star with earlier photometry by Kayser (1966) and van den Bergh and Humphreys (1979). Notice in addition to the normal red and blue supergiant branches, the very large number of stars populating the region normally occupied by the Hertzsprung gap. NGC 6822 is located at relatively low galactic latitude. Kayser (1966) showed these to be primarily foreground Galactic stars which complicate any analysis of the luminosity function. However, as discussed by Humphreys and Sandage (1980) if we restrict our attention only to the blue supergiants the foreground contamination over the magnitude range of interest is minimal. This is confirmed for background areas near NGC 6822 by Hoessel and Anderson (1985). In the discussion below only stars with B-V < 0.5 (i.e. G-R < 0.0) will be considered. These data were reduced using the RICHFLD point spread function fitting software at the KPNO IPPS. The data have been corrected for Galactic reddening assuming A_v = 0.90 mag.

In Figure 2 new CCD photometry is shown for two galaxies which have published CM diagrams, Sextans B and Leo A. Sandage and Carlson (1985a) produced a BV CM diagram for Sextans B, and Leo A was studied by Sandage and Tammann (1982), Demers et. al. (1984), and with new CCD data by Freedman (1984). The photographic data was calibrated with photoelectric sequences. Detailed star by star comparison of the new data by Hoessel and Danielson shown in Figure 2 with the historical record will be done in a later contribution. In Figure 3 new CCD photometry is shown for two galaxies previously unstudied. Based on its kinematics DDO 187 was shown to be nearby by Yahil et. al. (1977), perhaps a companion to G.R.8 (that is number 8 in the list of Reaves, 1956), which Hoessel and Danielson (1983) place near the outskirts of the local group. Here we will adopt a considerably larger distance from the brightest supergiant stars. A photograph of DDO 187 was published by Fisher and Tully (1979). According to Fisher and Tully (1975) DDO 168 is a kinematic member of the Canes Venatici I Cloud. Note the absence of any observed red supergiants in this galaxy. The photometry of these four galaxies was done with the aid of the DAOPHOT point spread function fitting software developed by P. Stetson.

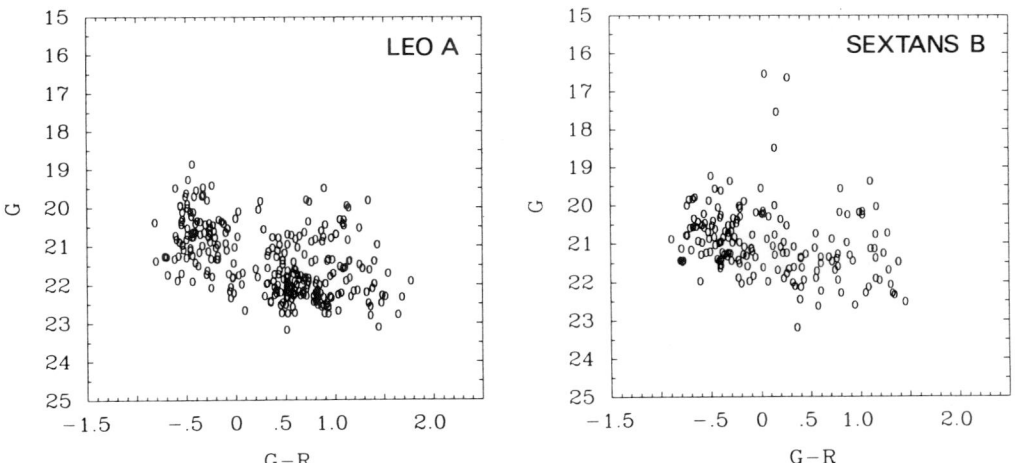

Figure 2--The color-magnitude diagram for Sextans B (DDO 70) and Leo A (DDO 69) from CCD observations reduced with DAOPHOT.

CCD photometry of luminous resolved stars is available in the literature for 7 other nearby dwarf irregulars. Pegasus was studied by Hoessel and Mould (1982). This galaxy also has a photographic CM diagram in Sandage and Tammann (1982). Hoessel and Danielson (1983) list photometry for G.R.8 which has also been surveyed by Hodge (1967). Photometry for stars in Holmberg I and Holmberg II was published by Hoessel and Danielson (1984) and for Sextans A by Hoessel, Schommer and Danielson (1983). Sextans A was studied photographically by Sandage and Carlson (1982) and with a CCD by Freedman (1984). A CCD CM diagram for Holmberg IX has been produced by Freedman (1984).

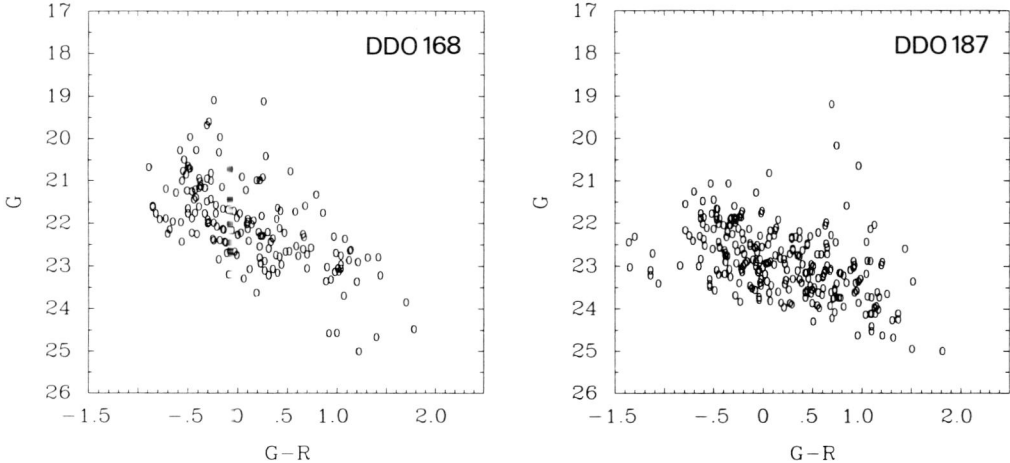

Figure 3--The color-magnitude diagram for DDO 168 and DDO 187 from CCD observations reduced with DAOPHOT. Note the absence of red supergiants in DDO 168.

Finally, two galaxies are included in the present study which as yet have only photographic photometry available for constructing luminosity functions. Sandage and Katem (1976) published a luminosity function for IC 1613. Recently Sandage and Carlson (1985b) have also produced photographic data for WLM. New CCD data have been obtained for both of these galaxies, but are not yet reduced.

Christian and Tully (1983) have shown that the local group irregular LGS-3 has no stars bright enough to be included in the present discussion. This may also be true for other very small stellar systems such as DDO 210, Phoenix, Sagittarius, Sculptor and UKS 2323-326. More detailed observations for each of these systems are necessary.

LUMINOSITY FUNCTIONS

In order to construct luminosity functions which can be directly compared to one another, two operations must be performed. First, the galactic foreground reddening must be removed. The data described above and shown below were all corrected via the Burstein-Heiles (1978) 21 cm method. Next a distance modulus for the galaxy is needed to set the absolute magnitude scale. Relatively accurate distances derived from Cepheid light curves exist for only five of the dwarf galaxies in the present sample. These are NGC 6822 (Kayser,1966, McAlary, et al 1983), IC 1613 (Sandage 1971, McAlary et al 1984), Sextans A and Sextans B (Sandage and Carlson 1985a), and WLM (Sandage and Carlson 1985b). A common distance of 3.2 Mpc is adopted for the three Holmberg galaxies which are in the NGC 2403-M81 group (Tammann and Sandage 1968, Madore 1976, McAlary and Madore 1984). For the remaining galaxies distance estimates from the literature were adopted except for DDO 187 and DDO 168, where the brightest red and blue star calibrations from Sandage and Carlson (1985b) have been adopted. Since DDO 168 has no observed red supergiants only the blue stars were used. Thus, for a majority of the nearby dwarf galaxies accurate distances are not available. This is the major source of uncertainty in interpreting the stellar content in these objects. Many of these galaxies are included in an ongoing program to find Cepheids. The adopted distance modulus for each galaxy is listed in Table 1.

Main sequence (that is, all stars with B-V < 0.5) differential luminosity functions are plotted in Figure 4 for the irregular galaxies discussed above except for Pegasus and G.R.8, where there are too few bright stars to justify a plot. The log number of stars per half magnitude bin is plotted as a function of absolute blue magnitude for the bin for each galaxy, with the number zeropoint shifted down one unit between each data set. Typical error bars, which are based only on the number of stars per bin are attached to two points on the NGC 6822 plot. The galaxies are roughly ordered from top to bottom in decreasing luminous blue star content. The effects of crowding and incompleteness have not yet been fully determined for these data.

The first result is that these plots look rather similar. The slopes at the bright end of the observed luminosity functions are consistent with being a constant value as found for generally more massive systems by Freedman (1984). The average logarithmic slope for the first three galaxies plotted, Holmberg II, NGC 6822, and Sextans A, is $d(\log N)/dM_B = 0.53$, but is not well determined. This value is less steep than that found by Freedman (1984). Both this new value for the luminosity function slope, and that found by Freedman (1984) are too shallow to fit with a purely statistical explanation of the correlation between the luminosity of the brightest blue stars and their parent galaxies (Sandage and Katem, 1976). If we assume continuous star formation this slope can be related to the IMF slope by using a mass-luminosity relation and including bolometric corrections. However, since many authors believe that star formation in these small systems occurs in bursts (Searle and Sargent, 1972, Gerola et al, 1980), interpretation of the bright-end slope is further complicated by the massive star lifetimes and the detailed chronology in each galaxy. Inspection of Figure 4 showns that smaller, less active systems also have somewhat steeper bright-end slopes as expected.

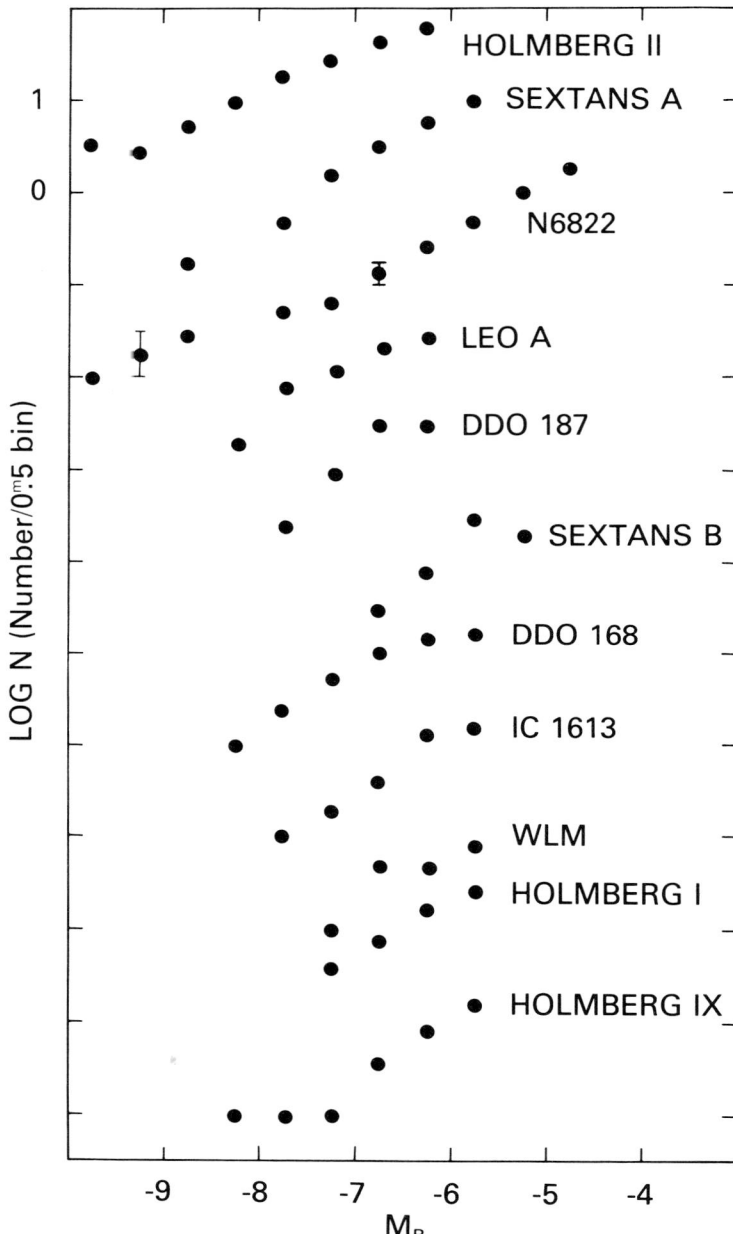

Figure 4--Luminosity functions for all stars with B-V < 0.5 for 11 dwarf irregular galaxies. The data have been corrected for foreground Galactic extinction and the absolute luminosities are based on the distances listed in Table 1. The log number of stars per half magnitude bin is plotted and the zeropoint of this scale is shifted one unit between each galaxy.

TABLE 1
Integrated Properties of Dwarf Irregular Galaxies

Name	Alias	m - M	M_H/L	ΔV	$\log M_H$	M_V
DDO 8	IC1613	24.5	-0.35	25	7.83	-15.1
DDO 50	Ho II	27.5	-0.07	57	8.92	-16.7
DDO 63	Ho I	27.5	0.12	27	8.06	-14.5
DDO 66	Ho IX	27.5	--	--	--	-13.5
DDO 69	Leo A	27.1	0.12	22	8.04	-14.7
DDO 70	Sex B	26.2	-0.20	39	7.87	-14.9
DDO 75	Sex A	26.2	0.03	48	8.17	-14.4
DDO 155	G.R.8	25.7	-0.05	30	6.56	-11.3
DDO 168		27.6	-0.11	63	8.29	-15.5
DDO 187		28.8	-0.18	35	8.01	-14.8
DDO 209	N 6822	23.5	-0.24	61	8.14	-15.5
DDO 216	Pegasus	25.3	-0.53	24	6.88	-13.3
DDO 221	WLM	25.0	-0.04	54	7.84	-13.9

LUMINOUS STARS AND INTEGRATED PROPERTIES

Searching for correlations between the properties of luminous stars and their parent galaxies could provide new clues concerning the star formation process and the evolution of galaxies. Several integrated properties of these dwarf irregular galaxies are listed in Table 1. The absolute magnitudes were obtained from the integrated photometry of de Vaucouleurs et al (1981) and using the listed distance modulus. The hydrogen mass to luminosity ratios were taken from Fisher and Tully (1975). The 21 cm widths are full width at 50% intensity (FWHM) and are from Huchtmeier and Richter (1985), as are the total hydrogen masses. Huchtmeier and Richter (1985) supply the line integrals which have been converted to total masses following the precepts in Fisher and Tully (1975) using distances listed here. In order to attempt to quantify the star formation activity in these galaxies we define a quantity N^* which is simply the total number of blue (B-V < 0.5) stars in the galaxy with M_B brighter than -6.0. In Figure 5 the log of the number of bright blue stars is plotted against the various integrated hydrogen properties of the objects. In the lower panel is the hydrogen mass to optical luminosity ratio which is a distance independent quantity. The central panel shows the bright star numbers in relation to the FWHM of the profile. In some of the larger objects this width indicates systematic rotation in the gas while in some of the smaller galaxies no systematic motion is observed indicating that the gas is turbulent. The upper panel shows the total neutral hydrogen mass as a function of massive star content. Holmberg IX is not included in these figures because its 21 cm data is badly confused with M 81. The filled circles indicate the existence of a Cepheid distance. The upper panels indicate a trend of increasing

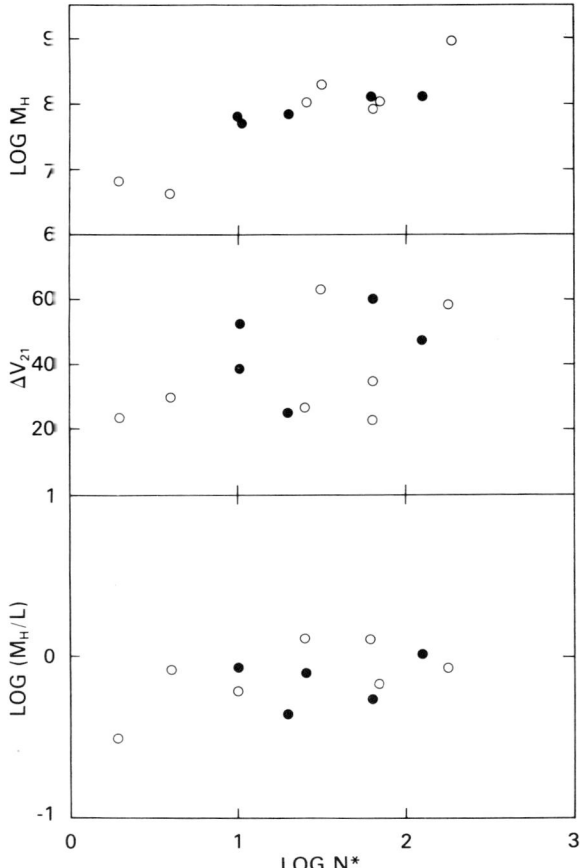

Figure 5—Relations between the number of bright blue stars (i.e. with $M_B < -6.0$ and $B-V < 0.5$) and the integrated neutral hydrogen properties are shown. Filled circles indicate that a Cepheid light curve distance is available. The lower panel contains the total hydrogen mass to luminosity ratio for the galaxy. The center panel shows the 21-cm. 50% intensity full width and the upper panel shows the total neutral hydrogen mass. These parameters are listed in Table 1. Only the total hydrogen mass is distance dependent.

number of bright stars with both larger galaxy dynamical and hydrogen mass. It should be noted that the veracity of these relations depends critically on galaxies with presently uncertain distances, particularly Pegasus and G.R.8. In the lower figure any trend of bright star population with the hydrogen to optical fraction is unclear, especially if the Pegasus point is ignored. Improved distances are necessary before any conclusions can safely be drawn on the relationship between luminous star content and galaxy hydrogen content.

The relationship between the number of bright blue stars and the total integrated visual magnitude for these galaxies is plotted in Figure 6. Although a trend exists in the sense that more luminous galaxies contain a larger number of massive stars, there is considerable scatter in the data. The luminous stars themselves contribute in general less than 5 percent (eg. Hoessel and Danielson, 1984) of the total visual luminosity of the galaxy. The scatter amounts to as much as a factor of ten in star number at a given galaxy

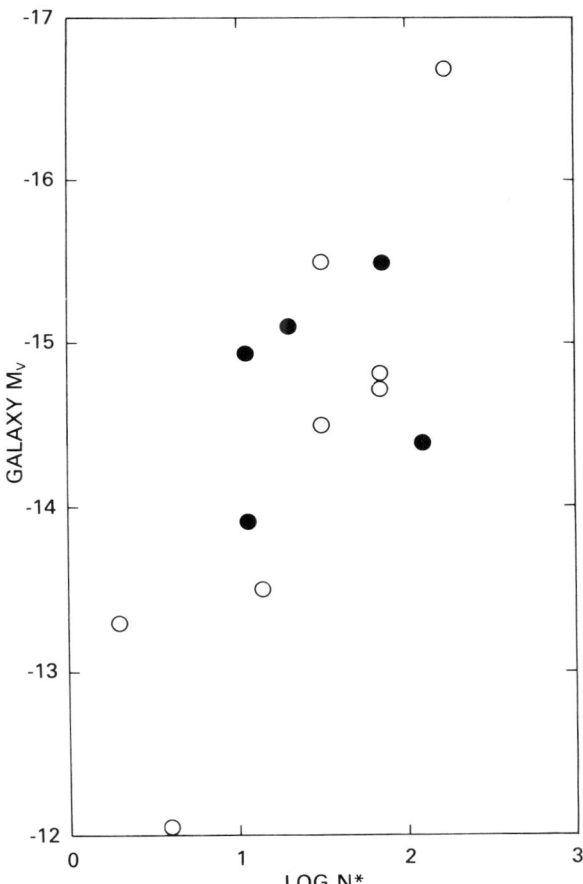

Figure 6--The relationship between the total visual absolute luminosity of the parent galaxy and the number of luminous stars is shown. Filled symbols indicate the existence of a Cepheid light curve distance.

absolute magnitude, or conversely up to 1.5 magnitudes in galaxy luminosity at a given absolute bright star population. There appear therefore to be substantial variations in the ratio of recent star formation activity to past average star formation rate.

FUTURE DIRECTIONS

In order to improve our understanding of the luminous stellar content in dwarf irregular galaxies new observations will be necessary. A first step is to obtain accurate distances to those galaxies with available photometry. A second step is to include new galaxies which will enlarge the range of integrated properties explored. Possible nearby systems include IC 10 in the northern hemisphere and NGC 3109, IC 5152, DDO 210, Sagittarius, Sculptor and Phoenix in the southern hemisphere. Galaxies in nearby groups such as the M 81 group, Sculptor group, and Canes Venatici I cloud will be at the resolution limit with the best ground based observing conditions. The Hubble Space Telescope will be a marvelous tool for observations of this sort. With its superior resolution and faint limiting magnitude the number of objects available for detailed study will be increased by an enormous factor. In a single orbit per color with the Wide Field Camera, luminosity functions similar to those shown in Figure 4 can be done at the distance of the Virgo Cluster. With HST most galaxies in the DDO catalog should be resolved.

Observations of the integrated properties of dwarf irregulars are in relatively good shape, except for their chemical composition which is only known for the very nearest objects. Perhaps the main impact of studies such as those reviewed here will eventually be that they provide finding lists for objects used in a variety of astrophysical studies.

I would like to thank Ed Danielson, Bob Schommer and Nelson Caldwell my collaborators on dwarf irregular galaxy projects. Jim Westphal and Jim Gunn have made excellent tools available for this work. I thank Holland Ford for a critical reading of an earlier version of this manuscript.

REFERENCES

Burstein, D. and Heiles, C.: 1978, Astrophys. J., **225**, 40.
Christian, C.A and Tully, R.B.: 1983, Astron. J., **88**, 934.
Demers, S., Kibblewhite, E.J., Irwin, M.J., Bunclark, P.S. and Bridgeland, M.T.: 1984, Astron. J., **89**, 1160.
deVaucouleurs, G., deVaucouleurs, A. and Buta, R.: 1981, Astron. J., **86**, 1429.
Fisher, J.R. and Tully, R.B.: 1975, Astron. Astrophys., **44**, 151.
Fisher, J.R. and Tully, R.B.: 1979, Astron. J., **84**, 62.

Freedman, W.L.: 1984, Ph.D. Thesis, University of Toronto.
Gerola, H., Seiden, P.E. and Schulman, L.S.: 1980, Astrophys. J., **242**, 517.
Hodge, P.W.: 1967, Astron. J., **148**, 719.
Hoessel, J.G. and Anderson, N.: 1985, preprint.
Hoessel, J.G. and Danielson, G.E.: 1983, Astrophys. J., **271**, 65.
Hoessel, J.G. and Danielson, G.E.: 1984, Astrophys. J., **286**, 159.
Hoessel, J.G. and Mould, J.R.: 1982, Astrophys. J., **254**, 38.
Hoessel, J.G., Schommer, R.A. and Danielson, G.E.: 1983, Astrophys. J., **274**, 577.
Huchtmeier, W.K. and Richter, O.G.: 1985, preprint.
Humphreys, R.M. and Sandage, A.: 1980, Astrophys. J. Suppl., **44**, 319.
Kayser, S.E.: 1966, Ph.D. Thesis, California Institute of Technology
Madore, B.F.: 1976, M.N.R.A.S., **177**, 157.
McAlary, C.W., Madore, B.F., McGonegal, R., McLaren, R.A. and Welch, D. L.: 1983, Astrophys. J., **273**, 539.
McAlary, C.W. and Madore, B.F.: 1984, Astrophys. J., **282**, 101.
McAlary, C.W., Madore, B.F. and Davis, L.E.: 1984, Astrophys. J., **276**, 487.
Reaves, G.: 1956, Astron. J., **61**, 69.
Sandage, A.: 1971, Astrophys. J., **166**, 13.
Sandage, A. and Carlson G.: 1982, Astrophys. J., **258**, 439.
Sandage, A. and Carlson, G.: 1985a, preprint.
Sandage, A. and Carlson, G.: 1985b, preprint.
Sandage, A. and Katem, B.: 1976, Astron. J., **81**, 743.
Sandage, A. and Tammann G.A.: 1982, in Astrophysical Cosmology, ed. H. A. Bruck, G.V.Coyne, and M.S.Longair (Pontificae Academiae Scientarum Scripta Varia 48), 23.
Searle, L. and Sargent, W.L.W.: 1972, Astrophys. J., **173**, 25.
Tammann, G.A. and Sandage, A.: 1968, Astrophys. J., **151**, 825.
Thuan, T.X. and Gunn, J.E.: 1976, P.A.S.P., **88**, 543.
van den Bergh, S. and Humphreys, R.M.: 1979, Astron. J., **84**, 604.
Yahil, A., Tammann, G.A. and Sandage, A.: 1977, Astrophys. J., **217**, 903.

Discussion : HOESSEL.

SHARA :

Can you do background subtraction using CCD frames of nearby pieces of sky? What are the problems associated with this technique?

HOESSEL :

This works statistically, but with very few background stars on a given frame. If your interest is specific stars, say red supergiants, then you must look at each one.

THE INITIAL MASS FUNCTION OF MASSIVE STARS IN GALAXIES: EMPIRICAL EVIDENCE

John M. Scalo
Department of Astronomy
University of Texas

ABSTRACT

Observational constraints on the form of the high-mass stellar IMF are reviewed. The evidence includes star counts in the solar neighborhood, individual and composite star clusters, and nearby galaxies, and arguments based on integrated light and chemical evolution modeling. There is no convincing evidence for any systematic variations of the shape of the high-mass IMF. However, the various determinations are very uncertain, and do not allow any firm estimate of the logarithmic slope of the upper IMF; the appropriate value is somewhere between -1.3 and -2.3, with region-to-region variations smaller than about ±0.5. A number of lines of evidence suggest that the lower mass limit or mode mass of the IMF increases with increasing star formation rate, reaching perhaps 10-15 $m_\odot$ in some starburst galaxies. It is also possible that the upper mass limit depends on metallicity, based on variations in excitation conditions of H II regions.

1. Introduction

The frequency distribution of stellar masses at birth, called the "initial mass function" (IMF) connects phenomena on the scale of stars with those on the galactic scale, and enters into nearly all studies of galactic evolution. The IMF of massive stars is especially important, since it is these stars that contribute most of the luminosity, ionization, and kinetic energy of star-forming regions in galaxies. Unfortunately, despite many independent studies using a variety of methods, our knowledge of the shape of the high-mass IMF remains very uncertain for a number of reasons, although substantial progress has been made in the past few years. The present paper is an attempt to briefly summarize the results and problems associated with a variety of approaches to the determination. A detailed review of this subject, as well as of the IMF of lower-mass stars, is given elsewhere (Scalo 1985).

For consistency's sake, we define the IMF as the number of stars formed at the same time in some volume of space per unit logarithmic mass interval; e.g. the classical Salpeter IMF would have a logarithmic slope of $\Gamma = -1.3$. It is useful to think of the IMF as a probability distribution. This definition assumes that the mechanisms responsible for the form of the IMF constitute a process that is statistically homogeneous in space and stationary in time. We also introduce the parameters m_l and m_u as the lower and upper mass limits of the IMF.

2. Star Counts in the Solar Neighborhood

The relative number of stars presently observed in each mass interval does not directly give the IMF because of the effects of stellar evolution; stellar lifetimes decrease with increasing mass, so the actual IMF is flatter than the presently observed mass distribution. For high mass stars the conversion does not depend on the history of the star formation rate (SFR) because their lifetimes are small compared to the timescale over which the global SFR has varied. Therefore the IMF can be constructed by dividing the presently-observed number of stars in each mass interval by the average stellar lifetime for that mass interval. In our galaxy this star counting technique can sample massive stars out to 1-3 kpc from the sun.

A number of recent papers have addressed this problem (Lequeux 1979; Garmany *et al.* 1982, GCC; Bisiacchi *et al.* 1983, BFS; Humphreys and McElroy 1984, HM; Van Buren 1984, Vanbeveren 1984; Scalo 1985), with differing results. Since masses cannot be directly observed, the present-day mass distribution must be derived either from the distribution of stars in the H-R diagram using theoretical evolutionary tracks and semi-empirical effective temperature and bolometric correction calibrations, or from the luminosity function (LF), using a mass-luminosity relation. All the studies listed above except that of Scalo (1985) use some variant of the former method. A major problem with all these studies is the question of completeness.

TABLE 1

Estimates of Γ for High-Mass Stars

Study	Γ	Notes
Lequeux	-1.3	
GCC	-1.4 to -1.7	a,b
BSF	-2.0	
HM	-2.2	c
VanBuren	-1.3	
Vanbeveren	-2.4(-1.9)	d
IMF(LF)	-1.5	e

[a]Range for 3 different sets of evolutionary tracks.
[b]Does not include B stars.
[c]Corrected for incompleteness in the 15-30 $m_\odot$ bin.
[d]For combined cluster and field star sample; value in parentheses was derived from Vanbeveren's results without weighting.
[e]From LF of GCC for m > 25, joined to adopted LF at smaller masses.

The results of the method which counts stars in various mass ranges in the H-R diagram are sensitive to the adopted evolutionary tracks and effective temperature scale. The evolutionary tracks present a particularly vexing problem since the quantitative effects and rates of mass loss, nonlocal convection, and internal rotational mixing are not well-understood. The same problems occur in using the mass-luminosity relation in conjunction with the luminosity function, although the effects are not as severe. On the other hand, the method which counts stars in the H-R diagram is in principle capable of higher accuracy in estimating an IMF slope because the steepness of the mass-luminosity relation means that a given uncertainty in the slope of the luminosity function translates into a much larger uncertainty in the IMF slope.

Another uncertainty, not very well recognized, enters when we divide the star counts by the stellar lifetimes. Main sequence lifetimes are uncertain by perhaps 30-40 percent, primarily due to our uncertainties concerning the importance of convective overshoot in the core and internal mixing due to rotational or other instabilities. Both of these effects increase the stellar lifetime.

Table 1 summarizes approximate logarithmic slopes of the high-mass IMF ($m \gtrsim 10 - 25\ m_o$) according to a number of investigations. The slope given for Lequeux (1979) assumes that none of the observed stars are stars from an evolved population. The slope given for HM does not include the data point at the highest mass, since the number of stars is small and the appropriate mass is uncertain. The slope given for Van Buren (1984) has been modified to account for the estimated effects of convective overshoot on the luminosities and lifetimes. The result derived from the luminosity function [denoted IMF(LF)] in Table 1 is based on the luminosity function of stars in the GCC catalogue, but joined to other luminosity function determinations at smaller luminosities in order to account for incompleteness at small luminosities. Notice that IMF(LF) is in reasonable agreement with the GCC estimates, suggesting that the two approaches are consistent.

A detailed discussion of all the causes of agreement and disagreement between the various determinations (lifetimes, evolutionary tracks, effective temperature scales, etc.) is beyond the scope of this review; such a discussion can be found in Scalo (1985). The major consideration worth mentioning is that the GCC catalogue included only O stars. The work of BSF, HM, and Vanbeveren (1984) clearly shows how the inclusion of later spectral types steepens the derived IMF, and, on this basis alone, we might conclude that the IMF slope of massive stars is probably around -2. However, Van Buren (1984) also included later spectral types, a large sample, and a very careful study of incompleteness corrections and reddening, and obtained a slope of -1.3 (or even flatter) for high-mass stars. I have been unable to resolve this discrepancy, and conclude that, without a more careful study and improvements in our knowledge of stellar evolution and the effective temperature scale, the existing studies provide little reason for choosing a particular value of the high-mass IMF slope. This is a disappointing but, I think, realistic conclusion.

3. Star Counts in OB Associations and Open Clusters

Additional information on the slope of the high-mass IMF comes from studies of individual (and composite) OB associations and young stellar clusters. Such studies are free from some of the problems mentioned above, but suffer from a number of new difficulties, such as mass segregation, small number statistics, and non-coeval star formation.

A number of papers have attempted to reduce the small-number fluctuations by combining the star counts for a fairly large number of clusters to obtain an average, or "composite" cluster IMF. The construction of such a composite function involves a number of difficulties which cannot be reviewed here. The only work which has attempted to address some of these problems is Taff (1974; see also Vanbeveren 1982).

The composite cluster IMFs derived from composite luminosity functions compiled by Taff (1974) and Burki (1977; large, medium and small diameter clusters shown separately) are shown in Figure 1 and compared with the IMF(LF) derived in

Scalo (1985). The normalization is arbitrary. For Taff's IMF, the IMF slope for 1 < m < 10 is about -1.8. The IMF appears to flatten for m ≳ 10, but the number of stars used is small. Burki's slopes between m = 7.5 and 25 $m_\odot$ (small clusters) or 50 $m_\odot$ (large clusters) are -1.7 (small), -1.5 (medium) and -1.2 (large). Perhaps the most important point is that these slopes are flatter than most of the studies of field stars using the H-R diagram reviewed earlier.

Studies of individual young clusters are problematic because of small number statistics and other difficulties, but a comparison of a number of careful studies (Scalo 1985, Sagar et al. 1985) suggests general agreement among themselves and with the composite functions. The IMF slope between about 3 and 10 $m_\odot$ appears to be around -1.5, but there is strong evidence for flattening at larger masses, especially in M17, NGC 654, and NGC 6611, in agreement with the effect noted above for Taff's (1974) composite cluster IMF.

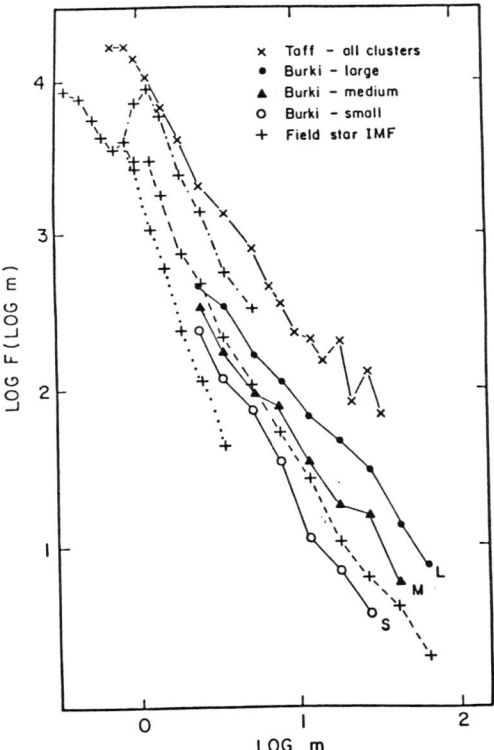

Figure 1. Estimates of composite open cluster IMFs by Taff (1974) and Burki (1977); the field star IMF constructed by Scalo (1985) is also shown. Normalization is arbitrary.

IMF variations between clusters are notoriously difficult to establish, so I will not attempt any review here, except to note the turnover in the NGC 3293 IMF at around 3 $m_\odot$ found by Herbst and Miller (1983) and the enormous variations in IMF slope found in the maximum likelihood fits of Tarrab (1982).

4. Star Counts in Nearby Galaxies

The shape of the IMF and its possible variations on scales $\gtrsim$ 1 kpc can be studied using star counts within and among nearby galaxies. After corrections for incompleteness, crowding, foreground stars, etc. (these are nontrivial problems), a luminosity function can be constructed. Comparisons of luminosity functions should give some idea as to whether galactic-scale IMF variations exist. (IMF studies based on counting stars in the H-R diagram will not be reviewed here; in my opinion they are of little use for extragalactic systems because of the bright limiting magnitude imposed by the necessity of obtaining spectra).

Several recent comparisons of this nature have been carried out. Scalo (1985) compared the LFs of six galaxies based on an inhomogeneous set of data culled from the literature, including both spirals (M33, M31) and irregulars (LMC, SMC, IC 1613, NGC 6822). Considering the variety of data sources, the agreement was remarkably good. Hoessel and coworkers (see Hoessel, this volume) have derived a homogeneous set of LFs for a number of dwarf irregulars; again, the agreement at the bright end, where incompleteness problems are minimized, is surprisingly good.

The most detailed study of galactic LFs is due to Freedman (see Freedman 1985, and this volume). Her study includes homogeneous CCD data for seven spiral and irregular galaxies, supplemented with data from the literature for three additional irregulars. The largest sample is for M33, with over 5000 stars. Freedman's papers should be consulted for details concerning automated star counting and image identification, crowding corrections, etc. The resulting LFs exhibit, again, a remarkable similarity with a mean slope dlog N/dM_v = 0.67 ± 0.03. If the LF slope differences estimated by Freedman are taken literally, they imply a variation in the IMF slope of less than ± 0.5, and probably much less, considering the uncertainties in the data. Furthermore, Freedman finds no significant variations in slope with position or metallicity in the galaxies studied, a conclusion in agreement with studies of the LMC LF at lower masses (see Scalo 1985).

One must conclude from these three comparisons that there is no evidence for high-mass IMF slope variations within or among nearby galaxies. This conclusion is especially striking because the galaxies vary greatly in morphology and metallicity.

5. Indirect Evidence: Integrated Light

I use the term "indirect" to refer to methods for estimating the IMF that do not employ star counts. The observable quantities used in such approaches include broad-band colors, mass-to-light ratios, spectral line strengths, and elemental abundances and abundance ratios. Most of these methods are very uncertain and ambiguous, but they can sample distant volumes of space and large numbers of objects which cannot be studied by star counts, and, taken together, can yield some clues to the gross properties of the IMF and especially its possible variations in space and time. In the following summary of some of the more recent work in this area, the reader should bear in mind the large uncertainties and ambiguities that accompany all such studies. The following discussion only includes methods which give information on the high-mass IMF, and even then the choice of topics is selective. A more comprehensive review is given elsewhere (Scalo 1985).

(a) Mass-to-Light Ratios

This ratio is roughly a measure of the ratio of the number of low-mass stars (and remnants and other dark matter), which contributed most of the mass, to the number of higher-mass stars, which contribute most of the light (see Faber and Gallagher 1979 for a review). Thus, M/L should be larger for a steeper IMF. In connection with the IMF, the M/L ratio has been used mostly as a probe in elliptical galaxies. Recently Terlevich and Melnick (1984) have interpreted an apparent correlation of metallicity Z with M/L in giant H II regions in galaxies and isolated intergalactic H II regions (as well as elliptical galaxies and globular clusters) in terms of an IMF slope which steepens with decreasing metallicity. They suggest the relation $\Gamma = -4.0 - \log Z$ for a pure power-law IMF. A problem is that their M/L values for H II regions are based on virial masses derived from emission line widths, and these masses may be inappropriate, as discussed by several authors. In addition, the suggested relation does not agree with results from star counts, at least for high-mass stars. Still, this evidence, along with additional correlation studies for elliptical galaxies, suggests at least that the ratio of high-to-low-mass stars may depend on metallicity, but the variation may not involve the slope of the high-mass IMF.

(b) Colors

Color variations among galaxies cannot in themselves be used to infer IMF variations because broad-band colors are very sensitive to the SFR and other parameters. The best illustration of this fact is provided in the figure presented by Larson and Tinsley (1978) and reproduced in Tinsley (1980) and Scalo (1985). The observed colors of galaxies are certainly consistent with a universal IMF, but they do not allow an estimate of the slope Γ and/or the upper mass limit m_u.

(c) Population Synthesis

Attempts to match observed galaxy colors, spectra, or line strengths by finding a "best" mixture of stars at various spectral types and luminosity classes is beset by problems concerning solution uniqueness and the choice of appropriate constraints, but, in principle, this method could yield an IMF estimate. A good example is provided by the paper of Ellis *et al.* (1982) for four spiral galaxies.

(d) Ultraviolet Luminosities

The UV flux spectrum from a young population depends on the IMF and SFR of massive stars. Use of the shape of the spectrum does not appear viable for IMF studies, but attempts to match the absolute UV luminosity are more promising. Donas and Deharveng (1984) have studied the UV fluxes of 40 spiral and irregular galaxies, and suggest $\Gamma = -2 \pm 0.5$ with the estimated limit on the dispersion based on the modest scatter in the derived SFR-gas mass correlation.

(e) Lyman Continuum Indicators

Probably the best indirect methods for studying the high mass IMF in galaxies involve quantities which allow an estimate of the rate at which photons are emitted by stars in the Lyman continuum, N(Lyc), since the Lyman continuum flux is dominated by stars with $m \gtrsim 20\ m_\odot$. The methods available include hydrogen line luminosities, which are proportional to N(Lyc), the thermal radio continuum emission from H II regions, and line intensity ratios like [O III]/Hβ, which give an estimate of an equivalent temperature usually denoted "T_{eff}". All these approaches suffer from severe uncertainties, e.g. the effects of dust, contamination by nonthermal radio emission, sensitivity to model atmospheres, etc. These difficulties are reviewed in Scalo (1985); here we will only summarize a few of the most interesting results, which should be regarded with caution.

(f) The W(Hα)-(B-V) Relation for Late-Type Spirals

The ratio of Hα or Hβ flux to the continuum flux just outside the line is a useful tool for studying IMFs and SFRs in galaxies. These ratios are usually referred to as the Hα or Hβ equivalent widths. The Hα or Hβ emission is due to massive O stars (through the Lyman continuum flux), while the continuum radiation comes mostly from B and A stars around Hβ and 1-3 $m_\odot$ red giants around Hα. Therefore, these quantities are measures of the ratio of $m \gtrsim 20\ m_\odot$ stars to intermediate-mass stars. However, they are also sensitive to the upper mass limit and SFR history.

Kennicutt (1983) has shown how the distribution of galaxies in the W(Hα)-(B-V) plane can be used to constrain the SFR histories and IMFs in late-type galaxies by comparing the observed distribution with theoretical galaxy evolution calculations. The comparison is reproduced in Figure 2. Each band corresponds to a different IMF for a range of SFR timescales, with the steepest IMF at the bottom. The best fit is found for the intermediate IMF, which is a power law with $\Gamma = -1.5$. Although the absolute value of this estimate is uncertain (e.g. it depends on the adopted upper mass limit m_u if $m_u \gtrsim 300\ m_\odot$, and rests on the validity of the theoretical evolution models), the modest spread does suggest that <u>variations</u> in Γ are probably less than about ± 0.3, in agreement with the star counts discussed earlier.

A related procedure involving W(Hβ)-color distributions for clusters in H II regions has been discussed by DeGioia-Eastwood (1984), and, although there are several problems, the method appears promising for future work.

The weak Hα emission in Sa and Sab galaxies has been interpreted as due to an IMF deficient in high-mass stars, for example by van den Bergh (1976); but the low excitation could just as well be due to small current SFRs. The question cannot be resolved without very accurate measurements of Hα equivalent widths in Sa and Sab galaxies. Further discussion of this question as it relates to anemic and smooth-armed spirals and the disk of M82 can be found in van den Bergh (1981) and Scalo (1985).

(g) Starburst Nuclei

"Starburst nuclei" refers to the central regions of several galaxies which have been found to be extremely luminous and interpreted in terms of very large SFRs, based on comparison with theoretical galaxy evolution models. These galaxies are often involved in a galaxy interaction, which is apparently able to trigger the burst by an unknown mechanism. The pioneering study of M82 and NGC 253 by Rieke *et al.* (1980) was important in establishing the starburst characteristics. More recent studies of IC 2153 by Olafsson *et al.* (1984), of Mk 171 (= NGC 3690 + IC 694) by Gehrz *et al.* (1983) and Augarde and Lequeux (1985), and of the extreme starburst galaxies Arp 220 and NGC 6240 by Rieke *et al.* (1985), as well as other less detailed studies of other galaxies, have for the most part substantiated the conclusions of Rieke *et al.* (1980) for M82 and NGC 253.

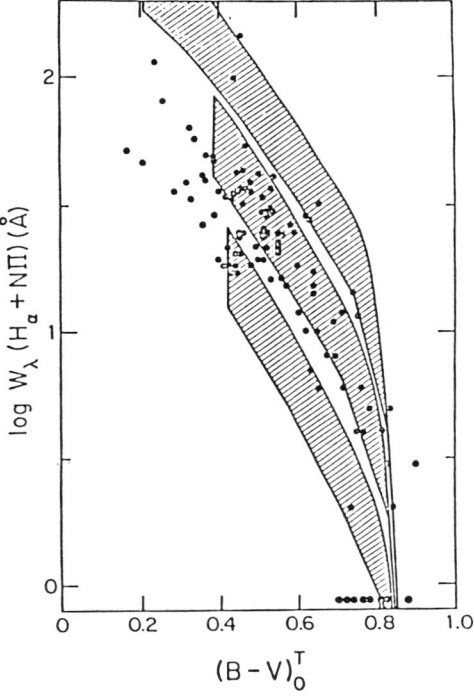

Figure 2. Hα equivalent width-(B-V) color diagram for late-type spirals from Kennicutt (1983). Dots are observed galaxies, bands are theoretical calculations for three IMFs with a range of star formation rate histories.

Concerning the IMF, the most striking conclusion is that it is difficult if not impossible to construct acceptable models unless the <u>lower mass limit is very large</u>, from around 3 m$_\odot$ (for M82 and NGC 253) to perhaps 20 m$_\odot$ for Mk 171. Apparently only high-mass stars are formed in these bursts. Although the result can be questioned on several grounds in individual cases (see Scalo 1985), the agreement of these independent studies based on differing constraints lead me to believe that the result is valid. The result is also consistent with other lines of evidence, discussed below, that in regions of large SFRs the lower mass cutoff <u>or</u> the mode (peak) of the IMF may be one to two orders of magnitude larger than in the solar neighborhood. In my opinion, these cases present the only convincing evidence for a significant variation in the large-scale IMF. However, there is as yet no compelling evidence that the <u>shape</u> of the IMF above the cutoff or mode varies.

(h) Blue Compact Galaxies and Related Objects

Several workers have attempted to infer the SFR histories of the blue compact galaxies, often referred to as extragalactic H II regions, using galactic spectrophotometric modeling. A secondary goal of these studies has been to provide constraints on the IMF. The IMF of these galaxies is especially interesting because of their large star formation rates and small metallicities. Unfortunately, as in all such investigations, it is extremely difficult to disentangle the history of the SFR from the form of the IMF, unless several observational constraints are used.

One of the most detailed studies is that of Viallefond and Thuan (1984) on I Zw36. The observational constraints included the Hβ equivalent width, related to the ratio of O stars to B and A stars, the "effective temperature" T_{eff}, which depends strongly on the high mass IMF slope Γ and the upper mass limit m_u, and ratios of continuum intensities at (mainly) two pairs of wavelength in the visual part of the spectrum, which are related to the IMF of stars of spectral type A to B2. For the high mass stars, the models require $\Gamma \approx$ -1.5 and $m_u \approx 100 \ m_\odot$. If the SFR has been constant with time, the lower mass limit must be about 4 $m_\odot$ to account for the large value of W(Hβ), a result similar to those found for interacting starburst galaxies. This again suggests m_l or the IMF mode value increases with SFR. However, if the burst is assumed to be instantaneous at some time in the past, smaller values of m_l would be allowed. A comparison for I Zw18 also gives $\Gamma \approx -1.5$ and $m_u \approx 100 \ m_\odot$, but only an upper limit on m_l of 2 $m_\odot$ could be determined.

In contrast, Viallefond and Thuan find that the giant H II region NGC 5471 in the spiral galaxy M101 can be modeled with $\Gamma \approx -2$, $m_u \approx 100 \ m_\odot$, and $m_l \approx 0.007$. They point out that the derived IMF slope is steeper in the more metal-rich region.

A similar metallicity-dependent IMF slope has been suggested for giant H II regions in nearby galaxies and blue compact galaxies by Terlevich and Melnick (1985) on the basis of correlations between Hβ strength, M/L_B, T_{eff}, and Z: As Z decreases, Hβ and T_{eff} increase while M/L_B decreases. But, as mentioned earlier, the correlations, except for the T_{eff}-Z relation, are based on assumed virial equilibrium of the regions studied, which has been called into question in several papers. Even if the result is valid, it only implicates a variation in the relative number of high and low mass stars (the IMF was assumed to be a pure power law from m_u to m_l), and may not reflect variations in IMF shape for the massive stars. As mentioned earlier, there is no evidence for a Z-dependent Γ for high mass stars from star counts in nearby galaxies, and the suggested Z-dependent Γ does not agree with determinations based on star counts of local stars (see Scalo 1985 for a detailed discussion).

The result of Terlevich and Melnick (1984) which *does* survive all this scrutiny is the anticorrelation between T_{eff} and Z, which has been known for some time (e.g. Alloin *et al.* 1979). However, with this relation alone it is not possible to decide whether the effect is due to a flatter IMF or a larger m_u at smaller Z. Further confusion arises because the definition of "T_{eff}" used by Terlevich and Melnick is inappropriate for a comparison with excitation models, and that the calculated T_{eff}, however it is defined, is very sensitive to the adopted model atmospheres, as pointed out by Lequeux (private communication). Because of the latter complication, as well as the presence of dust, it is possible that a Z-dependence of T_{eff} might exist which has nothing to do with the IMF. Obviously, the situation remains confused and deserves further study.

6. Indirect Evidence: Chemical Evolution Models

Attempts to constrain the IMF using chemical evolution models are very problematical, and so only a brief summary of the relevant arguments will be given here. Besides the uncertainty in elemental yields from massive stars as a function of mass, the conclusions concerning anomalous IMFs are in nearly all cases ambiguous; they cannot be disentangled from uncertainties in the chemical evolution model. There are additional problems concerning observational uncertainties and models for supernovae, which will be mentioned below. These issues are discussed in detail in Tinsley (1980) and Scalo (1985).

A well-known example concerns the metallicity distribution of disk dwarfs. A closed homogeneous chemical evolution model which starts with no metals and evolves with a constant IMF predicts too many low-Z stars. One solution is to postulate an IMF at early times which was enhanced in high-mass stars, using either pre-enrichment of the disk by an earlier generation of massive stars, or a disk IMF enriched in massive stars at early times. The IMF could have had a flatter slope or a larger value of m_l.

However, another possible resolution of the discrepancy is to relax the assumptions of the model. Examples include early enrichment of the disk by metals expelled from halo stars with a normal IMF (Ostriker and Thuan 1975) or later infall of metal-poor gas (Lynden-Bell 1975, Lacey and Fall 1983), both of which are consistent with Larson's (1976) dynamic models for disk galaxy formation. One could also speculate that the efficiency of star formation decreases strongly with decreasing metallicity (Talbot and Arnett 1973). At present there is no way to decide between these solutions and the assumption of a variable IMF.

Another argument for an early IMF enriched in massive stars, often mentioned in the literature, is that the O/Fe ratio is large in metal-poor stars, while C/Fe is roughly constant (see Sneden 1985 for a review of the data). Sneden et al. (1979) pointed out that, since the yields predicted by stellar evolution calculations predict that O/Fe increases with mass while C/Fe is relatively independent of mass, the observations suggest that the halo IMF was greatly enhanced in very massive stars relative to the disk IMF; according to Twarog and Wheeler (1981), the halo IMF would have had to contain most of its mass in stars with $m \gtrsim 50\ m_\odot$.

Although this conclusion may in fact be correct, it is not generally recognized that the argument depends crucially on the assumption that C, O, and Fe are produced in short-lived massive stars, an assumption which now appears doubtful. Explosions of accreting white dwarfs, which are currently the leading contenders for Type I supernovae, must eject large quantities of Fe (see Nomoto 1984, Woosley et al. 1984 and references therein), and possibly Si, S, and Ca (Nomoto et al. 1984). This would explain the behavior of the O/Fe ratio and possibly the observed constancy of S/Fe found by Clegg et al. (1983). In addition, single intermediate-mass stars exploding by carbon deflagration can also account for the observed O/Fe-Fe/H relation (Matteucci and Tornambe 1984). However, in order to escape the IMF argument, a long-lived source of C is required. The ability of exploding white dwarfs to produce significant C is uncertain at present. An alternate source of C is carbon stars, most of which have small masses (see Scalo 1981).

Evidently the use of abundance ratios of primary elements to constrain the IMF is not as clear-cut as is often supposed.

The existence of negative radial metallicity gradients in galaxies has also been invoked as an argument for a variable IMF, with a larger proportion of high-mass stars at small galactocentric distances, because the simple closed model for chemical evolution cannot account for such gradients. As pointed out below, there is no evidence for such an IMF slope gradient in M33 (Freedman 1985), which has a modest Z-gradient; nor is there any evidence from star counts for a Z-dependent IMF slope among other galaxies. Again, there exist alternative explanations involving modifications of the chemical evolution model. These include a gradient in the metallicity of infalling gas and/or the ratio of SFR to infall rate, both of which effects are found in Larson's (1976) dynamical models, and radial gas flows (Mayor and Vigroux 1981, Lacey and Fall 1985).

Perhaps the most promising explanation for the observed Z-gradients is that proposed by Güsten and Mezger (1983). Their model involves a variation in the IMF, but this variation is not supposed to be caused by Z-variations. Instead, Güsten and Mezger suggest that the IMF has the same shape everywhere, but the lower mass limit is large, m_l ~ 2-3 $m_\odot$, for stars formed in spiral arms, and small, $m_l \approx 0.1$ $m_\odot$, for stars formed in interarm regions. A Z-gradient arises because the SFR in arms at a given galactocentric distance R is assumed to be proportional to the rate at which interarm gas encounters arms, which is larger at smaller R because of differential galactic rotation. Therefore, the ratio of arm to interarm SFRs, and hence the ratio of stars with m ≳ 2-3 $m_\odot$ to stars with m ≳ 0.1 $m_\odot$, increases with R. Besides reducing uncomfortably large SFRs inferred for spiral galaxies, as explained in Güsten and Mezger's paper, the proposal that only high-mass stars form in spiral arms is consistent with analyses of surface photometry in nearby spirals by Jensen et al. (1981) and Bash and Visser (1981), and fits in well with the suggestion made above that m_l or the IMF mode increases in regions of large star formation rates.

Space precludes a discussion of the suggestion that the distribution of dwarf irregular and blue compact galaxies in the metallicity-gas fraction plane can be attributed to a dependence of the IMF on Z (see Scalo 1985 for a review). Suffice it to say that, besides the large uncertainties in the empirical gas masses, the observations can also be explained by large infall rates (perhaps implied by the prevalence of large H I halos in these galaxies), or galactic winds energized by supernova explosions, as shown in the thorough examination of the problem by Matteucci and Chiosi (1983).

In summary, every argument for a variable IMF based on chemical evolution arguments can be countered by a revision in the chemical evolution model or in the assumptions concerning the masses of stars which produce different elements. The observed metallicity distribution of dwarfs and the large O/C ratios in halo-population stars both suggest an IMF enriched in massive stars during the collapse of our galaxy, and such a conclusion is consistent with the suggestion already made several times above, that the lower mass limit or mode mass of the IMF increases with increasing SFR, assuming that our galaxy experienced a large global SFR during the protogalactic collapse phase. It should also be pointed out, however, that globular clusters, which presently contain only low-mass stars, should then contain huge numbers of stellar remnants and have very large M/L ratios. Larson (1985) has pursued this idea and suggests that these halo remnants (most of which would not presently reside in globular clusters, since most globular clusters are believed to have been disrupted by various processes during our galaxy's evolution) may account for the dark matter inferred to exist in the halo of our own and other galaxies.

7. Radial IMF Gradients in Galaxies

The question of systematic variations of the IMF with position in our own and other galaxies and its possible relation to spatial metallicity variations has received recurrent attention over the past ten years. Because the literature in this area is voluminous and the results often conflicting and/or ambiguous, only a short list of the major lines of evidence will be given here.

(a) Luminosity Functions

Burki (1977) found a correlation between upper LF slope and diameter for a sample of 27 young clusters, in the sense that the LF was flatter for larger clusters. Burki argued that, because mean cluster size increases with galactocentric distance, this result could be interpreted as a steepening of the IMF toward the galactic center. However, since Burki did not specifically group the clusters according to galactocentric distance, it is difficult to judge the validity of this conclusion. The young cluster IMF slopes estimated by Tarrab (1982) show no evidence for such a correlation (Scalo 1985), in agreement with the conclusion of Sagar et al. (1985) for a smaller number of better-studied clusters.

Garmany et al. (1982) claimed to have found a gradient in the IMF of O stars, with a flatter IMF toward the galactic center (opposite to the sense proposed by Burki). However, the LF of the sample does not show the effect, except for the faintest stars with $M_v > -5$, for which the counts appear incomplete (Scalo 1985). Humphreys and McElroy (1984) find no radial gradient in the bolometric LF of a sample which includes supergiants and B stars as well as the O stars, and show that the gradient proposed by Garmany et al. is probably due to a greater degree of incompleteness for stars toward the galactic center.

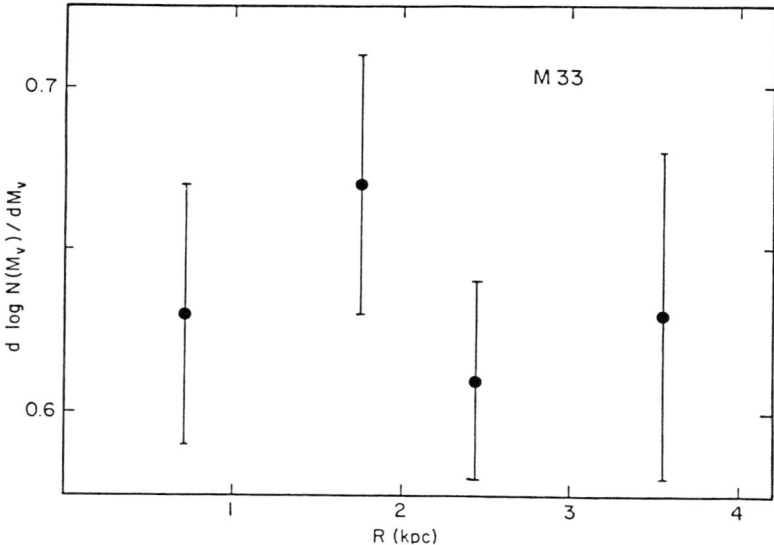

Figure 3. Logarithmic slope of the luminosity function as a function of galactocentric distance in M33, based on the results of Freedman (1985).

The available information on LFs at different positions within individual external galaxies gives no evidence for SMF variations. For the LMC, the studies of several regions outside the bar by Butcher (1977), Stryker and Butcher (1981), and Hardy (1979), and the study of a region near the northwest end of the bar by Hardy *et al.* (1984) all gave LFs in good agreement with each other and with the solar neighborhood function for various M_V ranges between -2 and +4, corresponding to a mass range from 8 $m_\odot$ to about 1.2 $m_\odot$. Freedman (1985) searched for radial variations in the LFs of several galaxies and found no such variations. The logarithmic slopes of the LFs for 4 difference radial intervals in M33 (the galaxy with the largest sample, over 5000 stars) are shown in Fig. 3. The estimated M33 metallicity gradient implies no dependence of LF shape on metal abundance over a factor of two change in metal abundance. In addition, Freedman found very similar LFs in the northern and southern regions, and in the arm and interarm regions. If the limits on LF slope variations are translated into power law IMF slope variations, the implied variation in the IMF slope is less than about ±0.3.

In summary, there is no convincing evidence for internal IMF variations in our own or other galaxies based on star counts.

(b) Infrared Excesses

The ratio of infrared flux to Lyman continuum flux, the so-called infrared excess or IRE, is sensitive to the IMF slope and m_u, but is difficult to interpret because of the unknown fraction of Lyc photons absorbed by dust.

Boisse *et al.* (1981) and Caux *et al.* (1985) found a gradient in the IRE of galactic H II complexes, with larger values of IRE toward the galactic center, although the gradient found by Caux *et al.* was much smaller than that of Boisse *et al.* These results suggest that the IMF is progressively deficient in massive stars toward the galactic center: either a steeper slope or a smaller m_u. However, Scoville *et al.* (1983) and Lester *et al.* (1984) have pointed out problems associated with the large beamsize of the IRE measurements which would reduce or eliminate the proposed gradient. Combined with the fact that variations in the IRE can be caused by variations in the dust optical depth and the stellar birthrate, I conclude that IRE variations which could be linked to IMF variations are not well-established. More high-resolution for IR measurements of distant H II regions are needed to improve the situation.

(c) Radial Excitation Gradients

Spiral galaxies exhibit radial gradients in excitation conditions as measured by the [O III/Hβ] ratio, excitation increasing with galactocentric distance. Some of the effect can be attributed to a metallicity gradient, but some of it may reflect a gradient in the effective temperatures of the exciting stars, possibly requiring more massive stars at larger galactocentric distances. Shields and Tinsley (1976) investigated the contribution of a dependence of m_u of Z (although their abstract states that consistency was found). However, as reviewed in Scalo (1985), even marginal consistency requires an extremely steep IMF with $\Gamma \lesssim -3$, which now appears unreasonable. Because of the sensitivity to various adopted scaling relations and other problems, it appears that the method used by Shields and Tinsley to test for a dependence of m_u on Z is currently indeterminate. At any rate, the scaling relations used by Shields and Tinsley and the empirical W(Hβ) gradient in M101 are inconsistent with any significant dependence of m_u on Z.

Panagia (1980) discussed effects which might account for the systematic decrease in the observed He^+/H^+ ratios in giant H II regions toward the galactic center, and concluded that the dominant effect is a dependence of the upper mass limit m_u on Z, with larger m_u at smaller Z. I have not attempted to review this argument and its associated uncertainties (e.g. dependence on stellar opacity details and the treatment of dust) in detail.

The matter of a dependence of m_u on Z as an explanation of ionization and excitation gradients remains open. An argument against such a Z-dependent m_u is that it would lead to an additional systematic effect in the correlation between luminosity of the brightest star in a galaxy and the total galaxy luminosity, an effect which is not seen (see Scalo 1985). On the other hand, there is an apparent anticorrelation between T_{eff} and Z among galaxies, suggesting an anticorrelation of either m_u or the IMF slope with Z; in my opinion, this anticorrelation is the only compelling evidence for a possible dependence of m_u on Z.

(d) Spiral Arm-Interarm Bimodality

There are three lines of evidence which suggest that stars less massive than about 1 to 3 $m_\odot$ do not form efficiently in spiral arms compared to more massive stars: The detailed analysis of UBVR Hα surface photometry of M83 by Jensen *et al* . (1981); the comparison of arm colors and widths predicted by a dynamical model with observations of M81 by Bash and Visser (1981); and the demonstration by Güsten and Mezger (1983) that the assumption of a lower mass limit $m_l \sim$ 2-3 $m_\odot$ in the arms and $m_l \sim$ 0.1 $m_\odot$ for stars formed outside the arms can account for the observed radial oxygen abundance gradient in our galaxy, as well as alleviate the uncomfortably small gas consumption timescales in our own and other spiral galaxies. Details of the arguments and their uncertainties can be found in Scalo (1985). Taken together, these independent studies strongly suggest that the lower mass limit or the mode mass of the IMF is much larger in spiral arms than in interarm regions.

This conclusion is consistent with the studies cited earlier that found evidence for only higher-mass star formation in starburst galaxies. The implication is that the lower mass limit or mode mass of the IMF increases with increasing star formation rate. Such an IMF is not intrinsically bimodal, but may appear bimodal or multimodal after a given period of time due to the superposition of time-dependent IMFs which arise as the SFR varies. A related interpretation, discussed by Larson (1985), is that the IMF is strictly bimodal and time-dependent. As shown by Larson, this hypothesis can account for a number of observed properties of galaxies. From a theoretical point of view, one idea which can account for the inferred IMF behavior is that discussed by Silk (1977), in which the enhanced cloud heating associated with an increasing number of high-mass stars increases the Jeans mass, and, hence, the characteristic mass. Any such coupling between the radiation field and fragment mass will result in the type of dependence of m_l or mode mass on SFR which is suggested by the observations.

8. Conclusions

All of the evidence summarized above is consistent with the hypothesis that the form of the massive star IMF, taken over scales $\gtrsim$ 1 kpc, is universal. If variations exist, then, for a power law IMF, the variations in logarithmic slope must be less than ± 0.5, and probably smaller. There is no compelling evidence for any systematic slope variations depending on metallicity, galactocentric distance, or galaxy morphology. Unfortunately, the uncertainties are so large that the appropriate value of the slope cannot be reliably estimated; values between -2.4 and -1.3 are possible for m $\gtrsim$ 15 $m_\odot$.

On the scales of open clusters, variations in the IMF form do seem likely. In particular, several open cluster IMFs appear much flatter than the field star IMF at large masses. Since most massive stars are probably formed in OB associations, this suggests that open cluster IMFs are flatter than association IMFs at large masses, although this conclusion must still be regarded as tentative.

More convincing arguments for IMF variations concern the upper and lower mass limits. There is some evidence, mainly from systematic variations in excitation conditions in H II regions, that m_u increases with decreasing metal abundance, but there is other evidence which conflicts with this idea. Several lines of argument suggest that the lower mass limit or mode mass of the IMF increases strongly with increasing star formation rate. The most extreme examples are certain starburst galaxies in which the lower mass limit or mode mass may be as large as 5-15 $m_\odot$.

References

Alloin, D., Collin-Souffrin, S., Joly, M., and Vigroux, L. 1979, *Astr. Ap.*, **78**, 200.
Augarde, R., and Lequeux, J. 1985, *Astr. Ap.*, in press.
Bash, F. N., and Visser, H. C. D. 1981, *Ap. J.*, **247**, 488.
Boisse, P., Gispert, R., Coron, N., Wijnbergen, J. J., Serra, G., Ryter, C., and Puget, J. L. 1981, *Astr. Ap.*, **94**, 265.
Burki, G. 1'977, *Astr. Ap.*, **57**, 135.
Butcher, H. 1977, *Ap. J.*, **216**, 372.
Caux, E., Puget, J. L., Serra, G., Gispert, R., and Ryter, C. 1985, *Astr. Ap.*, **144**, 37.
Clegg, R. E. S., Lambert, D. L., and Tomkin, J. 1983, *Ap. J.*, **250**, 262.
De Gioia-Eastwood, K. 1984, *P. A. S. P.*, **96**, 625.
Donas, J., and Deharveng, J. M. 1984, preprint.
Ellis, R. S., Gondhalekar, P. M., and Efstathiou, G. 1982, *M. N. R. A. S.*, **210**, 223.
Freedman, W. L. 1985, *Ap. J.*, in press.
Garmany, C. D., Conti, P. S., and Chiosi, C. 1982, *Ap. J.*, **263**, 777.
Gehrz, R. D., Sramek, R. A., and Weedman, D. W. 1983, *Ap. J.*, **267**, 551.
Güsten, R., and Mezger, P. G. 1983, *Vistas in Astronomy*, **26**, 159.
Hardy, E. 1979, *A. J.*, **83**, 319.
Hardy, E., Buonanno, R., Corsi, C. E., Jancs, K. A., and Schommer, R. A. 1984, *Ap. J.*, **278**, 592.
Herbst, W., and Miller, D. P. 1982, *A. J.*, **87**, 1478.
Humphreys, R. M., and McElroy, D. B. 1984, *Ap. J.*, **284**, 565.
Jensen, E. B., Talbot, R. J., and Dufour, R. J. 1981, *Ap. J.*, **243**, 716.
Kennicutt, R. C. 1983, *Ap. J.*, **272**, 54.
Lacey, C. G., and Fall, S. M. 1983, *M. N.*, **204**, 791.
Lacey, C. G., and Fall, S. M. 1985, *Ap. J.*, **290**, 154.
Larson, R. B. 1976, *M. N. R. A. S.*, **176**, 31.
Larson, R. B. 1985, preprint.
Larson, R. B., and Tinsley, B. M. 1978, *Ap. J.*, **219**, 46.
Lequeux, J. 1979, *Astr. Ap.*, **80**, 35.
Lester, D. F., Dinerstein, H. L., Werner, M. W., and Harvey, P. M. 1985, *Ap. J.*, in press.
Lynden-Bell, D. 1975, *Vistas in Astronomy*, **19**, 299.
Matteucci, F., and Tornambé, A. 1984, IAU Symp. No. 105, in *Observational Tests of Stellar Evolution Theory*, eds. A. Maeder and A. Renzini (Dordrecht: Reidel), 577.

Mayor, M., and Vigroux, L. 1981, *Astr. Ap.*, **98**, 1.
Nomoto, K. 1984, in *Stellar Nucleosynthesis*, eds. C. Chiosi and A. Renzini (Dordrecht: Reidel), 205, 238.
Nomoto, K., Thielemann, F.-K., and Yokoi, K. 1985, submitted to *Ap. J.*
Olafsson, K., Bergvall, N., and Ekman, A. 1984, *Astr. Ap.*, **137**, 327.
Ostriker, J. P., and Thuan, T. X. 1975, *Ap. J.*, **202**, 353.
Panagia, N. 1980, in *Radio Recombination Line*, ed. P. A. Shaver (Dordrecht: Reidel), 99.
Rieke, G. H., Cutri, R. M., Black, J. H., Kailey, W. F., McAlary, C. W., Lebofsky, M. J., and Elston, R. 1985, *Ap. J.*, **290**, 116.
Rieke, G. H., Lebofsky, M. J., Tompson, R. I., Low, F. J., and Tokunaga, A. T. 1980, *Ap. J.*, **238**, 24.
Sagar, R., Piskunov, A. E., Myakutin, V. I., and Joshi, V. C. 1985, preprint.
Scalo, J. M. 1981, in *Physical Processes in Red Giants,* eds. I. Iben and A. Renzini (Dordrecht: Reidel), 77.
Scalo, J. M. 1985, *Fund. Cosmic Phys.*, in press.
Scoville, N. Z., Becklin, E. E., Young, J. S., and Capps, R. N. 1983, *Ap. J.*, **271**, 512.
Shields, G. A., and Tinsley, B. M. 1976, *Ap. J.*, **203**, 66.
Silk, J. 1977, *Ap. J.*, **214**, 718.
Sneden, C. 1985, in *ESO Workshop on Production and Distribution of C, N, O Elements,* in press.
Sneden, C., Lambert, D. L., and Whitaker, R. W. 1979, *Ap. J.*, **234**, 964.
Stryker, L. L, and Butcher, H. R. 1981, in IAU Colloq. No. 68, eds. A. G. Davis Philip and D. S. Hayes (Schenectady: Davis), p. 255.
Taff, L. G. 1974, *A. J.*, **79**, 1280.
Talbot, R. J., and Arnett, W. D. 1973, *Ap. J.*, **186**, 69.
Tarrab, I. 1982, *Astr. Ap.*, **109**, 285.
Terlevich, R., and Melnick, J. 1985, *M. N. R. A. S.*, in press.
Tinsley, B. M. 1980, *Fund. Cos. Phys.*, **5**, 287.
Twarog, B. A., and Wheelerf, J. C. 1981, *Ap. J.*, **261**, 636.
Vanbeveren, D. 1982, *Astr. Ap.*, **115**, 65.
Vanbeveren, D. 1984, *Astr. Ap.*, **139**, 545.
Van Buren, D. 1985, *Ap. J.*, in press.
van den Bergh, S. 1976, *A. J.*, **81**, 797.
van den Bergh, S. 1981, *P. A. S. P.*, **93**, 712.
Viallefond, F., and Thuan, T. X. 1984, *Ap. J.*, **269**, 444.
Woosley, S. E., Axelrod, T. S., and Weaver, T. A. 1984, in *Stellar Nucleosynthesis,* eds. C. Chiosi and A. Renzini (Dordrecht: Reidel).

Discussion   :   SCALO.

ZINNECKER :

A comment regarding the upper stellar mass limit : it seems to me that on the basis of theoretical considerations collapsing protostars cannot be more massive than 100-150 M_o. However, one possible way to overcome this limit is the merging of two or several massive stars (see the related poster paper).

LUMINOUS CONTENT OF GALAXIES: INFERENCES FROM SUPERNOVAE

J. Craig Wheeler
Department of Astronomy
University of Texas

Abstract - Some luminous stars undoubtedly explode as supernovae, but it is not yet certain that fate awaits them all. The connection between luminous stars and supernovae is reviewed in terms of the statistical rates of supernovae and pulsars, the constraints of nucleosynthesis, and the various classifications of supernovae by their spectra and light curves, including a newly confirmed class of peculiar Type I supernovae.

1. Introduction

There are a few simple statements about supernovae that reflect the common understanding of them. The rate of explosion of Type I supernovae (SNI) is about the same as that of Type II supernovae (SNII) in our Galaxy. The rate of explosion of SNII is about the same as the birthrate of stars with mass in excess of $10\,M_\odot$ and of the birthrate of pulsars. The Crab nebula is an obvious example of a supernova which produced a pulsar. Massive stars account for the synthesis of the heavy elements.

Based on these popular notions, one might conclude that all stars with $M > 10\,M_\odot$ explode as SNII, produce pulsars, and generate the heavy elements. There are numerous unanswered questions, however, that suggest that few of these issues are settled. The evolution of massive stars will be briefly summarized in Section 2 to set the stage. The information, uncertainties and open questions involved with the study of supernova statistics in various galaxy types will be addressed in Section 3. The constraints from nucleosynthesis are discussed in Section 4, and the different categories of supernovae classified by spectra and light curves and some speculations on those differences are presented in Section 5.

2. Evolution

Figure 1 gives a summary in the $\log \rho_c$, $\log T_c$ plane of the evolution of massive stars. Stars with mass less than about $8\,M_\odot$ form dense, degenerate carbon/oxygen cores. If they proceed to carbon ignition, they are believed to explode by means of a subsonic deflagration process (Mazurek, Meier, and Wheeler 1977) and leave no compact remnant. Most of these stars, however, are destined to eject their envelopes as planetary nebulae

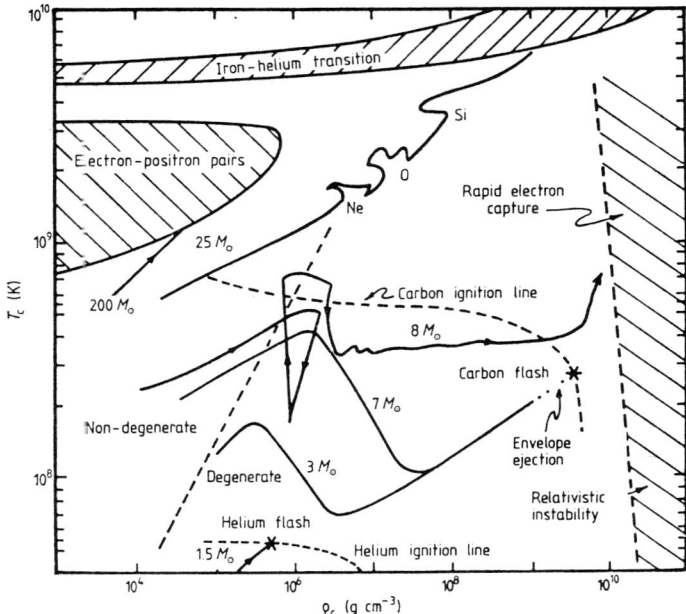

Figure 1. Evolutionary tracks are given in the plane of central density and central temperature. Most stars with mass less than about 8 solar masses produce white dwarfs after envelope ejection, but some may explode by degenerate carbon deflagration. Above 8 solar masses most stars are predicted to undergo core collapse. (From Wheeler 1981).

and leave the cores to from cooling stable white dwarfs. We do not know if some fraction of these stars near the upper mass limit retain their envelopes and proceed to explode. The currently most popular model for SNI involves the delayed deflagration of such a C/O core left behind in the first stage of evolution which is later rejuvenated by mass transfer from a companion (Sutherland and Wheeler 1984).

Evolutionary models of stars with mass just in excess of about 8 $M_\odot$ up to about 10 - 12 $M_\odot$ show that carbon ignites off-center and burns non-degenerately. A degenerate core of O/Ne/Mg forms and collapses from the effects of rapid electron capture (Nomoto 1984). The resulting collapse is hindered, but not reversed by oxygen burning. This and the steep density profile which accelerates the shock enhance the probability that a rebound shock from neutron star formation will generate an explosion (Hillebrandt, Nomoto, and Wolff 1984).

Above 10 - 12 $M_\odot$, all burning occurs quasistatically and the star forms an iron core. This core photodisintegrates and collapses. There is considerable question whether an explosion ensues in this case by the direct action of a bounce shock, but heating by neutrinos leaking from the neutronized core on timescales approaching 1 second may boost a standing, stagnated shock into an outward propagating explosion (Wilson 1984).

Stars of this mass, which are thought to be prime candidates for the synthesis of heavy elements, may explode, but some may collapse to produce black holes.

Masses in excess of about 100 $M_\odot$ are thought to collapse due to the formation of electron/positron pairs. This collapse is reversed by oxygen ignition and leads to total disruption unless the mass exceeds about 300 $M_\odot$, in which case the collapse proceeds to completion resulting in the formation of a black hole (Wheeler 1977, Woosley and Weaver 1982).

3. Statistics

Tammann (1982) gives the rate of explosion of SNI and SNII in the Galaxy as 1 per 36 years and 1 per 44 years, respectively, based on interpolation from extragalactic rates. A comparison of the rate of occurrence of SNII with the birthrate of massive stars (Wheeler 1981) shows that if stars with mass significantly less than $10 M_\odot$ explode, there would be more SNII than observed. On the other hand, statistics do not determine whether stars with mass much in excess of 20 $M_\odot$ explode, because such explosions are expected to be rare and lost in the noise. Nucleosynthesis provides some constraint on these massive stars, as will be argued in section 4.

Figure 2 shows that there are uncertainties in the ratio of SNI to SNII in spiral galaxies. This figure presents raw number counts from the magnitude limited sample of Oemler and Tinsley (1979). One cannot prove, based on this data, that the ratio of SNI to SNII is not unity, but neither is there a strong basis for assuming that in our Galaxy (Sbc) that the ratio is unity. The net supernova rate can also be estimated from the occurrence of historical supernovae. Breaking down the rate by type is more difficult. Several are suspected of being SNI, but there is no known SNII in that sample. The Crab nebula event might have been a SNII, but its light curve is ill determined, and it has many features which are simply unorthodox, and difficult to categorize. In any case, despite these uncertainties, there is little evidence which would support the notion that SNII derive from stars with original mass much less than 10 $M_\odot$.

The rate of pulsar formation, assuming only about 20 percent are observed due to beaming, is about one per 20 -50 years (Lyne 1982). This rate is crudely consistent with the rate of formation of SNII. A troubling fact is that evidence for neutron stars is observed only in filled center or composite type supernova remnants (SNR), not in the shell- type remnants which are 3/4 of the sample (Helfand 1984a). With the advent of the Einstein Observatory this search can be made not only in the radio which may be beamed, but by searching for X-ray synchrotron nebulae which should be isotropic emitters. Since the rate of formation of filled shell remnants which show evidence for pulsars is low, and the majority of SNR show no such evidence, the question is raised as to whether there must be a source of pulsars with no associated supernova remnant.

Helfand (1984b) has raised another interesting possibility. He questions whether the pulsar radiation is, in fact, not beamed, so the directly observed rate, about one pulsar per 200 years, which agrees with the rate of formation of filled center SNR, is the appropriate one. By this hypothesis, pulsars would be produced in all, but only in, the filled center SNR. The shell type SNR would be supposed to come from some event, perhaps SNI, which left no neutron star.

Another view on this problem is provided by a statistical study of the angular distribution of SNR (Li, Wheeler, and Bash, 1984; Li, Wheeler, Bash, and Jefferys 1985). These studies examine the distance independent angular distribution of the SNR as compared to Monte Carlo models which vary the proportion of SNR in the spiral arms and

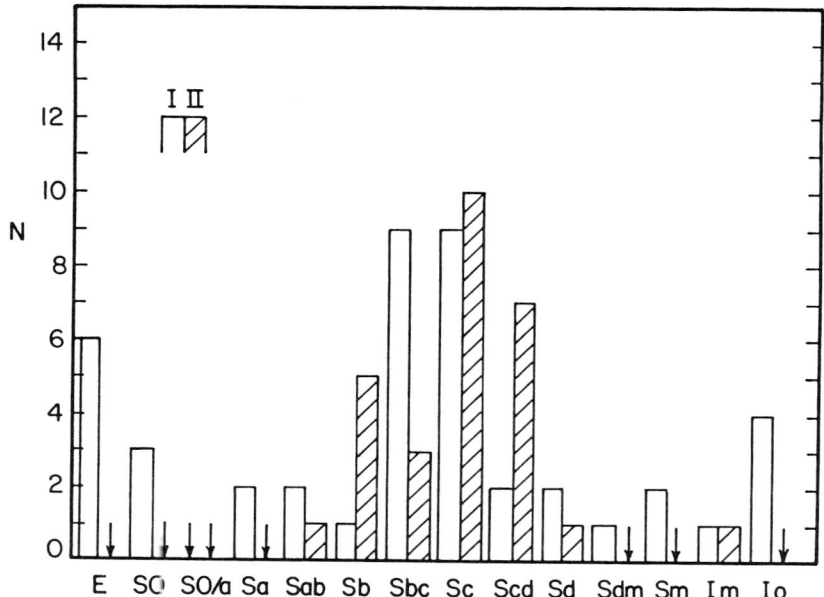

Figure 2. The number of supernovae identified from their spectra as SNI or SNII is presented as a function of galaxy type from the magnitude limited sample of Oemler and Tinsley (1979). Two of the supernovae in the I0 category were in NGC 5253 which should perhaps be reclassified as Im. (From Wheeler and Wheeler 1983).

the exponential disk. These studies reject the hypothesis that the shell type SNR are a disk population as would be expected if they were entirely from SNI. The model which best fits the data is one in which all the SNR reside in spiral arms, implying that they had massive star progenitors. The questions of why SNI do not seem to contribute to the sample of old SNR (perhaps because with large scale height they expell their ejecta from the galaxy) and why the shell SNR show no sign of pulsars remain open.

Models of SNII light curves, particularly the type with a distinct plateau (Falk and Arnett 1977; Litvinova and Nadyozhin 1983), can be reproduced with an explosion in an extended envelope. This strongly suggests that such supernovae are not merely massive stars, but likely to be red supergiants. An interesting question then arises as to why there is an apparent paucity of such events in late-type galaxies.

Figure 2 shows the extreme example of this phenomenon are the I0 galaxies. As for elliptical galaxies, these galaxies have been observed to have only SNI, no SNII. For ellipticals the standard argument is that there is no recent star formation, so all the supernovae must be old, but I0 galaxies typically have dark dust lanes, a typical sign of on-going star formation. Oemler and Tinsley (1979) have used such evidence to argue that SNI come from intermediate mass stars and that ellipticals must have a current level of star formation. An interesting example of a galaxy which is typically, if not completely accurately, labeled an I0 is M82, a site of particularly active star formation which apparently has generated a number of deeply buried radio SNR (Kronberg et al 1985). The type of these events is, of course, unknown.

The paucity of SNII is not complete in other late type galaxies, but perhaps just as mystifying. Oemler and Tinsley present rates for galaxies of different morphological type corrected for various selection effects. For Sc and Scd galaxies, the most prolific supernova producers by number, they give the rates of SNI and SNII as 0.40 and 0.35, respectively, in units of 10^{12} $L_\odot$-yr. For the category including Sd, Sdm, Sm, and Im they give 0.48 and 0.14 in the same units, showing the marked paucity of SNII. They give 1.7 and 0 for I0 galaxies. The rate of production of SNI in I0 galaxies is based on only four supernovae, the two events in NGC 5253, SN1985b and SN1972e and two others. Campbell and Terlevich (1984) have shown, however, that NGC 5253 has an abundance of red supergiants and it is probably nearer to an Im in classification, than an I0. Thus the rate of SNI in I0 galaxies should be only 0.8 based on only two objects, and hence uncertain, although interestingly large. The rate of SNI in the Sd-Im category should probably likewise be increased slightly, enhancing the relative paucity of SNII. The Sd through Im galaxies are characterized by abundant star formation, O stars, and red supergiants. Where, then, are the SNII? Are they hidden by dust as in M82? Is the paucity a temporary thing, caused by observing a starburst at the wrong phase? Does bi-model star formation short change some galaxies of massive stars? These questions are worthy of serious investigation.

4. Nucleosynthesis

An important development in the long study of nucleosynthesis of the heavy elements in massive stars has been the recent upward revision of the rate of the $^{12}C(\alpha, \gamma)^{16}O$ reaction by a factor of 3 to 5 (Kettner et al 1982). This revision changes the systematics all along the alpha chain and gives a particularly satisfying fit of the isotopic and elemental *ratios* to the cosmic values for elements from carbon to the iron peak in the model of a star of 25 $M_\odot$ (Woosley 1985). The increased rate also changes the *absolute* abundances. The mass fraction of the ^{16}O increases by a factor of 1.5 to 2 (for an increase in the rate a factor of 3 and 5, respectively) for models with mass in the range 10 to 30 $M_\odot$ (Thielemann, 1984). This raises the possibility for an interesting constraint. Stars of about 20 $M_\odot$ are argued to be typical agents of nucleosynthesis since they are near the peak of the curve weighting ejected mass with the stellar mass function. Twarog and Wheeler (1982) pointed out, however, that for standard yields and mass functions, there was a danger that if all the massive stars (M > 12 $M_\odot$) exploded, an overproduction of various elements, particularly oxygen, would occur. The overproduction was at least a factor of 1.5 for a reasonably steep mass function ($dN/dM \propto M^{-3}$), and worse for shallower mass functions which imply greater numbers of massive stars. With the new rates, the overproduction would be at least a factor of 2 to 3, and that is far larger than the scatter in the observed values at the current epoch.

There are three possible conclusions that might be reached from this discussion. Perhaps the rate for $^{12}C(\alpha, \gamma)^{16}O$ is not enhanced. This helps the overproduction of oxygen, but restores old problems with the ratios of certain species. For instance Ne/O and Mg/O are typically too high by a factor of about four. A second possibility is that the rate is higher and the bulk production of elements like oxygen is predicted to be too large. This may imply that some, perhaps a majority, of the massive stars that were presumed to be the major source of nucleosynthesis collapse rather than explode. Alternatively, the excess ejecta may be expelled from the Galaxy, a notion previously raised for SNI. A third possibility is that we simply do not yet know the nuclear physics and the evolution of massive stars sufficiently well to make precise comparisons with observations. In this case, any conclusion concerning the fate of massive stars, that they explode and provide the site of synthesis of heavy elements, or that they collapse and do not, is premature. The fate of Wolf-Rayet stars which are thought to be representative of this class of massive stars is likewise uncertain.

5. Supernova Classifications

Recent developments have suggested we are on the verge of a period of new understanding and reclassification of supernovae. Despite the suggestion of five classes by Zwicky (1965), supernovae have been typically characterized by their spectra as either SNI (no apparent hydrogen) thought to be caused by a thermonuclear explosion or SNII (with detectable hydrogen) thought to arise from core collapse. Two recent supernovae, SN 1983n and SN1984l, have confirmed that there is a class of supernovae which blur the typical lines of empirical classification and provoke new questions about the nature of supernovae.

SN1983n and SN1984l had spectra which were virtually identical at maximum light (Wheeler and Levreault 1985). They are hydrogen deficient, and certainly do not resemble the spectra of SNII (nor any other of Zwicky's types). The spectra qualitatively resemble those of SNI , but there are distinct differences, and hence these events have been called peculiar SNI. The Doppler broadening and hence the ejecta velocity is about the same as for a SNI. The shape of the light curve is also very similar. This implies that the mass of the ejecta, which determines the thermal diffusion time and the width of the light curve peak, is comparable for the classical and peculiar SNI. The peculiar events seem to be associated with Population I stellar environments, HII regions, spiral arms, and bars of barred spiral galaxies, environments distinctly eschewed by classical SNI. SN1983n was a radio supernova (Sramek et al 1984) and SN1984l may be as well (Sramek 1985). No classical SNI has been observed in the radio.

The peculiar SNI appear to be dimmer at maximum light than the classical variety by a factor of order four. In the absence of a hydrogen envelope, the luminosity is presumed to come from the radioactive decay of ^{56}Ni, as for classical SNI. Less luminosity implies less nickel ejected, however, and hence a weaker thermonuclear explosion. Since the peculiar events seem to have the same mass and velocity and hence kinetic energy as the classical variety, the implication is that the kinetic energy can not arise purely from thermonuclear burning, but must rely on some other process. The only other obvious candidate is core collapse, as for SNII.

The ejecta mass must be of order 2 - 3 $M_\odot$ from the similarity of the light curve shapes. If to the ejecta mass one adds the mass of a neutron star, of order 1 $M_\odot$, then the mass of the immediate progenitor of the peculiar SNI is of order 3- 4 $M_\odot$. If this is interpreted as the mass of an evolving helium core, then the original main sequence mass would be of order 10 - 20 $M_\odot$, basically the same mass range as that which produces the bulk of the SNII. This implies that while the true nature of the peculiar SNI is still uncertain, there is a strong suggestion that the physical mechanism of their explosion is more like that of a SNII than a SNI, despite the spectral similarity to the latter. The relatively small mass deduced for the progenitor stars argues that the peculiar SNI are not directly related to Wolf-Rayet stars despite the suggestive nature of the hydrogen deficiency and the association with Pop I. Note, however, that Cahen (1985) argues that

Wolf-Rayet stars can be as bright as normal supernovae despite their lack of extended envelopes if they eject of order 0.1 $M_\odot$ of ^{56}Ni, and that the light curve decline is enhanced by the rapid recombination wave in the helium envelope.

SNII also come in two categories, characterized by their light curves, those which show a plateau, and those which show a linear decline. Recently Doggett and Branch (1985) have noted that whereas the majority (2/3 of the sample) plateau events are about 1.5^m dimmer than SNI, the linear events (1/3 of the sample) are nearly as bright, and that when plotted on the same scale, the light curves of the linear SNII qualitatively resemble those of classical SNI. This leads to the following speculation. Perhaps of the events which show hydrogen in the spectrum the plateau SNII are core collapse events from stars with $M > 8\ M_\odot$ and the linear variety are the result of thermonuclear explosions of stars with degenerate C/O cores and mass slightly less than 8 $M_\odot$. Of the supernovae with no apparent hydrogen, the classical SNI are thermonuclear explosions of C/O cores which have been laid bare in binary systems and then rekindled, whereas the peculiar events arise from the collapse of more massive stars which have lost their envelopes by mass transfer or winds.

6. Conclusion

There is no question that some supernovae come from luminous stars. The question of which luminous stars produce supernovae is more difficult to answer. Statistics suggest that stars with $M > 10\ M_\odot$ (or perhaps a little less) produce SNII. Current evolutionary theory demands that stars above this limit collapse to form neutron stars, but the question of the progenitors of pulsars and of SNR of various morphology is still very uncertain. Some galaxies with an abundance of luminous stars nevertheless seem not to produce a corresponding number of SNII. Constraints from nucleosynthesis hint that not all massive stars should eject their full complement of heavy elements. Some may collapse to produce black holes or evolve in ways or environments we have yet to fathom.

This research was supported in part by NSF grant 8413301.

References

Cahen, S. 1986, Proceedings of the Fifth Moriond Conference on Astrophysics.
Campbell, A. W., and Terlevich, R. 1984, MNRAS, **211**, 15.
Doggett, J. B., and Branch, D. 1985, preprint.
Falk, S. W., and Arnett, W. D. 1977, Astrophys. J. Suppl., **33**, 515.
Helfand, D. J. 1984a, in Supernova Remnants and Their X-Ray Emission, ed. P. Gorenstein and J. Danziger (Dordrecht: Reidel).
Helfand, D. J. 1984b, private communication.
Hillebrandt, W., Nomoto, K. and Wolff, R. G. 1984, Astr. and Ap., **133**, 175.
Kettner, K. U., Becker, H. W., Buchmann, L., Görres, J., Kräwinkel, H., Rolfs, C., Schmalbrock, P., Trautvetter, H. P., Vlieks, A. 1982, Z. Physik, **A308**, 73.
Kronberg, P. P., Biermann, P., and Schwab, F. R. 1985, Ap. J., **291**, 693.
Li, Z., Wheeler, J. C., and Bash, F. N. 1984, in Proceedings of the Erice Workshop on Stellar Nucleosynthesis, ed. C. Chiosi and A. Renzini.
Li, Z., Wheeler, J. C., Bash, F. N., and Jefferys, W. H. 1985, in preparation.
Litvinova, I. Yu., and Nadyoozhin, D. K. 1983, Ap. and Space Sci., **89**, 89.

Lyne, A. G. 1982, in Supernovae: A Survey of Current Research, ed. M. J. Rees and R. J. Stoneham (Dordrecht: Reidel), p. 405.
Mazurek, T. J., Meier, D. L., and Wheeler, J. C. 1977, Astrophys. J., **215**, 518.
Nomoto, K. 1984, in Proceedings of the Erice Workshop on Stellar Nucleosynthesis, ed. C. Chiosi and A. Renzini, in press.
Oemler, A., and Tinsley, B. M. 1979, A. J., **84**, 985.
Sramek, R. A., Panagia, N., and Weiler, K. W. 1984, Ap. J. Letters, **285**, L63.
Sramek, R. A. 1985 private communication.
Sutherland, P. G., and Wheeler, J. C. 1984, Ap. J., **280**, 282.
Tammann, G. A. 1982, in Supernovae: A Survey of Current Research, ed. M. J. Rees and R. J. Stoneham (Dordrecht: Reidel), p. 371.
Thielemann, F. K. 1984, private communication.
Twarog, B. A., and Wheeler, J. C. 1982, Ap. J., **261**, 636.
Wheeler, J. C. 1977, Astrophys. Space Sci., **50**, 125.
Wheeler, J. C. 1981, Reports on Progress in Physics, **44**, 85.
Wheeler, J. C. and Levereault, R. 1985, Ap. J. Letters, in press.
Wheeler, J. C., and Wheeler, J. A. 1983, in Science Underground, ed. M. M. Nieto, W. C. Haxton, C. M. Hoffman, E. W. Kolb, V. D. Sandberg, and J. W. Toevs (New York: American Institute of Physics), p. 214.
Wilson, J. R. 1984, preprint.
Woosley, S. E. 1986, Proceedings of the Fifth Moriond Conference on Astrophysics.
Woosley, S. E., and Weaver, T. A. 1982, in Supernovae: A Survey of Current Research, ed M. J. Rees and R. J. Stoneham (Dordrecht: Reidel).
Zwicky, F. 1965, in Stars and Stellar Systems, ed. L. H. Aller and D. B. McLaughlin (Chicago, University of Chicago Press), p. 367, vol. VIII.

Discussion   :   WHEELER.

FEAST :

Do you believe that the statistics of runaway stars gives one any information on the masses of supernovae?

WHEELER :

I have never been convinced that runaway O, B stars had anything to do with supernovae. The recent work of Gies, referred to by de Loore, does nothing to dissuade me of that notion.

SREENIVASAN :

Is there a good theroretical reason why WR stars should expell more than 2-3 M_o in a SN event?

WHEELER :

Observations suggest that Wolf-Rayet stars have masses exceeding 8 M_o. I believe the theory is consistent with that, but it is possible that a final very rapid, very short lived mass loss phase could bring the mass down.

DE JAGER :

1) Theoretical research seems to show that Wolf-Rayet stars become supernovae. But in view of the low birth rate of WR's the WR supernovae must constitute a small fraction of the supernovae. Since WR's are H-poor they cannot yield type II SN's, but they cannot be type I either because WR's are extreme population I objects. So, WR supernovae, if they exist, constitute another class of fairly rare supernovae.

With regard to Niemela's interesting observations I think it is not sure that the progenitor was a WR; the WR characteristics may just have been a consequence of the explosion just having started.

2) the Cas A progenitor must have been faint at maximum, M_V about - 12. How do we reconcile this with the claim that the progenitor was massive?

WHEELER :

As I said, I think there is doubt that all Wolf-Rayet stars necessarily explode even on the basis of the theoretical models. I argued that constraints based on nucleosynthesis may cut either way. Models for the core-collapse induced explosion are still ambivalent. One can understand why Cas A may have been -12^m at peak in terms of an explosion in a rather compact object like a Wolf-Rayet star so that much of the shock energy is dissipated in adiabatic expansion before the ejecta become sufficiently optically thin to radiate, as has been argued by Chevalier. The work of Cahen, Schaeffer and Cass{ et al. suggests that if about 0.1 M_o of radioactive material is ejected, the event will have a normal Type II magnitude, so Cas A must not have ejected such an amount of nickel.

CAMPBELL :

I would just like to say something about NGC 5253. It's a lenticular galaxy with a smooth red population, well fit by an $R^{1/4}$ law, which has a starburst nucleus. Why not describe it as an S0 galaxy with a nuclear starburst, just as for spirals?

SANDAGE :

The morphological definition of an S0 is that it doesn't have dust lanes or emission lines. NGC 5253 is amorphous like M82 and therefore should be classified as IrII.

WHEELER :

But it is not amorphous- for example, Jedrjewski's thesis shows that it is a lenticular except right at the nucleus.

KAFATOS :

Craig, could you tell us something about the number statistics of these peculiar type I supernovae?

WHEELER :

Over the last few years during which these SN have been observed the number seems to be about 10% of all supernovae. Branch has argued that they may be selected against because of the fainter magnitudes compared to classical type I events.

GRAHAM to NIEMELA :

Is it not true that in Niemela's work with Maria Teresa Ruiz and Mark Phillips they saw a WR type spectrum in a recent SN before maximum and does this not therefore support the idea that some WR stars turn into supernovae?

NIEMELA :

Yes, we observed a WN6 type spectrum about 10 days before the maximum light. However, we do not know whether this WR spectrum was caused by the explosion, or if it was the WN star that exploded. In any case, the spectrum we saw was N enriched (ref. Ap.J. 289, 52, 1985).

SESSION 8.

LUMINOUS STELLAR CONTENT OF THE GALAXY-INTEGRATED PROPERTIES-II.

Chairman : M.FEAST.

1. P.MEZGER: Luminous Stellar Content of the Galaxy: Inferences from Radio and Infrared Data.

LUMINOUS STELLAR CONTENT OF THE GALAXY: INFERENCES FROM RADIO AND INFRARED DATA

P. G. Mezger
Max-Planck-Institut für Radioastronomie
Auf dem Hügel 69
5300 Bonn 1, F.R.G.

ABSTRACT. Lyman continuum (Lyc) photon production rates can be estimated from radio free-free emission and used to estimate the star formation rate (SFR) of O stars. If this SFR is linked to the total SFR through a constant IMF (m $\gtrsim$ 0.1 $m_\odot$) one derives for our Galaxy a present-day SFR of $\sim$10 $m_\odot$ yr^{-1}, which is close to the average SFR over the age of the galactic disk. This is difficult to reconcile with a formation law of the form SFR $\Psi \propto M_{gas}^k$ with k>0 which yields SFRs which decrease with time. Even more severe is the fact that the mass distribution of the galactic disk cannot be reproduced by the present-day SFR with a constant IMF. Bimodal star formation, however, reduces the rate at which matter is permanently locked up in low mass and dead stars by nearly a factor of three, and gets reasonable agreement between the present-day distribution of stellar mass and lock-up rate. Bimodal star formation means that stars with m >0.1 $m_\odot$ form in the interarm region while in spiral arms induced star formation produces only stars with m >m_c $\sim$2-3 $m_\odot$.

Our Galaxy emits about one third of its stellar luminosity as IR emission from dust. Warm (30-40 K) dust is heated by OB stars. Most of the hot (250-500 K) dust is heated by M giants with heavy mass loss, whose progenitors appear to be stars with $\sim$2-8 $m_\odot$, and which have a similar radial distribution as OB stars. From an estimate of their present-day luminosity we arrive at the conclusion that the SFR of stars with m $\gtrsim m_c$ was about constant over the past 10^8 yr.

Observations of radio recombination of H and He show that the Lyc photon radiation field gets softer with decreasing galactocentric distance. This effect is linked to an increase of the metal abundance towards the galactic center. This can be attributed to the combined effect of UV line blanketing, increase of stellar radius and decrease of the upper mass limit of the IMF.

Stellar parameters and observable radio and infrared characteristics of HII regions.

Practically all radiation of a massive, hot and luminous star surrounded by gas and dust of sufficiently high column density is absorbed and re-emitted as radio and infrared (IR) radiation. The number of Lyman continuum (Lyc) photons, N'_{Lyc}, absorbed per second by the gas is related to the free-free flux density S_ν (in Jy) at a frequency ν (in GHz) high enough so that opacity effects are negligible, by

$$N'_{Lyc} = a(\nu, T_e)(1+y^+)^{-1}\, 4.76\, 10^{48}\, T_e^{-0.45}\, \nu^{-0.1}\, S_\nu D^2 \qquad (1a)$$

Here D is the distance in kpc between the sun and the HII region, T_e in degree K is the electron temperature, y^+ the number abundance of ionized helium and $a(\nu, T_e)$ a factor of order unity. The Lyc photon production rate, N_{Lyc}, is related to N'_{Lyc} by

$$N'_{Lyc} = (1-f_{esc})f_{net}\, N_{Lyc} \qquad (1b)$$

with f_{esc} the fraction of Lyc photons which escape from the HII region and $(1-f_{net})$ the fraction of Lyc photons which are directly absorbed by dust. The total radio and IR spectrum of a typical radio HII region (i.e. a relatively young HII region with central emission measures of $\gtrsim 10^4$ pc cm^{-6} and angular diameters of some arc minutes or less) is shown in Fig. 1.

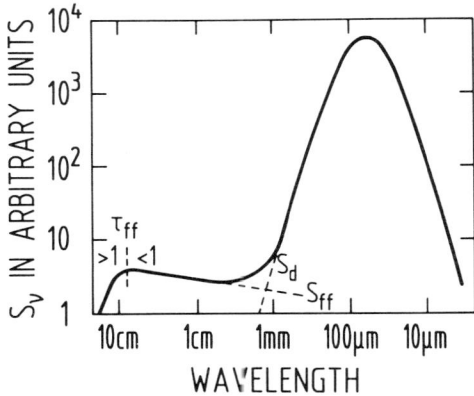

Fig. 1: *Typical spectrum of a radio HII region. At higher frequencies ($\tau_{ff} < 1$) $S_{ff} \propto \nu^{-0.1}$. To a first approximation the spectrum of the dust emission can be approximated by a modified Planck function $S_d \propto \nu^m B_\nu(T_c)$.*

The steep increase of the spectrum at wavelengths $\lesssim 1$ mm is due to thermal emission from dust grains. This part of the spectrum can usually be approximated by the modified Planck-function $\nu^m B_\nu(T_d)$, with m ~ 2 and T_d a mean temperature of the dust grains. The integrated spectrum yields the IR luminosity, L_{IR}, which is related to the total stellar luminosity of the ionizing star(s), L_*, by

$$L_{IR} = f_d L_*$$

with f_d the fraction of the stellar luminosity which is absorbed and reradiated by dust.

Finally, the ratio of the integrated line intensities of two adjacent radio recombination lines of H and He yields immediately the number abundance of ionized helium

$$y^+ = n(He^+)/n(H^+) \qquad (3)$$

averaged over the HII region. Knowing the true He-abundance, y = n(He)/n(H), one can infer the ratio of He ionizing Lyc photons to the total number of Lyc photons of the ionizing star(s), (see, e.g. Mezger, 1979).

The derivation of star formation rates (SFR) from Lyc photon production rates.

This procedure is straight forward and has most recently been discussed by Güsten and Mezger (1983; hereafter referred to as Paper I). In a first step one separates thermal (free-free) and nonthermal (synchrotron) emission using radio continuum surveys made at two widely separated frequencies. This yields N'_{Lyc} (see eq. (1a)) and - with appropriate corrections - via eq. (1b) the total Lyc photon production rate N_{Lyc} of all O stars in the region under consideration.

The formation rate of O stars is linked to the total SFR by the Initial Mass Function (IMF). As usual we separate the creation function, i.e. the number of stars formed per unit time in the mass interval m, m+dm, into a time-dependent SFR $\Psi(t)$ (in units of $m_\odot$ yr^{-1}) and a time-independent initial mass function (IMF) $\phi(m)$, with upper and lower mass limits m_u and m_L, respectively. The IMF is normalized in the usual way, viz.

$$\int_{m_L}^{m_u} \phi(m) m \, dm = 1$$

In the following discussion we consider the IMF derived by Salpeter (1955; $\phi(m) \propto m^{-2.35}$) and by Miller and Scalo (1979; $\phi(m) \propto m^{-3.62}$ for $m \gtrsim 30\, m_\odot$) as limiting cases for O stars. With the assumption that the present-day SFR, $\Psi(t_o)$, did not change during the main sequence (MS) lifetime of O stars, τ_{MS}, we derive in Paper I the following relation between Lyc photon production rate $N_{Lyc}(t_o)$ and SFR $\Psi(t_o)$

$$N_{Lyc}(t_o) = \Psi(t_o) \int_{m_L}^{m_u} \phi(m) dm \int_0^{\tau_{MS}(m)} N_{Lyc}(m,\tau) d\tau \qquad (4)$$

The ratio of Lyc photon luminosity to SFR can be computed for a given IMF $\phi(m)$ from stellar model atmospheres. In Fig. 2 is shown the product $\phi(m) N_{Lyc}(m)$ for the two IMFs. If multiplied with $\tau_{MS}(m)$ the maximum of the corresponding curves is shifted towards stars with masses $m = 30-40\, m_\odot$. This means that stars of spectral type O7-O6 are the main contributors to the Lyc photon production rate in the galactic plane. Contributions of stars with $m \lesssim 10\, m_\odot$ (i.e. sp. type B2 and later) are negligible. Hence we have to extend the integration only over $m \geq 10\, m_\odot$. Convenient numerical expressions for the quantity

$$\frac{\langle N_{Lyc}(t,\phi)\rangle}{\Psi(t)} = \phi(1)\cdot P(\phi) = \left\{\int_{m_L'}^{m_u}\phi(m)mdm\right\}^{-1}\int_{10m_\odot}^{m_u}\phi(m)dm\int_0^{\tau_{MS}(m)}N_{Lyc}(m,\tau)d\tau \quad (5a)$$

are given in Paper I. Note that (5a) depends on the lower mass limit m_L' through the normalization factor of $\phi(m)$, which is

$$\int_{m_L'}^{m_L}\phi(m)mdm = \begin{cases} 1 & \text{for } m_L' = m_L \\ 1-\Delta(m_c) & \text{for } m_L' = m_c > m_L \end{cases} \quad (5b)$$

with $\Delta(m_c) = \int_{m_L}^{m_c}\phi(m)mdm$ and $1-\Delta(m_c) = 0.80\, m_c^{-0.4} - 0.23$ (5c)

for the Miller-Scalo IMF, $m_L = 0.1\, m_\odot$ and $1 \leq m_c/m_\odot \leq 6$. All numerical values given in the following relate to this IMF.

The SFR which corresponds to a given Lyc photon production rate at a time t is

$$\Psi(t) = N_{Lyc}(t)/\phi(1)P(\phi) \quad (5d)$$

of which the fraction $\frac{dM_*}{dt} = (1-r)\Psi$ is permanently locked up in stars. $\phi(1)P(\phi) = 2.5\, 10^{52}$ Lyc photons $s^{-1}/m_\odot$ yr^{-1} is the number of Lyc photons produced per sec per solar mass per year of gas converted into stars (Paper I), $r = 0.42$ is the "instantaneous return rate".

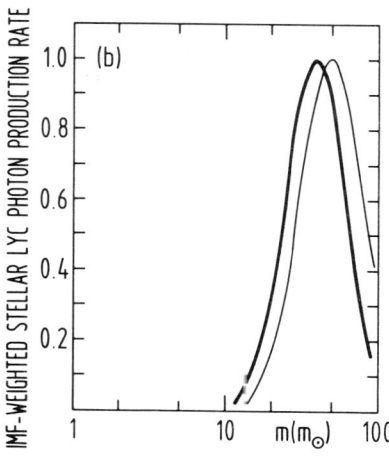

Fig. 2: The IMF weighted stellar Lyc photon production rate (Cox et al., 1985). The thin solid curve relates to the Salpeter IMF, the heavy solid curve relates to the Miller-Scalo IMF.

Global present-day SFRs in the Galactic Disk derived for a constant IMF

In Paper I we estimate a Lyc photon production rate in the galactic disk of

$$\sum_{R=2}^{10} N_{Lyc}(R) = 2.6 (^{+0.8}_{-0.5}) \, 10^{53} \, s^{-1} \tag{6a}$$

Substitution in eq. (5d) yields a present-day SFR in the galactic disk of

$$\sum_{R=2}^{10} \Psi(t_o, R) \sim 10.4 \, m_\odot \, yr^{-1} \tag{6b}$$

(The SFR in the central region $R < 2$ kpc amounts to $\sim 10\%$ of the above value; that outside $R > 10$ kpc amounts to less than 10%).
This value has to be compared to the SFR $\langle\Psi(t)\rangle$, averaged over the age of the galactic disk, τ_{Disk}. With a total stellar mass of the galactic disk of $\sum_{R=2}^{10} M_*(R) = 6.1 \, 10^{10} \, m_\odot$, an "instantaneous return rate" $r \sim 0.42$ of the gas transformed into stars and an estimated age of the galactic disk of $\tau_{Disk} \sim 10^{10}$ yr, this average value is

$$\langle\Psi\rangle = \frac{1}{(1-r)} \frac{\Sigma M_*}{\tau_{Disk}} \sim 10.3 \, m_\odot \, yr \tag{7a}$$

If star formation is a continuous process and if stars and interstellar matter in the galactic disk form a closed system (i.e. negligible infall of gas from the halo or radial gas flow), then a ratio $\Psi(t_o)/\langle\Psi\rangle \gtrsim 1$ is difficult to reconcile with a power-law dependence of the SFR on the total mass of gas available to be transformed into stars. A relation of the type $\Psi \propto M_{gas}^k$ yields for a closed system ratios of present-day to average SFRs of

$$\frac{\Psi(t_o)}{\langle\Psi(t)\rangle} = \begin{cases} 1 & \text{for } k = 0 \\ \frac{\mu \ln \mu^{-1}}{1-\mu} = 0.134 & \text{for } k = 1 \end{cases} \tag{7b}$$

which means that for $k = 1$ (and quite generally for $k > 0$) the ratio $\Psi(t)/\langle\Psi(t)\rangle$ decreases with time, since $\mu = M_{gas}/M_{tot}$ decreases with time. The numerical value of $\sim 13\%$ for $k = 1$ holds for $\mu \sim 0.04$, the present-day gas-to-total mass ratio averaged over the galactic disk ($R = 2-10$ kpc), as inferred from observations. A present-day SFR of

∼10 $m_\odot$ yr^{-1} would mean a time-independent SFR; it would also mean that the remaining ISM of the galactic disk would be locked up in stars within some 10^8 years.

If one accepts the much more likely situation of a SFR which continuously decreases with time one has to face the fact that for a present-day SFR $\Psi(t_o) < <\Psi(t)>$ there are too many Lyc photons produced. To reconcile the observed high Lyc photon production rate with a reasonably low SFR we suggest that the local IMF, determined from star counts in the solar vicinity, does not hold throughout the galactic disk. We further suggest that spatial bimodal star formation yields the appropriate modification of the IMF.

SFRs of O stars.

From observations we know that O stars form out of molecular clouds both in the spiral arms (sa) and in the interarm region (ia). The surface density of O stars in spiral arms is much higher than in the interarm region. While the O star SFR in the interarm region appears to be proportional to the amount of ISM contained in molecular clouds to some power k, $\Psi_{OB}^{ia} \propto M_{H_2}^k$, additional O star formation is induced if the ISM flows into spiral arms. We assume that there the O star SFR is proportional to the amount of ISM that flows per unit time through the spiral arms. This amount is determined by the orbital velocity of the ISM $R\Omega_R$, relative to the pattern speed, $R\Omega_P$, of the spiral arms. Integrated over a concentric annulus of radius R and width $\Delta R = 1$ kpc the O star SFR is

$$\Psi_{OB}(R) = \Psi_{OB}^{ia} + \Psi_{OB}^{sa} \propto M_{H_2}^k(R)[1+\alpha\nu(R)] \tag{8a}$$

Here $\alpha = \Psi_{OB}^{sa}(R_\odot)/\Psi_{OB}^{ia}(R_\odot)$ is the ratio of O stars formed in spiral arms and in the interarm region, respectively, at the distance of the solar circle, $R_\odot$, while

$$\nu(R) = \frac{\Omega_R - \Omega_P}{\Omega_\odot - \Omega_P} \tag{8b}$$

is the increase of induced star formation in spiral arms at distance R relative to that at $R = R_\odot$. For a derivation of these relations and convenient numerical approximations see Paper I. $\nu(R)$ is given in column (8) of Table 1.

With $N_{Lyc} \propto \Psi_{OB}$ we can check the validity of the adopted law of the formation rate of massive stars. In Fig. 3 are plotted as a histogram the "observed" Lyc photon production rates $N_{Lyc}(R)$ from Table 1, column (5). Substitution of M_{H_2} (column (2c)) in eq. (8a) and normalization to $\Sigma[\text{const } M_{H_2}^k(1+\alpha\nu)] = 2.52\ 10^{53}$ s^{-1} yields the points plotted for k = 1, 1.5 and 2, respectively. We see that eq. (8a) with $1 \leq k \leq 1.5$ fits the observed Lyc photon production rate reasonably well. In the following we adopt, for convenience, k = 1. Observations suggest that at $R_\odot = 10$ kpc equal amounts of O stars are formed in spiral arms and in the interarm region, i.e. $\alpha = 1$. Integrated over the galactic disk (R = 2-10 kpc) this means $\psi^{sa}/(\psi^{ia}+\psi^{sa}) \sim 0.7$, i.e. 70% of all O stars are formed in spiral arms.

Table 1: Star Formation Rates $\psi(t_0)$ and Rates at which Matter is permanently locked up in Low Mass Stars and Stellar Remnants, (dM_*/dt), for Constant (Columns (6) and (7)) and Bimodal (Columns (9) and (10)) Star Formation and Related Quantities

R (kpc)	M_{ISM} ($10^8\ m_\odot$)	M_{HI}	M_{H_2}	M_* ($10^9\ m_\odot$)	$\mu = M_{ISM}/M_*$	N_{Lyc} ($10^{52}\ s^{-1}$)	$\psi(t_0)$ ($m_\odot\ yr^{-1}$)	dM_*/dt ($m_\odot\ yr^{-1}$)	$\nu(R)$	$\psi^{bm}(t_0)$ ($m_\odot\ yr^{-1}$)	$(dM_*/dt)^{bm}$ ($m_\odot\ yr^{-1}$)	$(dM_*/dt)^{pred}$ ($m_\odot\ yr^{-1}$)	
												$k=0$	$k=1$
(1)	(2a)	(2b)	(2c)	(3)	(4)	(5)	(6)	(7)	(8)	(9)	(10)	(11)	(12)
2 – 3	0.6	0.24	0.36	5.6	0.011	1.01	0.40	0.23	5.41	0.17	0.05	0.53	0.03
3 – 4	1.0	0.34	0.66	7.3	0.014	1.84	0.74	0.43	4.05	0.31	0.11	0.73	0.04
4 – 5	2.3	0.74	1.56	8.6	0.027	5.47	2.18	1.26	3.17	0.99	0.37	0.86	0.09
5 – 6	3.5	0.84	2.66	9.6	0.036	6.81	2.72	1.58	2.57	1.33	0.52	0.96	0.12
6 – 7	3.3	0.76	2.54	8.8	0.038	3.60	1.44	0.84	2.10	0.75	0.31	0.88	0.11
7 – 8	3.9	1.95	1.95	8.2	0.048	3.53	1.41	0.82	1.74	0.77	0.33	0.82	0.13
8 – 9	3.3	1.95	1.35	6.7	0.049	2.61	1.04	0.60	1.43	0.60	0.27	0.67	0.10
9 – 10	3.4	2.28	1.12	6.4	0.053	1.31	0.52	0.30	1.14	0.32	0.15	0.64	0.11
$\Sigma(R=2-10)$	21.3	9.10	12.20	60.9		26.2	10.4	6.1		5.24	2.11	6.1	0.73

*) M_{ISM}, M_{HI} and M_{H_2} relate to the total mass of interstellar matter, i.e. the contribution of heavy elements is included.

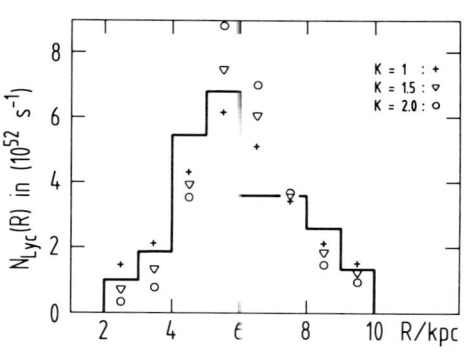

Fig. 3: The Lyc photon production rate in the galactic disk as derived from radio free-free emission (histogram) is compared to an O star formation rate of the form $\Psi_{OB}(R) \propto M_H^k (1+\alpha\nu)$. (See text).

Spatial bimodal star formation.

Spatial bimodal star formation means that in the interarm region stars form according to the IMF derived for the solar vicinity, i.e. in the total mass range m_L, m_u while only stars above a critical mass $m \geq m_c$ form by induced star formation such as in spiral arms. This was first suggested by Mezger and Smith 1977) based on a review of relevant observations. It was put into a quantitative form by Güsten and Mezger (1983; Paper I) to explain the existence of abundance gradients in the galactic disk as a consequence of chemical evolution. In the following we adopt $m_c = 3\ m_\odot$. It appears that in small molecular clouds in the interarm region low mass stars $\gtrsim 0.1\ m_\odot$ form first and OB stars form last, while in spiral arms (or - more generally - wherever star formation is induced by large-scale compression of molecular clouds) the massive stars form at about the same time and suppress the formation of low mass stars. Physical explanations for this process are summarized by Silk (this symposium).

With ϕ the IMF derived for the solar vicinity the bimodal IMF assumes the form

$$\phi^{bm}(R,\alpha,m_c) = N^{-1} [\phi^{ia} + \alpha\nu(R)\phi^{sa}] \qquad (9a)$$

with $\phi^{ia} = \phi(m \geq m_L)$, $\phi^{sa} = \phi(m \geq m_c)$. The normalization factor is

$$N = \int_{m_L}^{m_u} \phi^{bm}(R,\alpha,m_c) m\,dm = 1 + \alpha\nu(R)[1-\Delta(m_c)] \qquad (9b)$$

$1-\Delta(m_c)$, as given by eq. (5c), yields 0.29 for $m_c = 3\ m_\odot$. The fraction of matter permanently locked up in stars is

$$(1-r)^{bm} = \frac{1}{1+\alpha\nu(R)[1-\Delta(m_c)]} \left\{ (1-r)^{ia} + \alpha\nu(R)(1-r)^{sa} \right\} \qquad (10a)$$

with

$$(1-r)^{sa} = [0.6 \int_{m_c}^{6} \phi(m)dm + 1.4 \int_{6}^{m_u} \phi(m)dm] = 0.136\, m_c^{-1.4} + 0.008 \quad (10b)$$

for $1 \leq m_c/m_\odot \leq 6$.

In the derivation of eqs. (10) it is assumed that the MS lifetime $\tau_{MS}(m \leq 1\, m_\odot) \gtrsim \tau_{Disk}$, that stars with $1 \leq m/m_\odot \leq 6$ evolve to white dwarfs with an average mass of $\sim 0.6\, m_\odot$, and more massive stars evolve to neutron stars with an average mass $\sim 1.4\, m_\odot$. For $m_c = 3\, m_\odot$, $m_u = 60\, m_\odot$ the fraction of gas transformed into stars which is permanently locked up in low mass and dead stars is $(1-r)^{ia} = 0.58$ in the interarm region and $(1-r)^{sa} = 0.04$ in spiral arms.

SFRs for spatial bimodal star formation

Substitution of ϕ^{bm} (eqs. 9a,b) in eq. (4) yields the Lyc photon production rate per solar mass of gas transformed per year into stars in the case of spatial bimodal star formation

$$\frac{<N_{Lyc}>}{\psi^{bm}} = \phi(1)P^{bm}(\phi) = \phi(1)P(\phi)\, \frac{1+\alpha\nu(R)}{1+\alpha\nu(R)[1-\Delta(m_c)]} \quad (11)$$

In the derivation of eq. (11) use has been made of eqs. (5a,b). $\phi(1)P(\phi) = 2.5\, 10^{52}\, s^{-1}/m_\odot\, yr^{-1}$ applies for the interarm region, $\phi(1)P(\phi)/[1-\Delta(m_c)]$ applies for spiral arms.

For a given Lyc photon production rate $N_{Lyc}(R)$ the corresponding SFR is

$$\psi^{bm}(R) = \frac{N_{Lyc}(R)}{\phi(1)P(\phi)} \left\{ \frac{1}{1+\alpha\nu(R)} + \frac{\alpha\nu(R)[1-\Delta(m_c)]}{1+\alpha\nu(R)} \right\} \quad (12a)$$

and the fraction permanently locked up in low mass and dead stars is

$$\frac{dM_*^{bm}}{dt} = \psi^{bm}(1-r)^{bm} = \frac{N_{Lyc}(R)}{\phi(1)P(\phi)} \left\{ \frac{(1-r)^{ia}}{1+\alpha\nu(R)} + \frac{\alpha\nu(R)(1-r)^{sa}}{1+\alpha\nu(R)} \right\} \quad (12b)$$

The first and second term in eqs. (12) relate to star formation in interarm and spiral arm region, respectively.

Dividing the Lyc photon production rates $N_{Lyc}(R)$ given in column (5) of Table 1 by $\phi(1)P(\phi) = 2.5\, 10^{52}\, s^{-1}/(m_\odot\, yr^{-1})$ yields the SFR $\Psi(t_o,R)$ for constant IMF listed in column (6), and the lock-up rates of matter, $dM_*(R)/dt = 0.58\Psi(R)$, listed in column (7). The corresponding values for spatial bimodal star formation are given in column (9) and (10).

Nearly all the matter that is permanently locked up in low mass and dead stars (viz. 1.82 $m_\odot$ yr^{-1}) comes from star formation in the interarm region. This is the reason why $(dM_*(R)/dt)^{bm}$, shown in Fig. 4 by open triangles, is much flatter than the corresponding rate obtained for a constant IMF (shown by open circles), where the induced SFR in spiral arms leads to an enormous lock-up rate in low mass and dead stars in the range between R = 4 and 8 kpc. For comparison purposes are also shown in Fig. 4 the predicted lock-up rates assuming a closed system and a power-law dependence of SFR on the mass of interstellar matter.

$$(dM_*/dt)^{pred} = \frac{M_*(R)}{\tau_{Disk}} \begin{cases} 1 \text{ for } k = 0 \\ \\ \dfrac{\mu \ell n \mu^{-1}}{1-\mu} \text{ for } k = 1 \end{cases}$$

Here we substituted for $\tau_{Disk} = 10^{10}$ yr and for M_* and μ the values given in columns (3) and (4) of Table 1. While the lock-up rates averaged over the galactic disk, $\sum_{R=2}^{10}(dM_*/dt)$, derived for constant IMF and predicted for constant SFR (k=0) (lowest line of columns (7) and (11)) agree well their radial variations, as shown in Fig. 4, are incompatible. The present lock-up rate predicted for k=1, 0.73 $m_\odot$ yr^{-1}, (column (12)), is by about a factor of three lower than the value derived for the specific model of spatial bimodal star formation to which the values in column (10) relate. Remember that we use here a model developed in Paper I to explain abundance gradients. The above discrepancy is partly resolved if we use a more realistic model with a gradual build-up of the galactic disk by infalling halo gas. For an exponentially decreasing accretion rate ($\propto \exp\{-t/\tau\}$, $\tau \sim 5$ Gyr) Vader and de Jong (1981) have shown that even for k = 1 values $\langle\Psi\rangle/\Psi(t_o) > 1$ are possible.

Much more important than the total SFR integrated over the galactic disk is the rate at which matter is permanently locked up in low mass stars and stellar remnants and especially its radial dependance which – integrated over the age of the disk – determines the present-day mass distribution of the disk. Bimodal star formation leads to a radial dependence of the present-day lock-up rate $(dM_*/dt)^{bm}$ which agrees in shape reasonably well with that predicted from the present-day mass distribution $M_*(R)$. To demonstrate this agreement we have arbitrarily multiplied $(dM_*/dt)^{pred}_{k=1}$ by a factor of three (dotted line in Fig. 4). A justification for such an increase is given above by gradual accretion of the disk.

Note that eqs. (7b and 13) hold for a closed system with $\mu = M_{ISM}/M_{tot}$ and M_{ISM} the total mass of interstellar matter, while eq. (8a) relates the SFR to the mass of hydrogen in molecular form. Formally this can be justified if eq. (8a) is rewritten as a relation for the lock-up rate

$$(dM_*/dt)^{bm} \sim (dM_*/dt)^{ia} \propto M_{ISM}^k \qquad (13)$$

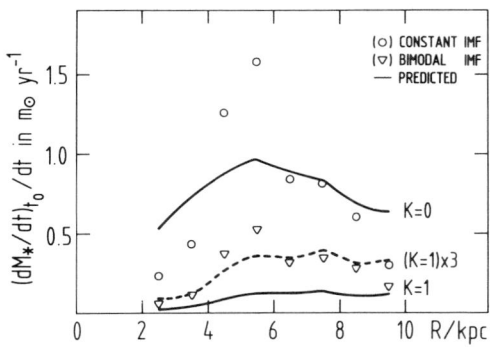

Fig. 4: *The rate at which ISM is locked up in low mass and dead stars. Solid and dotted curves: Predicted from the observed mass distribution in the galactic disk and an adopted relation $(dM_*/dt) \propto M_{ISM}^k$ with $k = 0$, 1. Dots and triangles: Lock-up rates derived from observed Lyc photon production rates and with a constant and bimodal IMF, respectively.*

since matter is permanently locked up in low-mass stars predominantly in the interarm region. This allows us to use solutions for the closed system with $M_{ISM} + M_* = M_{tot}$ to compute the "predicted" lock-up rates given in columns (11) and (12) of Table 1 for k = 0 and 1, respectively, and shown in Fig. 4. In the case of spatial bimodal star formation the star formation in spiral arms appears like a (compared to the galactic rotation time) short firework, which leaves most of the ISM transformed into stars as ashes in the form of ISM enriched with heavy elements, while only a small fraction (viz. ∼4%) remains permanently locked up in stellar remnants. In the interarm regions, on the other hand, about equal amounts of the ISM transformed into stars are returned "instantaneously" to interstellar space and permanently locked up predominantly in low mass stars. Hence it is star formation in the interarm region that accounts for the present-day stellar mass distribution in the galactic disk.

Other applications of bimodal star formation

It should be stressed that spatial bimodal star formation was initially introduced to explain the existence of element abundance gradients in the galactic disk. The different return rates in spiral arms and interarm regions, respectively, of gas processed in intermediate and massive stars, together with the increasing fraction of intermediate and massive stars, formed in spiral arms with decreasing galactocentric distance R, accounts for the variable yield required for an explanation of the observed abundance gradients. (Paper I and Güsten, 1985).

With a generalized concept of bimodal star formation, i.e. whenever star formation is induced by some large-scale effects such as in starbursts (observed in the central regions of some galaxies and in galaxies with extreme infrared emission), the derived extremely low mass-to-luminosity ratios can be explained (Silk, this symposium).

The secondary maximum, which appears in IMFs derived from observed luminosity functions with SFRs decreasing with time, find a natural explanation by spatial bimodal star formation (see Scalo, this symposium).

Larson (1985) has constructed a star formation model that is bimodal in time, which can account for all of the unseen mass in the solar neighbourhood in the form of stellar remnants, as well as for many other observed characteristics of spiral galaxies.

Luminous MS stars and infrared emission from warm dust

Dusty galaxies emit a large fraction of their stellar luminosity as thermal infrared emission from dust. The IR spectrum of the inner part of our Galaxy (R ≲8 kpc), as taken from Cox, Krügel and Mezger (1985; hereafter referred to as Paper II) is shown in Fig. 5. The spectrum can be decomposed into contributions of cold (10-25 K) dust, warm (30-40 K) dust and hot (250-500 K) dust. The total IR luminosity is $\sim$1.5 10^{10} $L_\odot$ or $\sim$30% of the total stellar luminosity; the luminosity of warm and hot dust are $\sim$5.7 10^9 $L_\odot$ and $\sim$1.7 10^9 $L_\odot$, respectively.

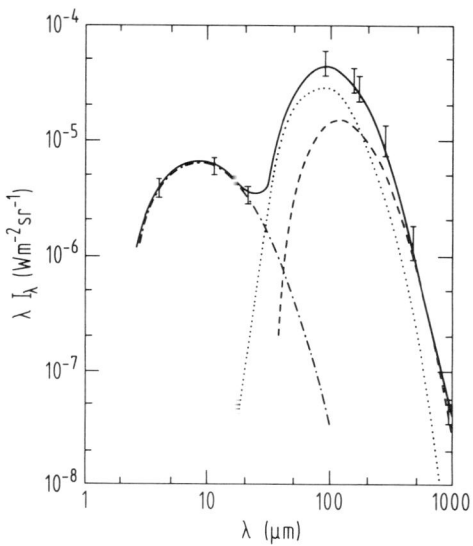

Fig. 5: Spectrum of the dust emission from the inner part (R ≲8 kpc) of our Galaxy averaged over galactic longitudes 3-35° and latitudes |b| >1°. Our analysis in Paper II shows that the spectrum can be decomposed in contributions from: i) cold (10-25 K) dust associated with atomic and molecular hydrogen heated by the general ISRF (dashed curve); ii) warm (30-40 K) dust heated by OB stars (dotted curve); and iii) hot (250-500 K) dust heated by M giants with heavy mass loss.

For each of these three dust components a dominant heating source has been identified (Paper II and references therein): Cold dust is heated by the general interstellar radiation field, warm dust is heated by O and B stars, hot dust is heated by OH/IR stars, i.e. M giants with heavy mass loss. Hence, it is only warm dust which traces luminous MS stars. As can be seen from Fig. 5 the spatial distribution of warm dust emission should be well represented by emission at $\lambda \sim$60 μm, where IRAS data will soon become available.

With eq. (2) one can formally write a relation between warm dust luminosity and SFR which is analogous to eq. (4):

$$L_{IR}^{wd}(t_o) = \Psi(t_o) \int_{m_L}^{m_u} \phi(m) f_d(m) dm \int_0^{\tau_{MS}(t)} L_*(m,\tau) d\tau \qquad (14)$$

In Fig. 6 is shown the product $\phi(m)L_*(m)$. The two curves behave qualitatively like the curves $\phi(m)N_{Lyc}(m)$ shown in Fig. 2, but peak at lower stellar masses, viz. $\sim$13 $m_\odot$ (spectr. type B1) and $\sim$28 $m_\odot$ (O7), respectively. However, when multiplied with the main sequence lifetime $\tau_{MS}(m)$, the stellar luminosity curves diverge, while the maxima of the Lyc photon luminosity curves are only shifted toward lower stellar masses.

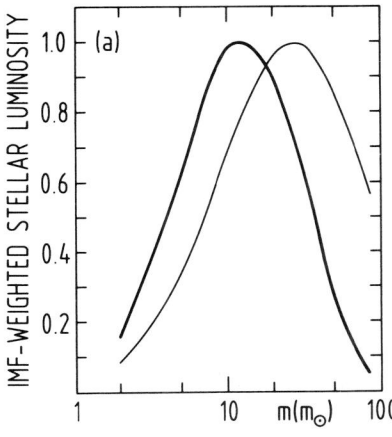

Fig. 6: The IMF weighted stellar luminosity. The thin solid curve relates to the Salpeter IMF, the heavy solid curve relates to the Miller-Scalo IMF. (Cox et al., 1985).

This means that the factor $f_d(m)$, i.e. the fraction of stars of a given stellar mass which contribute to the heating of warm dust, governs the computed ratio of warm dust luminosity of the SFR. Since this fraction is very difficult to estimate for stars later than OB, SFRs derived from IR luminosities are rather unreliable.

For our galaxy we know for R $\leq$10 kpc both the total Lyc photon luminosities, $N_{Lyc} = 2.6 \; 10^{53}$ s^{-1} and the warm dust luminosity, $L_{IR}^{wd} = 7.3 \; 10^9$ L$_\odot$ (Paper II), so that

$$\frac{\langle N_{Lyc} \rangle}{\langle L_{IR}^{wd} \rangle} = 3.56 \; 10^{43} \; s^{-1} \; L_\odot^{-1} \tag{15a}$$

On the assumption that this ratio is typical for galaxies with a high formation rate of luminous stars one can derive an empirical factor for converting observed warm dust IR luminosities into SFRs, which is

$$\Psi(t) = \frac{3.5 \; 10^{43} \; L_{IR}^{wd}}{\phi(1)P(\phi)} \; [1-\Delta(m_c)] = 1.4 \; L_{IR,9}^{wd} \; [1-\Delta(m_c)] \tag{15b}$$

Here $\Psi(t)$ is in m$_\odot$ yr^{-1}, $L_{IR,9}$ in 10^9 L$_\odot$ and $\Delta(m_c)$ is given by eq. (5c). As an example consider a star-burst galaxy with $L_{IR} \sim 10^{11}$ L$_\odot$. If stars of all masses were produced in the star burst a SFR of 140 m$_\odot$ yr^{-1} would be required to sustain this luminosity, of which 81 m$_\odot$ yr^{-1} would be permanently locked up in low mass stars and stellar remnants. If only stars above $m_c \sim 3$ m$_\odot$ were formed the corresponding rates are Ψ = 39 m$_\odot$ yr^{-1} and dM^*/dt = 5.6 m$_\odot$ yr^{-1}.

Luminous post MS stars and infrared emission from hot dust

The dust emission spectra of our Galaxy (Fig. 5) and of external galaxies show a secondary maximum around $\lambda \sim 10$ μm. It corresponds to hot dust in the temperature range 250-500 K. The hot dust luminosity in our Galaxy is $L_{IR}^{hd} \sim 1.7 \; 10^9$ L$_\odot$ or $\sim$10% of the total IR luminosity. There are two competing explanations for this "M(=middle) IR shoulder".
i) The existence of very small grains, made of polycyclic aromatic molecules, which have spectral features between λ = 3.7 and 11.9 μm.

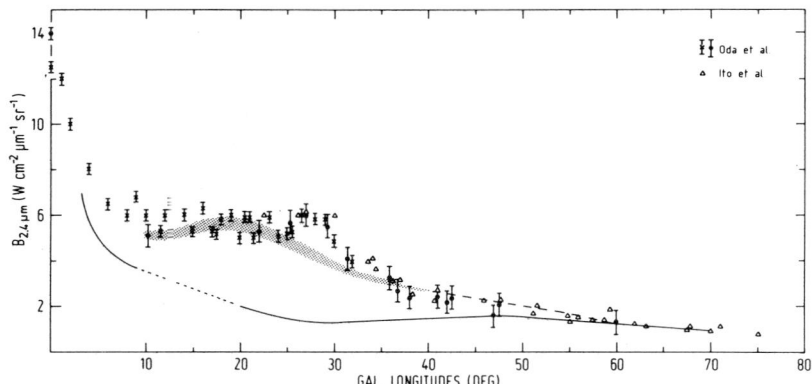

Fig. 7: The ridge line intensity at $\lambda 2.4\,\mu m$ as observed by Oda et al. (1979) and Ito et al. (1977). The solid curve relates to model computations, where the space density of M giants in the solar vicinity is scaled with the total mass distribution and dust opacity at $\lambda 2.4\,\mu m$ is considered. The shaded curve relates to a model, where an excess component of medium mass M giants is superimposed on the distribution of old disk population M giants. The radial distribution of this excess component is similar to that of the present-day O stars. This excess component finds a natural explanation if i) spatial bimodal star formation applies, ii) all stars in the mass range $m_c \sim 3\,m_\odot \lesssim m \lesssim 8\,m_\odot$ evolve to M giants, and iii) the SFR in the galactic disk was approximately constant over the past 10^8 yr (Güsten and Mezger, 1983; Paper I).

When heated by absorption of an energetic photon to temperatures of ~ 600 K these grains emit predominantly in the MIR range and thus may account for the MIR shoulder (Léger and Puget, 1984); ii) the presence of some 10^4 M giants in the galactic plane with heavy mass loss ($\dot{M} \gtrsim 10^{-5}\,m_\odot\,yr^{-1}$) and typical luminosities of some $10^4\,L_\odot$ which emit all their radiation in the MIR. (Paper II).

In Paper II we conclude that both explanations apply and that the observed MIR shoulder in our and external galaxies is due to a superposition of the two mechanisms. The small particles explain observations of the MIR shoulder in dust heated by the general ISRF, e.g. in the IRAS "cirrus clouds" at high galactic lattitudes. The M giants with mass loss and MIR emission from surrounding dust shells, observationally known as OH/IR stars, explain the strong MIR emission close to the galactic plane.

Owing to 2.4 µm surveys of the galactic plane especially by Japanese groups (see, e.g. Hayakawa et al., 1977) it is well known that there exists a large number of M giants, which account for a total luminosity of $\sim 2 \cdot 10^{10}\,L_\odot$ (see, e.g. Mathis et al., 1983). The ridge line intensity of the 2.4 µm emission is shown in Fig. 7. It consists of contributions from old disk population M giants (shown in Fig. 7 as (interrupted) solid curve) and an excess component of M giants which accounts for a total luminosity of ~6 10^9 $L_\odot$. In Paper II it is shown that this excess component can be explained if all stars with m >m_c ~2-3 $m_\odot$ evolve to M giants and if their distribution in the galactic plane is similar to that of O stars. Remember that m_c is the critical lower mass in spatial bimodal star formation.

The inference of the NIR ($\lambda 2.4$ μm) emission is that its "excess component" relates to post MS stars in the mass range m ~3-8 $m_\odot$, which have evolved to M giants. The inference of the MIR (λ~10 μm) emission is that it relates to a subgroup of these medium mass M giants viz. those, which are in a (necessarily relatively short-living) phase of heavy mass loss. Now remember that the MS lifetime of a 3 $m_\odot$ star is ~3 10^8 yr. With a reasonable estimate of luminosity integrated over post MS lifetimes of intermediate mass stars we arrive at the conclusion that the present-day SFR is compatible with the number of intermediate mass M giants observed today through their NIR emission. On this basis we suggest that the SFR of luminous stars in the galactic disk was rather constant at least during the last galactic rotation (Paper I).

Ionization state of HII regions and the upper mass limit of the IMF

While the ^{4}He abundance in the galactic disk appears to increase towards the galactic center the ionized helium abundance observed in giant HII regions actually starts to decrease at galactic radii R ~8-9 kpc. (For the latest review of radio recombination line observations see Mezger and Wink, 1983).

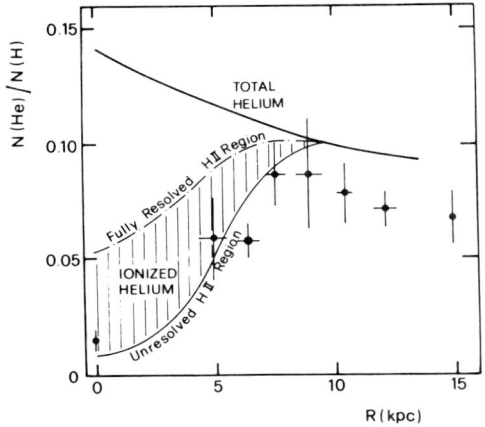

Fig. 8: Observed He^+ abundances (-•-) are compared to model computations. Heavy solid curve: Adopted He abundance. Thin solid curves: Results of model computations where $<T_{eff}>$ decreases with R (Panagia, 1979).

This means that with decreasing galactocentric distance the He^+ Strömgren sphere shrinks relative to that of ionized hydrogen. Panagia (1979) interprets this effect as a decrease of the average effective temperature, which characterizes the Lyc radiation field in giant HII regions, according to

$$<T_{eff}> = 3.1\ 10^4 \exp\{R_{kpc}/50\} \text{ in K} \qquad (16)$$

Panagia links this decreases of $<T_{eff}>$ to the increase of the metal abundance Z with decreasing R. Blanketing of UV lines with an increase of the stellar radius on the one hand, and a decrease of the upper IMF mass limit on the other hand should contribute about equally to this effect. Such a decrease of $m_u \propto Z^{-0.5}$ was in fact predicted by Kahn (1974) based on an accretion theory for the formation of massive stars.

REFERENCES

Cox, P., Krügel, E., Mezger, P.G. (1985) Astron. Astrophys. (in print) (Paper II)
Güsten, R. (1985) "The Chemical Evolution of the Milky Way", Erice Lectures, preprint
Güsten, R., Mezger, P.G. (1983) Vistas in Astronomy 26, 159 (Paper I)
Hayakawa, S., Ito, K., Matsumoto, T., Ugama, K. (1977) Astron. Astrophys. 58, 325
Ito, K., Matsumoto, T., Uyama, K. (1977) Nature 256, 517
Kahn, F.D. (1974) Astron. Astrophys. 37, 149
Larson, R.B. (1985) "Bimodal star formation and remnant-dominant galactic models", preprint, subm. to MNRAS
Léger, A., Puget, J.L. (1984) Astron. Astrophys. 137, L5
Mathis, J., Mezger, P.G., Panagia, N. (1983) Astron. Astrophys. 128, 212
Mezger, P.G. (1979) in "Radio Recombination Lines", (P.A. Shaver, ed.), D. Reidel Publ. Co., p. 81
Mezger, P.G., Wink, J. (1983) Proc. ESO Workshop on "Primodial Helium", (Shaver, Kunth, and Kjär, eds.), p. 281
Miller, G.E., Scalo, J.M. (1979) Astrophys. J. Suppl. 41, 513
Oda, N., Maihara, T., Sugiyama, T., Okuda, H. (1979) Astron. Astrophys. 72, 309
Panagia, N. (1979) in "Radio Recombination Lines", (P.A. Shaver, ed.), D. Reidel Publ. Co., p. 99
Salpeter, E.E. (1955) Astropyhs. J. 121, 161
Vader, J.P., de Jong, T. (1981) Astron. Astrophys. 100, 124

Discussion : MEZGER.

ZINNECKER :

The Lyman continuum production rate does not depend on the upper mass limit using the Miller and Scalo IMF but peaks for $M \sim 30 M_o$ you said. How is it possible then to get a handle on the upper mass limit as a function of metallicity?

MEZGER :

What counts for the He ionization structure is not the Lyc photon production rate but rather the ratio of the ionizing photons to all Lyc photons, a quantity which depends strongly on $\langle T_{eff} \rangle$ and hence on the upper mass limit M_u.

SILK :

I do not think you are on very secure ground in appealing to the theoretical prediction that the upper mass limit depends on metallicities in a specific way in order to account for the radial gradient in the hardness of the ionizing photon flux. For example, the theoretical calculations assume spherical symmetry, yet the situation studied in which radiation from the accreting core interacts with an infalling shell is Rayleigh-Taylor unstable. I suspect that this can only allow additional accretion : in any case, additional simulations are needed. Is there any other viable mechanism for giving a gradient in selective absorption of Lyman continuum photons as the metallicity changes, for example due to line blankting or to small grain properties in HII regions?

MEZGER :

Maybe Kahn's theory oversimplifies the actual situation. But model fitting to the observed IR/submm spectra of compact HII regions shows that the inner region is heavily depleted in dust but that dust cocoons form at distances of .. 10^{17} cm from the surface of O stars, probably as a result of radiation pressure; hence, radiation pressure may, in fact, limit the mass of the most massive stars and this effect should certainly depend on the dust-to-gas ratio and thus on the metal abundance.

POSTER PAPERS 4.

Chairman : A. MAEDER

1. A.CAMPBELL and L.SMITH:
    A search for WR Stars in Giant Extragalactic Bursts of Star Formation.

2. M.KAUFMAN, R.C.KENNICUTT and R.N.BASH:
    Giant HII Regions in M81.

3. J.MELNICK, R. TERLEVICH and M. MOLES:
    Warmers: Massive Stars in the Nuclei of Galaxies.

4. G.BERTELLI, A.BRESSAN, C.CHIOSI, E.NASI and L.PIGATTO:
    Convective Overshooting: New Integrated Colours vs. Age Relations for Star Clusters.

5. C.CHIOSI and L.PIGATTO: The Distance Modulus of LMC.

6. Y.H.CHU: NGC 2070 and NGC 3603.

A SEARCH FOR WOLF-RAYET STARS IN GIANT EXTRAGALACTIC BURSTS OF STAR FORMATION

Alison W. Campbell
Institute of Astronomy
The Observatories
Cambridge CB3 OHA
U.K.

Linda J. Smith
Dept. of Physics & Astronomy
University College London
Gower St.
London WC1E 6BT
U.K.

It is well known that some giant extragalactic star-forming regions contain WR stars. D'Odorico, Massey, Rosa and coworkers found many examples in nearby galaxies of giant HII regions whose spectra show that they contain WN, and occasionally, WC stars. The dwarf emission-line galaxies He 2-10 (Allen et al. 1976) and Tol 3 (Kunth & Sargent 1981) have a strong broad emission feature near HeII 4686Å; in the latter object ∿150 WN stars are required to explain the observed equivalent width.

Giant star-forming regions ($\gtrsim 10^3$ O stars) are ideal laboratories in which to study the evolution of the most massive stars. We have begun a detailed optical/infrared investigation of a sample of "HII galaxies" (galaxies experiencing a relatively very luminous burst of star formation) with the objective of studying the occurence and evolution of WR stars as functions of the age, mass and abundance of the the ionizing cluster. Our sample consists of 15 HII galaxies observed at the AAT and a further 3 from the sample of Campbell et al. (1985), obtained at LCO: we have moderate dispersion, high (typically 30) continuum S/N spectrophotometry covering the wavelength range 3500-7000Å for a total of 20 starburst regions in these objects. We report here on our initial findings and concentrate on the four objects in which we have detected strong WR features.

The spectra are characterised by broad emission near HeII 4686Å and are shown in Fig. 1. The most conspicous WR features are seen in Mi 499 (= NGC 4385) which also shows broad NIII 4640Å emission of comparable strength ($W_\lambda \sim 4$Å) and width (FWHM ~ 20Å) to HeII 4686Å. This feature is very similar to that discovered in NGC 300(7) by D'Odorico et al. (1983) and indicates a predominantly late WN population. The WC feature CIV 5800 is just detected at the 2σ level with a FWHM ~ 40Å, indicating that early WC stars may also be present. Assuming a distance of 50 Mpc (from the emission-line redshift; $H_0 = 50$ km s^{-1} Mpc^{-1}), we estimate that $\sim 1.5 \times 10^4$ late WN stars are required to produce the HeII emission flux of 1.5×10^{-14} ergs s^{-1} cm^{-2} in Mi 499.

The other three HII galaxies shown in Fig. 1 have a broad HeII 4686 feature (FWHM∿30Å; $W_\lambda \simeq 3$Å) but no NIII 4640Å emission, indicating that

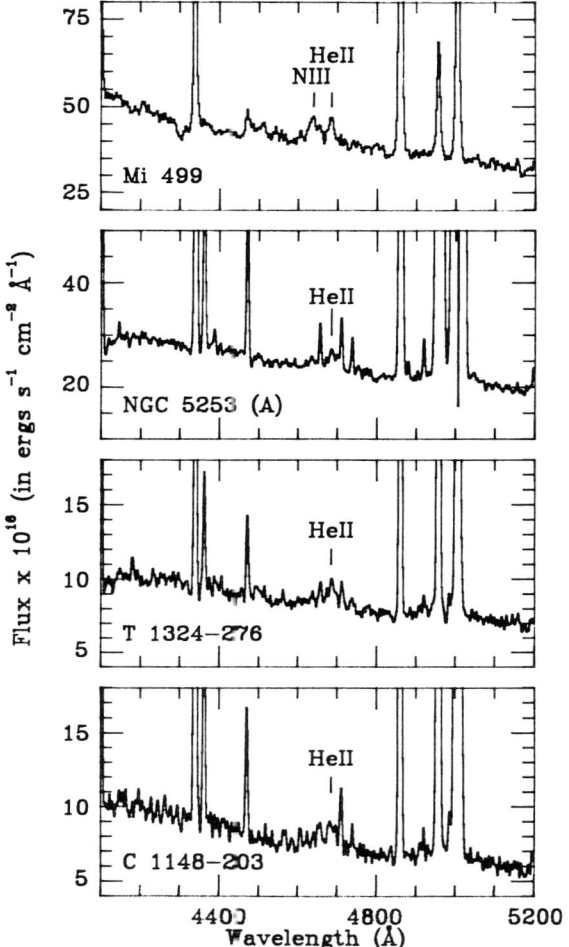

Fig 1: The 4 spectra containing WR stars

the WR stars are probably WN4-6, since this line is about a factor of 10 weaker than HeII 4686 in early WN stars. In addition, the CIV 5800 feature is not observed in these objects.

To begin to relate the presence of these WR features to physical properties of the HII galaxy, we have derived the gaseous O abundance relative to that in the solar neighbourhood (sn) (Shaver et al. 1983), using the measured T_e and a typical N_e of 100 cm^{-3}. Where [OIII] 4363 is not observed (in regions of moderate age or high O abundance), T_e cannot be measured, and we were not able to determine O/H in such objects. The O abundance may be estimated indirectly (for O/H≳0.70/H$_{sn}$) from the flux of [OIII]4959, 5007Å relative to that of Hβ; using the calibration of Edmunds & Pagel (1984 and references therein), we derive an abundance for Mi 499 of ~ 1.50/H$_{sn}$. We have divided the objects arbitrarily into 3 abundance groups, as indicated by the symbols in Fig. 2. The WR containing objects have abundances throughout the range (0.150/H$_{sn}$<O/H<1.50/H$_{sn}$) covered by our sample. The only object which may contain WC stars, Mi 499, has an unusually high gaseous abundance.

Fig. 2 shows the distribution of the objects in the log (4*[OIII] 4959/ [OII] 3727), log ($W_{Hβ}$) plane. As the ionizing cluster evolves and its integrated spectral energy distribution changes, an HII galaxy moves from upper right to lower left in this diagram, thus providing an age estimator (Terlevich et al. 1985). The WR-containing objects are grouped in a narrow age range (~3-4 Myr) with the exception of Mi 499, whose position in the diagram indicates an age of ≳ 6Myr.

We have both broad-band (JHK) and narrow-band (2.3μ, CO index) infrared photometry for the 2 bursts in Tol 3, the brightest burst in

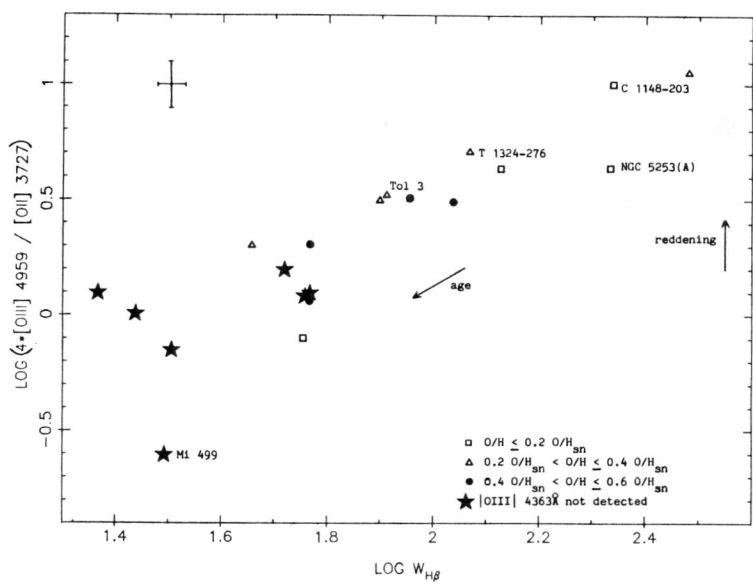

Fig. 2: The distribution in the log $|OIII|/|OII|$, log $W_{H\beta}$ plane.

T 1457-262 and the 2 brightest knots in the nucleus of NGC 5253. A large CO index or, less directly, aperture photometry, indicates that red supergiants (RSG) are present in NGC 5253 (B). T1457-262 (A) and Tol 3 (SE) (Campbell & Terlevich 1984). Similarly, RSGs are not present in NGC 5253 (A) and Tol 3 (NW). The 4 objects in Fig. 1, Mi 499, C 1148-203, NGC 5253 (A) and T 1324-276, together with Tol 3 (NW), are known to contain WN stars. At comparable continuum S/N, no WR stars are detected in Tol 3 (SE), NGC 5253 (B) and T 1457-262 (A). The data therefore suggest that, as found in the Galaxy (e.g. Maeder et al. 1980), RSG and WR stars do not occur (at the same time) in the same star-forming regions. A more detailed analysis, now underway, is required to explain how this phenomenon is related to the physical properties of the cluster such as age, mass and abundance.

REFERENCES

Allen,D.A., Wright,A.E. & Goss,W.M., 1976, Mon.Not.R.astr.Soc.,177, 91
Kunth,D. & Sargent,W.L.W., 1981, Astron.Astrophys.,101, L5
Campbell,A.W. & Terlevich,R.J., 1984, Mon.Not.R.astr.Soc.,211, 15
Campbell,A.W., Terlevich,R.J. & Melnick,J., 1985, in preparation.
D'Odorico,S., Rosa,M. & Wampler,E.J., 1983, Astron.Astrophys.Suppl.,53, 97
Edmunds,M.G. & Pagel,B.E.J., 1984, Mon.Not.R.astr.Soc.,211, 507
Maeder,A., Lequeux,J. & Azzopardi,M., 1980, Astron.Astrophys.,90, L17
Shaver,P.G., McGee,R.X., Danks,A.C. & Pottasch,S.R., 1983,
        Mon.Not.R.astr.Soc.,204, 53
Terlevich,R.J., Melnick,J. & Campbell,A.W., 1985, in preparation

GIANT H II REGIONS IN M81

Michele Kaufman, Ohio State University

R. C. Kennicutt, University of Minnesota

F. N. Bash, University of Texas

Giant HII regions are important tracers of recent star formation in distant galaxies. For a selection of HII regions in our galaxy where the exciting stars can be identified, Rumstay (1985) finds that the measured Hα and radio continuum luminosities of an HII region correlate with the stellar ionizing flux derived from model atmospheres and the known exciting stars. Therefore, we use flux measurements of giant HII regions as an index of the distribution of O stars in M81.

Using the VLA, Bash and Kaufman (1985) have mapped M81 at λ6 and 20 cm with a resolution of 10" (190 pc, if the distance of M81 is 4 Mpc). About 40 of the radio continuum sources coincide with Hα sources observed by Hodge and Kennicutt (1983). Certain studies (Visser 1980; Leisawitz and Bash 1982) of spiral structure in M81 have compared predicted star formation rates with the global distribution of optical HII regions detected by Connolly et al. (1972). However Connolly et al. do not distinguish bright HII regions from faint ones. The HII regions detected in our VLA survey are giant HII regions with excitation parameter $U > 170$ pc cm^{-2}. These giant radio HII regions are more concentrated in a two-armed pattern and show a more narrow distribution as a function of galactocentric distance R than the HII regions plotted in Connolly et al. As indicated in Fig. 1 the distribution of giant radio HII regions in the plane of M81 exhibits a strong peak near R = 300". Thus M81 may show the same phenomenon that Rumstay and Kaufman (1983) find in M83 and M33; namely, the set of high luminosity HII regions is more sharply peaked in azimuthal and in radial distribution than the set of low luminosity HII regions.

We also compare the Hα and the radio continuum fluxes of the giant HII regions in M81. For each region we convolve the Hα and the radio data to the same resolution, integrate over the same coordinates, and correct both the radio and optical data for local background in the same way. We restrict here to the HII regions where the determination of the local background is the least ambiguous and assume an electron temperature of 10^4 K. For these regions the average radio spectral index is $- 0.09 \pm 0.06$. For 25 sources, the 20 cm and Hα data yield a mean value of 1.0 ± 0.15 magnitudes for the visual extinction A_V(20 cm). For 19 sources, the 6 cm and Hα data yield a mean value for A_V(6 cm) of 1.0 ± 0.16 magnitudes. The resulting distribution of extinction as a function of galactocentric R (see Fig. 2) shows considerable scatter at any given R and no strong trend.

REFERENCES

Bash, F.N., and Kaufman, M. 1985, Bull. A.A.S., 16, 976.
Connolly, L.P., Mantarakis, P.Z., and Thompson, L.A. 1972, Publ. A.S.P. 84, 61.
Hodge, P.W., and Kennicutt, R.C. 1983, Astron. J. 88, 296.
Leisawitz, D., and Bash, F. 1982, Astrophys. J. 259, 133.
Rumstay, K.S. 1985, preprint.
Rumstay, K.S., and Kaufman, M. 1983, Astrophys. J. 274, 611.
Visser, H.C.D. 1980, Astron. Astrophys. 88, 159.

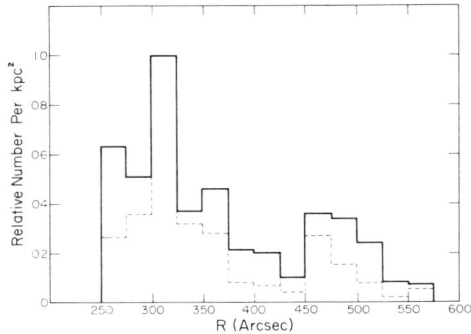

Fig. 1. Radial distribution of giant radio HII regions in the plane of M81. The solid histogram gives the number of giant HII regions per kpc^2; the dashed histogram, the 20 cm flux per kpc^2 from giant HII regions. Both histograms are normalized to one for the 300-325" annulus.

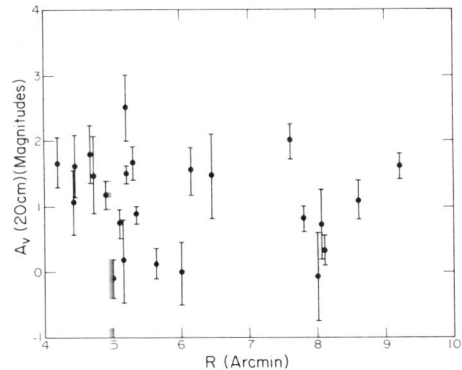

Fig. 2 Radial distribution of visual extinction in the plane of M81; A_V(20cm) is the visual extinction determined from the 20 cm and Hα observations of HII regions.

WARMERS: MASSIVE STARS IN THE NUCLEI OF GALAXIES

Jorge Melnick
European Southern Observatory

Roberto Terlevich
Royal Greenwich Observatory

Mariano Moles
Instituto de Astrofísica de Andalucía

1. ACTIVE GALACTIC NUCLEI

A very small minority of all galaxies exhibit activity in their nuclear regions. The best known and most studied signature of nuclear activity is the presence of emission lines in the optical spectrum. In fact, active galaxies are classified as Seyferts, Liners or Starburst galaxies according to the strengths, widths and intensity ratios of their nuclear emission lines (figure 1).

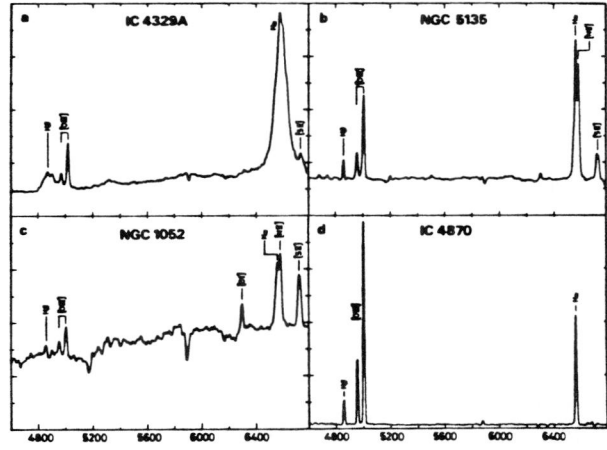

Figure 1: Typical examples of active galaxies taken from the Vérons' catalogue.
a) Seyfert 1
b) Seyfert 2
c) Liner
d) HII region.

There are two competing scenarios to explain nuclear activity: Starburst models, that postulate activity to be caused by one or several violent star formation events in the central regions of galaxies, and Nonthermal or Monster models that assume nuclear activity to be powered by a compact nonthermal powerhouse (the Monster) lodged in the nuclei of galaxies. Whilst many of the properties of active galaxies are well explained by Starburst models, conventional starburst photoionization models fail to reproduce the morphology of the emission line spectra of Seyferts and Liners.

On the other hand, nonthermal power-law photoionization models fit reasonably well the observed spectra of active nuclei but fail to explain why some galactic nuclei harbor active star formation while others contain luminous nonthermal sources or why only a small fraction of all galaxies display nuclear activity.

2. STARBURSTS OR MONSTERS?

According to Heckman et al. (1983; A.J. 88, 1077) the starburst theory can be consistent with the optical, radio and X-ray observations of Seyferts and Liners only if,

a) Nuclear starbursts differ in several fundamental ways from extranuclear starbursts; and

b) The type of optical emission line spectrum produced by the nuclear starburst is strongly correlated with both the radio properties of the starburst and the nature of the parent galaxy.

While these two conditions are not inconsistent with observation they were regarded by Heckman et al. to be ad-hoc, to have neither empirical nor theoretical justification and therefore to argue strongly against the starburst scenario.

However, new developments in evolutionary theories for massive stars lend theoretical support to these two conditions. Here we will present the empirical evidence that is obtained by reanalyzing the observations of Seyferts and Liners in the light of these new theories.

3. WARMERS

The models presented by André Maeder in this Symposium predict that massive stars with strong mass-loss will spend a significant fraction of their He-burning phase to the left of the ZAMS on the HR diagram (figure 2). Thus, the ionizing spectrum of a young cluster of massive stars will be dramatically influenced by these extremely hot and luminous stars that we will refer to as WARMERS (observationally young Warmers may correspond to early WC and WO stars).

We have computed evolutionary models for the emission line spectra of constant density, spherically symmetric nebulae photoionized by coeval clusters of massive stars with solar-neighborhood IMF (Terlevich and Melnick, 1985, M.N. 213, 481).

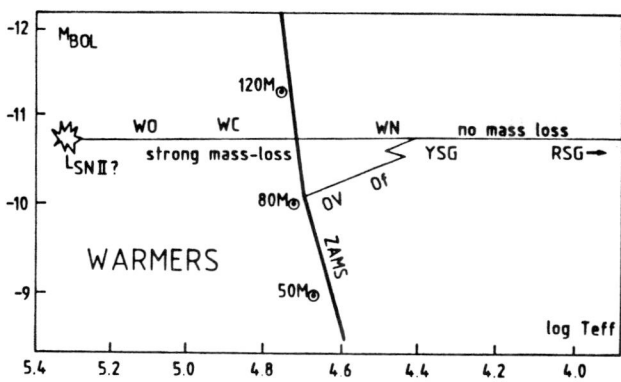

Figure 2: Schematic illustration of the effect of mass-loss on the evolution of massive stars.

Figure 3 presents some results from these models. After 2-3 million years, Warmers begin to appear in the cluster and the nebular spectrum suddenly changes from a normal, high excitation HII region (typical of extranuclear starbursts) to a Seyfert and/or Liner spectrum (figure 3).

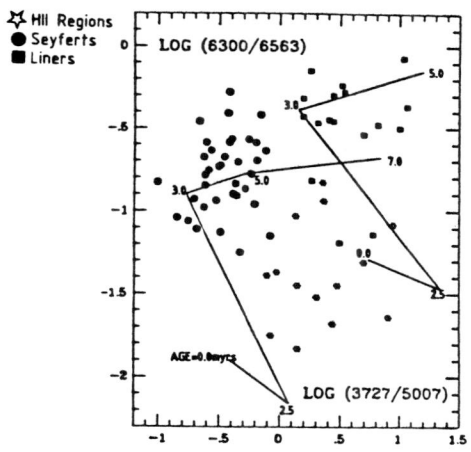

Figure 3: Starburst photoionization models with Warmers.

These models naturally explain the different types of active nuclei as an evolutionary sequence determined by the total mass of the ionizing cluster (or starburst) and the mass-loss rates of the ionizing stars. The models predict two parallel evolutionary sequences in the case of strong mass-loss,

    Large Starbursts: HII regions → Seyferts → (Blue) Liners

    Small Starbursts: HII regions → (Red) Liners.

If mass-loss is not strong, massive stars end their evolution as Red Supergiants or late WN stars but not as Warmers. Therefore metal poor starbursts will not be expected to develop Seyfert or Liner characteristics.

4. NUCLEAR VS. EXTRANUCLEAR STARBURSTS: MASS-LOSS

There is no complete theory of mass-loss but most theoretical scenarios predict a strong correlation between mass-loss rate and metallicity. There is substantial, albeit indirect, observational evidence for such a correlation (e.g. the dependence of Wolf-Rayet populations of parent galaxy composition). Most spiral galaxies show abundance gradients such that their nuclear regions are more metal rich. Thus, the Warmers models naturally explain the observed differences between nuclear and extranuclear starbursts as a metallicity effect through its influence on the stellar mass-loss rates.

In addition, for a given Hubble type extranuclear starbursts are on average considerably less massive than nuclear ones and therefore are not expected to contain very massive stars. For example, the most massive star in the Orion nebula has only 30 $M_\odot$.

5. NUCLEAR ACTIVITY AND GALAXY MORPHOLOGY

Figure 4 presents a histogram of the frequency of different types of nuclear activity as a function of the Hubble type of the parent galaxy. This distribution was drawn from the volume-limited sample of Keel (1983; Ph.D. thesis UCSC) which contains 93 galaxies out of which 30 have HII-region nuclei and 52 have Liners and 5 Seyferts.

The histogram shows that HII-region nuclei occur predominantly in spiral galaxies with small bulges while Liners are mostly found in

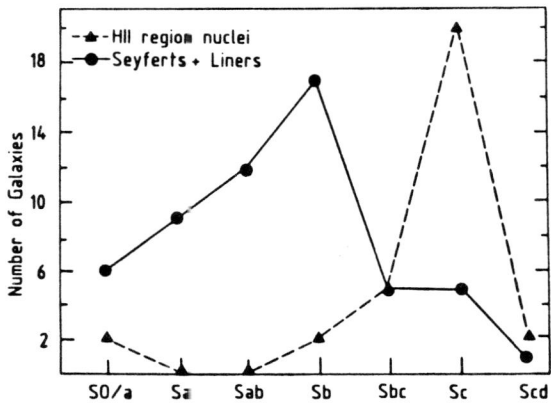

Figure 4: Distribution of nuclear morphology versus parent galaxy type.

early type spirals. The work of Cowley, Crampton and McClure (1982; Ap.J. 263, 1) shows that there is a good correlation between the luminosity of the bulge component of spiral galaxies and the chemical composition and/or the age of the stellar population such that luminous bulges are metal rich and old. According to this work and the calibration of Hubble (or de Vaucouleurs) morphological type as a function of bulge luminosity by Simien and de Vaucouleurs (in IAU Symp. 100, p. 375) the bulges of spirals of de Vaucouleurs type 4 or Hubble type Sbc have approximately solar abundances.

Spirals later than Sbc have bulge abundances lower than solar while the bulges of early type spirals have over-solar metallicities. Thus, Warmer models explain the Hubble type distribution of galaxies with active nuclei as a metallicity effect.

EPILOGUE: THE UNDERLYING CONTINUUM

Starburst models with Warmers explain, at least as well as Monsters, the observational properties of active galaxies. For reasons of space we cannot discuss here any of these properties but one: The morphology of the optical continuum in Seyferts and Liners.

After deconvolving the bulge component, the (underlying) continuum of Seyfert nuclei is featureless and is roughly a power-law of slope 1. Since a power-law ionizing spectrum of approximately the same slope is used to fit the emission-line spectrum of active galaxies, the morphology of the optical continuum is generally considered to be one of the strongest lines of evidence in favor of the Monster hypothesis. This interpretation, however, fails to consider reddening. After correction for reddening the mean observed slope of the continuum is -1, totally inconsistent with the ionizing power-law. On the other hand, the observed continuum of young, high excitation HII regions (figure 5) is also featureless and has the correct slope!

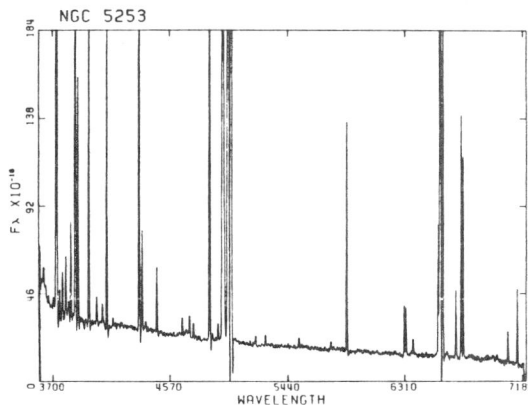

Figure 5: The featureless continuum of the nucleus of NGC 5253.

To conclude, although starburst models will certainly not fit the observations of all active galaxies, they explain the activity observed in the vast majority of spiral galaxies.

CONVECTIVE OVERSHOOTING: NEW INTEGRATED COLOURS VS AGE RELATIONS FOR STAR CLUSTERS

G. BERTELLI, A.G. BRESSAN, C. CHIOSI, E. NASI, L. PIGATTO

Institute of Astronomy, Padova, Italy

If the integrated colours of a star cluster mainly depend on chemical composition and age, then theoretical calibrations of colours as function of age for different chemical compositions are very useful to obtain quantitative determinations of the age and composition of individual clusters, and thus to trace the chemical history of nearby galaxies. Several calibration curves exist in the literature which rest on the standard theory of stellar evolution. However, a growing amount of observational evidence seems to indicate that overshooting from convective cores may be an important phenomenon in stellar evolution. In fact models computed with overshooting are significantly different from the standard ones. The aim of this preliminary investigation is to study the effects of convective overshooting on the integrated colours of clusters whose turnoff mass is in that range in which convective overshooting is effective.

The Results

We have computed (B-V):(U-B) colours as a function of age adopting the following theoretical background:
a) Evolutionary tracks with overshooting parameter $\lambda=1$ and chemical composition $X=0.700$ and $Z=0.02$ (Bertelli et al 1985)
b) Relation between Mbol, LogTeff and BC, (B-V), (U-B) from model atmospheres collected from many sources and amalgamated in homogeneous manner
c) The mass loss rate during the AGB phases as in Reimers (1975) with $\eta=1$.
The separate contributions of main sequence, main sequence + red giant, main sequence + red giant + AGB phases are shown in fig 1 for (B-V) only. In the same figure the new relations are compared with those derived from classical models. Two points are evident: a) the small effect on the colour by including the AGB phase, while the contribution of the red giant phase is dominant, b) the sudden change in (B-V) slope at age $\simeq 10^9$ yr, which does not occur with other relationships. The calibration curves of fig 1 rest on the hypothesis that the clusters possess an infinite number of stars. On the other hand it is easy to understand that stochastic fluctuations in the populations of rare but luminous stars may determine a dispersion in the colours of real clusters, and that the amplitude of the dispersion depends on the total number of evolved stars. To test the point

and to ascertain the amount of dispersion, a large number of Montecarlo simulations of real clusters have been performed. The dispersion in (B-V) as a function of different number of evolved stars is shown by the vertical bars along which are annotated the numbers of evolved stars per cluster. In order to test the reliability of the new calibration we have also plotted several LMC clusters, whose colours are from van den Bergh (1981) and ages from Chiosi et al (1985) (with overshooting). The points populate in a satisfactory way the dispersion band predicted by a finite number of stars. If the old calibration curve and the ages from Hodge (1983) are used (no overshooting) the agreement is not as good as in the previous case. New calibration curves with different chemical composition (Z=0.001 and X=0.700) are in progress on the aim of exploring if the sudden change in the slope of (B-V)-age relation at $t \simeq 10^9$ yr, together with dispersion effects due to the finite number of evolved stars, and differences in chemical composition might be responsible of the discontinuity shown by LMC clusters in the relation (B-V) versus cluster type of the Searle et al (1980) classification.

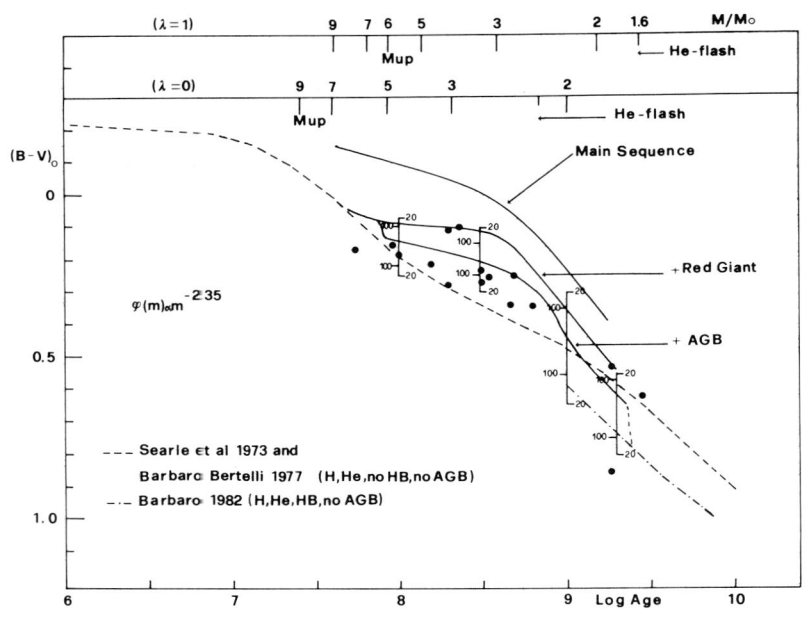

Fig. 1

References

Barbaro, G.,1982, Astrophys. Space Sci.,83,143
Barbaro, G., Bertelli, G.,1977, Astron. Astrophys.,54,243
Bertelli, G., Bressan, A.G., Chiosi, C.,1985,Astron. Astrophys., in press
Chiosi, C., Bertelli, G., Bressan, A.G., Nasi, E.,1985, preprint
Hodge, P.W.,1983, Astrophys. J.,264,470
Reimers, D.,1975, Mem. Soc. Roy. Sci.,Liège,8,369
Searle, L., Sargent, W.L.W., Bagnuolo, W.G.,1973,Astrophys. J.,179,427
Searle, L., Wilkinson, A., Bagnuolo, W.G.,1980, Astrophys. J.,239,803
van den Bergh, S.,1981, Astron. Astrophys. Suppl. Ser.,46,79

THE DISTANCE MODULUS OF LMC

Cesare Chiosi[1] and Luisa Pigatto[2]
1 Istituto di Astronomia, Università di Padova
2 Osservatorio Astronomico di Padova
Vicolo dell'Osservatorio,5 - I-35122 Padova
Italy

Deep CCD photometry of the star clusters NGC2162 and NGC2190 in LMC presented by Schommer et al.(1984) is used togheter with new evolutionary models computed by Bertelli et al.(1985a) which take into account overshooting from convective cores, to derive the clusters ages and the distance modulus of LMC. A preliminary analysis of the two clusters indicates that NGC 2162 and NGC 2190 belong to the same class of clusters discussed by Barbaro and Pigatto(1984). In fact, for the turn-off mass estimated by means of classical models (<2.2 $m_\odot$), these clusters should possess an extended red giant branch and a bimodal distribution of red stars(cifr.Fig.2). On the contrary they show a clump of red stars. This means that ages and other properties derived from classical models for this range of masses, may not correspond to reality. With the new models, stars of mass as low as 1.6 $m_\odot$, ignite helium in non degenerate conditions, avoid the long lived RG phase, and burn helium as more massive stars. As consequence of it, a clump of red giants is expected. In Fig.1, we show new isochrones(Bertelli et al.1985b) derived from models with overshooting, overlaid to the CM diagram of NGC 2162. Theoretical luminosities and T_{eff}'s are converted into M_V:(B-V)$_0$ plane by means of T_{eff}:(B-V):BC scales based on models atmospheres collected from several authors(Chiosi,1985). At any given age, the new isochrones run brighter than those of Ciardullo and Demarque(1977). By means of the luminosity function, a method more objective(Paczsynski,1984) than the standard one of ZAMS and/or isochrone fitting, with a reddening of E(B-V)=0.06 and chemical composition X=0.700 and Z=0.02, we find ages of 1 10^9yr and a true distance modulus of $(m-M)_0$=18.6 instead of 18.2±0.2 mag given by Schommer et al.(1984). Fig.2 shows the theoretical luminosity function at age 1 10^9yr, (age preliminarly assigned to the clusters by isochrone fitting) for main sequence and red giant stars obtained with Salpeter's IMF(top panel), compares it with the correspondent one of Ciardullo and Demarque(1977), and finally shows the observational LF we derive from stars counts(bottom panel) for NGC 2162. By imposing coincidence between theoretical and observational LF's at the side of main sequence fall-off and rising of the red giant clump, we derive the distance modulus $(m-M)_0$=18.6. In conclusions, models with overshooting not only interpret the morphology of this class of clusters, but assign LMC a distance modulus in agreement with other independent determinations(Walker,1984; Visvanathan,1985).

REFERENCES

Barbaro,G.,and Pigatto,L.1984,Astron.Astrophys.,136,355
Bertelli,G.,Bressan,A.,Chiosi,C.1985a,Astron.Astrophys.,in press
Bertelli,G.,Bressan,A.,Chiosi,C.,Nasi,E.,Pigatto,L.1985b,in preparation
Chiosi,C.1985,in preparation
Ciardullo,R.B.,Demarque,P.1977,Yale Trans.Vol.33
Paczynski,B.1984,Astrophys.J.,288,182
Schommer,R.A.,Olszewski,E.W.,Aaronson,M.1984,Astrophys.J.(Letters)285,L.53
Visvanathan,N.1985,Astrophys.J.,288,182
Walker,A.R.1985,Monthly Notices Roy.Astr.Soc.,212,343

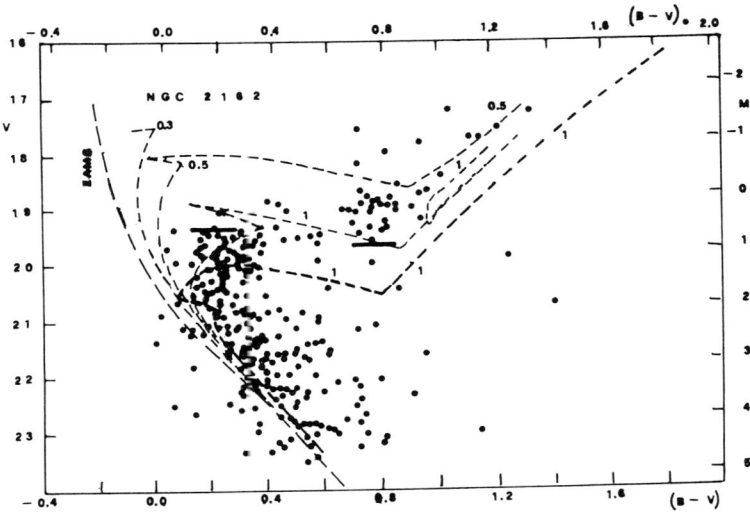

Figure 1. CM diagram of NGC 2162 from Schommer et al. (1984). The dashed lines indicate new isochrones for $Z = 0.02$ as described in the text. The heavy dashed line shows the isochrone of $1 \ 10^9$ yr of Ciardullo and Demarque(1977) with $Z=0.01$. The heavy horizontal lines indicate the caracteristic luminosity used to derive the main sequence and red giant star luminosity function.

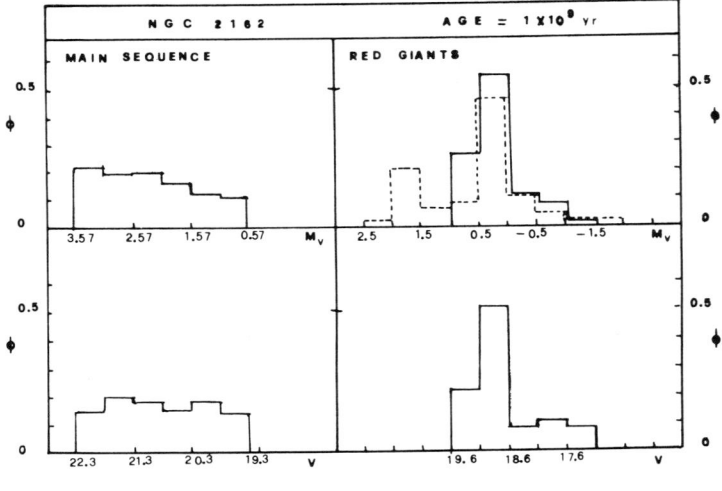

Figure 2. Theoretical luminosity functions (top panel) for main sequence and red giant stars along the isochrone $1 \ 10^9$ yr. The red giant luminosity function derived from Ciardullo's and Demarque's isochrone with age $1 \ 10^9$ is shown by the dashed line. Observational luminosity functions (bottom panel) for NGC 2162.

NGC 2070 and NGC 3603

You-Hua Chu
University of Illinois at Urbana-Champaign
Patrick Seitzer
Kitt Peak National Observatory
Nolan Walborn
Space Telescope Science Institute

Giant H II regions are usually so distant that only the integrated properties of their stellar contents can be studied. However, the nearby giant H II regions 30 Doradus in the Large Magellanic Cloud and NGC 3603 in the Galaxy offer a rare opportunity for a spatially resolved study of their central stellar associations - NGC 2070 and NGC 3603. We have obtained 4m prime focus plates with interference filters centered on the He II 4686 line ($\Delta\lambda = 65$ A) and the adjacent blue continuum at 4765 A ($\Delta\lambda = 70$ A) to identify the Wolf-Rayet (WR) stars in their cores. We have also obtained CCD images in UBVRI for NGC 2070, and BVRI for NGC 3603 to determine their color-magnitude diagrams and luminosity functions. Since the WR stars in the core of 30 Dor have been reported by Chu, Cassinelli, and Wolfire (1984) and Moffat, Seggewiss, and Shara (1985), and the CCD photometry of NGC 2070 and 3603 is still preliminary, we will discuss in these proceedings only the WR stars resolved in the core of NGC 3603.

The core of NGC 3603, HD 97950, was resolved into ABCDEFG components (Walborn 1973; see Fig. 1 for their positions). The A component was recently further resolved into 3 components by speckle interferometry (Weigelt, Baier, and Ladebeck 1985). Moffat, Seggewiss, and Shara (1985) have determined that there is more than one WR star in HD 97950. The resolution of their CCD images is unfortunately limited by the pixel size of 0.6 arcsec, so that the individual WR stars are not resolved.

Based upon our on-line, off-line pair of two-minute exposure prime focus plates, the C component is clearly a WR star, and the B component of the AB group is probably the main contributor of the He II line emission. (The He II line emission components are best illustrated in the false color contour plots as shown in the poster. A black and white photocopy is presented in Fig. 2.) The excellent seeing also allows the D component to be resolved into two stars (see Fig. 1).

The unresolved cores of giant H II regions have been suggested to contain peculiar objects, e.g. supermassive stars. The WR stars resolved in our data and the stars resolved by speckle interferometry (Weigelt, Baier, and Ladebeck 1985) for NGC 2070 and 3603 demonstrate that peculiar objects may not be necessary.

References:
Chu, Y.-H., Cassinelli, J. P., Wolfire, M. G. 1985, Ap. J., 283, 560.
Moffat, A. F. J., Seggewiss, W., and Shara, M. M. 1985, Ap.J., in press.
Walborn, N. R. 1973, Ap. J. Letters, 182, L21.
Weigelt, G., Baier, G., and Ladebeck, R. 1985, ESO Messenger, No. 40.

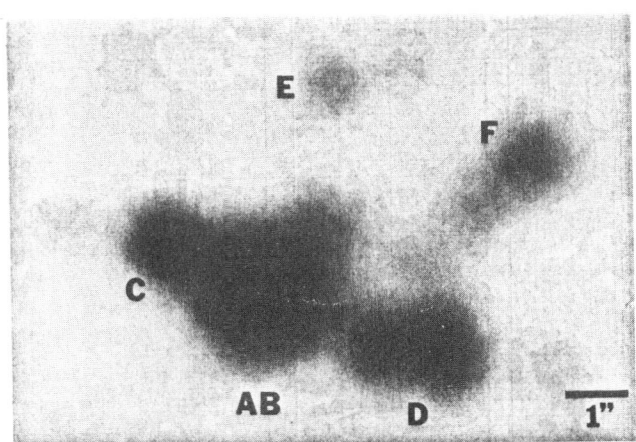

Fig.1. HD 97950, the core of NGC 3603. North is at top, and east is at left.

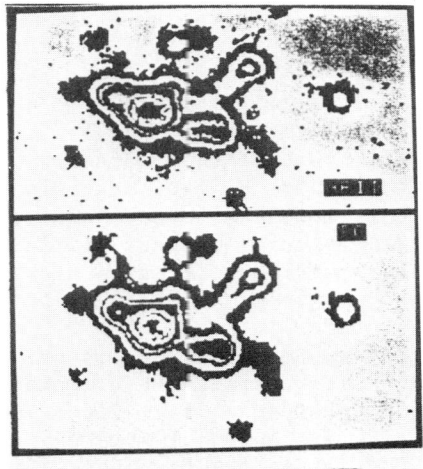

Fig.2. Black and white reproduction of the false color contour plots of HD 97950. The upper panel is for the He II image, and the lower panel for the blue continuum image. The scale below represents plate densities of 0 to 5 from left to right. The C component is brighter in the He II than the blue continuum by about 0.3 in density. The He II image of AB is denser than the blue continuum counterpart, especially at the southeast where the B component is located.

POSTER PAPERS - SESSIONS 7 and 8.

Chairman : A.MAEDER.

Maeder:

There are very few posters to be discussed today, and before going these I thought I would show some "nice" comments pronounced during the meeting:

(i) Conti: "I know we are in a spiral galaxy, but I have a lot of trouble in seeing spiral arms"
(ii) Concerning agreement between data and theory; Kudritzki has said: "you cannot see it because the agreement is so good"
(iii) Joe Silk: "the fact we don't know anything does not restrain theorists from studying star formation"
(iv) Peter Conti: "how do we make a superstar by Hans Zinnecker"
(v) Hodge: "when you look to associations in other galaxies you must have good reasons for making your life more difficult"
(vi) Hoessel: "the facts are just clear enough for publication" - I wonder whether the observations are also so clear!
(vii) Wheeler: "this is the total number of supernovae in units of a few".

Turning now to the posters, we have a series of papers from Chiosi and co-workers discussing the effects of overshooting on the morphology of the HRD and integrated colours of clusters in the range of intermediate mass stars together with the distance modulus of the LMC where they get a better fit with overshooting and age estimates of various star clusters in the LMC where again the inclusion of overshooting provides better results.

Feast:

If I remember correctly the paper of Chomer et al. also did a comparison with observed HR diagrams. Does that fit with your changes resulting from the different theory?

Chiosi:

The major difference is in the following. Those particular clusters belong to a particular class of clusters which show, in our opinion an atypical morphology of the HRD, since on the basis of standard theory they should develop a red giant branch whereas the red stars lie in a clump. This is also evident looking at the red giant luminosity function. This means that the theory is not applicable in this domain of cluster ages. In the paper of Bressan, et al., on the minimum mass and maximum mass for stars undergoing a helium burning phase, they have found this value, instead of 2.2 M_o in classical models, is now down to 1.6 M_o. Actually they are able to confine the mass limits in a very narrow band. This means that stars as old in mass as 1.6 M_o will not develop a red giant phase but will burn helium in the core as for massive stars. This leads to the fact that the red giant luminosity function is a single sequence as indicated by the observed data. As a consequence of this result the new distance modulus comes out immediately in the sense that overshooting models at any given age are brighter than standard models, so for a given apparent absolute magnitude it is obvious that if they found a distance modulus of 18.2 with the new models we have to find a slightly higher distance, namely 18.6.

Maeder:

I would like to mention the paper dealing with luminous stars in SMC-associations, where, if I understand correctly the results are very preliminary.

Kaufman:

What are the stellar types in the associations?

Kontizas E.:

From the diagrams we show, there are mainly early-type stars. There seems to be a lack of late-type supergiants, but this is preliminary. We cannot distinguish B and A stars, so we call them BA stars.

Maeder:

One has to be very careful about overshooting to get just the right amount of this physical effect. I think that one needs to compute the details of the internal chemistry in order to compare with evolutionary observational results in order to confirm or reaffirm the existence of this process in massive stars. Further progress including turbulent diffusion, overshooting and so on will slightly modify the result, but presently we come out with 40 to 50 M_o.

Maeder:

The next poster, deals with a search for WR stars in giant extragalactic regions by Campbell, Smith and Terlevich. They have found

DISCUSSION

spectra of four HII galaxies showing WR features and additionally a large CO index showing evidence of red supergiants which anticorrelate with the WR features.

Zinnecker:

How secure are the CO indices; they are very difficult to measure. One problem for instance is that a measurement of the CO index distinguishing between giants and supergiants is difficult. Maybe you have an old underlying population of giants.

Campbell:

You have to do very accurate photometry, needing about 5 to 10 observations or better. We do not really have enough observations and our first results thus need to be confirmed. However, I believe that our technique is a powerful one for detecting red supergiants unambiguously.

Massey:

In his thesis, Howard French studied a few HII galaxies and reported HeII 4686 emission, but he sort of missed the point and simply assumed it was nebular. Did you do any of the galaxies he did, and what was the HeII 4686 due to?

Campbell:

We do have some common objects. The 4686 in our sample is very narrow HeII emission; using ordinary O stars you would need temperatures greater than say 150,000 K to produce a narrow HeII line, but I am not sure whether they can do this without also producing other, possibly broad, features which are not observed. This HeII emission is always observed in the very high excitation, low-abundance, regions and we do not know how it is produced.

Maeder:

Finally we come to the paper dealing with "Warmers" by Melnick and Terlevich where the emphasis is put on the star-burst model. They show that the contribution of WC and WO stars, which they call "warmers" might just give the emission spectra observed in active nuclei, which would appear after two or three million years according to this work.

Silk:

I think this is a very exciting idea. However, it seems to me that on the basis of the model, you would get Seyfert-like features in HII regions where one expects a weak metallicity.

Campbell:

I can perhaps answer for Terlevich. The reason you do not see these features in HII regions is simple. Firstly, you need an extremely massive burst of star formation to produce these stars massive enough. Secondly you need a very high density and you probably need solar abundances.

Silk:

Surely there must be a giant HII region somewhere with about solar abundances?

Campbell:

There is not an HII galaxy like that.

Melnick:

There are some nuclei that have intermediate characteristics and there are some giant HII regions (cf. the spectrum of NGC 5253 I showed in my talk) that have nebular HeII 4686 that you do not expect from a normal population of O stars).

Maeder:

Perhaps starburst models are slightly more conservative than "monsters" and this is clearly a very interesting line of research. We have a little time for any further discussion of today's reviews.

Zinnecker:

A brief comment ont the question of mass at which the bi-modal star formation is cut into two pieces with a separating mass of 3 M_o. I think that mass can be justified to some extent, as was mentioned, but I would like to emphasise the point, by the near-IR, 4 micron, observations. The fact is that we need to explain this IR excess in the 5 kpc arm by lots of giants, but at the same time you must satisfy the requirement that the scale height of these stars is not too high as the

DISCUSSION

observations show. Now, in order to do that you must not allow the stars any time to disperse in this direction and that puts a limit on their age and that is another reason, I think, why the limit cannot be around 1 M_o, but must be somewhere around 2 to 3 M_o so that the lifetime we are talking about is around 10^8 years.

SUMMARY TALK: WHAT IS HAPPENING TO MASSIVE STARS IN GALAXIES?

J. A. Graham
Cerro Tololo Inter-American Observatory
National Optical Astronomy Observatories
Casilla 603, La Serena, Chile.

Taft E. Armandroff
Astronomy Department
Yale University
Box 6666, New Haven, CT 06511

ABSTRACT. Highlights of the IAU Symposium 116 are reviewed. Some of the general themes running through the meeting are identified. These include:i) the fruitful interaction between observation, laboratory work and theory. ii) the need for understanding and, if possible, correcting for the effects of incompleteness and bias in observing lists. iii) the importance of the Magellanic Clouds, as the nearest independently evolving stellar systems, in the study of massive star formation and evolution in galaxies.

It can be hard sometimes, to make our colleagues believe that gatherings at places as beautiful as Porto Heli can serve a useful purpose. Yet, by getting together at the right location for a week-long interchange of ideas, results and information, we can set the stage for new and often pivotal advances in a subject. As we listened to the papers during the last week, many of us felt that this was indeed a good time for a symposium about luminous stars and associations in galaxies. If one were to choose a major theme of our meeting it might well have been the fruitful interaction between observation, laboratory work and theory at this present stage of our understanding. The recent burst of progress has come from a confluence of these three streams, each itself newly enriched by significant new advances. First, we now have at hand an impressive array of new instrumentation on our telescopes which allows observers to measure basic parameters for stars in external galaxies to a precision hitherto unavailable. Second, there has been a lot of new laboratory work, both in the measurement of atomic data and in the development of new techniques for the measurement and analysis of the raw observational material. Third, but of equal significance, there have been advances in theory, helped on by the enormous increase in computing power which has arrived on the scene. In the future we all look forward to the Hubble Space Telescope. Many here have already tasted what is to come from their experience with

IUE. A still greater advance in computing power will follow from the appearance of supercomputer facilities which will not only allow theoreticians to carry out numerical calculations previously thought impossible but will also enable observers to handle promptly the huge data blocks which come out of modern digital recording devices.

Kudritzki and Hummer got our symposium off to a good start in telling us about the resolution of some inconsistent basic data for the brightest O star, ζ Pup. This was achieved by careful attention to the combined effects of electron scattering, gravity and helium abundance. In this first talk, the term windblanketing came up, a term new to many of us but one which has turned out to be very important owing to the dominating effect which strong stellar winds and subsequent mass loss seem to have in the evolution of early-type stars. Kudritzki pointed out that a consistent theory of stellar winds is still necessary. Improvements, such as the taking into account of the finite size of the stellar disk, are now possible with increased computing power. Some of the strongest constraints on models presently come from the observed CNO abundances.

One thing that was missing at this meeting, to which we were all looking forward, was the cheerful, rational presence of Katy Garmany. Peter Conti filled in by telling us about her recent work on the Initial Mass Function (IMF) for massive stars in the Galaxy and the Magellanic Clouds. There continues to be some debate about completeness at the faint end of the solar-neighborhood O star sample. One of us (JAG) continues to worry about how many late O and early B-type stars might still be hidden from visual observation by their placental dust clouds. The weaker radiation pressure of these stars might not allow them to disrupt their surrounding clouds as rapidly as the early O stars and could lead to a skewing of the IMF. Garmany, Conti and Massey have embarked on a major project to extend the study to the Large and Small Magellanic Clouds. This requires good quality spectra of every O star candidate and good atmospheric models to analyse them. The paper by Gehren et al. also gave an excellent discussion of these needs.

Allan Sandage treated us to a masterly survey showing the power of luminous stars as distance indicators. We all saw how the bold approaches initiated by Hubble have succeeded in a way that would have amazed him. Distance calibrations of the first 3 blue and red stars in a galaxy (a statistical procedure urged in a poster paper by Laura Greggio) have been performed using Cepheid distances for 16 nearby galaxies including Sextans A and B and the WLM system. The ultimate aim is not only to fix the Universal distance scale but to map the local velocity perturbations. Some of this will have to wait for the revolution brought about by Space Telescope but much can be done from the ground. Sandage pointed out that the discovery and prompt follow-up of supernova of type I, for example, and their subsequent folding into the distance calibration has become a major observational challenge. Later in the week, in another invited talk, Craig Wheeler enlarged on the astrophysical opportunities which currently exist in supernova research.

Roberta Humphreys presented a paper about cooler supergiants as crucial signposts in the life cycles of massive stars. New results have been obtained for a number of galaxies including M31, M81, M101 and NGC 2403. Here, another theme of the meeting became prominent. This is the importance of understanding bias and selection effects and, as well, completeness in forming our lists of stars. Interloping foreground stars must be removed from our lists ruthlessly! This can be done either from the measurement of proper motions or from spectroscopic criteria such as those outlined in the paper by Graham and Humphreys. Completeness in our lists of Galactic red supergiants is being improved by work such as that reported by MacConnell et al. Wendy Freedman gave us a grand talk about the stellar luminosity function with results for several external galaxies. She showed how the huge data sets obtained from CCDs and automatic measuring machines could be handled to give results of high statistical weight. In a related paper Tom Manley demonstrated the promise and potential of the Minnesota Automatic Plate Scanner, while John Hoessel's talk on Friday illustrated further the capabilities of CCD detectors for studies of this type as applied to dwarf galaxies.

Poster papers in many ways represented the core of the new results at the meeting. We had a rich array over our four days at Porto Heli. Two papers on the massive G-type hypergiant HD 217476 by Smolinski et al. and by Zsoldos introduced us to the photometric and spectroscopic complexity of this star. De Jager et al. showed that the rate of stellar mass loss depends almost entirely on effective temperature and luminosity for stars of types O through M. Only WR and C stars do not seem to fit. Among the other papers, the Magellanic Clouds were well represented. Two surveys of emission line stars were presented by Azzopardi and Meyssonnier (SMC) and Bohannan (LMC). Dubois showed us that there are still problems in the use of hydrogen lines as distance indicators for the blue supergiants in the SMC. Chiosi's group applied state-of-the-art theory to the interpretation of color-magnitude diagrams. The new evolutionary tracks not only provide a better morphological fit for intermediate age clusters in the LMC but support a distance modulus close to $18^{m}\!.6$ rather than the smaller values which have recently been suggested by Aaronson and others.

Nolan Walborn's paper was officially concerned with the broad topic of spectral morphology of O stars but it was made especially memorable by his review of the stellar populations associated with the core object R136a in the 30 Dor complex of the LMC. There has been a substantial advance in the imagery and detailed spectroscopy of this object. Particularly impressive is the new holographic speckle interferometry which has been carried out by Weigelt and Baier. This shows that R136a has 8 stellar components, the three most luminous of which are similar in brightness. There are now 28 resolved stars in what was originally labelled as the star R136. In its immediate surroundings, numerous early O and WR stars have been found which can easily account for the ionization of the 30 Dor nebula. There now seems no identifiable core object which could conceivably have a mass higher than about 250 solar masses. The poster paper by Lortet et al. further emphasized that there are many bright unresolved multiple systems in the Magellanic Clouds which are only now being identified as such.

"How massive can a stable star be?" was the question addressed on Tuesday morning by Immo Appenzeller "and what are the main sources of instability?" He considered the equilibrium states of stellar models and the effects on these of various perturbations. Vibrational instability is potentially one of the more effective ones but it is thermal instability which ultimately plays the most important role because of the strong temperature dependence of the interior nuclear reactions. 90 $M_\odot$ seems to be an upper limit for a stable star on the zero-age main sequence but a star with a mass up to 200 $M_\odot$, while unstable, could evolve quickly into an observable, almost stable, post-main sequence object. Shock wave induced mass ejection seems to rule out as violently unstable any stars as massive as 1000 $M_\odot$. Observations of stars near the upper mass limit were reviewed by Bernard Wolf and Henny Lamers. Between them, they covered S Dor, Hubble-Sandage and P Cyg stars. During the outburst states we generally see an accentuated rate of mass loss up probably by a factor of 100 from that observed in the quiescent stage. Observed velocities are not high but most of the significant mass loss in the star's lifetime occurs during these events. Interestingly, their bolometric magnitude is probably not changing much. It is mostly just a redistribution of the total stellar flux which gives rise to the optical variability.

On Tuesday afternoon, Peter Conti brought us up to date on the astrophysical parameters of Wolf Rayet stars. Especially timely was his discussion based on eclipsing binary studies of an upward revision in the effective temperatures of these stars to about 100,000K. An investigation of V444 Cyg reported by Pauldrach et al. reached similar conclusions. Conti also noted that the effect of back-scattering from stellar winds, discussed by Kudritzki, would be very important in WR atmospheres. He presented new radio-derived mass loss rates by Abbott. These showed a good correlation between $\dot{M}$ and M indicating like several other papers at this meeting that mass loss is radiatively driven. Finally Conti addressed the question of the WR/O star number ratio and showed how the new tracks by Maeder result in greatly improved theoretical and observational accord. Phil Massey spoke about the use of WR stars as tracers of the massive star population in nearby galaxies. He stressed the observational difficulty of isolating a sample of unevolved O stars using photometry alone. WR stars, on the other hand, are easily detected and very effectively trace concentrations of massive stars in a galaxy. He reviewed the various techniques for discovering WR stars emphasizing the problems of bias and completeness. Neither the WR density or the WC/WN number ratio appears to be a function of metallicity alone and it may be necessary to invoke IMF variations between galaxies. Finally, a preliminary survey of M31

WHAT IS HAPPENING TO MASSIVE STARS IN GALAXIES?

for WR stars by Armandroff, Massey and Conti was summarised. Another recent survey of M31 reported by Shara and Moffat has produced quite a different result which can probably be attributed to varying degrees of completeness at small equivalent widths. Moffat, however, argued in discussion that the two surveys were complementary. The Shara and Moffat survey might be less complete but it does cover the whole galaxy. Armandroff et al.'s survey may be biased by the small size of the areas searched.

Thursday morning brought a related theoretical session. Maeder emphasized how mass loss determines the course of evolution through the WR phase. Mass loss may have little effect on the central regions of the star but it radically changes the abundances at the stellar surface. Rotational mixing can lead to material diffusion and may be the agent for producing massive, nitrogen-rich blue straggler stars, such as those observed in Per OB1. Sreenivasan and Wilson demonstrated how spinning models also lead to the prediction of high effective temperatures for WR stars. Bert de Loore and colleagues presented a comprehensive review, with new numerical results, of the evolution of massive binary stars. A large number of evolutionary possibilities exist depending on whether mass is lost or accreted through Roche lobe overflow or through stellar winds. Wolf Rayet stars can be produced in both ways depending on the circumstances. Cases relevant to the formation of X-ray binaries have been investigated and were reported by Hellings.

Chiosi's review concentrated on recent improvements to the theoretical models of massive star evolution, especially those incorporating convective overshooting. He showed how convective overshooting can reconcile the distribution of stars on the AGB with turnoff ages for Magellanic Cloud clusters. This can prove important for studies of the brightest stars in these clusters such as those reported at this meeting by the Kontizases and co-workers. Further application of convective overshooting, this time to the upper mass limit for stars undergoing the He flash, was presented by Bressan and Bertelli. Chiosi also reminded us of the importance of stochastic effects in determining the integrated colors of a star cluster.

Joe Silk gave us a view into the likely processes of massive star formation. He favored the concept of fragmentary molecularclouds being swept up into giant molecular clouds through galactic density waves. Star formation is probably bimodal. An extended period of low-mass star formation precedes the eventual triggering of massive star birth which, by heating the environment, suppresses star formation almost completely for a while. This talk led naturally to the observational papers by Michael Rosa and Paul Hodge in the afternoon. Rosa considered the stellar knots within giant HII regions in external galaxies where a large fraction of the most massive stars are usually found. Since individual stars generally can no longer be resolved the interpretation of the observations is far from simple. To discriminate between age and IMF effects, for example, observations must be made in the far

UV as well as in the optical and infrared. Wolf Rayet stars make a substantial contribution to the observed spectrum which often shows strong variations over distances of only 1". New observations were reported of stellar knots in M101, NGC 55 and NGC 5128. Wolf Rayet features have been identified here and as well, at the sites of 3 supernovae in M83. Paul Hodge described to us the many difficulties in defining stellar associations in nearby galaxies. Here, bias problems again become substantial and consequently, integrated magnitude measurements of stellar associations are probably not good distance indicators. Nevertheless, it does seem that the size or mass distribution of associations cannot be very different from one galaxy to another. Hodge showed part of his new atlas of stellar associations in the Small Magellanic Cloud. New surveys of M31 and M33 were also displayed by Nikolov, Ivanov and collaborators at this time.

While consensus and agreement were reached in several problem areas at this meeting, new questions arose and we are all going to be kept busy in the future searching for their solution. Some flavor of this activity in the field came through after Rosa's paper on Thursday when a spirited discussion ensued on which type of Wolf Rayet stars one would expect to see in the cores of giant extragalactic HII regions. Theorists, for the most part, felt that young clusters of massive stars should contain mostly WC stars, since these are thought to originate from more massive progenitors than WNs. Walborn, on the other hand, stated that in the 30 Dor complex, WR stars are mostly of WN type and that a previous study by Roberts had found that Galactic associations contain mainly WN types. Massey countered that the Roberts study was based on data of 1950s vintage: more complete surveys are now available and, in M33 at least, WC stars are as common as WN stars in the large OB associations. Conti pointed out that, although we know that all WC stars have gone through a WN stage, it is not at all clear that all WN stars continue on to become WCs and that the relative lifetimes are very uncertain. Someone asked Rosa just how he expected to form stars in these HII region cores. Rosa didn't know; another symposium in 3-6 years time might be necessary to answer this one.

Peter Conti was clearly thinking aloud already of possibilities. "A meeting in Java, perhaps?" he suggested.

SUBJECT INDEX

30Dor, 185,186,189,191,217,235,274,285,286,307,313,316,317,356,359,401,
    426,434,515

abundance, 350
abundance-gradient, 13,70
accretion, 141
AGB, 3,324,326,327,332,334
association, 25,27,215,223,224,242,243,317,319,369,370,371,372,373,
    374,375,376,377,387,390,391,399,409,419,425,434,439,453,518

backscattering, 203
Blue compact galaxy, 459
Be-stars, 109,158,197,227,228,267,268,269
beer, 155
bimodality, 308,310,317,464,479,486,487,489,521
bimodal-star-formation, 70
birthrate, 469
blanketing, 6,7
blocking, 7
blue-stragglers, 302
burst, 311,355,356,358

C/N, 293
C12, 125
calibration, 31,32,33,46,133,134
carbon-stars, 404
Carina, 22
Carson, 125,126
CASPEC, 13,237,250,411,430
CCD, 235,236,332,399,411,439,440,441,442,455,513,515
cenA, 364
cepheid, 32,33,40,335,336,447
champagne, 155
chemical-evolution, 460
cloud, 306,309,311,410
cloud-mass, 305
clump, 308,370
cluster, 274,297,319,323,324,331,336,359,377,405,406,407,408,423,
    451,454
CNO, 12,15,175,320
collapse, 305,306,316
collapsing-clouds, 304

completeness, 22,30

Davidson-limit, 180
De Jager-limit, 58,155,180
deflagration, 467,468
diffusion, 346
distance-indicator, 37
distance-modulus, 224,235,513,518
distances, 32
dredge-up, 292
dust, 61,123,130,131,248,250,251,260,269,273,303,367,401,479
dust-lane, 393,431
dwarf-irregular, 445,448,461

Eddington-limit, 35
Eta-Car, 155,156,251,252,253,254,280,295,436
evolution, 129,133
Evolutionary, 132

filaments, 251
fragmentation, 303,304,306,307,308,316

galaxy, 46,101,159
galaxy-morphology, 506
globular-cluster, 325,407,461
gradient, 126
grains, 308

H-R, 20
H-S, 154
HeI4471, 229,237
HeII4686, 192,233,235,243,362,367,497
HI-shell, 391
HII, 185,212,218,235,273,303,304,309,317,355,357,361,364,367,376,377,
    389,393,395,427,433,434,435,451,493,497,501,502
HII-region, 65,70,80,505,515,520
horizontal-branch, 3
HS-variables, 149
Hubble-Sandage, 139,151,180,195,251
Humphreys-Davidson-limit, 58
Humphreys-limit, 155,180
Hyades, 323
hypergiant, 89,130,149,158,172,176,260

IC1613, 63,216,242,376,377,439,442,455
IMF, 19,20,67,79,101,240,243,303,304,306,307,310,313,316,356,361,362,
    363,391,392,451,452,453,454,456,458,460,462,464,465,481,482,483,
    484,486,491
incompleteness, 65,123,332
instability, 139,142,143,144,146,147,149,257
IRAS, 91,31,231,234,245,296

SUBJECT INDEX

isochrone, 297,514

Leo, 441
liners, 505
LMC, 195
Local-Group, 36,50,51,242
LTE, 4
luminosity-function, 64,65,67,68,79,324,453

M101, 33,41,53,58,59,68,135
M31, 52,58,63,68,74,134,136,151,157,159,216,217,218,233,241,242,243,
      371,372,373,374,387,455
M33, 33,41,51,52,53,63,65,67,68,74,79,80,151,154,157,159,215,216,219,
      242,243,389,419,427,433,455,463
M81, 33,40,41,53,58,63,68,135,501
M83, 362
mass-distribution, 452
mass-function, 377
mass-loss, 9,46,49,55,109,121,125,128,143,146,150,153,155,158,165,170,
      174,179,199,205,229,232,251,254,258,263,271,278,289,290,291,292,
      319,320,324,325,328,330,332,341,342,345,348,363,397,412,452,506
mass-loss-rate, 18,46,129,132,152,167,226,435
mass-ratio, 342,347
mass-transfer, 397
merger, 273,431,432
metallicity, 25,49,51,63,70,101,134,217,233,272,344,345,413, 421, 430,
      460,461
Milky-Way, 51
mixing, 289
molecular-cloud, 303,308,309,317,395,486
monsters, 504
MS-widening, 291,323
multiple-system, 191,399,400,426
MWC300, 281

NGC55, 358
NGC-206, 136,216
NGC1866, 323,331,332
NGC2264, 310
NGC2403, 33,40,41,53,58,63,83
NGC300, 53,63,103
NGC3603, 235
NGC5128, 393,431,432
NGC5253, 476
NGC588, 218
NGC6128, 439
NGC6822, 51,63,216,242,243,376,440,455
NGC7331, 374
nucleosynthesis, 350,467,471

O/N, 294

O16, 125
Oe-stars, 197
opacity, 7,121,125,126,180,210,320,325,326,330,335
open-cluster, 453,465
oscillation, 396
overshooting, 10,67,121,125,160,239,240,289,296,297,319,320,325,326,
  328,330,332,334,335,346,397,412,421,422,509

parent-galaxy, 403,445
P-Cygni, 55,157,158,160,169,170,172,173,229,247,249,254,267,271,272,
  284
photodisintegration, 468
planetary-nebula, 435,467
population-deconvolution, 367
population-synthesis, 363,364,456
protostars, 306,307
pulsar-formation, 469
pulsations, 127,131

R136, 143,185,186,191,198,278,307,308,313,357,399
radiation-pressure, 130
reddening, 135
Reimers, 129
ring-nebulae, 246
Roche-lobe, 353,383
rotation, 93,177,298,299,302,303,385
rotational-mixing, 452
Roxburgh, 348
runaway, 209,255,344,475

S-Dor, 146,147,149,151,152,153,155,156,157,160,169,176,177,230,237,249
Schwarzschild, 343
segregation, 407,408
Sextans, 51,133,441
Seyfert, 505
SFR, 20
shell, 171,181,195
SN, 3,4,347,349,353,361,383,467,477
SNI, 469,470,471,472,473
SNII, 469,470,471,472,473
Space-Telescope, 254
speckle, 186
spiral-arm, 419,484,486
star-burst, 503
star-cluster, 509
star-counts, 455
star-formation, 3,25,61,71,72,73,101,233,243,273,303,305,306,308,310,
  311,312,313,355,365,392,393,394,395,409,410,431,432,439,445,
  460,468,479,481,486,487,497,501,520
star-formation-rate, 19,79,451,452,457,459,465
starburst, 310,505

SUBJECT INDEX

starburst-nuclei, 457
superbubbles, 409
supergiant, 33,40,45,130,165,167,176,177,179,181,182,185,198,229,230,
    235,250,290,291,295,296,321,385,389,430,440
supermassive-star, 399,516
supershell, 409,410
superstar, 428
synthetic HR diagrams, 125

T-Tauri, 426
turbulence, 258
turnoff, 333,334,336,421,509,513

V729Cyg, 398,428
Virgo, 36,37,374

warmers, 503,504,505,520
wind, 139,149,151,195,197
wind-blanketing, 112

X-ray, 231,251,261,341,343,349,350,383,384,469

Zeta-Pup, 4,5,6,11,112